Shell Structures in Civil and Mechanical Engineering

Institution of Civil Engineers publishing

Shell Structures in Civil and Mechanical Engineering

Theory and analysis

Second edition

Alphose Zingoni
BSc (Eng), MSc (Eng), PhD, DIC, PrEng, CEng, FSAAE, FIABSE, FIStructE
Professor of Structural Engineering & Mechanics
Department of Civil Engineering, University of Cape Town, South Africa

Published by ICE Publishing, One Great George Street, Westminster, London SW1P 3AA.

Full details of ICE Publishing representatives and distributors can be found at: www.icebookshop.com/bookshop_contact.asp

Other titles by ICE Publishing:

Advanced Structural Mechanics, Second edition.
D. Johnson. ISBN 978-0-7277-2860-9
Finite-element Design of Concrete Structures.
G. Rombach. ISBN 978-0-7277-3274-3
Geotechnical Finite Element Analysis.
A. Lees. ISBN 978-0-7277-6087-6

www.icebookshop.com

A catalogue record for this book is available from the British Library

ISBN 978-0-7277-6028-9

ICE Publishing is a division of Thomas Telford Ltd, a wholly owned subsidiary of the Institution of Civil Engineers (ICE).

Commissioning Editor: Rachel Gerlis
Development Editor: Inês Pinheiro
Production Editor: Madhubanti Bhattacharyya
Market Development Executive: Elizabeth Hobson

FSC
www.fsc.org
MIX
Paper from responsible sources
FSC® C019520

Typeset by Out of House Publishing
Index created by Cathryn Pritchard
Printed and bound by NOVOPRINT S.A., Spain

To
Ratidzo, Tafadzwa and Simbarashe

Contents

Preface

Shell structures find application in many fields of engineering, notably civil, mechanical, marine and aeronautical disciplines. Since about the beginning of the twentieth century, considerable effort has been expended on the development of rigorous theories – both general and specialist – to describe the behaviour of shells in the elastic range as realistically as possible. The problem with such rigorous formulations has been that, in almost all cases, they have led to sets of differential equations for which exact mathematical solutions cannot be found, or for which the mathematical solutions that are obtained are not very practical from an engineering point of view. Not surprisingly, the development of numerical formulations over the past 60 years, notably the finite-element method, has seen the gradual decline of attempts to find analytical solutions to rigorous mathematical formulations of shell behaviour.

The finite-element method is a general numerical procedure that can be traced back to the early 1960s. It can be used to tackle any shell problem to any desired degree of accuracy; one needs only to model the behaviour of the basic individual elements making up a structure (rather than the behaviour of the whole structure), and then assemble the relevant structural properties by enforcing compatibility conditions between adjacent elements, conditions of equilibrium of forces and moments, as well as boundary conditions at the edges or supports of the structure. In this way, the overall behaviour of the complete structure is approximated. The finite-difference method is another popular numerical approach, where the exact differential equations of the problem are approximated with equivalent finite-difference equations, which can then be solved more conveniently as linear algebraic equations. Numerical modelling is particularly useful in those instances where the mathematical formulation results in sets of differential equations that are difficult or impossible to solve analytically, particularly where non-linearities in material and/or geometric behaviour have to be accounted for.

However, for situations where the mathematical formulation results in sets of differential equations that can be solved analytically, the numerical solution approach, while still effective as a tool for obtaining actual values of stresses and displacements in a structure, is not necessarily the most efficient for the purposes of studying the effect of varying a single parameter (such as the thickness of a shell) during the course of design, since each time such a change is made, the entire analysis procedure has to be carried out, albeit by a computer. Numerous runs of the computer program may be necessary in order to obtain a comprehensive understanding of the effect (on the stresses and deformations in the shell) of varying a parameter of interest through the full range of possibilities. Thus the process of preliminary design can easily become very expensive if several structural parameters need to be studied.

It is in situations such as these that analytical formulations can be extremely useful. In fact, we may justify the pursuit of exact mathematical solutions to the differential equations of the shell, in an era where numerical solution approaches have become very popular (owing to their ease of programming), on the grounds that analytical solutions: (*a*) provide valuable insights on the key parameters that govern the behaviour of the shell; (*b*) are more amenable to systematic parametric study than numerical formulations; and (*c*) provide a means (sometimes the only means) for checking the correctness of numerical solution procedures.

In many situations, particularly where the aim is simply to calculate the stresses and deformations in the shell under normal service conditions, the assumption of linear-elastic material and geometric behaviour is usually made right from the outset. This certainly simplifies the problem from a mathematical point of view. Unfortunately for many types of shells, linear-elastic assumptions still result in sets of differential equations that are very difficult to solve exactly. To be able to obtain analytical solutions for such problems, we may either (*a*) make further simplifying assumptions at the beginning of our formulation, or (*b*) introduce simplifying approximations in the rigorously obtained sets of governing differential equations. Either way we end up with differential equations that, while not exact, have the merit of having known mathematical solutions. Provided the simplifying assumptions are good, this approach can afford reasonably accurate closed-form analytical solutions to a wide range of practical shell problems. It is an effective strategy that we will adopt throughout this book.

Closed-form analytical solutions are highly suitable for practical calculations, enabling quantities that are of interest to a designer (such as stresses and deformations) to be computed with ease. General expressions (in terms of the parameters defining the geometry of the shell and the nature of the loading) can easily be programmed on portable scientific calculators or small computers, permitting the designer to rapidly evaluate stresses in a shell structure of a given set of parameters, or to obtain important design insights by simply examining the form of the mathematical expressions, or to obtain a deeper understanding of the effect of varying a particular parameter by simply plotting the results with respect to that parameter.

This book is intended to be both a textbook for senior undergraduate and postgraduate students of civil and mechanical engineering, and a reference book for practising engineers in these disciplines. It is also intended to serve as a useful resource for researchers involved with the development of more effective methods of shell analysis, as well as software developers. As with the first edition, the distinguishing features of the book are as follows:

(*a*) emphasis is placed on the presentation of simplified shell theory for which closed-form analytical solutions can be obtained;
(*b*) for a wide range of shell problems, closed-form results for stresses and deformations are derived and presented in a way that allows convenient evaluation of these quantities by the designer, and that permits any desired parametric studies to be conducted. For many problems, the results of such parametric studies, drawn from the author's previous work, are presented and design implications pointed out, enhancing the reference value of the book.

The main differences between the first edition and the present edition are (*i*) the inclusion of new sections on non-conventional shell forms for liquid-containment applications in Chapters 3, 7 and 8, based on recent work of the author, and (*ii*) the addition of three new chapters covering buckling of thin shells, finite-element analysis and design considerations for shells. Although the main emphasis of this book remains the development of simplified mathematical theory and presentation of closed-form analytical solutions for a wide variety of shell problems, the last three chapters have been included for the sake of completeness; it is acknowledged that no treatment of shells would be regarded as complete without some consideration of the important topics of buckling, finite-element modelling and design.

In shells where there is a significant amount of compression or shear, and the shell happens to be very thin, there is always the possibility of failure by buckling rather than material crushing or rupture. It is therefore important to be able to predict the critical loading that will trigger failure by buckling, and be able to design against this instability phenomenon. Also, where a shell problem cannot be solved analytically owing to the complexity of the differential equations or boundary conditions, resort is usually made to finite-element analysis, so there is merit in understanding how the finite-element method works, and how it can be implemented in the context of shell analysis. Ultimately, all analysis serves the purpose of informing design, and the connection between the two is important. Thus an overview of design considerations for shell structures seems an appropriate way of concluding a book of this type.

Chapter 1 explains the essential properties of shell structures and the manner in which shells resist loads. It traces the historical development of shell theory and analytical solution procedures, and discusses the assumptions and limitations of the membrane hypothesis in shell analysis. Chapters 2 to 4 develop the membrane theory of shells of revolution, and present membrane solutions for pressure vessels, liquid-containment vessels, domes, roofs and cooling towers. Chapters 5 and 6 present the theories of the axisymmetric bending of cylindrical shells and general shells of revolution. Chapter 7 explains the flexibility formulation for shell structures, and applies this to shell-ring problems and multi-shell assemblies. Chapter 8 features a series of parametric studies of spherical tanks and domes, as well as feasibility studies of certain non-conventional shapes for sludge digesters. The use of shell theory as a tool for exploring alternative new shell forms is also illustrated.

Chapter 9 presents the membrane theory of general cylinders and its application to liquid-containment vessels and barrel shell roofs of circular, elliptic and parabolic cross-sections. The membrane theory of shells of arbitrary shape is developed in Chapter 10, and applied to the analysis of elliptic and hyperbolic paraboloidal shell roofs. An approximate theory for the linear elastic buckling of cylindrical shells is presented in Chapter 11, and results for critical buckling loads derived for the cases of axial loading, uniform lateral pressure and torsion. Results for other common shell geometries (spherical, conical, ellipsoidal and toroidal shells) are also presented. The finite-element formulation for shells is outlined in Chapter 12, with considerations covering static, dynamic and stability behaviour. Various types of shell elements are discussed, and the process of assembly and solution of system equations explained.

The last chapter (Chapter 13) is devoted to design considerations for concrete and steel shells. For steel vessels, Eurocode design criteria and guidelines of the American Petroleum Institute are reviewed, while for concrete shells and thin shell roofs in particular, the key recommendations of the American Concrete Institute and of the International Association for Shell and Spatial Structures are summarised.

While every effort has been made to ensure that all mathematical expressions are correct, it is possible a few errors have escaped detection, given the large number of equations and mathematical results that are presented. The author welcomes all feedback from readers in this regard. It is hoped the book will succeed in serving its intended purpose.

Alphose Zingoni
Cape Town, 2017

Acknowledgements

The author dedicates this book to his wife Lydia and children Ratidzo, Tafadzwa and Simbarashe, for their patience, understanding and constant encouragement throughout the writing stages. The author has continued to draw inspiration from the generations of prominent scholars who have laid the foundations and extended the boundaries of shell theory in the last century, including the late Professor Milija Pavlović, his PhD supervisor at Imperial College London, who first introduced him to the study of shells.

The author also wishes to acknowledge the input of various international experts who made some useful comments on the content, and of his PhD students, who assisted with some of the numerical calculations. This second edition has benefitted from the constructive feedback received from many readers across the world, since the appearance of the first edition in 1997. The author is most grateful for this. The many illustrations in the book were prepared by Angus Rule of the University of Cape Town, to whom the author owes deep gratitude.

The author was a recipient of a grant of the National Research Foundation of South Africa during the writing of this book, which helped in meeting some of the costs of preparing the manuscript. The author is also grateful for the 6 months' sabbatical leave granted by the University of Cape Town in early 2017, for the purposes of completing the book.

Without the constant encouragement of Inês Pinheiro, Development Editor at ICE Publishing, this edition might have taken a little longer to complete. The professionalism of the production department at ICE Publishing is gratefully acknowledged.

Alphose Zingoni

About the author

Alphose Zingoni PhD, DIC, PrEng, CEng, FSAAE, FIABSE, FIStructE is Professor of Structural Engineering and Mechanics in the Department of Civil Engineering at the University of Cape Town. He earned a PhD degree from Imperial College London in 1992, for a thesis on the bending of non-shallow spherical shells. He served as Dean of the Faculty of Engineering at the University of Zimbabwe from 1997 to 1999, and as Head of the Department of Civil Engineering at the University of Cape Town from 2008 to 2012. A past recipient of a Research Fellowship of the Royal Commission for the Exhibition of 1851 (1992–1994), his current research interests encompass shell structures, structural vibration analysis, finite-element modelling, studies of symmetry and applications of group theory to computational problems in structural mechanics. He has also written a book on vibration analysis, published by CRC Press (Taylor & Francis) in 2015. He serves on the editorial boards of several international journals, and is registered as a Professional Engineer with the Engineering Council of South Africa, and as a Chartered Engineer with the Engineering Council of the UK. Among his most recent honours is the award of a Fellowship of the University of Cape Town in 2016, in recognition of 'original distinguished academic work'.

Shell Structures in Civil and Mechanical Engineering
ISBN 978-0-7277-6028-9

http://dx.doi.org/10.1680/ssicame.60289.001

Chapter 1
Introduction

1.1. Overview of shell applications

1.1.1 Early shell structures

Man-made shell structures have been in existence for many centuries. One of the earliest applications of the shell as a structural form is the dome. Since the times of the ancient civilisations of Greece and Rome, many domes have been constructed for the purposes of providing roofing for temples, cathedrals, monuments and other buildings. Notable historical examples include the Pantheon of ancient Rome, built around two thousand years ago; the Hagia Sophia of Constantinople (now Istanbul), which is approximately one thousand five hundred years old; St Peter's Basilica in Rome, a Renaissance structure whose dome, dating back to about 1590, was designed by Michelangelo; St Paul's Cathedral in London, designed by Sir Christopher Wren and built around three hundred years ago; and the Taj Mahal of India, built in the seventeenth century by the Mogul Emperor, Shah Jahan.

1.1.2 Modern shell applications

The generally high strength-to-weight ratio of the shell form, combined with its inherent stiffness, has formed the basis of modern applications of shell structures. As may be seen in more detail in the historical account of Sechler,[1] developments in various sectors of human activity, since the beginning of the Industrial Revolution in Europe more than two hundred years ago, have opened up numerous fields of application for shells. Among these are thin concrete shell roofs, of which many interesting architectural shapes were constructed in the middle part of the twentieth century. A relatively recent example is the elegant roof of the central building of the L'Oceanogràfic aquarium in Valencia (Spain), opened in 2003. Shaped like a water lily and with a thickness of only 60 mm, the shell roof (Figure 1.1) was designed by the architect Felix Candela and structural engineers Carlos Lázaro and Alberto Domingo.

Apart from roofing, shells find application in many other areas. Figure 1.2 shows a hyperbolic cooling tower under construction at the Niederaussem Power Station in Germany. With an eventual height of 200 m, this structure became the tallest cooling tower in the world at the time of completion in 2000. Cylindrical silos for the storage of grain (Figure 1.3) represent another application of shells, as do elevated water towers, which come in all sorts of shapes (Figure 1.4).

Large storage vessels for oil and industrial chemicals[2,3] are usually constructed in thin steel plate, to give a variety of containment shell forms that include cylindrical, conical, spherical and ellipsoidal tanks (Figure 1.5). In the manufacturing and process industries, boilers, pressure vessels and associated piping are further examples of shell structures in thin metal construction. Figure 1.6 shows an 80 m-long catalytic-distillation steel vessel being transported to its place of deployment in Malaysia. Hollow members of large industrial steel structures, such as offshore oil platforms, are another example of shell applications, as are bodies of transportation structures such as motor vehicles, ships, submarines, aircraft, missiles and spacecraft.

Figure 1.1 Thin concrete shell roof of the Oceanogràfic aquarium building in Valencia, Spain: (a) completed shell before fitting of glass walls (photo supplied by Dr Carlos Lázaro); (b) view of shell roof after opening of aquarium (reproduced from Wikimedia under a Creative Commons license). The shell has a thickness of 60 mm and is made of steel-fibre reinforced concrete.

(a)

(b)

Figure 1.2 Niederaussem Cooling Tower in Germany nearing completion (photo supplied by Krätzig & Partners, courtesy RWE Energie AG). With a height of 200 m, this concrete shell became the tallest cooling tower in the world at the time of completion in 2000.

Figure 1.3 Grain silos in South Africa (courtesy NWK Ltd).

Figure 1.4 Examples of elevated water towers: (a) conical tower in Johannesburg, South Africa (reproduced from Wikimedia under a Creative Commons license, courtesy NJR ZA); (b) spheroidal tower in Barnesville, Minnesota, USA (reproduced from Wikimedia under a Creative Commons license, courtesy JP Salvatore).

(a)

Figure 1.4 (*Continued*)

(b)

Within civil engineering, the great diversity of shell applications is also seen in the curved panels of the world-famous Thames Barrier in London. Another innovative application of shells has been in the area of wastewater treatment, where giant egg-shaped sludge digesters (Figure 1.7) have provided energy-efficient and highly effective solutions in many countries around the world, including the USA, Germany, the UK, Australia, China and South Africa.

The essential geometrical property of all the above shell structures that distinguishes them from other structural forms is, as Calladine[4] points out, the possession of both 'surface' and 'curvature'. It is, in fact,

Figure 1.5 Petroleum storage steel vessels: (a) large cylindrical oil tanks at Stolthaven Houston Terminal, Texas, USA (photo supplied by Dr Luis A. Godoy); (b) spherical vessels for liquefied petroleum gas at MiRO petroleum refinery, Karlsruhe, Germany (reproduced from Wikimedia under a Creative Commons license).

(a)

(b)

this combination of 'surface' and 'curvature' that endows shells with their characteristic strength and stiffness.

A shell may be defined as a relatively thin structural element in which the material of the element is bound between two curved surfaces a relatively small distance apart. The structural behaviour of the shell is usually modelled on the basis of its middle surface (alternatively referred to as 'midsurface'), which is the locus of interior points equidistant from the two bounding surfaces of the shell.

Figure 1.6 An 80 m-long chemical reaction steel vessel (catalytic distillation column) under transportation in Malaysia (courtesy deugro Group, Germany).

Figure 1.7 Egg-shaped concrete digesters under construction at Darvill Wastewater Treatment Works in South Africa. Photo supplied by Group Five (contractor) with the permission of Umgeni Water (owner). The digesters were designed by Hatch Africa.

1.2. Shell action in relation to beam, arch and plate actions

1.2.1 General comparisons

The beam, a structural element with a straight-line idealisation (Figure 1.8(a)), mainly resists any applied transverse loading by bending. On the other hand, the arch, a structural element with a curved-line idealisation (Figure 1.8(b)), will resist such a loading primarily by internal thrust action, assuming of course that the ends are adequately restrained (i.e. the supports of the arch must be capable of providing force reactions to the thrust transmitted to them).

The plate, a structural element that may be idealised as a flat surface (Figure 1.8(c)), has to bend in order to carry applied transverse loading. In contrast, the shell, a structural element that may be idealised as a curved surface (Figure 1.8(d)), if suitably supported, is quite capable of resisting smoothly varying transverse surface loading through the action of internal forces in the tangential plane of the shell's middle surface at any given point; such forces will henceforth be referred to as 'in-plane' actions.

From these simple considerations, it is evident that the property that makes both arches and shells capable of resisting external transverse loadings by 'extensional' action (as opposed to 'flexural' action) is *curvature*. Here, the term 'extensional' refers to any in-plane direct forces, whether tensile or compressive. Now, structural materials are generally more efficient in carrying loads when they are resisting the applied loading by extensional action rather than flexural action. It follows that arches and shells, when suitably supported, are stronger and stiffer than beams and plates of the same cross-sectional area or thickness.

To illustrate this last point, consider Figure 1.9, which shows a straight prismatic bar of length l and rectangular cross-section of constant width b and constant depth d (so that the cross-sectional area $A = b \times d$). The dimension l is assumed to be much larger than the dimensions b and d. The bar is required to support a weight W. In Figure 1.9(a), the bar is vertical and supported at the top, with the weight W hung from the lower (free) end of the bar. In Figure 1.9(b), the bar is horizontal and simply supported at its ends, with the weight W applied as a vertical point load at midspan. In Figure 1.9(c), the bar is horizontal and rigidly built into a vertical wall at the left end, while the right end is free and subjected to the weight W applied as a vertical point load. Figure 1.9(d) shows a cross-section of the horizontal bar in the bending mode, with $N - N$ denoting the neutral axis for bending.

Let us denote the yield stress of the material of the bar by σ_y, and furthermore, let us assume this to be the same in tension and in compression. From elementary considerations of direct stress in axially loaded members and members subjected to pure bending, we can easily calculate the maximum weights W_{max} that can be supported by the arrangements of Figures 1.9(a), (b) and (c); the results are $W_{max} = \sigma_y A$, $(W_1)_{max} = (2/3)(d/l)\sigma_y A$ and $(W_2)_{max} = (1/6)(d/l)\sigma_y A$, respectively. Thus, the vertical bar will support in direct tension a weight hung at the lower end of magnitude $1.5\, l/d$ times greater than the weight it can support at midspan as a simply supported horizontal beam, and $6\, l/d$ times greater than the weight it can support at the free end as a horizontal cantilever.

Since $l/d >> 1$, we conclude that the extensional mode of load resistance is much more efficient than the bending mode, implying that arches and shells, with their capability to mobilise extensional action, are inherently more efficient than beams and plates of the same proportions, which rely on bending action only.

In beams and plates, transverse external loads do not give rise to any appreciable extensional action within these elements, provided that the displacements during deformation remain relatively small in relation to the transverse dimension of the elements (i.e. depth in the case of a beam, and thickness in the case of a plate).

Figure 1.8 Transverse loadings on: (a) beam; (b) arch; (c) plate; (d) shell.

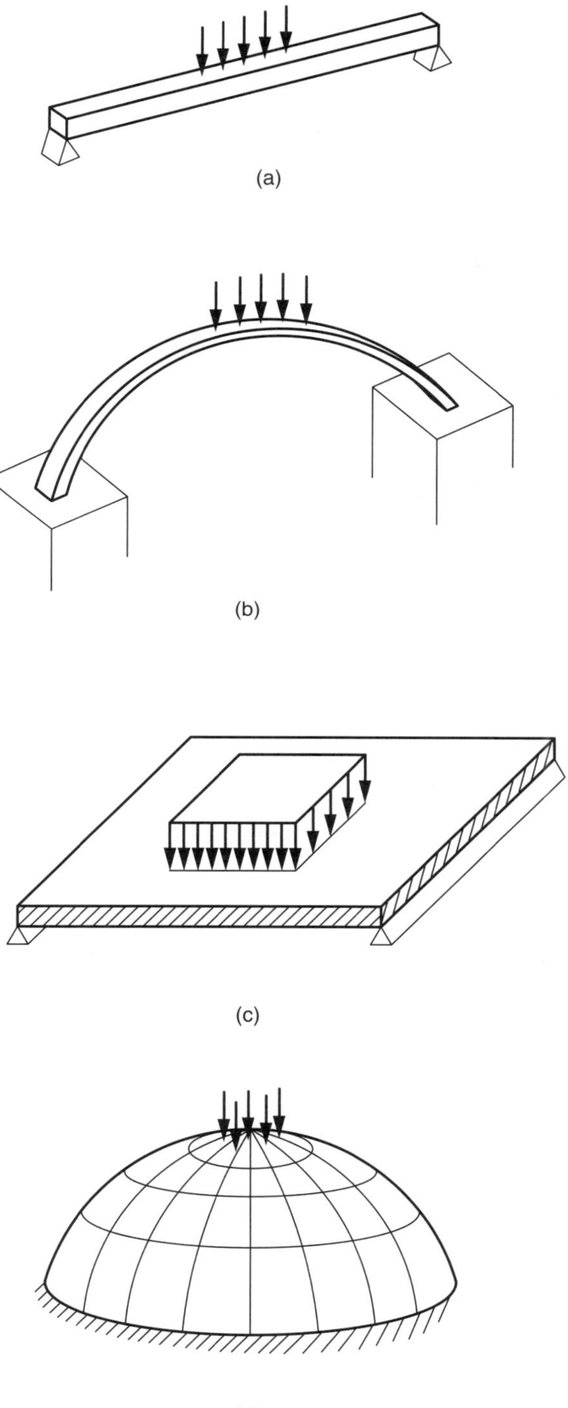

Figure 1.9 Comparison of maximum weights that can be supported by a bar of uniform cross-section in direct tension and in bending: (a) bar in direct tension: $W_{max} = \sigma_y A$; (b) bar in bending as a simply supported beam: $(W_1)_{max} = (2/3)(d/l)\sigma_y A$; (c) bar in bending as a cantilever beam: $(W_2)_{max} = (1/6)(d/l)\sigma_y A$; (d) cross-section of beam: $N - N$ denotes neutral axis for bending. Parameters A (= bd) and σ_y denote cross-sectional area of bar and yield stress of material respectively.

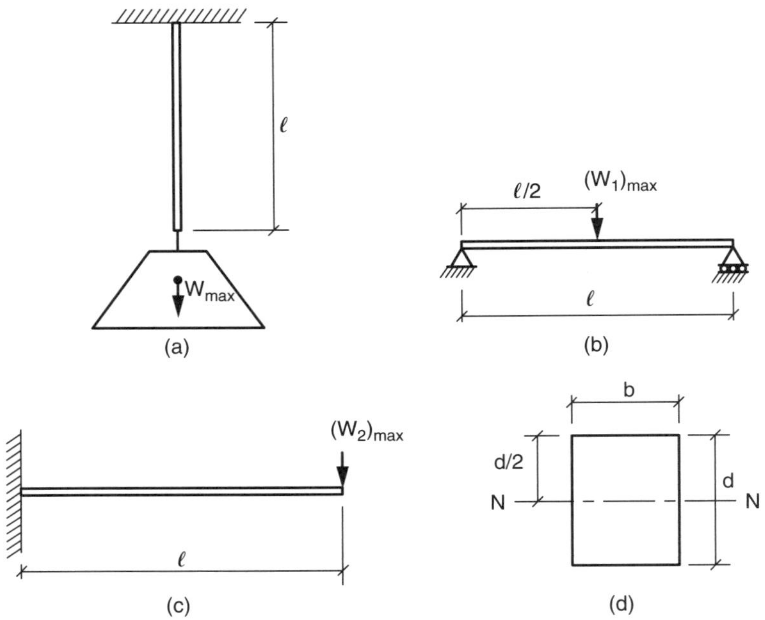

On the other hand, arches and shells can resist distributed transverse applied loads almost wholly by extensional action provided that, in the case of arches, the loading pattern is of a certain specific type and suitable boundary conditions are provided, and in the case of shells, suitable support conditions are provided and certain smoothness criteria for variations of loading and shell geometry are met.

In practice, a mixture of extensional and flexural actions usually occurs in arches and shells, since the above ideal conditions are not exactly satisfied; for these structural elements, good design consists in minimising the flexural actions.

1.2.2 Distribution of extensional and flexural effects in shells

For shells, the relative proportions of extensional and flexural effects at a given point depend on several factors such as the type of shell surface, support conditions, loading configuration, and the proximity of edges and certain discontinuities. A useful parameter in the classification of shell surfaces is the *Gaussian curvature*. This is defined as the product of the two principal curvatures (i.e. the maximum and minimum curvatures) at a given point of a curved surface. Principal curvatures exist in two planes called the principal planes, which are perpendicular to each other. Based on Gaussian curvature, three types of shell surfaces are distinguished in Figure 1.10.

Synclastic surfaces are those with positive Gaussian curvature. As is clear from Figure 1.10(a), for a synclastic surface, the centres of maximum and minimum curvature at a given point lie on the same side of the surface, so that the surface, when suitably oriented in space, has the capability to hold a small amount of liquid around the point.

On the other hand, anticlastic surfaces possess negative Gaussian curvature, the centres of maximum and minimum curvatures at any given point of the surface lying on opposite sides of the surface, so that the surface is 'saddle shaped' and therefore incapable of holding a small amount of liquid around the point no matter how the surface is oriented in space (Figure 1.10(b)).

Developable surfaces, as the name implies, are those that can be flattened into a plane surface, either directly or after making a single straight-line cut in the surface (Figure 1.10(c)); they are characterised by zero Gaussian curvature, and as such, are also not capable of holding a small amount of liquid around an arbitrary point on the surface (the liquid would simply run off along the direction of zero curvature!).

Figure 1.10 Types of shell surfaces: *synclastic*: (a) paraboloid of revolution; (b) elliptic paraboloid; *anticlastic*: (c) hyperboloid of revolution; (d) hyperbolic paraboloid; *developable*: (e) circular cylinder; (f) circular cone.

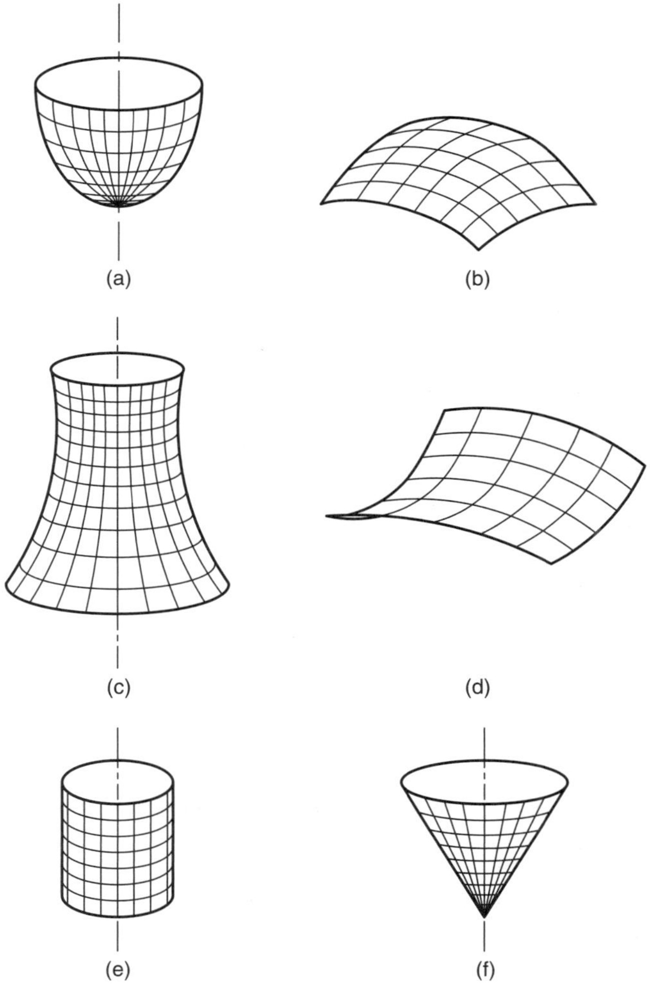

For synclastic shells, if both the shell geometry (shell thickness, midsurface slope, principal radii of curvature, etc.) and surface-loading components are smoothly varying (i.e. exhibiting no discontinuities), then extensional effects generally prevail in the interior regions of the shell, while a mixture of extensional and flexural effects occurs in the edge zones bordering the supports, for most practical constructions of supports. Such localisation of bending effects to the boundary zones of the shell is also noted, but to a somewhat reduced extent, in developable shells. In anticlastic shells, bending emanating from the edges of the shell may propagate along certain directions deeply into the interior of the shell, a behaviour not unlike the bending in beams and plates.

1.3. Thin-shell theories in relation to more general theories

As already indicated, a shell may be defined as 'a three-dimensional structure bounded primarily by two arbitrary curved surfaces a relatively small distance apart'.[5] Criteria for thinness vary, depending on the magnitude of errors that can be tolerated in regarding a particular shell as thin, and hence suited to study by thin-shell theories rather than by the more general (but less convenient) thick-shell theories that, by their nature, are closer to the general theory of elasticity.

Assuming a permissible error of 5% for ordinary engineering computations, Novozhilov[6] proposes $t/r_m = 1/20$ as the cut-off point between thin and thick shells, where t is the thickness of the shell and r_m is the minimum radius of curvature of the midsurface of the shell. He notes that a large number of shells encountered in practical applications lie in the range $1/1000 \leq t/r_m \leq 1/50$, and therefore belong to the class of thin shells. Vlasov[7] adopts a more stringent criterion for thinness: $t/r_m \leq 1/30$.

The results of Goldberg *et al.*,[8] who compared the thick- and thin-shell theories on the basis of a circular-cylindrical shell built-in at both ends and subjected to a uniform internal pressure, suggest that less stringent criteria may be possible for particular shell geometries, loading configurations and boundary conditions. In this study, an error of only 6% was observed for a t/a ratio (where a is the radius of the cylinder) as large as 1/10. This reduced to a mere 4% when the t/a ratio became 1/15.

Being a 3-dimensional body (no matter how thin it might actually be), any shell could, in principle, be tackled using the general theory of elasticity. However, owing to the complexities of such an approach, the shell is almost invariably regarded as a 2-dimensional entity characterised by the coordinates of the middle surface. In principle, the simplified theory for such a 2-dimensional approximation could be deduced directly from the more general 3-dimensional theory of elasticity as a limiting case of the latter when the thickness dimension becomes very small in comparison to the other two dimensions. As pointed out by Mollmann,[9] however, this method of derivation of a theory of shells leads to contradictions in seeking to relate the two theories (i.e. the 2-dimensional shell theory and 3-dimensional elasticity theory). Consequently, shell theories have mostly been derived independently of the 3-dimensional theory of elasticity, by adopting at the outset a set of clearly defined assumptions from which to proceed.

1.4. Historical developments in linear shell theory and analytical solutions

1.4.1 Historical note

An investigation of the general theory of shells, based on the Kirchhoff hypothesis concerning the deformation of plates,[10] was first attempted by Aron in 1874.[11] This development, however, did not present the theory satisfactorily. On the basis of the same Kirchhoff assumptions, Love derived, in 1888, the basic equations that govern the behaviour of thin elastic shells.[12,13] Although a theory of thin elastic shells had emerged from Love's considerations, the primary concern of that work in the first instance had been the search for a theory of the vibration of bells.[12] A fascinating account of the controversy surrounding the beginnings of the theory of thin shells as postulated by Love may be seen in papers by Calladine,[14,15] in which important contributions of Love's contemporaries, namely Rayleigh, Bassett and Lamb, are also discussed.

1.4.2 The Love–Kirchhoff assumptions and Love's approximations

The simplifying assumptions first adopted by Kirchhoff for plates and later extended to shells by Love are sometimes referred to as the Love–Kirchhoff assumptions. These simplifications may be stated as follows:

(*a*) *The shell thickness is negligibly small in comparison with the least radius of curvature of the shell middle surface.* This implies that the shell is thin, and hence is the most fundamental of the assumptions. Criteria for thinness have been discussed in Section 1.3.

(*b*) *Strains and displacements that arise within the shell are small.* This implies that products of deformation quantities occurring in the derivation of the theory may be neglected, ensuring that the system is described by a set of geometrically linear equations. This also makes it possible to formulate the equilibrium conditions of the deformed middle surface with reference to the original position of the middle surface prior to deformation.

(*c*) *Straight lines that are normal to the middle surface prior to deformation remain straight and normal to the middle surface during deformation, and experience no change in length.* This assumption is analogous to Navier's hypothesis for beams, which states that plane sections before bending remain plane after bending, leading to the Bernoulli–Euler elementary theory for beams.[10,11] It implies that the direct strain in the direction normal to the middle surface, and the shearing strains in planes perpendicular to the middle surface and due to transverse shear forces, are all zero.
From this assumption alone, which is a statement about deformations and not stresses, it cannot be concluded that the direct stress in the direction normal to the middle surface, and the transverse shearing stresses acting on planes perpendicular to the middle surface, are also zero. It must follow, therefore, that the shear moduli corresponding to transverse shearing in the planes perpendicular to the middle surface are infinite; however, Young's modulus in the direction normal to the middle surface need not be regarded as infinite in view of the next and last assumption.

(*d*) *The direct stress acting in the direction normal to the shell middle surface is negligible.* This assumption is justified because the shell has already been assumed to be thin. However, the assumption is not valid in the vicinity of concentrated transverse loads, even if the shell is thin.

The theory of thin shells derived on the basis of all the above assumptions is referred to in the literature as 'Love's first approximation'. Love later made efforts to lift some of these restrictions from the theory (by no longer regarding the length of a normal to the middle surface before deformation as invariant during deformation, while still preserving the assumption that it remains straight and normal to the middle surface throughout deformation). In so doing, he obtained another version of shell equations, now referred to as 'Love's second approximation'.

1.4.3 Refinements and the question of consistency

Flügge in 1934[16] and Byrne in 1944[17] independently developed a more general set of shell equations by retaining all of the first-approximation assumptions of Love except (*a*) on thinness. These investigators retained, in the stress-resultant and the strain-displacement relations, terms of the order z^2 / r^2 compared to unity, where z ($\leq t/2$) is the perpendicular distance of a point from the middle surface and, as before, r is the radius of curvature of the middle surface. E. Reissner later pointed out, in 1952,[18] that such a correction to Love's first approximation amounted to the same order of magnitude as that which would be obtained by taking into account transverse normal stresses and transverse shear deformations, and therefore argued that assumptions (*a*), (*c*) and (*d*) above must be abandoned simultaneously for consistency.

Further refinements in the theory, in the period 1948–1958, included the incorporation of the effects of transverse normal stress and transverse shear deformations.[18–21] These additional refinements are justifiable only in the case of thick shells. Love's second approximation, the Flügge–Byrne theory, and the refinements abandoning all of the Love–Kirchhoff assumptions except the one on smallness of displacements are often collectively referred to as 'higher-order approximations' because they suspend (or, rather, attempt to remove) one or more of the assumptions fundamental to the first approximation of Love.

In 1963, the refined theories of E. Reissner[18] and Naghdi[21] were made invariant under different coordinate systems by Naghdi,[22] this being more of a mathematical refinement of the formulation than an extension of the physical validity of the theory. An account of the various theories of thin elastic shells that had been put forward by 1955 may be seen in the survey by Naghdi.[23]

Returning to the first-approximation theories that retain all of the assumptions (*a*)–(*d*), it is noted that in 1941 E. Reissner[24] had given a new derivation of the equations for the deformation of elastic shells, by taking the lines of curvature on the middle surface as the coordinate curves. In 1956, Knowles and E. Reissner[25] generalised this derivation of a first-approximation theory by adopting an arbitrary orthogonal system of coordinates on the middle surface of the shell. A yet more general derivation of Love's first-approximation theory, which is valid for non-orthogonal coordinate curves on the middle surface, is contained in the 1954 work of Green and Zerna.[26] This derivation made use of the powerful tool of tensor notation.

Not surprisingly, the increasing number of apparently different versions of the linear theory of elastic thin shells that were being put forward as 'improvements' of Love's first-approximation theory soon became a cause for concern, giving rise to the question as to which of the various sets of equations were most accurate and most consistent. The studies of Koiter, published in 1959,[27] clarified the question of accuracy, and confirmed the suspected view that there was very little difference in the accuracy of the many sets of derivations existing at that time (in a review paper published in 1967, Koiter[28] re-examines this fundamental issue from a somewhat broader perspective).

In 1959, Sanders[29] improved Love's first-approximation theory, while retaining the four basic assumptions, by removing certain inconsistencies that were embedded in Love's theory. This work was based on lines-of-curvature coordinates. Four years later, Budiansky and Sanders[30] derived what they termed a 'best' first-order linear shell theory, using general tensor notation in terms of an arbitrary coordinate system for the middle surface of the shell. This theory, the authors hoped, would be accepted as standard, since it embodied all the desirable characteristics of a good theory as defined by the authors.

Today's version of Love's first-approximation theory is referred to, quite appropriately, as the Sanders–Koiter set of equations, since it was they who had finally settled the fundamental questions of consistency and accuracy.

1.4.4 Developments towards the solution of equations

Owing to the complexity of the eighth-order system of the differential equations of the first-approximation theory of shells, it became necessary to simplify the equations by neglecting or adding certain minor terms in such a way as to make the resulting equations more readily reducible to solvable forms.

One of the methods proposed for this purpose is that of complex transformation. This technique, first suggested for the simplification of shell equations by Novozhilov (see the 1962 paper of Novozhilov[31] for a history of the development of complex transformation), involves formulating the general equations of the theory in complex form. In the first instance, Novozhilov reduced the equations of shell theory to complex form for the particular case of zero Poisson's ratio, without discarding any terms in the equilibrium and compatibility equations.

Novozhilov was able to extend the complex transformation to cases of arbitrary values of Poisson's ratio only after discarding certain terms of the order t / r in the equilibrium and compatibility relations, a simplification he justified on the grounds that it leads to errors of the same magnitude as those inherent in the fundamental assumptions of the first-approximation theory of shells.

In 1967, Sanders[32] achieved complex transformation by a method that avoided the necessity of approximating the equilibrium and compatibility equations, or any of the other relations except the constitutive equations. This was effected at the expense of introducing a secondary set of equations for certain 'error terms' in the constitutive relations.

An advantage of the method of complex transformation is that the order of the equations is reduced by a factor of two, and they become 'less cumbersome'.[31] However, the accuracy of the resulting equations may become questionable for certain classes of problems. In 1961, Koiter[33] had pointed out that Novozhilov's equations did not adequately describe the axisymmetric deformation of a helicoidal shell. Even Sanders' somewhat more accurate complex reduction[32] was also subject to this particular shortcoming.

The set of 12 equilibrium, compatibility and constitutive relations comprising the Sanders–Koiter first-approximation theory was simplified, without any appreciable loss of accuracy, by Simmonds in 1970.[34] These 12 equations, when specialised to shells of constant *average* midsurface curvature, were reduced to eight equations by adding certain small terms (on the justification that such alterations introduce inaccuracies of the same order of magnitude as are inherent in the unaltered relations) to the stress-strain relations. For shells of non-zero Gaussian curvature, these simplified equations were further reduced to four coupled equations.

In 1975, Simmonds[35] reduced, through extensive use of order-of-magnitude arguments in neglecting certain terms, again without any significant loss of accuracy, the Sanders–Koiter equations for shells of constant thickness with a smooth (but otherwise arbitrary) midsurface, to a pair of coupled fourth-order partial differential equations.

Drawing on the characterisation of Koiter,[33] Bouma, in a 1962 review of approximate methods of shell analysis,[36] had highlighted a systematic scheme for the simplification of the sets of differential equations of the general theory of shells. Koiter's scheme was based on a classification of possible states of stress in the shell. Combining three cases of deformation modes (predominantly extensional; predominantly flexural; mixed) with three ranges of the wavelength of the deformation pattern of the shell relative to a characteristic dimension of the shell (long; medium; short) resulted in seven possible simplification approaches of the general shell equations (corresponding to the equations of (*i*) plane stress, (*ii*) plate bending, (*iii*) membrane theory, (*iv*) inextensional bending, (*v*) superposition of plane stress and plate bending, (*vi*) superposition of membrane theory and inextensional bending, and (*vii*) shallow-shell theory), leaving two exceptional cases.

1.4.5 Specialised theories

1.4.5.1 Spherical shells, general shells of revolution and shallow shells

The differential equations of the theory of elastic thin shells as given by Love[12,13] were not in a form suitable for the practical analysis of shells, since they needed to be solved first. With the development of automated procedures for numerical integration in later years, the need to obtain analytical solutions of such equations,

Figure 1.11 Actions on an element of an axisymmetrically loaded shell of revolution: N_ϕ and N_θ are direct forces per unit length in the meridional and hoop directions respectively; M_ϕ and M_θ are bending moments per unit length in the meridional plane and the hoop plane of the shell midsurface respectively; Q_ϕ is the transverse shear force per unit length in the meridional plane; p_ϕ and p_r are external load components per unit area of shell midsurface in the direction of the tangent to the meridian and the normal to the shell midsurface respectively; r_1 and r_2 are the principal radii of curvature of the shell midsurface in the first principle plane (i.e. the meridional plane) and the second principal plane respectively; ϕ and θ are angular coordinates along a meridian and a hoop circle respectively.

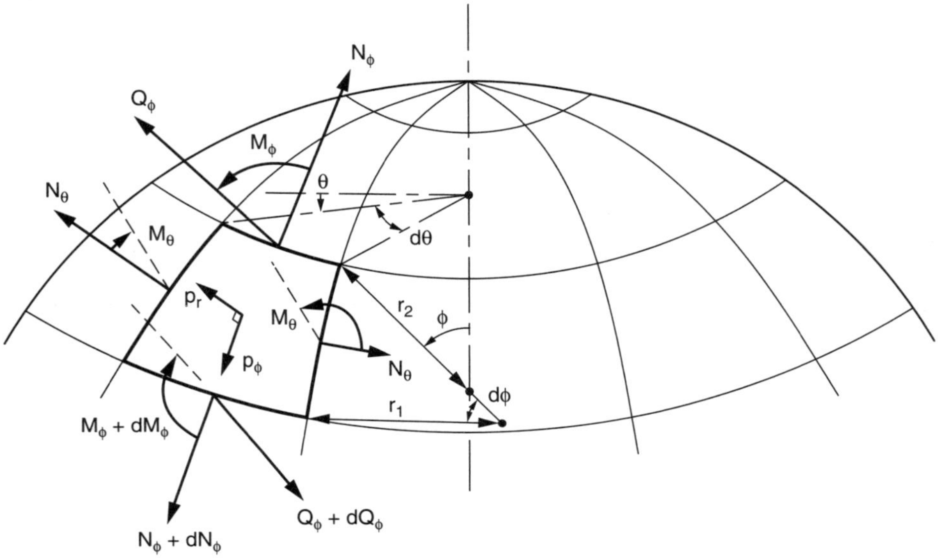

for the primary purpose of evaluating the internal actions and displacements of the shell, was to eventually diminish.) Specialisation of the first-approximation differential equations to shells possessing axial symmetry (i.e. shells of revolution) and, furthermore, loaded symmetrically with respect to their axis of symmetry, was expected to render these equations more amenable to reduction to forms for which solutions could be found. Thus, around 1910, there emerged some efforts towards the development of usable solutions for the spherical shell, the circular-cylindrical shell, and shells of revolution in general. However, the reduction of the equations of the first-approximation theory for more arbitrary shells was to take longer to achieve (as noted in the previous section), and only then after further minor simplifications.

In 1912, H. Reissner[37] had managed to reduce the differential equations for the axisymmetric deformation of a spherical shell to two simultaneous ordinary second-order differential equations in the variables Q_ϕ and V. In the present and subsequent considerations, Q_ϕ is the transverse shear per unit length of a shell element cut out by two adjacent meridians and two adjacent circles of latitude of the shell, acting on the parallel edges of the circles of latitude. Figure 1.11 shows the Q_ϕ forces and all other actions associated with an element of an axisymmetrically loaded shell of revolution. The variable V is the rotation of a meridian as seen in the plane containing the meridian and the axis of revolution of the shell.

H. Reissner discovered that a good approximation could be achieved by expressing the complete solution of these equations as the sum of the membrane solution for the distributed surface loads, and the bending

Figure 1.12 Geometrical parameters of the spherical shell: a is the shell radius, t the shell thickness, ϕ the meridional angle, and ϕ_e the value of ϕ at the edge.

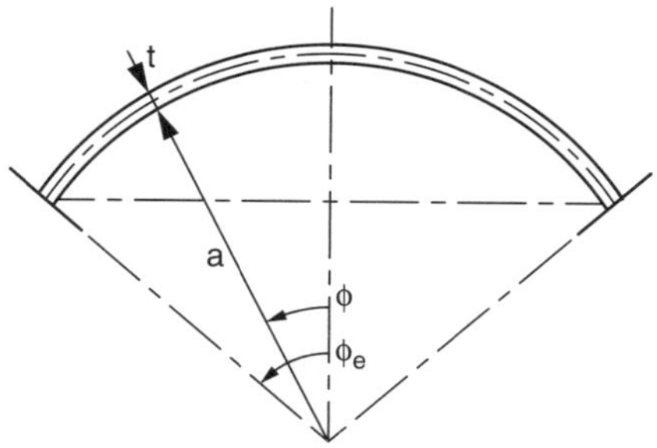

solution due to uniformly distributed shearing forces and bending moments applied axisymmetrically along the circular edge of the shell (which lies in a plane perpendicular to the axis of revolution of the shell). This was deduced on the basis of the observation that the effect of the latter system of (edge) loads was confined to a narrow edge zone. As will be explained in more detail in due course, the membrane solution is obtained purely from equilibrium considerations of the shell, assuming that the applied surface loading is wholly resisted by in-plane actions; bending and twisting actions are neglected altogether. H. Reissner gave a proof of the validity of this approximation for the case of the spherical shell subjected to self-weight loading.

The possibility of obtaining the homogeneous solution (i.e. the solution corresponding to the application of axisymmetric edge loads, in the absence of surface loads) as an asymptotic series was also indicated by H. Reissner. In 1912, Blumenthal[38] obtained such an asymptotic series for the axisymmetrically loaded spherical shell of constant thickness. In the following year, Meissner[39] extended H. Reissner's reduction to general shells of revolution, and noted that the equations were of the hypergeometric type. In 1915, Meissner[40] gave the exact solution for the spherical shell in terms of the hypergeometric series. The reduced pair of ordinary second-order differential equations in the variables Q_ϕ and V, for axisymmetrically loaded shells of revolution, is appropriately referred to as the 'Reissner–Meissner equations'.

The advantage of representing the homogeneous solution for the spherical shell as an asymptotic-series expansion rather than as a hypergeometric series is that, while the latter (mathematically exact) solution requires the retention of a relatively large number of terms of the series for sufficient computational accuracy in practically all cases of a / t (spherical shell's radius-to-thickness ratio – see Figure 1.12), the former, though always generally remaining an approximate solution regardless of how many terms of the associated infinite series are taken, usually requires, in practice, the retention of only the first few terms of its series for reasonable accuracy. Moreover, the accuracy of that retained portion increases rapidly with increasing a / t. In sharp contrast to this, while the hypergeometric-series results would become truly exact for all a / t if all the terms of the infinite series could be taken, their accuracy generally *decreases* with increasing a / t for a fixed finite number of terms of the series, owing to the slower convergence of the series for higher values of a / t.

It is worthwhile at this stage to digress temporarily from the present historical account in order to take note of other investigators' views on, and experiences with, the hypergeometric-series solution and its variants. Novozhilov[6] argues that, in the derivation of the Reissner–Meissner pair of reduced equations for the spherical shell, errors of order t / a in comparison with unity are involved, so that there is no point in retaining in the solution to the equations terms of order smaller than t / a, as appears to be done when the hypergeometric series is adopted. He concludes that the simplifying asymptotic-series method must, therefore, be employed irrespective of the difference between the solutions obtained by this method and the hypergeometric series, especially as the convergence problems associated with the mathematically exact hypergeometric solution hamper the use of that solution for practical engineering calculations. As pointed out earlier, these convergence problems are greater the higher the a / t ratio of the shell. Timoshenko and Woinowsky-Krieger[41] point out that, even for an a / t ratio as low as 30, as many as 10 terms are required in order to obtain sufficient accuracy for the hypergeometric-series solution; for $a / t = 62.5$ – still a relatively low ratio of shell slenderness – they indicate that no fewer than 18 terms of the series are necessary. In considering a relatively thick shell of $a / t = 15$, Flügge[42] also reports problems with the convergence of the hypergeometric series.

In 1950, Hildebrand[43] extended the asymptotic-series solution originated by Blumenthal[38] (for the analysis of the spherical shell) to general shells of revolution. He obtained a homogeneous solution in the form of a product of $e^{\lambda\phi}$ and a series of inverse powers of λ, where $\lambda = [3(1 - \nu^2)\, r_1^4 / (r_2^2 t^2)]^{1/4}$ is a shell-slenderness parameter, and ϕ is a meridional-angle coordinate. (In the expression for λ, ν is Poisson's ratio, t is the shell thickness, r_1 is the meridional principal radius of curvature, and r_2 is the second principal radius of curvature of the shell of revolution – refer to Figure 1.11 for r_1 and r_2.) Considering only the leading term of the series, he further obtained an approximate asymptotic solution, for which he gave closed-form expressions for interior actions and deformations, as well as for edge deformations.

In the same paper, Hildebrand advocated the use of the membrane solution as an approximate particular solution of the general equations for a shell subjected to distributed surface loading, and outlined the conditions necessary for this approximation to be valid. Should it be decided to use the approximate asymptotic homogeneous solution stemming from the retention of only the leading term of the asymptotic series, Hildebrand argued that, for consistency, the membrane-solution meridional rotation (we will denote this by V^m), as well as the bending moments calculated from the membrane solution (we will denote these by M_ϕ^m for the meridional bending moment per unit length and M_θ^m for the hoop bending moment per unit length), must all be neglected if the surface-loading intensity does not vary appreciably over a distance of $\sqrt{rt}$. The calculation of V^m, M_ϕ^m and M_θ^m will be dealt with in Section 1.5.3.

Another version of the asymptotic solution, applicable to constant-thickness spherical shells that are not too shallow (i.e. where the angle ϕ_e in Figure 1.12 is not too small), had already been given by Hetényi back in 1938.[44] This solution does not involve any series, and all the quantities of interest can be obtained in closed form. Using the same transformation of the Reissner[37] pair of simultaneous second-order ordinary differential equations in Q_ϕ and V as originally employed by Blumenthal,[38] Hetényi found that the terms containing the first derivatives of Q_ϕ and V vanished, leaving only terms in the second derivatives of Q_ϕ and V, and terms in the undifferentiated variables Q_ϕ and V themselves. Neglecting the terms in the undifferentiated variables Q_ϕ and V in comparison with the terms in the second derivatives of these variables, and combining the ensuing pair of simplified equations, Hetényi obtained a simple fourth-order ordinary differential equation in one dependent variable Q_ϕ and having constant coefficients. The general solution of any differential equation of this form was well known, enabling him to give explicit closed-form results for interior actions and deformations of interest.

The classical method of asymptotic integration (as applied by Hildebrand[43]) and its variants are not valid around the apex of the shell of revolution – that is, for $\phi \to 0$. (ϕ is the angular coordinate of a point in the meridional plane of a shell of revolution; it is the angle between the axis of revolution of the shell and the normal to the shell midsurface at the point in question, as depicted in Figure 1.11.) The method of Hetényi,[44] being in essence a simpler version of Hildebrand's asymptotic procedure, is subject to the same limitation. One way of bypassing this limitation has been to replace $\cot\phi$ in the basic second-order ordinary differential equation in Q_ϕ (or V) by ϕ^{-1} (see, for example, Ref. 42). In 1935, Langer[45] had developed a different type of asymptotic solution, which, unlike the classical method of asymptotic integration, is valid for all values of ϕ.

In 1960, in discussing asymptotic solutions for the spherical shell subjected to axisymmetric loading, Leckie[46] concluded, on the basis of a comparison between the classical method of asymptotic integration and the Langer method, that the method of Hetényi[44] is valid provided that $\lambda\phi > 3\sqrt{2}$.

In 1946, Hetényi[47] had also arrived at another set of slightly different and somewhat less accurate expressions for the quantities of interest in an axisymmetrically loaded non-shallow spherical shell of constant thickness, on the basis of the theory of a beam on an elastic foundation in place of shell theory. Hetényi showed that the agreement between the results of either of his methods and those of the exact hypergeometric-series solution is excellent for the combination $a/t = 30$ and $\phi_e = 35°$, from which we can conclude that the shell-theory version of the Hetényi approximation[44] is practically exact for all spherical shells for which $a/t > 30$ and $\phi_e > 30°$.

Earlier in 1926, Geckeler[48] had achieved an even more drastic but more general simplification than that of Hetényi,[44] by dropping, from the Reissner–Meissner pair of equations for arbitrary shells of revolution subjected to axisymmetric loading, the terms in the two dependent variables Q_ϕ and V, and the terms in the first derivatives of these variables. He noted that the rapidly decaying character of the edge disturbance means that the first derivatives of these variables are large compared with the variables themselves, and the second derivatives are, in turn, large compared with the first, provided that the shell is sufficiently thin and not too shallow. Retaining only the second-derivative terms of the Reissner–Meissner equations, and then eliminating one of the two variables (Q_ϕ or V), Geckeler obtained a fourth-order ordinary differential equation in Q_ϕ or V. If the shell has a constant slenderness parameter λ (as in the case of a constant-thickness spherical shell), then Geckeler's equation has constant coefficients, and the solution is readily obtained in simple closed form.

It is worthwhile pointing out at this stage that Geckeler's equation, which, as already implied, is a good approximation for thin shells of revolution with sufficiently steep sides at the opening, and which happens to be exact for circular-cylindrical shells under axisymmetric edge loads, is identical in form to the well-known differential equation of bending of a *straight* beam lying on a Winkler-type elastic medium: comparing the Geckeler equation with the actual beam-on-elastic-foundation equation,[47] the transverse-shear variable Q_ϕ in the former equation is seen to correspond to the transverse deflection (w, say) of a beam uniformly supported on a Winkler foundation, while the shell slenderness parameter λ corresponds to the beam-foundation interaction parameter $\{k/(4EI)\}^{1/4}$, where k is the modulus of the foundation and EI the flexural rigidity of the beam. It is also worth noting that, for spherical shells, Geckeler's approximation is less accurate than Hetényi's set of shell equations derived on the basis of the beam-on-elastic-foundation analogy,[47] as the latter approach adopts the more precise model of a *curved* beam.

Geckeler's approximation is the most practical of all the methods discussed in this section so far. From the point of view of simplicity, and the rapidity with which a stress analysis can be performed, the Geckeler approximation is undoubtedly the most convenient analytical tool available for obtaining edge-zone stresses in axisymmetrically loaded spherical shells. It is most suited to such shells if, in addition, they are sufficiently non-shallow, sufficiently thin, and of constant thickness. As implied earlier, the Geckeler approximation is also applicable, but somewhat less conveniently, to other non-shallow thin shells of revolution of arbitrary meridional shape, provided of course that the loading and boundary conditions are axisymmetric. For such shells, if the parameter λ varies along the meridian in accordance with a certain specific law, a closed-form solution is also possible.[42]

Finally, it may be noted that spherical caps (and caps in the form of other surfaces of revolution) may be analysed using shallow-shell theory, if the angle of opening is too small for either the Geckeler or the Hetényi approximation to be applicable. This approximate theory is particularly relevant in the context of shell roofs, and applicable when the rise of the shell is small in comparison with the smaller plan dimension.

Figure 1.13 Vlasov's criterion for shallowness, illustrated by the spherical shell of constant plan diameter D: shell B (with centre O_B) lies in the non-shallow range; shell C (with centre O_C) lies in the shallow range; shell A (with centre O_A and $\phi_s = 44°$) shows the cut-off rise of $H = D/5$.

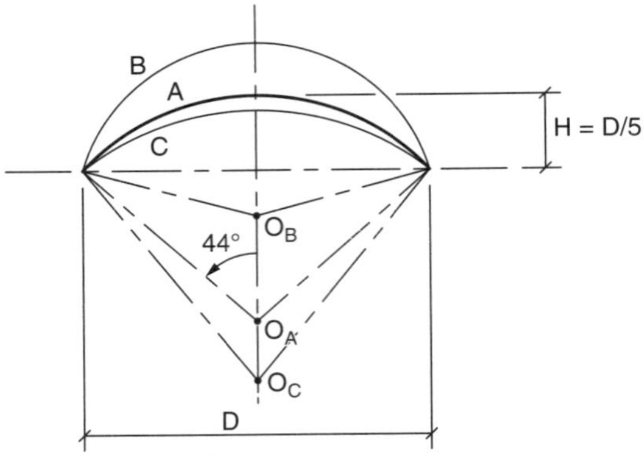

A general theory of shallow shells was given in 1938 by Marguerre.[49] In 1944, Vlasov[7] independently developed a bending theory applicable to shallow shells. According to Vlasov,[50] a shell may be regarded as shallow if the ratio of the rise to the shorter side (for a shell of rectangular plan) or to the diameter (for one of circular plan) is less than 1/5. Such a criterion for shallowness corresponds to a ϕ value at the shell edge of approximately 44° in the case of spherical shells, for which the plan shape is circular (see Figure 1.13).

Now, to ensure reasonable accuracy, the Geckeler approximation requires the angle ϕ_e ($\leq 90°$ for caps and domes) to be at least 25° within the a/t range of thin shells, henceforth arbitrarily taken as $a/t \geq 30$. Thus, according to Vlasov's criterion, the domain of validity of shallow-shell theory for spherical caps overlaps considerably with that of the Geckeler approximation.

For a shell of revolution with a circular plan, E. Reissner gave, in a paper published in 1946,[51] a more restrictive criterion for shallowness requiring that the ratio of rise to plan diameter be less than or equal to 1/8. For a spherical shell, this criterion corresponds to a ϕ value at the shell edge of approximately 28°, which, fortunately, still overlaps (albeit slightly) with the range of applicability of the Geckeler approximation, ensuring that all possible cases of thin spherical shells subjected to axisymmetric surface loads and boundary conditions can be analysed using the Geckeler approximation (or the method of Hetényi if more accuracy is required) if they are non-shallow, or by shallow-shell theory if they are otherwise, without the necessity of resorting to the cumbersome exact hypergeometric-series solution. Moreover, the results obtained from shallow-shell theory will often be applicable to shells that are not shallow if the loads are such that the stresses are effectively restricted to shallow zones.[51]

1.4.5.2 Cylindrical shells

The general theories of cylindrical shells had their origins in the period 1930–1940, the stimulus for their development having stemmed mainly from the need to construct reinforced concrete shell roofs. The earliest of the several rigorous and approximate bending theories put forward during this period was due to Finsterwalder.[52,53] The characteristic feature of Finsterwalder's theory consisted in neglecting the longitudinal bending moment M_x per unit length, the longitudinal transverse shear force Q_x per unit length, and the twisting moment $M_{x\phi}$ per unit length. Figure 1.14 shows all the actions associated with an element of a cylindrical shell.

Figure 1.14 Actions on an element of a cylindrical shell: ϕ is the angular coordinate along the circumferential direction and x is the distance coordinate along the axis of the cylinder; N_ϕ and N_x are direct forces per unit length in the circumferential (ϕ) and axial (x) directions respectively; $N_{\phi x}$ and $N_{x\phi}$ are in-plane shearing forces per unit length acting on the same edges of the cylindrical element as N_ϕ and N_x respectively; Q_ϕ and Q_x are transverse shear forces per unit length acting on the same edges of the cylindrical element as N_ϕ and N_x respectively; M_ϕ and M_x are bending moments per unit length acting on the same edges of the cylindrical element as N_ϕ and N_x respectively; $M_{\phi x}$ and $M_{x\phi}$ are twisting moments per unit length acting on the same edges of the cylindrical element as N_ϕ and N_x respectively; p_ϕ, p_x and p_r are external load components per unit area of shell midsurface, in the direction of the tangent to a circumferential arc (i.e. the ϕ direction), the direction parallel to the x axis, and the direction normal to the shell midsurface; a is the radius of the circular cylinder and l its length. The incremented actions on the edges $\phi + d\phi$ and $x + dx$ are not labelled to avoid clutter in the diagram.

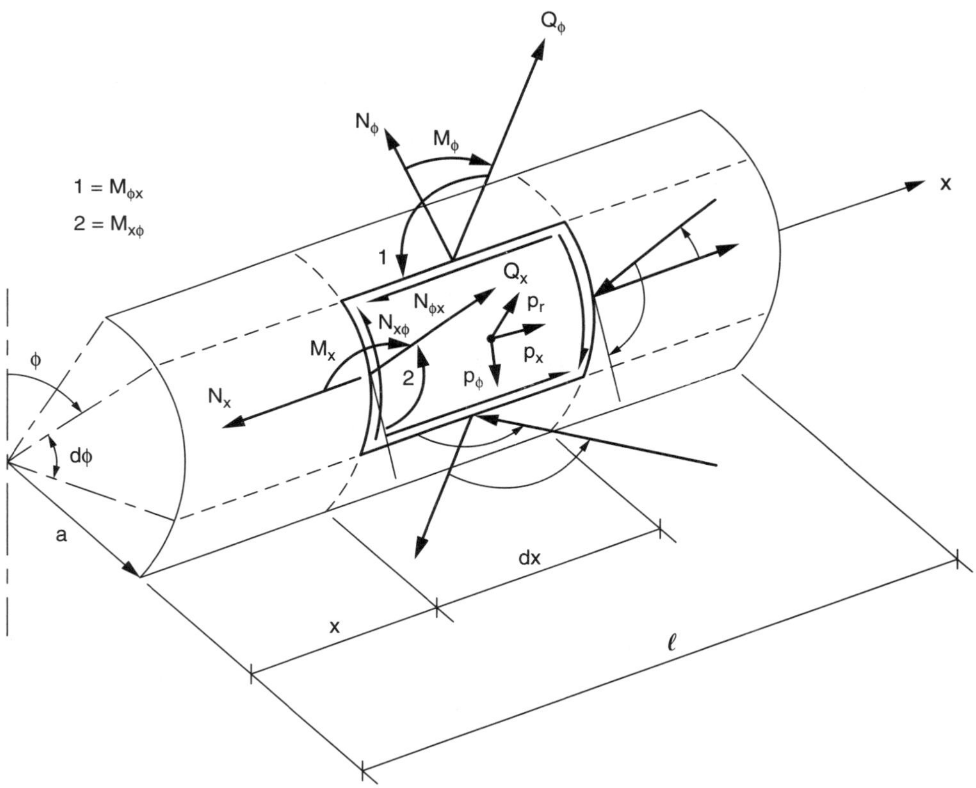

Like most of the bending theories for cylindrical shells, Finsterwalder's derivation culminates in a single eighth-order partial differential equation, which can be transformed into an eighth-degree polynomial equation (the characteristic equation) having eight roots. The Finsterwalder theory is applicable only to long shells.

In his publication of 1934, Flügge[16] was the first to present the theory of the cylindrical shell in a rigorous and concise form. His theory led to three simultaneous partial differential equations in the displacement variables u, v and w (being displacements in the x (axial) direction, the ϕ (circumferential/tangential) direction, and the direction normal to the shell midsurface – see Figure 1.14). These equations led to a characteristic equation that may be treated as exact and used as a standard to assess the accuracy of other theories. Various efforts have been made to obtain approximate but accurate roots of Flügge's characteristic equation.[54–56]

In 1935, Dischinger[57] also derived a fairly rigorous and concise theory that is nearly as accurate as Flügge's theory, and had the added advantage that it was more suitable for design. Although Dischinger arrived at the same three simultaneous differential equations in u, v and w as Flügge obtained, the characteristic equations of the two theories differ, albeit only in certain minor terms.

In the same year, Schorer[58] achieved a drastic simplification by assuming, like Finsterwalder, that $M_x = Q_x = M_{x\phi} = 0$, but went further to assume that the circumferential strain ε_ϕ and the shear $\gamma_{x\phi}$ are both small and may be neglected in relation to the longitudinal strain ε_x. The Schorer theory leads to an eighth-order partial differential equation of relatively simple form. In considering barrel-roof type cylindrical shells, Schorer's approach for dealing with the bending disturbances originating at the straight edges is rather analogous to the assessment of the edge effect in open spherical shells by the Geckeler approximation. However, as Bouma[36] points out, the approximate character of Schorer's solution is more pronounced than that of the Geckeler approximation, as the limits of applicability of the former depend on more factors (in particular, Schorer's theory gives reliable results only in the case of long shells). The range of validity of Schorer's method is practically the same as that of the Finsterwalder theory: $l / a \geq \pi$, where l is the length of the shell (in the axial direction) and a is the radius of the circular cylinder of which the shell is a part.

Around the same time, in work primarily concerned with the elastic stability of shells, Donnell[59,60] developed a theory that was more simplified than the theories of Flügge and Dischinger, but more general than those of Finsterwalder and Schorer. In the Donnell theory, the displacements u, v and w appear in uncoupled form. In arriving at these equations, non-essential terms were discarded on the basis of order-of-magnitude considerations. The Donnell equations are much easier to solve than Flügge's equations. For the special case of a circular cylindrical shell, the completely general theory of shells of arbitrary shape that Vlasov later gave in 1944[7] degenerates into the Donnell theory. The procedure employed by Kármán and Tsien in 1941,[61] in their investigations on the buckling of cylindrical shells, and the theory given by Jenkins in 1947,[62] presented in a form suitable for the analysis of cylindrical shell roofs, are essentially the same as the Donnell theory.

The accuracy of the Donnell equations was studied by Hoff in a paper published in 1955,[63] using Flügge's theory as a standard. In 1959, Morley[64] achieved an improvement in the Donnell approximation while essentially retaining its simplicity, and demonstrated the effectiveness of his modification by applying it to the solution of an infinitely long thin-walled circular cylinder subjected to equal and opposite concentrated radial loads.[65] In 1974, Wang *et al.*[66] proposed a theory claimed to be more accurate than Donnell's theory and yet simpler than Morley's, on the assumption that the cylinder is circumferentially inextensible. In general, Donnell's theory is best suited for short shells ($l / a \leq 1.6$).

Various other theories have been put forward. For example, Parme introduced a new formulation as a basis for a manual of the American Society of Civil Engineers published in 1952,[67] while another was proposed by Holand in 1957.[68] Finally, it is useful to note that an extremely simplified theory called *the beam method* had been devised by Lundgren in 1949,[69] in response to the need for an elementary method suited to practical design. In general, the beam method of cylindrical-shell analysis, which combines a beam analysis (considering the shell as a beam of curved cross section) and an arch analysis, should only be used for very long shells ($l / a > 5$), or moderately long shells whose cross-sections are prevented from distortion during deformation.

1.5. General aspects of the membrane theory of shells

1.5.1 Introduction

The membrane theory of shells had its origins in the work of Lamé and Clapeyron, who, in 1828, had considered shells of revolution loaded symmetrically with respect to their axes.[11] Beltrami and Lecornu are generally credited with having first established the general form of the equations of the membrane theory in the second half of the nineteenth century.[6]

The theory is applicable to either completely flexible membranes (e.g. an inflated tyre tube or toy balloon), which have negligible bending stiffness, or shells with finite bending rigidity but in which the moments that are developed are so small as to be negligible (i.e. the state of stress is essentially *momentless*[6]), owing to the geometry of the shell, the nature of the boundary conditions at the shell edges, and the manner in which the applied loading is distributed.

For shells with finite bending rigidity, which are the sole concern of this book, the membrane solution may serve as a convenient and satisfactory approximation to the actual state of stress in the shell. In general, the exact state of stress in the shell, within the confines of shell theory rather than the more general theory of elasticity, can only be determined on the basis of the *bending theory* of shells, which is more general than the membrane theory. The bending theory of shells, with which most of the developments outlined in Section 1.4 were concerned, takes into account both extensional (i.e. in-plane) effects and flexural (i.e. bending, twisting and shearing) effects within the shell material. Unfortunately, and as already seen in Section 1.4, the equations of the general bending theory are generally very difficult to solve. That is why it is important for the analyst to know the conditions under which the membrane solution constitutes a satisfactory approximation, in order to take advantage of this much simpler approach.

As will be elaborated in due course, the membrane solution may also serve as an approximate particular integral to the general bending-theory equations, to be added to the homogeneous solution of the equations. Such a role of the membrane solution saves the analyst the generally very difficult task of searching for the exact particular solution corresponding to a given surface loading.

1.5.2 Momentless states of stress in shells versus arches

In Section 1.2.1, some comparisons of shells, plates, arches and beams were made. Whereas then mechanisms of resisting loads were compared in a general manner, the considerations in this section concern only the *momentless* states of stress in arches and shells of finite bending rigidities, structural forms that were noted in Section 1.2.1 to be capable of substantial extensional action under transverse applied loads.

Drawing on the examples of Heyman[70] and Novozhilov,[6] a flexible chain or cable suspended from two points lying in a horizontal plane will sag between these supports along a catenary profile, the self-weight of the chain causing purely tensile forces to be transmitted along its profile. Inversion of such a profile gives a catenary arch, in which the self-weight is resisted by purely compressive internal forces acting along the profile. Thus, in order for self-weight loading to be resisted purely by compressive in-profile forces, an arch of uniform thickness must be given a catenary profile. Similarly, for an imposed loading uniformly distributed over the horizontal projection of an arch to be resisted by essentially compressive internal forces without the necessity of mobilising the flexural rigidity of the arch, the profile must be parabolic.

As pointed out in Section 1.2.1, it is this ability of arches to support certain transverse loadings without any appreciable bending (by virtue of their curvature) that underlies their greater structural efficiency over straight beams. Since a different profile of the arch is required to resist each particular distribution of transverse loading without bending, it means that, if more than one loading type is anticipated (for instance, in the case of an arch bridge where the self-weight of the arch and the superimposed loads may be equally significant), then considerable arch thickness may be required in order to allow all the individual *momentless profiles* (alternatively visualised as *compressive-stress paths*) associated with the various load configurations to fit within the bounds of the arch,[70] thereby minimising bending actions in the real structure.

The essential difference between arch action and membrane shell behaviour is that, although the arch and the shell both possess the property of *curvature* essential for the in-plane internal resistance of transverse external loadings, the shell possesses the additional property of *surface* or, as Calladine[4] terms it, *continuity*. Thus, while a slender arch of a given shape will support only one type of loading without appreciable bending, a thin shell of given shape has, as a rule, the same property for a wide range of loads that satisfy only very general requirements, provided its edges are suitably supported.[6] The shell structure, therefore, does not have to be given a substantial thickness simply to accommodate different loads without significant bending, although, as Heyman points out,[70] a certain minimum thickness may be necessary to safeguard against local instability.

1.5.3 Applicability of the membrane hypothesis

As pointed out above, all the shells dealt with in this book are assumed to possess finite bending rigidity so that, in the most general circumstances, both flexural and extensional effects occur. Furthermore, the existence of the momentless state of stress in such shells is subject to the provision of suitable boundary conditions, as well as the fulfilment of certain smoothness conditions regarding shell geometry and loading distribution.

Guidelines regarding the various locations at which the membrane hypothesis for shells (i.e. the postulation that moments in the shell are negligible) fails outright to fulfil the interior conditions of equilibrium and continuity of the middle surface, or the external conditions of equilibrium and kinematic fixities of the shell edge, will be discussed in due course. For the moment, it will suffice to note that these locations include external supports, discontinuities in loading or shell geometry, and concentrated line loadings. All these sites, interior and external alike, at which the breakdown in the membrane solution is manifested by its failure to fulfil actual boundary conditions, will be referred to in this discussion by the general term *edges*. At such edges (whether within the physical extremities of the shell, or at the actual physical boundaries of the shell), transverse shearing forces, bending moments and twisting moments are, in general, required in order to satisfy the boundary conditions, so that, quite clearly, a momentless state of stress cannot exist in their vicinity.

At locations that are remote from edges, the membrane hypothesis may still be unsatisfactory, though we may not observe the obvious breakdown in its validity that characterises edges. The absence of 'tell-tale' incompatibilities (such as occur at edges) in locations that are remote from edges necessitates a formal procedure for assessing the performance of the membrane solution. These locations will be referred to as *interior regions*, to distinguish them from the vicinities of edges, which will be referred to as *edge zones*. By implication, interior regions are essentially free from the so-called edge effects associated with localised bending, although, in general, they may still be subject to some relatively small bending actions. For present purposes, an interior region may be considered to extend right up to a bounding edge if that edge happens to be free of edge effects; otherwise, an interior region generally extends to only a certain distance of the bounding edge in question, remaining separated from that edge by the edge zone in which edge effects occur. For interior regions, in the absence of detailed knowledge of shell behaviour, the logical approach is to regard the momentless assumption as only a hypothesis, whose truth must be checked after the membrane-solution results have been obtained.[4]

For the important class of shells of revolution loaded symmetrically with respect to the axis of rotation, a procedure for checking the validity of the membrane hypothesis in interior regions has been suggested by

Zingoni.[71,72] Instead of the standard procedure usually indicated for such purposes,[41] where displacements v (in the direction of the tangent to a meridian) and w (in the direction of the normal to the midsurface) need to be calculated as an intermediate step, we may adopt a more direct approach that does not involve the explicit determination of displacements. As will be seen in Chapter 2, the general solution for the displacement component v features a constant of integration that requires a kinematic boundary condition for its evaluation, before v and w can be obtained explicitly. By using an approach that does not require the determination of displacements v and w, we bypass the need to know the kinematic conditions at the boundaries of the shell; in fact, these conditions are not necessary.

For a physical interpretation of the actions referred to in this discussion, reference should be made to Figure 1.11 if necessary. In the context of this discussion, we will adopt the superscript m for the key variables, to emphasize the fact that quantities under consideration relate to the membrane solution, or are calculated on the basis of the results of the membrane solution.

The membrane stress resultants in the shell, N_ϕ^m and N_θ^m (direct forces per unit length, in the directions of the tangent to a meridian and the tangent to a circle of latitude, respectively), are first derived from equilibrium considerations, as will be shown in detail in Chapter 2. After expressions for these stress resultants are available, the meridional rotation V^m due to membrane action is written down directly from the general relation (see Chapter 2)

$$V^m = \frac{1}{r_1}\left[\frac{\cot\phi}{Et}\left\{\left(r_1 + \nu r_2\right)N_\phi^m - \left(r_2 + \nu r_1\right)N_\theta^m\right\} - \frac{d}{d\phi}\left\{\frac{r_2}{Et}\left(N_\theta^m - \nu N_\phi^m\right)\right\}\right] \tag{1.1}$$

where E is Young's modulus of elasticity of the shell material, ν is its Poisson's ratio, and the other symbols are as previously defined. The bending moments M_ϕ^m and M_θ^m associated with the rotation V^m of the membrane solution then follow directly from the relations (see Chapter 6)

$$M_\phi^m = -D\left\{\frac{1}{r_1}\frac{dV^m}{d\phi} + \frac{\nu}{r_2}V^m \cot\phi\right\} \tag{1.2a}$$

$$M_\theta^m = -D\left\{\frac{\cot\phi}{r_2}V^m + \frac{\nu}{r_1}\frac{dV^m}{d\phi}\right\} \tag{1.2b}$$

If the membrane hypothesis is valid, the extreme-fibre flexural stresses associated with M_ϕ^m and M_θ^m (these are $\pm 6M_\phi^m / t^2$ and $\pm 6M_\theta^m / t^2$ respectively, the upper and lower signs of the symbol $\pm$ referring to the inner and outer shell surfaces respectively) must turn out to be negligible in comparison with the purely extensional stresses associated with N_ϕ^m and N_θ^m (these are N_ϕ^m / t and N_θ^m / t respectively, uniform across the shell thickness).

If $V^m(\phi)$ as given by equation (1.1) turns out to be zero throughout an interior region of the shell, then M_ϕ^m and M_θ^m must, from equations (1.2), also both be zero. This indicates that the membrane hypothesis, which inherently assumes that the state of stress in the shell is momentless, is completely free of self-contradiction, and must therefore represent the actual state of stress in the shell *exactly*. Indeed, M_ϕ^m and M_θ^m can only be zero simultaneously for arbitrary ν and ϕ if and only if $V^m(\phi)$ is zero. If $V^m(\phi) \neq 0$, as is often the case, then $M_\phi^m(\phi)$ and $M_\theta^m(\phi)$ cannot simultaneously vanish over a region; this result contradicts the basic membrane-hypothesis assumption of momentlessness.

The ratio of the magnitude of the stresses associated with M_ϕ^m and M_θ^m to that of the stresses associated with N_ϕ^m and N_θ^m is a relative measure of the degree of self-contradiction of the membrane hypothesis. Though quantitatively derived, this ratio is merely a *qualitative* measure of the departure of the actual state of stress (in an interior region of the shell) from the calculated membrane state of stress. In this regard, we must note that the parameters M_ϕ^m and M_θ^m are *merely calculated* from the membrane solution, and, in general, bear no definable relationship to the true flexural actions actually occurring in the interior region of the shell in question, except, of course, insofar as their magnitude gives some indication of the magnitude of the flexural effects ignored at the outset by the membrane hypothesis.

To distinguish them from the true flexural actions in the shell, M_ϕ^m and M_θ^m will be termed *membrane bending moments*, a description that sounds self-contradictory but is intended to underscore the pseudo-character of M_ϕ^m and M_θ^m. However, as already pointed out, the vanishing of the membrane bending moments does indicate quite conclusively that no flexural actions occur in the shell, so that the state of stress then is truly momentless.

The procedure that has been outlined above does not, of course, reveal the presence of edge effects. In order to check whether or not a given location is a source of edge effects, and hence an *edge* in the sense defined earlier, we may proceed by first deriving the membrane stress resultants N_ϕ^m and N_θ^m, and then evaluating the membrane deformations V^m and δ^m (displacement in the direction perpendicular to the axis of revolution of the shell) at the location in question, or at positions immediately on either side of the location. While V^m is given by equation (1.1), δ^m is computed from the expression

$$\delta^m = \frac{1}{Et}\left(r_2 \sin\phi\right)\left(N_\theta - \nu N_\phi\right) \tag{1.3}$$

Note that both the deformations V^m and δ^m are fully determinate once expressions for the stress resultants N_ϕ^m and N_θ^m are known.

If the given location is indeed a source of edge effects, this will be manifested as a failure of the membrane solution to satisfy the known conditions of continuity, equilibrium and/or kinematic constraints at that location.

Gol'denveizer[73] terms these sources of edge effects *lines of distortion*. So far we have conceptually referred to these locations as simply *edges*, whether they are at the physical edge of the shell or at some interior location. To facilitate their identification, Gol'denveizer categorised *lines of distortion* as

(*a*) the physical edges of a shell
(*b*) lines along which discontinuities of the components of the external surface loads or of certain of their derivatives occur
(*c*) lines along which the middle surface of a shell has a break, or the curvature of the middle surface changes abruptly
(*d*) lines along which the rigidity of a shell or its thickness undergoes sudden changes.

Figure 1.15 illustrates some axisymmetric lines of distortion for shells of revolution. This enabled Gol'denveizer to state the conditions *for the negligibility of edge effects* (which are different from the generally weaker and unlocalised flexural effects associated with the distributed surface loads, whose negligibility should be separately assessed via the earlier-given procedure) in a given part of the shell in terms of these lines of distortion.

Figure 1.15 Some axisymmetric 'lines of distortion' for shells of revolution: D_1 is a discontinuity in the radial loading component; D_2 is a discontinuity in the first derivative of the radial loading component with respect to distance along the meridian; D_3 is a discontinuity in the shell thickness; D_4 is a discontinuity in the first derivative of the shell thickness with respect to distance along the meridian; D_5 is a discontinuity in the meridional radius of curvature; D_6 is a discontinuity in both the meridional slope and the meridional radius of curvature; D_7 is a concentrated line load; D_8 is a line of support.

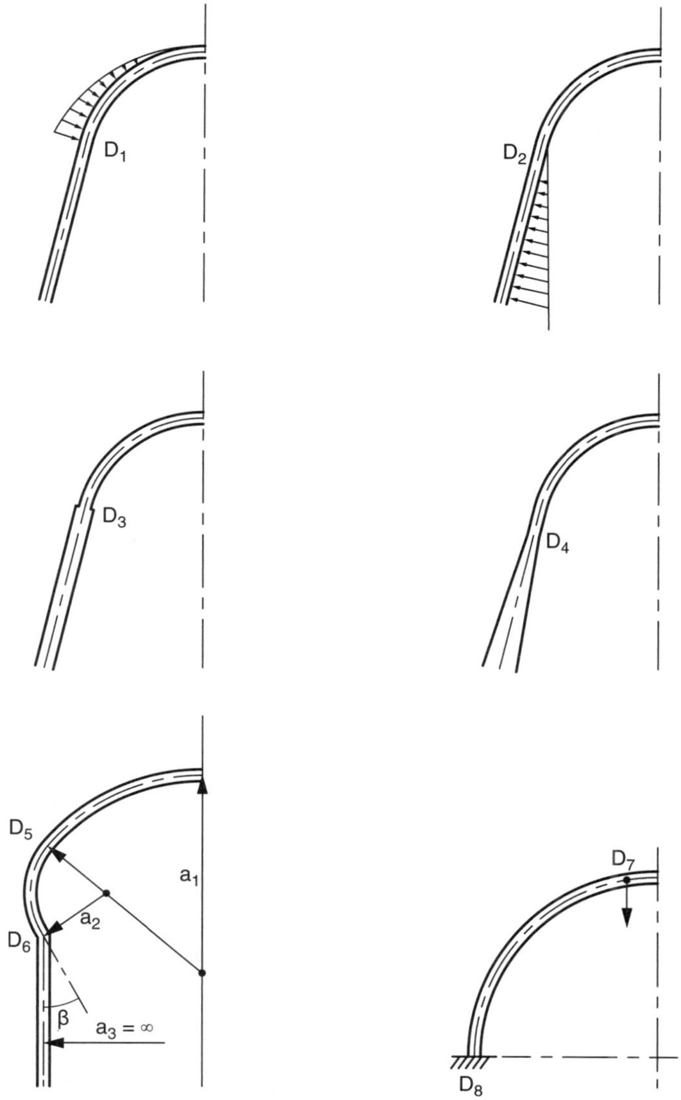

For instance, Gol'denveizer's requirement that '*lines of distortion do not form too tight a net on the middle surface*' (p. 474) implies that a part of the shell that satisfies this condition may be analysed on the basis of the membrane solution alone; the obtained results would be valid everywhere on that part of the shell except in the zone(s) adjacent to the line(s) of distortion. This assumes that the momentless hypothesis, as tested in accordance with the procedure that was described earlier, is reasonably satisfied over such an interior region.

In their considerations of the axisymmetric deformation of shells of revolution, Novozhilov[6] and Nielsen[74] discuss these conditions from a somewhat more practical perspective than does Gol'denveizer. Novozhilov notes that, at certain locations in the interior of the shell, the continuity of the middle surface appears to be violated when one uses the membrane hypothesis alone; however, since the middle surface is in reality physically continuous, it follows that there must exist at these locations transverse shearing forces and moments to satisfy the conditions of continuity.

For axisymmetric shells of revolution, locations at which such phenomena occur are those where the following quantities are discontinuous along a meridian:

(*a*) meridional loading component, radial loading component, and the first derivative of the radial loading component with respect to distance along the meridian
(*b*) shell thickness, and the first derivative of the shell thickness with respect to distance along the meridian
(*c*) meridional radius of curvature, and its first derivative with respect to distance along the meridian.

Nielsen identifies locations associated with the severest flexural effects in axisymmetric shells of revolution as those of discontinuous (*a*) radial loading component, (*b*) shell thickness, and (*c*) meridional radius of curvature, as well as those at which concentrated line loadings are applied.

The avoidance of the above discontinuities in loading components and shell geometry ensures that the continuity of the shell midsurface is not violated through the use of the membrane hypothesis on its own. These conditions are not sufficient, since so far we have not said anything about the type of supports that are consistent with the momentless assumption. Assuming the membrane hypothesis has already been shown, through the earlier-given test, to be adequate in regions that are free of edge effects, we now need to know the conditions for its sufficiency in the vicinity of the line of support of the shell, which is a special type of edge or line of distortion.

Since any component of the force reaction at the support, in the direction normal to the shell midsurface, would induce transverse shear in the shell (which the membrane hypothesis assumes is absent), it follows that the force reaction at the support must be purely tangential to the shell midsurface, if the membrane hypothesis is to be valid in the vicinity of the line of support of the shell. Clearly, the construction of the support must be such that it cannot provide any moment reaction upon the shell (implying a pinned, hinged or simply supported edge), as any moment reaction of the support (and hence moment action in the shell) would violate the momentless assumption outright.

As regards kinematic constraints, the support must permit the shell to move freely in the direction normal to the shell midsurface, and to rotate freely in the meridional plane. In other words, the boundary conditions must be such that w (the displacement in the direction normal to the shell midsurface) and V (meridional rotation) cannot be specified. Note that these kinematic variables are similar in sense to the static variables Q_ϕ (transverse shear) and M_ϕ (meridional moment) that characterise axisymmetric bending in shells of revolution. Just as only tangential force reactions are allowed at a supported edge, only tangential displacement components may be specified there; otherwise an edge effect, which the membrane hypothesis cannot take into account, will occur. Figure 1.16 depicts membrane-type support conditions for an axisymmetric shell of revolution.

Figure 1.16 Membrane-type support conditions for an axisymmetric shell of revolution: inclined rollers permit free movement w of the shell edge in the radial direction, and provide a force reaction F_R tangential to the shell midsurface; a hinge or pin permits free rotation V of the shell edge in the meridional plane, and eliminates moment reaction (i.e. $M_R = 0$).

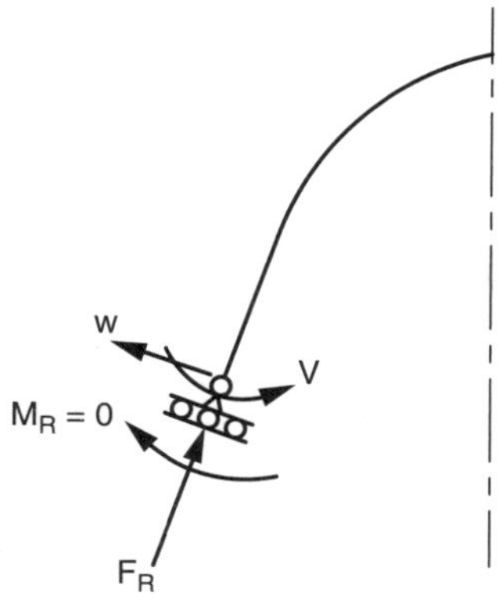

1.5.4 The membrane solution as a particular integral of the general bending-theory equations

The final aspect of the membrane solution that merits attention is its widespread use as an approximate particular integral of the bending-theory equations. Where the membrane solution fails to satisfy the conditions of midsurface continuity, shell equilibrium and/or support constraints, it is usual practice, certainly in the case of axisymmetric shells of revolution of zero or positive Gaussian curvature, to apply uniformly distributed shearing forces and bending moments along all such edges so as to restore the correct boundary conditions. This is illustrated in Figure 1.17.

The homogeneous solution, corresponding to the system of axisymmetric shearing forces and bending moments applied along each edge of the shell in the absence of surface loads, is simply superimposed with the membrane solution, just as if the membrane solution were the actual particular integral of the complete bending-theory equations. In fact, a true particular solution is rarely sought at all, since the membrane solution, which is generally much easier to obtain than the true particular solution, is in most cases a very good approximation to the latter. The homogeneous solution, which is that part of the total solution that is associated with the edge effect, has enough free constants to fulfil the arbitrary boundary conditions at the edges. The particular solution, unique to the specific type of surface loading to which the shell is subjected, does not necessarily have to satisfy the prescribed boundary conditions of the problem.

Let us limit present considerations to axisymmetrically loaded shells of revolution, for simplicity. Disregarding edge effects, or considering locations that are remote from such effects, the membrane solution must give the exact picture of the true state of stress in the shell if it is completely free of self-contradiction (i.e. if $V^m = 0$ so that the membrane bending moments M_ϕ^m and M_θ^m vanish), as explained in Section 1.5.3. On the other hand, the exact particular solution must always give the true picture of the state of stress in the shell, where edge effects are absent or totally insignificant. It must follow, therefore, that the membrane solution and the true particular solution coincide if and only if the membrane-solution parameter V^m is zero throughout a region of the shell in question. Thus, the condition $V^m(\phi) = 0$ shows that the membrane solution is indeed an exact particular integral of the full bending-theory equations, a

Figure 1.17 The homogeneous component of the total solution for axisymmetrically loaded shells of revolution: uniformly distributed shearing forces H_e and bending moments M_e applied along the shell-edge circumference.

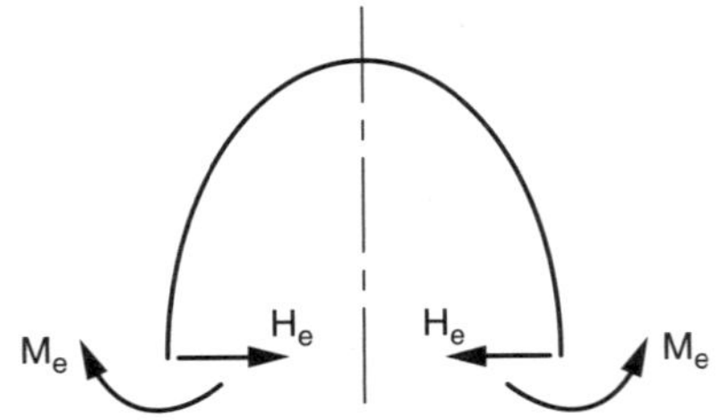

result that can also be arrived at from an inspection of the complete differential equations of bending that take into account surface loads.

In Section 1.5.3, a distinction was made between the localised flexural effects associated with the edge zones, and the relatively feeble flexural effects that pervade all regions of the shell if $V^m \neq 0$. It was pointed out that the latter flexural actions cannot, in general, be precisely quantified from the calculated membrane bending moments M_ϕ^m and M_θ^m. In fact, they are a function of the applied surface loading, and can only be quantified by the exact particular solution.

Flügge[42] and Seide[5] obtain exact particular solutions for the constant-thickness spherical shell subjected to two loading cases for which $V^m \neq 0$, namely self-weight and hydrostatic internal pressure. The difference between the direct stresses associated with the exact particular solution and those calculated from the membrane hypothesis is of the order of t^2 / a^2 in comparison with unity, while the flexural stresses added by the exact particular solution to these direct stresses are of the order of t / a in comparison with unity. Thus, for the cases where the membrane solution is not an exact particular integral of the complete equations of bending (i.e. if $V^m \neq 0$), the accuracy of its use as an approximate particular integral is generally satisfactory, provided, of course, that $V^m(\phi)$ and its derivative with respect to ϕ are relatively small throughout (see equations (1.2)). The accuracy becomes better as a shell of given radius becomes thinner.

From the above, it is evident that the test procedure that was suggested in Section 1.5.3 for assessing the validity of the membrane hypothesis in interior regions of the shell may also be used in a wider sense to check the suitability of the membrane solution as a particular integral throughout the entire shell.

With reference to axisymmetric deformations of shells of revolution, Hildebrand[43] states the following conditions for the validity of adopting the membrane solution as an approximate particular integral:

(*a*) the shell should be thin, its meridian should be smooth (i.e. with no discontinuities in the middle-surface profile nor in the slope of the meridian), and the principal radii of curvature r_1 and r_2 as well as the shell thickness t should vary continuously and smoothly (but not too rapidly) along the meridian

(*b*) the surface-loading components in both the meridional direction and in the direction of the normal to the middle surface should vary continuously and smoothly along the meridian, but must not change appreciably over a distance of $\sqrt{rt}$.

Hildebrand also points out that it is indeed consistent to adopt the membrane solution as an approximate particular integral when using the asymptotic-series solution, in view of the simplifications inherent in the method of asymptotic integration, provided that the above conditions are satisfied. For spherical shells, for instance, the approximations of Geckeler[48] and Hetényi[44] are both inherently less accurate than the asymptotic-series solution that Hildebrand discusses, so that the use of the membrane solution as an approximate particular integral for such problems is the only reasonable way to proceed whenever the homogeneous solution is determined on the basis of either the Geckeler approximation or the Hetényi improvement. Similar arguments hold for cylindrical, conical and other shells of revolution.

For examples of stress investigations based on the above approach, the reader is referred to some relatively recent articles on the feasibility of various liquid-containment shells of revolution encompassing 'egg-shaped' digesters,[75,76] conical assemblies[77] and multi-segmented spherical vessels.[78]

References

1. Sechler E.E. The historical development of shell research and design. In *Thin Shell Structures*. Prentice-Hall, Englewood Cliffs (New Jersey), 1974, 3–25.
2. Tooth A.S. Storage vessels. In *Developments in Thin-Walled Structures* (Vol. 1). Applied Science Publishers, London, 1982, 1–52.
3. CBI. *Hortonsphere pressure vessels*. Chicago Bridge & Iron Co., Oak Brook (Illinois), 1980.
4. Calladine C.R. *Theory of Shell Structures*. Cambridge University Press, Cambridge, 1983.
5. Seide P. *Small Elastic Deformations of Thin Shells*. Noordhoff, Leyden, 1975.
6. Novozhilov V.V. *Thin Shell Theory*. Wolters-Noordhoff Publishing, Groningen, 1970 (translated from the 2nd Russian edition by P.G. Lowe).
7. Vlasov V.Z. Osnovnye differentsialnye uravnenia obshche teorii uprugikh obolochek. *Prikladnaia Matematika i Mekhanika*, 1944, **8**, 109–140. (English translation: Basic differential equations in general theory of elastic shells. *NACA Technical Memorandum No. 1241*. National Advisory Committee for Aeronautics, Washington D.C., 1951.)
8. Goldberg J.E., Korman T. and Baluch M.H. On the use of various shell theories in the analysis of axisymmetrically loaded circular cylinders. *Nuclear Engineering and Design*, 1974, **30**, 88–92.
9. Mollmann H. *Introduction to the Theory of Thin Shells*. John Wiley & Sons, New York, 1981.
10. Todhunter I. and Pearson K. *A History of the Theory of Elasticity and of the Strength of Materials* (2 vols). Cambridge University Press, Cambridge, 1886 and 1893.
11. Timoshenko S.P. *History of Strength of Materials*. McGraw-Hill, New York, 1953. (Also published by Dover Publications, New York, 1983.)
12. Love A.E.H. The small free vibrations and deformations of a thin elastic shell. *Philosophical Transactions of the Royal Society of London* (Series A), 1888, **179**, 491–546.
13. Love A.E.H. *A Treatise on the Mathematical Theory of Elasticity*. Dover Publications, New York, 1944.
14. Calladine C.R. The theory of shell structures: Aims and methods. *International Journal of Mechanical Sciences*, 1982, **24**, 219–230.
15. Calladine C.R. The theory of thin shell structures 1888–1988 (Love Centenary Lecture). *Proceedings of the Institution of Mechanical Engineers*, 1988, **202** (42).
16. Flügge W. *Statik und Dynamik der Schalen*. Springer-Verlag, Berlin, 1934 (alternatively, see 3rd edition of 1962).
17. Byrne R. Theory of small deformations of a thin elastic shell. *University of California Publications in Mathematics* (New Series), **2**, 103–152.
18. Reissner E. Stress strain relations in the theory of thin elastic shells. *Journal of Mathematics and Physics*, 1952, **31**, 109–119.
19. Hildebrand F.B., Reissner E. and Thomas G.B. Notes on the foundations of the theory of small displacements of orthotropic shells. *NACA Technical Note No. 1833*. National Advisory Committee for Aeronautics, Washington D.C., 1949.

20. Green A.E. and Zerna W. The equilibrium of thin elastic shells. *Quarterly Journal of Mechanics and Applied Mathematics*, 1950, **3**, 9–22.
21. Naghdi P.M. On the theory of thin elastic shells. *Quarterly of Applied Mathematics*, 1957, **14**, 369–380.
22. Naghdi P.M. Foundations of shell theory. In *Progress in Solid Mechanics* (Vol. 4), North-Holland Publishing Co., Amsterdam, 1963, 1–90.
23. Naghdi P.M. A survey of recent progress in the theory of elastic shells. *Applied Mechanics Reviews*, 1956, **9**, 365–368.
24. Reissner E. A new derivation of the equations for the deformation of elastic shells. *American Journal of Mathematics*, 1941, **63**, 177–184.
25. Knowles J.K. and Reissner E. A derivation of the equations of shell theory for general orthogonal coordinates. *Journal of Mathematics and Physics*, 1956, **35**, 351–358.
26. Green A.E. and Zerna W. *Theoretical Elasticity*. Oxford University Press, Oxford, 1954.
27. Koiter W.T. A consistent first approximation in the general theory of thin elastic shells. In *Proceedings of the IUTAM Symposium on the Theory of Thin Elastic Shells* (Delft, 1959). North-Holland Publishing Co., Amsterdam, 1960, 12–33.
28. Koiter W.T. Foundations and basic equations of shell theory: A survey of recent progress. In *Proceedings of the IUTAM Symposium on the Theory of Thin Shells* (Copenhagen, 1967). Springer-Verlag, Berlin, 1969, 93–105.
29. Sanders J.L. An improved first approximation theory for thin shells. *NASA Report No. 24*. National Aeronautics and Space Administration, Washington D.C., 1959.
30. Budiansky B. and Sanders J.L. On the 'best' first-order linear shell theory. In *Progress in Applied Mechanics* (Prager Anniversary Volume). Macmillan, New York, 1963, 129–140.
31. Novozhilov V.V. Development of the method of complex transformation in the linear theory of shells during the last fifty years. In *Proceedings of the Fourth All-Union Conference on Shells and Plates* (Everan, 1962). NASA Technical Translation F341.
32. Sanders J.L. On the shell equations in complex form. In *Proceedings of the IUTAM Symposium on the Theory of Thin Shells* (Copenhagen, 1967). Springer-Verlag, Berlin, 1969, 135–156.
33. Koiter W.T. A systematic simplification of the general equations in the linear theory of thin shells. *Proceedings of Nederlandse Akademie van Wetenschappen*, 1961, **64** (B), 612–619.
34. Simmonds J.G. Simplification and reduction of the Sanders–Koiter linear shell equations for various midsurface geometries. *Quarterly of Applied Mathematics*, 1970, **28**, 259–275.
35. Simmonds J.G. Reduction of the linear Sanders–Koiter shell equations for nondevelopable midsurfaces to two coupled equations. *Transactions of the American Society of Mechanical Engineers: Journal of Applied Mechanics*, 1975, **42**, 511–513.
36. Bouma A.L. On asymptotic methods of shell analysis: A general survey. *Proceedings of the World Conference on Shell Structures* (San Francisco), 1962, 475–486.
37. Reissner H. Spannungen in Kugelschalen. *Müller-Breslau Festschrift*, Leipzig, 1912, 181–193.
38. Blumenthal O. Über asymptotische Integration von Differentialgleichungen mit Anwendung auf die Berechnung von Spannungen in Kugelschalen. *Proceedings of the Fifth International Congress of Mathematicians* (Cambridge) (Vol. 2), 1912, 319–327.
39. Meissner E. Das Elastizitätsproblem für dünne Schalen von Ringflächen-, Kugel- und Kegelform. *Physik. Zeitschrift*, 1913, **14**, 343–349.
40. Meissner E. Über Elastizität und Festigkeit dünner Schalen. *Vierteljahrschrift der Naturforschenden Gesellschaft in Zürich*, 1915, **60**, 23–47.
41. Timoshenko S.P. and Woinowsky-Krieger S. *Theory of Plates and Shells*. McGraw-Hill, New York, 1959.
42. Flügge W. *Stresses in Shells*. Springer-Verlag, Berlin, 1973.

43. Hildebrand F.B. On asymptotic integration in shell theory. In *Proceedings of the Symposium in Applied Mathematics* (Vol. 3). McGraw-Hill, New York, 1950, 53–66.
44. Hetényi M. Spherical shells subjected to axial symmetrical bending. *Publications of the International Association for Bridge and Structural Engineering*, 1938, **5**, 173–185.
45. Langer R.E. On the asymptotic solution of ordinary differential equations with reference to the Stokes' phenomenon about a singular point. *Transactions of the American Mathematical Society*, 1935, **37**, 397–416.
46. Leckie F.A. Asymptotic solutions for the spherical shell subjected to axially symmetric loading. In *Nuclear Reactor Containment Buildings and Pressure Vessels*. Butterworths, London, 1960, 286–297.
47. Hetényi M. *Beams on Elastic Foundation*. University of Michigan Press, Ann Arbor (Chicago), 1946.
48. Geckeler J.W. Über die Festigkeit achsensymmetrischer Schalen. *Forschungsarb-eiten auf dem Gebiet des Ingenieurwesens*, 1926, **276**, 1–52.
49. Marguerre K. Zur Theorie der gekrümmten Platte grosser Formänderung. *Proceedings of the Fifth International Congress of Applied Mechanics* (Cambridge, Massachusetts), 1938, 93–101.
50. Vlasov V.Z. *Obshchaya teoriya obolochek i yeye prilozheniya v tekhnike*. Gosudarstvennoye Izdatel'stvo Tekhniko-Teoreticheskoy Literatury, Moscow/Lenin-grad, 1949. (English translation: *General Theory of Shells and Its Applications in Engineering*. NASA Technical Translation F-99, 1964.)
51. Reissner E. Stresses and small displacements of shallow spherical shells. *Journal of Mathematics and Physics*, 1946, **25**, 80–85 and 279–300.
52. Finsterwalder U. Die Theorie der zylindrischen Schalengewölbe System Zeiss-Dywidag und ihre Anwendung auf die Grossmarkthalle in Budapest. *International Association for Bridge and Structural Engineering*, 1932, **1**, 127–150.
53. Finsterwalder U. Die Querversteiften Zylindrischen Schalengewölbe mit Kreissegmentförmigem Querschnitt. *Ing.-Arch.*, 1933, **4**, 43–65.
54. Holand I. Characteristic equations in the theory of circular cylindrical shells. *The Aeronautical Quarterly*, 1962 (Feb), 88–91.
55. Rish R.F. The roots of Flügge's cylindrical shell equation. *ASME Journal of Applied Mechanics*, 1968 (March), 172–173.
56. Colbourne J.R. Approximate roots of Flügge's characteristic equation for the closed cylindrical shell. *ASME Journal of Applied Mechanics*, 1969 (June), 352–355.
57. Dischinger F. Die strenge Theorie der Kreiszylinderschale in ihrer Anwendung auf die Zeiss-Dywidag-Schalen. *Beton u. Eisen*,1935, **34**, 257–264; 283–294.
58. Schorer H. Line load action on thin cylindrical shells. *Proceedings of the American Society of Civil Engineers*, 1935, **61**, 281–316.
59. Donnell L.H. Stability of thin-walled tubes under torsion. *NACA Report No. 479*, National Advisory Committee for Aeronautics, Washington D.C., 1933.
60. Donnell L.H. A new theory for the buckling of thin cylinders under axial compression and bending. *Transactions of the American Society of Mechanical Engineers*, 1934, **56**, 795–806.
61. Kármán T. Von and Tsien H.S. The buckling of thin cylindrical shells under axial compression. *Journal of Aeronautical Science*, 1941 (June), **8**, 303.
62. Jenkins R.S. *Theory and Design of Cylindrical Shell Structures*. Ove Arup & Partners, London, 1947.
63. Hoff N.J. The accuracy of Donnell's equations. *ASME Journal of Applied Mechanics*, 1955 (Sept), **22** (3), 329–334.
64. Morley L.S.D. An improvement on Donnell's approximation for thin-walled circular cylinders. *Quarterly Journal of Mechanics and Applied Mathematics*, 1959, **12** (1), 89–99.
65. Morley L.S.D. The thin-walled circular cylinder subjected to concentrated radial loads. *Quarterly Journal of Mechanics and Applied Mathematics*, 1960, **13** (1), 24–37.
66. Wang Y., Lee P.C.Y. and Billington D.P. A simplified theory of thin cylindrical shells. *ASCE Journal of Engineering Mechanics*, 1974 (Aug), **EM4**, 719–735.

67. ASCE. *Design of Cylindrical Concrete Shell Roofs*. American Society of Civil Engineers, Manual of Engineering Practice No. 31, New York, 1952.
68. Holand I. *Design of Circular Cylindrical Shells*. Oslo University Press, Oslo, 1957.
69. Lundgren H. *Cylindrical Shells*. The Danish Technical Press, The Institution of Danish Civil Engineers, Copenhagen, 1949.
70. Heyman J. *Equilibrium of Shell Structures*. Oxford University Press, Oxford, 1977.
71. Zingoni A. *Bending-disturbance considerations in the axisymmetric response of non-shallow spherical shell structures*. PhD Thesis, University of London, 1991.
72. Zingoni A. and Pavlovic M.N. Some observations on the exactness of the membrane hypothesis in the bending analysis of symmetrically-loaded shells of revolution. *International Journal of Mechanical Engineering Education*, 1991, **19** (4), 305–311.
73. Gol'denveizer A.L. *Theory of Elastic Thin Shells*. Pergamon Press, Oxford, 1961.
74. Nielsen T.B. *Axisymmetric Bending of Shells*. Danish Technical Press, The Institution of Danish Civil Engineers, Copenhagen, 1962.
75. Zingoni A. Stresses and deformations in egg-shaped sludge digesters: Membrane effects. *Engineering Structures*, 2001, **23** (11), 1365–1372.
76. Zingoni A. Stresses and deformations in egg-shaped sludge digesters: Discontinuity effects. *Engineering Structures*, 2001, **23** (11), 1373–1382.
77. Zingoni A. Discontinuity effects at cone-cone axisymmetric shell junctions. *Thin-Walled Structures*, 2002, **40** (10), 877–891.
78. Zingoni A., Mokhothu B. and Enoma N. A theoretical formulation for the stress analysis of multi-segmented spherical shells for high-volume liquid containment. *Engineering Structures*, 2015, **87**, 21–31.

Shell Structures in Civil and Mechanical Engineering
ISBN 978-0-7277-6028-9

http://dx.doi.org/10.1680/ssicame.60289.035

Chapter 2
Membrane theory of shells of revolution

2.1. General shells of revolution under axisymmetric loading
2.1.1 Introduction

A *surface of revolution* is generated by rotating a plane curve through 360° about a straight line (called the axis of revolution) in the plane of the curve. Any given position of the plane curve as it is moved around the axis of revolution constitutes a *meridian* of the generated surface of revolution.

In considerations of shells of revolution, it is convenient to regard the axis of revolution of the shell midsurface as vertical, so that the parallel '*circles of latitude*' of the shell, which are the intersections of the shell midsurface with planes perpendicular to its axis of revolution, lie in horizontal planes. The generally non-parallel meridians therefore lie in vertical planes containing the meridian in question and the vertical axis of revolution of the shell. These planes will be referred to as *meridional planes*. The direction of a tangent to a meridian at a given point on the shell midsurface will be referred to as the *meridional direction*. Circles of latitude will alternatively be referred to as *hoop circles*, and the direction of a tangent to a hoop circle will be referred to as the *hoop direction*. Hoop circles and meridians form an orthogonal net on the shell midsurface, which for a spheroidal shell has the appearance of the lines of latitude and longitude on a globe map of the earth.

Shells of revolution find application in the design of pressure vessels, liquid-filled tanks, roof domes and cooling towers. For these structures, the principal loading conditions are usually axisymmetric, so that every meridian of the shell of revolution deforms in the same manner; there is no relative shearing between adjacent portions of the shell when viewed in the plane of a circle of latitude, and all circles of latitude remain perfectly circular and centred about the axis of revolution after deformation.

Consider a shell element bounded by two adjacent circles of latitude and two adjacent meridians, as shown in Figure 2.1. Two angular coordinates may be adopted to describe the position of any point on the shell midsurface. The coordinate ϕ is the angle measured from the axis of revolution of the shell to the normal to the shell midsurface at the point in question; this will be referred to as the *meridional angle*. The coordinate θ is the angle measured in the horizontal plane of a circle of latitude, and defines the position of a given meridian relative to some reference meridian (just as the position of any meridian on a 3-dimensional surface map of the earth is usually defined by an angle measured relative to the Greenwich Meridian).

Consistent with the notation introduced in Chapter 1, r_1 will be used to denote the principal radius of curvature of the shell midsurface at a given point as seen in the meridional plane, while r_2 denotes the second principal radius of curvature of the shell midsurface, given by the distance between the point in question and the intercept of the normal to the shell midsurface at that point and the axis of revolution of the shell. The radius of the circle of latitude through the point will be denoted by R (see Figure 2.1).

Figure 2.1 Element of an axisymmetrically loaded shell of revolution.

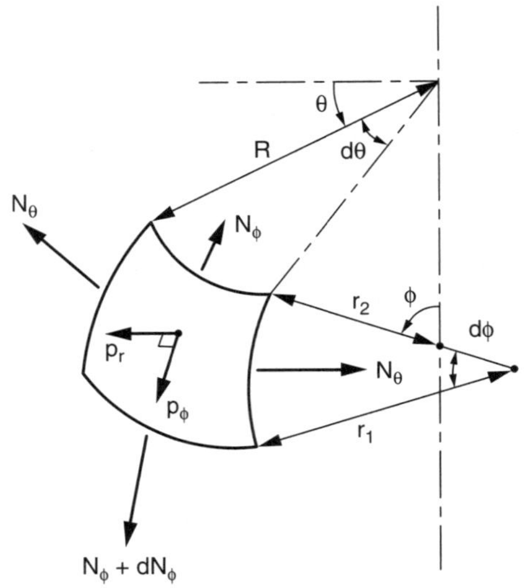

In the membrane theory of shells, and as explained in Chapter 1, we assume that no bending moments, twisting moments and transverse shearing forces exist in the shell. For shells of revolution subjected to axisymmetric loading and boundary conditions, in-plane shearing forces cannot exist, so that the only internal actions in the shell are N_ϕ (the direct force per unit length in the meridional direction) and N_θ (the direct force per unit length in the hoop direction), which we will refer to as the meridional and hoop stress resultants, respectively. We will consider N_ϕ and N_θ as positive if tensile (i.e. tending to stretch the element). Note that N_θ does not vary with respect to θ (owing to axisymmetry), but N_ϕ will generally vary with respect to ϕ.

Axisymmetric external loading may be resolved into two components (refer to Figure 2.1): p_ϕ (load per unit area of the shell midsurface, in the meridional direction) and p_r (load per unit area of the shell midsurface, in the direction of the normal to the shell midsurface). In the more general circumstances of non-axisymmetric loading, there is a third loading component p_θ (load per unit area of the shell midsurface, in the hoop direction), but under axisymmetric loading conditions, this is zero. We will consider p_ϕ as positive if pointing in the direction of increasing ϕ, and p_r as positive if pointing away from the axis of revolution of the shell.

2.1.2 Geometric preliminaries

From Figure 2.1, the radius R of the circle of latitude at the coordinate ϕ is given by

$$R = r_2 \sin\phi \tag{2.1}$$

The edges of the shell element in the two meridional sections are each of arc length $r_1\, d\phi$. The upper horizontal edge of the element is of arc length $R\, d\theta$, but the lower horizontal edge of the element is of increased arc length $(R + dR)d\theta$ since R has increased with respect to ϕ.

Although the shell element is curved and trapezoidal in projection, its surface area may be approximated as $R\, d\theta \times r_1\, d\phi$ (as if the area was a plane rectangle of sides $R\, d\theta$ and $r_1\, d\phi$).

Figure 2.2 Derivation of the governing equations of equilibrium for an element of an axisymmetrically loaded shell of revolution.

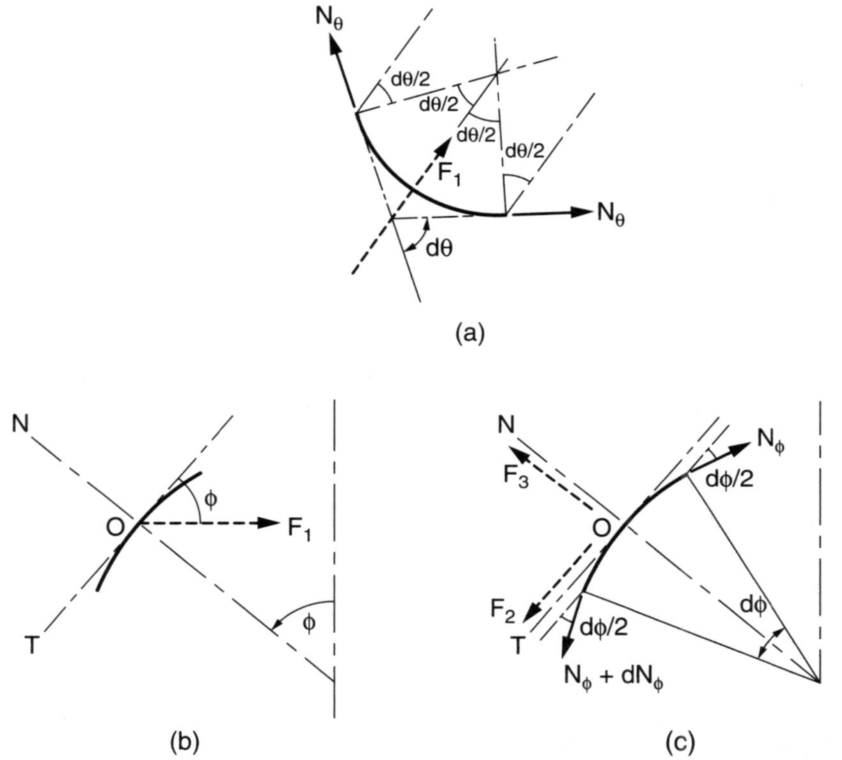

2.1.3 Governing equations of equilibrium

The shell element is in equilibrium under the external applied loading p_ϕ and p_r and the internal stress resultants N_ϕ and N_θ of the shell (now exposed as edge forces when the element is isolated as a free body). Force equilibrium requires that the sum of the forces acting upon the shell element, resolved in the directions of the tangent to a meridian and of the normal to the shell midsurface, must in both cases be zero. We will make use of the sketches in Figure 2.2, which show various sections of the shell element.

Let us first consider the components in the positive direction of the tangent to a meridian (i.e. the direction of increasing ϕ) of all forces acting upon the shell element. The stress resultants N_ϕ and N_θ, as well as the external loading component p_ϕ, all contribute a force component in this direction.

Figure 2.2(a) shows a section of the element in the horizontal plane of a circle of latitude. The forces acting at the ends of the element are given by $N_\theta \times$ (length of the edge over which N_θ acts) – that is, $N_\theta\, r_1\, d\phi$. The resultant of these two end forces is a horizontal force F_1 directed towards the axis of revolution of the shell, and given by

$$\begin{aligned} F_1 &= N_\theta\, r_1\, d\phi \left(\sin \frac{d\theta}{2} \right) + N_\theta\, r_1\, d\phi \left(\sin \frac{d\theta}{2} \right) \\ &\approx N_\theta\, r_1\, d\phi\, d\theta \end{aligned} \tag{2.2}$$

Since $(\sin\beta) \approx \beta$ for very small angles β.

Figure 2.2(b) shows a section of the element in the vertical plane of a meridian. Resolving F_1 (as obtained in the last step) along OT (the positive direction of the tangent to the meridian), we obtain the contribution of N_θ to the force resultant along OT as

$$-F_1 \cos\phi$$

which, upon substituting the expression for F_1 (equation 2.2), becomes

$$-N_\theta\, r_1 \cos\phi\, d\phi\, d\theta \tag{2.3}$$

Figure 2.2(c) depicts a section of the shell element in the vertical plane of a meridian, with the meridional end forces shown. The total meridional force acting at the upper edge is $N_\phi \times$ (length of the upper edge) – that is, $N_\phi\, R\, d\theta$. The total meridional force acting at the lower edge is $(N_\phi + dN_\phi) \times$ (length of the lower edge) – that is, $(N_\phi + dN_\phi)(R + dR)d\theta$. Resolving these two forces along OT, we obtain their resultant F_2 as

$$\begin{aligned} F_2 &= (N_\phi + dN_\phi)(R + dR)d\theta \times \cos\frac{d\phi}{2} - N_\phi\, R\, d\theta \times \cos\frac{d\phi}{2} \\ &\approx N_\phi\, dR\, d\theta + R\, dN_\phi\, d\theta \\ &= d(RN_\phi)\, d\theta \end{aligned} \tag{2.4}$$

since $(\cos\beta) \approx 1$ for very small angles β; the term containing $dN_\phi\, dR\, d\theta$ has also been neglected in comparison with the other terms, as it is much smaller.

Finally, the contribution of the external loading to the force resultant along OT is simply given by $p_\phi \times$(surface area of the shell element), that is,

$$p_\phi\, r_1 R\, d\phi\, d\theta \tag{2.5}$$

Summing up the contributions (2.3)–(2.5), we may write

$$d(RN_\phi)d\theta - N_\theta\, r_1 \cos\phi\, d\phi\, d\theta + p_\phi\, r_1 R\, d\phi\, d\theta = 0$$

Dividing throughout by $d\theta$, we obtain

$$d(RN_\phi) - r_1 N_\theta \cos\phi\, d\phi = -\, r_1 R\, p_\phi\, d\phi \tag{2.6}$$

Next, let us find the components of all the forces in the direction ON, which is the positive direction of the normal to the shell midsurface. With reference to Figure 2.2(b), F_1 (stemming from N_θ) has the component

$$-F_1 \sin\phi$$

which, upon substitution of the expression for F_1, becomes

$$-\, r_1 N_\theta \sin\phi\, d\phi\, d\theta \tag{2.7}$$

From Figure 2.2(c), the resultant of the N_ϕ forces along ON, which we will denote by F_3, is given by

$$F_3 = -N_\phi\, R\, d\theta \sin\frac{d\phi}{2} - \left(N_\phi + dN_\phi\right)\left(R + dR\right) d\theta \sin\frac{d\phi}{2}$$

$$\approx -RN_\phi\, d\theta\, d\phi \tag{2.8}$$

noting that $\sin\left(d\phi/2\right) \approx \left(d\phi/2\right)$, and ignoring the relatively small quantities

$$N_\phi\, dR\, d\theta \frac{d\phi}{2};\ R\, dN_\phi\, d\theta \frac{d\phi}{2}\ ;\ dN_\phi\, dR\, d\theta \frac{d\phi}{2}$$

The resultant of the external loading along the direction ON is simply

$$p_r\, r_1 R\, d\phi\, d\theta \tag{2.9}$$

Summing up all the contributions (2.7)–(2.9), we have

$$-r_1 \sin\phi\, N_\theta\, d\phi\, d\theta - RN_\phi\, d\phi\, d\theta + r_1 R\, p_r\, d\phi\, d\theta = 0$$

Upon dividing throughout by $d\phi\, d\theta$, replacing R with $r_2 \sin\phi$, and finally dividing throughout by $\sin\phi$ and rearranging, we obtain

$$r_2 N_\phi + r_1 N_\theta = r_1 r_2 p_r \tag{2.10}$$

This expression, when divided throughout by $r_1 r_2$, assumes the form

$$\frac{N_\phi}{r_1} + \frac{N_\theta}{r_2} = p_r \tag{2.11}$$

2.1.4 General solution

From equation (2.11), we may write

$$N_\theta = r_2 p_r - \frac{r_2}{r_1} N_\phi \tag{2.12}$$

Substituting the above expression into equation (2.6), we obtain

$$d\left(RN_\phi\right) - r_1\left(r_2 p_r - \frac{r_2}{r_1} N_\phi\right)\cos\phi\, d\phi = -r_1 R p_\phi\, d\phi$$

Dividing throughout by $d\phi$ leads to the expression

$$\frac{d\left(RN_\phi\right)}{d\phi} + r_2 N_\phi \cos\phi = r_1 r_2 p_r \cos\phi - r_1 R p_\phi$$

which, upon multiplication throughout by $\sin\phi$, becomes

$$\left[\frac{d(RN_\phi)}{d\phi}\right]\sin\phi + (r_2\sin\phi)N_\phi\cos\phi = r_1r_2p_r\cos\phi\sin\phi - r_1Rp_\phi\sin\phi \tag{2.13}$$

Replacing ($r_2\sin\phi$) in the second term of the left-hand side of the above equation by R, we may write the left-hand side as

$$\left[\frac{d(RN_\phi)}{d\phi}\right]\sin\phi + RN_\phi\cos\phi$$

which, by applying the product rule of differential calculus in reverse, reduces to

$$\frac{d}{d\phi}(RN_\phi\sin\phi)$$

Thus, expression (2.13) may now be written as

$$\frac{d}{d\phi}(RN_\phi\sin\phi) = r_1r_2p_r\cos\phi\sin\phi - r_1Rp_\phi\sin\phi \tag{2.14}$$

Integrating both sides of equation (2.14) with respect to ϕ, we obtain

$$RN_\phi\sin\phi = \left[\int(r_1r_2p_r\cos\phi\sin\phi - r_1Rp_\phi\sin\phi)d\phi + k\right]$$

We may now replace R with ($r_2\sin\phi$) to obtain

$$(r_2\sin^2\phi)N_\phi = \left[\int r_1r_2(p_r\cos\phi - p_\phi\sin\phi)\sin\phi\,d\phi + k\right]$$

for which the general solution for N_ϕ can finally be written as

$$N_\phi = \frac{1}{r_2\sin^2\phi}\left[\int r_1r_2(p_r\cos\phi - p_\phi\sin\phi)\sin\phi\,d\phi + k\right] \tag{2.15}$$

in which k is a constant of integration to be obtained from an appropriate boundary condition. In the case of a shell of revolution with two axisymmetric openings or edges, once k has been evaluated by specifying a boundary condition for N_ϕ at one edge of the shell, we no longer have the flexibility of adjusting N_ϕ to suit the boundary condition at the other edge, and we simply have to accept the value yielded at the second edge on the basis of specifying a boundary condition at the first edge. If the so-obtained value of N_ϕ at the second edge does not match the real boundary condition for N_ϕ there, it implies that a homogeneous correction will be required in order to adapt the total solution (i.e. membrane solution + homogeneous solution) to the actual boundary condition, in accordance with the considerations of Chapter 1.

Once N_ϕ has been obtained, N_θ then follows from equation (2.12).

2.2. Special cases of axisymmetric shells of revolution of zero Gaussian curvature

2.2.1 General remarks

In the special cases of the circular cylindrical shell and the circular conical shell, the generating meridian of the midsurface is a straight line, parallel to the axis of revolution in the case of the cylinder and inclined to the axis of revolution in the case of the cone. For both these geometries, r_1 is infinite, while ϕ is a constant at all points along the meridian and can no longer be used as a coordinate.

For the circular cylindrical shell, it is convenient to adopt, in place of ϕ, a distance coordinate x measured along the longitudinal axis of the cylinder, with the origin being at one circular end of the cylindrical shell. For the circular conical shell or its frustum, the coordinate s is normally adopted, this being the distance along a meridian from the vertex of the cone (i.e. the point on the axis of rotation where all meridians converge) to the point in question.

While it is quite possible to deduce the results for the circular cylindrical shell and for the circular conical shell as limiting cases of those already obtained for general shells of revolution, a return to first principles is preferred here, as this approach is simpler and sheds more insight on the influence of curvature on the load-carrying mechanism of the shell.

2.2.2 The circular cylinder

Let us take the longitudinal axis (i.e. axis of revolution) of the cylinder to be vertical, and the x axis of the coordinate system of the cylinder to be coincident with its axis of revolution. As before, the coordinate θ will be used to define the position of a vertical meridional section of the cylinder relative to some datum vertical plane containing the axis of revolution of the cylinder. Let the radius of the cylinder be denoted by a. Figure 2.3 shows an element of such a cylindrical shell. The vertical edges of the element are of length dx, while the curved horizontal edges are of arc length $a\,d\theta$. Thus the element has a surface area of $a\,d\theta\,dx$.

Resolving forces in the positive x direction, we obtain, by reference to Figure 2.3(a),

$$(N_x + dN_x)a\,d\theta - N_x a\,d\theta + p_x\,a\,d\theta\,dx = 0$$

which simplifies to

$$a\,dN_x\,d\theta = -\,ap_x\,dx\,d\theta$$

giving the result

$$dN_x = -p_x\,dx \tag{2.16}$$

Integrating both sides with respect to x, we obtain

$$N_x = \int -p_x\,dx + k \tag{2.17}$$

where k is a constant of integration to be determined from a boundary condition at one horizontal circular edge of the cylindrical shell. As pointed out earlier for the general shell of revolution, one constant of integration allows us to specify N_x at only one circular edge of the cylinder, and we have to accept whatever value ensues at the other end; if the ensuing value of N_x at the second circular edge is not consistent with the actual boundary condition at that edge, then a homogeneous solution will be required to rectify the inconsistency.

Figure 2.3 Element of an axisymmetrically loaded circular cylinder.

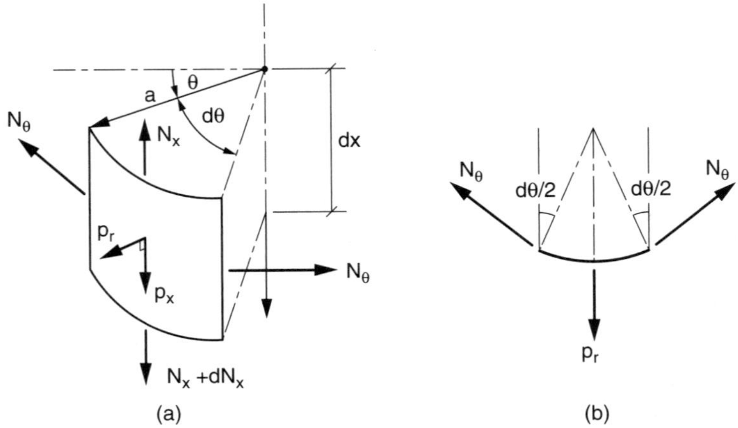

Resolving forces in the positive direction of the normal to the shell midsurface, we obtain, by reference to Figure 2.3(b),

$$2N_\theta\, dx \sin\frac{d\theta}{2} = p_r\, a\, d\theta\, dx$$

Making use of the approximation $\sin(d\theta/2) \approx (d\theta/2)$, we obtain the result

$$N_\theta = ap_r \tag{2.18}$$

2.2.3 The circular cone

Figure 2.4 shows an element of an axisymmetrically loaded circular cone in the $\{s, \theta\}$ coordinate system applicable to conical shells.

The inclined straight edges of the conical element in Figure 2.4(a) are each of length ds, while the upper and lower curved edges of the element (which both lie in horizontal planes) are of length $R\, d\theta$ and $(R + dR)d\theta$ respectively, where, as for general shells of revolution, R is the radius of the horizontal circle containing the upper curved edge of the element. The surface area of the curved element is approximately $R\, d\theta\, ds$. The angle of inclination of all meridians of the cone (relative to the horizontal plane) is a constant value of ϕ, which we will denote by α.

By reference to Figures 2.4(b) and (c), which correspond to Figures 2.2 (a) and (b) for the general shell of revolution, the horizontal resultant F_1 of the hoop actions N_θ is given by

$$F_1 = 2N_\theta\, ds \sin\frac{d\theta}{2} \approx N_\theta\, ds\, d\theta \tag{2.19}$$

which, upon resolving in the direction OT, gives the meridional component

$$-F_1 \cos\alpha = -\, N_\theta \cos\alpha\, ds\, d\theta \tag{2.20}$$

Figure 2.4 Element of an axisymmetrically loaded circular cone.

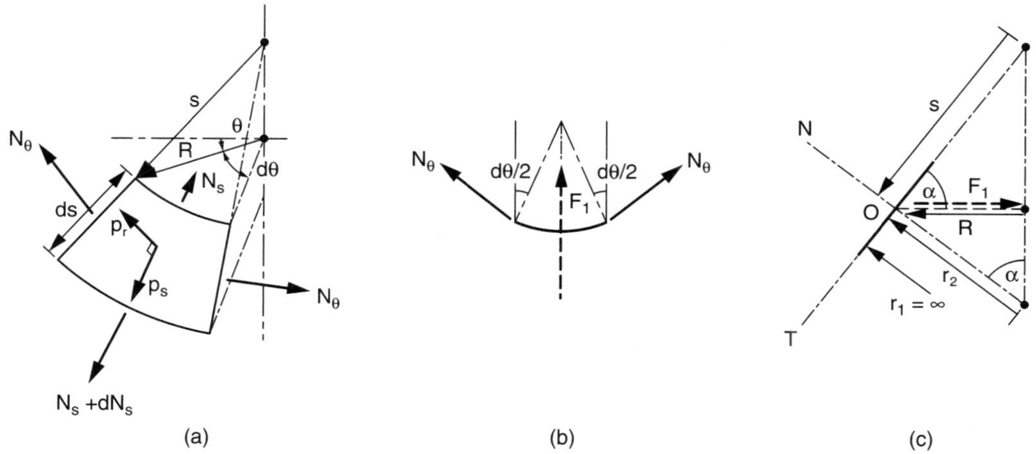

The resultant of the N_s actions in the direction OT is simply

$$(N_s + dN_s)(R + dR)d\theta - N_s R\, d\theta$$

$$\approx (N_s\, dR + R\, dN_s)d\theta \text{ (ignoring the small term)}$$

$$= d(RN_s)\, d\theta \tag{2.21}$$

while the external loading contributes

$$p_s\, R\, d\theta\, ds \tag{2.22}$$

in the direction OT.

Collecting results (2.20)–(2.22) together, we have

$$d(RN_s)\, d\theta - N_\theta \cos\alpha\, ds\, d\theta + p_s\, R\, d\theta\, ds = 0$$

– that is,

$$d(RN_s) - N_\theta(\cos\alpha)\, ds + p_s\, R\, ds = 0 \tag{2.23}$$

In the direction ON (normal to the shell midsurface), the N_s actions do not contribute a resultant, owing to the lack of curvature of the shell in the meridional direction. (In the case of the doubly curved general shell of revolution, the corresponding N_ϕ actions did contribute a component along ON by virtue of the shell having curvature in the meridional section.) Only the N_θ actions and, of course, the p_r external loading, enter into the equilibrium equation for the direction ON. By reference to Figures 2.4(b) and (c), the resultant of the N_θ actions in the direction ON is given by

$$-F_1 \sin\alpha = -\, N_\theta(\sin\alpha)\, ds\, d\theta \tag{2.24}$$

while the external-load contribution is

$$p_r\ R\, d\theta\, ds \tag{2.25}$$

Summing all elemental forces in the direction ON, we obtain

$$-N_\theta \left(\sin\alpha\right) ds\ d\theta + p_r\ R\ d\theta\, ds = 0 \tag{2.26}$$

from which

$$N_\theta = \frac{R}{\sin\alpha}\ p_r$$

$$= s\left(\frac{\cos\alpha}{\sin\alpha}\right) p_r \ \text{(since, from Figure 2.4(c), } R = s\cos\alpha\text{)}$$

$$= s\ (\cot\alpha)\ p_r \tag{2.27}$$

Substituting this result into equation (2.23), we obtain

$$d\left(RN_s\right) - s\ (\cot\alpha)\ (\cos\alpha)\ p_r\ ds + Rp_s\ ds = 0$$

leading to the result

$$d\left(RN_s\right) = s\cos\alpha\left(p_r\cot\alpha - p_s\right) ds \tag{2.28}$$

recalling that $R = s\cos\alpha$.

Integrating both sides of equation (2.28) with respect to s, we obtain

$$RN_s = \int s\cos\alpha\left(p_r\cot\alpha - p_s\right) ds + k'$$

giving

$$N_s = \frac{1}{s\cos\alpha}\left[\int s\cos\alpha\left(p_r\cot\alpha - p_s\right) ds + k'\right]$$

$$= \frac{1}{s}\left[\int\left(p_r\cot\alpha - p_s\right) s\ ds + k\right] \tag{2.29}$$

where, as before, k is a constant of integration to be determined from a suitable boundary condition.

2.3. General shells of revolution under non-axisymmetric loading

2.3.1 Introductory remarks

Shells of revolution may be subject to non-axisymmetric loadings, such as wind loading on domes, elevated water towers and cooling towers. Lateral inertial forces arising from earthquake accelerations also constitute a particular type of non-axisymmetric loading. Circular cylindrical vessels for the storage or transportation of oil and industrial chemicals are often supported with their longitudinal axis horizontal, so that the hydrostatic pressure acting on the inside surface of the cylindrical portion, and the vertical self-weight of the cylinder, are both non-axisymmetric with respect to the axis of the cylinder. The end closures of such vessels are usually in the form of a spherical cap, a torispherical cap (i.e. a toroidal portion surmounted by a spherical cap) or an ellipsoidal cap; the hydrostatic pressure acting on these end shells, as well as their self-weight, also constitute non-axisymmetric loadings.

For all the above reasons, a study of the non-axisymmetric response of shells of revolution is in order.

Non-axisymmetrically loaded circular cylindrical shells, such as the liquid-containing horizontal vessels mentioned above, and barrel shell roofs (where the axis of the parent cylinder is horizontal, while self-weight and superimposed dead loads are, of course, vertical), will be dealt with separately in Chapter 9. In the present section, we aim to develop a general approach for arbitrary shells of revolution subjected to non-axisymmetric loading.

2.3.2 Governing equations of equilibrium

The difference between non-axisymmetric and axisymmetric stress conditions is evident from a comparison of the element in Figure 2.5 versus the earlier one of Figure 2.1. For non-axisymmetric loading, there is now an additional loading component p_θ in the positive hoop direction. As a result of this, in-plane shear stress resultants $N_{\phi\theta}$ and $N_{\theta\phi}$ (per unit length) now exist along the edges of the element. Moreover, stress resultants now generally vary as one goes around a circle of latitude, so that the values of N_θ and $N_{\theta\phi}$ at the right-hand meridional edge of the element are incrementally larger than those at the left-hand meridional edge. Since quantities now vary with respect to both ϕ and θ, we must replace total derivatives with partial derivatives. It is now necessary to consider force equilibrium in three perpendicular directions, namely the directions of p_ϕ, p_r and p_θ.

Let us begin by considering equilibrium in the direction of the tangent to a meridian. The contributions stemming from the N_θ actions, N_ϕ actions and the p_ϕ loading component are as for the axisymmetric loading situation if, (*a*) in the derivation for the N_θ contribution, the small quantity $\left(\partial N_\theta / \partial\theta\right) d\theta\, r_1\, d\phi \sin\left(d\theta/2\right)$ now

Figure 2.5 Element of a non-axisymmetrically loaded shell of revolution.

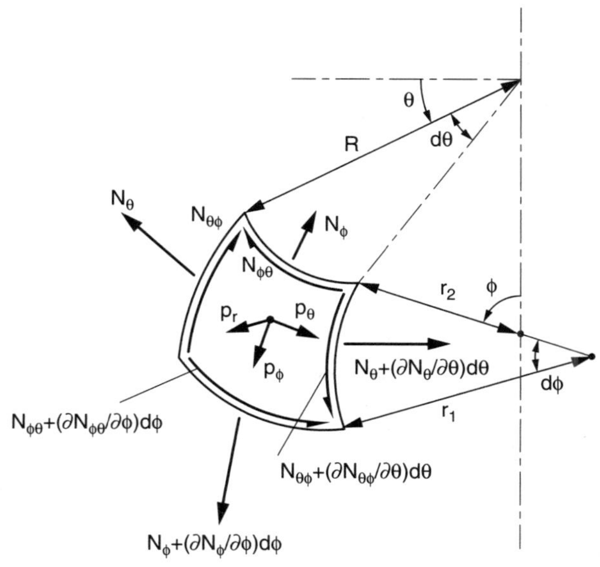

contributing to F_1 (see equation (2.2)) is ignored, and (*b*) in the derivation for the N_ϕ contribution (i.e. the resultant F_2 of equation (2.4)), dN_ϕ is replaced with $\left(\partial N_\phi / \partial\phi\right)d\phi$ and dR is replaced with $\left(\partial R / \partial\phi\right)d\phi$, so that

$$F_2 = \left[N_\phi \frac{\partial R}{\partial \phi} + R\frac{\partial N_\phi}{\partial \phi}\right] d\phi\, d\theta$$

$$= \frac{\partial}{\partial \phi}\left(RN_\phi\right) d\phi\, d\theta$$

There is now an additional contribution to the net force in the direction OT (see Figure 2.2 for this direction), due to the shear stress resultants along the meridional edges of the element. Figure 2.6(a) shows a direct view of the element along the normal to the shell midsurface, towards the axis of revolution of the shell. When extrapolated, the non-parallel sides of the trapezoidal projection intersect at V, a point on the

Figure 2.6 Derivation of the governing equations of equilibrium for an element of a non-axisymmetrically loaded shell of revolution.

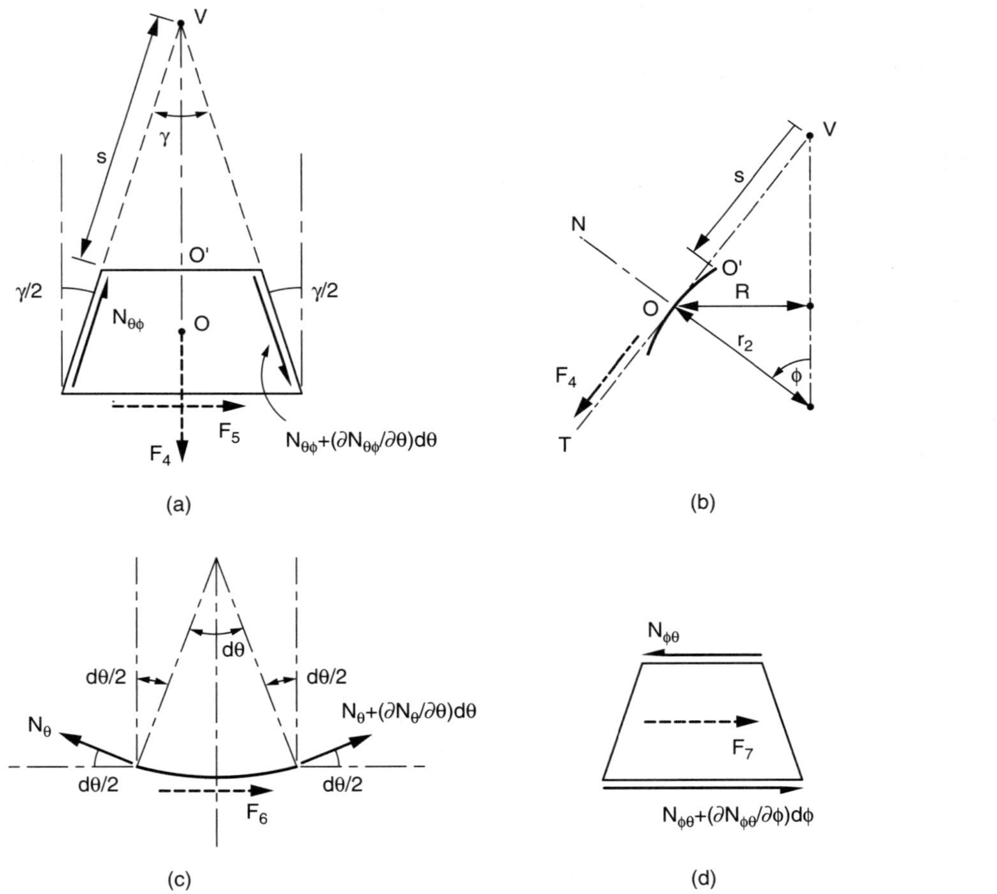

axis of revolution of the shell. This is clearer by reference to Figure 2.6(b), showing a view of the element in the meridional section through its centre. The upper horizontal edge of the element is of length $R\,d\theta$ as before.

In Figure 2.6, the distance s is very large in comparison with the dimensions of the element, so that $s \approx VO' \approx VO$. The small angle γ is therefore given by

$$\gamma \approx \frac{R\,d\theta}{s} = \frac{r_2 \sin\phi}{r_2 \tan\phi}\,d\theta = \cos\phi\,d\theta$$

since, from Figure 2.6(b), $R = r_2 \sin\phi$ and $s \approx r_2 \tan\phi$.

The net resultant F_4 (in the direction OT) of the shear actions shown in Figure 2.6(a) is given by

$$F_4 = \left(N_{\theta\phi} + \frac{\partial N_{\theta\phi}}{\partial\theta}\,d\theta\right) r_1\,d\phi\left(\cos\frac{\gamma}{2}\right) - N_{\theta\phi}\,r_1\,d\phi\left(\cos\frac{\gamma}{2}\right)$$

$$\approx r_1 \frac{\partial N_{\theta\phi}}{\partial\theta}\,d\phi\,d\theta \qquad (2.30)$$

noting that $\cos(\gamma/2) \approx 1$.

Summing up all force components in the direction OT, we obtain

$$\frac{\partial}{\partial\phi}\left(RN_\phi\right)d\phi\,d\theta - r_1 N_\theta \cos\phi\,d\phi\,d\theta + r_1\frac{\partial N_{\theta\phi}}{\partial\theta}\,d\phi\,d\theta + r_1 R p_\phi\,d\phi\,d\theta = 0$$

– that is,

$$\frac{\partial}{\partial\phi}\left(RN_\phi\right) + r_1\frac{\partial N_{\theta\phi}}{\partial\theta} - r_1 N_\theta \cos\phi = -r_1 R p_\phi \qquad (2.31)$$

Considering next the equilibrium of the element in the direction of the tangent to a circle of latitude (i.e. the positive hoop direction, which is the direction of increasing θ), the resultant F_5 of the $N_{\theta\phi}$ actions is given by (refer to Figure 2.6(a))

$$F_5 = \left(N_{\theta\phi} + \frac{\partial N_{\theta\phi}}{\partial\theta}\,d\theta\right)\left(r_1\,d\phi\right)\left(\sin\frac{\gamma}{2}\right) + N_{\theta\phi}\left(r_1\,d\phi\right)\left(\sin\frac{\gamma}{2}\right)$$

$$\approx r_1 N_{\theta\phi}\gamma\,d\phi$$

$$= r_1 N_{\theta\phi}\cos\phi\,d\phi\,d\theta \qquad (2.32)$$

recalling that $\gamma = (\cos\phi)\,d\theta$.

Figure 2.6(c) shows the hoop actions N_θ in a horizontal section of the element. Their resultant F_6 is simply

$$F_6 = \left(N_\theta + \frac{\partial N_\theta}{\partial\theta}\,d\theta\right)\left(r_1\,d\phi\right)\left(\cos\frac{d\theta}{2}\right) - N_\theta\left(r_1\,d\phi\right)\left(\cos\frac{d\theta}{2}\right)$$

$$\approx r_1\frac{\partial N_\theta}{\partial\theta}\,d\phi\,d\theta \qquad (2.33)$$

The same view of the element shown in Figure 2.6(a) appears in Figure 2.6(d), but with only the in-plane horizontal shear actions $N_{\phi\theta}$ shown. Their resultant F_7 in the positive tangential hoop direction is given by

$$
\begin{aligned}
F_7 &= \left(N_{\phi\theta} + \frac{\partial N_{\phi\theta}}{\partial \phi}\, d\phi\right)\left(R + \frac{\partial R}{\partial \phi}\, d\phi\right) d\theta - N_{\phi\theta} R\, d\theta \\
&\approx N_{\phi\theta} \frac{\partial R}{\partial \phi}\, d\phi\, d\theta + R \frac{\partial N_{\phi\theta}}{\partial \phi}\, d\phi\, d\theta \\
&= \frac{\partial}{\partial \phi}\left(R N_{\phi\theta}\right) d\phi\, d\theta \qquad (2.34)
\end{aligned}
$$

The resultant of the external loading in the positive tangential hoop direction is simply

$$p_\theta\, R\, d\theta\, r_1\, d\phi \qquad (2.35)$$

Summing up the terms in expressions (2.32)–(2.35), we obtain

$$\frac{\partial}{\partial \phi}\left(R N_{\phi\theta}\right) d\phi\, d\theta + r_1 \frac{\partial N_\theta}{\partial \theta}\, d\phi\, d\theta + r_1 N_{\theta\phi}\left(\cos\phi\right) d\phi\, d\theta + r_1 R p_\theta\, d\phi\, d\theta = 0$$

– that is,

$$\frac{\partial}{\partial \phi}\left(R N_{\phi\theta}\right) + r_1 \frac{\partial N_\theta}{\partial \theta} + r_1 N_{\theta\phi} \cos\phi = -r_1 R p_\theta \qquad (2.36)$$

Since the in-plane shears do not have a resultant in the direction normal to the shell midsurface, equilibrium of the element in this direction yields the same expression as was obtained for axisymmetric loading, when small second-order terms are neglected. Thus, the third equation of equilibrium remains

$$\frac{N_\phi}{r_1} + \frac{N_\theta}{r_2} = p_r \qquad (2.37)$$

Regarding the special cases of shells of revolution of zero Gaussian curvature mentioned earlier, we have already indicated that the non-axisymmetrically loaded circular cylinder will be covered in Chapter 9. Owing to the relatively simple geometry of general cylinders (of which the circular cylinder is a special case), the differential equations of equilibrium for arbitrarily loaded general cylinders (see Chapter 9) are of much simpler form than the above equations for arbitrarily loaded general shells of revolution (i.e. equations (2.31), (2.36) and (2.37)). (A general cylinder is a *shell of translation* – as opposed to a shell of revolution – for which the midsurface is generated by translating a straight line around a closed or open smooth curve of generally non-zero curvature throughout, while keeping the straight line always parallel to its initial orientation, and perpendicular to the plane of the curve.) For this reason, we have chosen to cover the arbitrarily loaded circular cylinder in Chapter 9, as a special case of arbitrarily loaded general cylinders, rather than as a special case of arbitrarily loaded shells of revolution of present considerations.

As for the arbitrarily loaded circular cone, we may deduce its governing equations from those of the general shell of revolution in the following way. We first divide equations (2.31) and (2.36) by r_1 throughout, and then replace the ensuing $r_1\, \partial\phi$ in the denominator of the first term on the left-hand side of each equation by ∂s, at the same time replacing, wherever they occur in either equation, N_ϕ by N_s, $N_{\theta\phi}$ by $N_{\theta s}$, $N_{\phi\theta}$

by $N_{s\theta}$, $\cos\phi$ by $\cos\alpha$, R by $s\cos\alpha$ (see Figure 2.4(c)), and p_ϕ by p_s. Finally, dividing throughout by $\cos\alpha$, we obtain

$$\frac{\partial}{\partial s}(sN_s) + \frac{1}{\cos\alpha}\frac{\partial N_{\theta s}}{\partial\theta} - N_\theta = -sp_s \tag{2.38}$$

$$\frac{\partial}{\partial s}(sN_{s\theta}) + \frac{1}{\cos\alpha}\frac{\partial N_\theta}{\partial\theta} + N_{\theta s} = -sp_\theta \tag{2.39}$$

Equation (2.37) is adapted for the circular cone by putting $r_1 = \infty$ (so that the first term of the left-hand side vanishes), and $r_2 = s\cot\alpha$ (see Figure 2.4(c)). The result is

$$N_\theta = s(\cot\alpha)p_r \tag{2.40}$$

which, as expected, is the same result as for axisymmetric loading (see equation (2.27)).

2.3.3 Solution approach

From equation (2.37), we have

$$N_\theta = r_2 p_r - \frac{r_2}{r_1}N_\phi \tag{2.41}$$

Let us first consider equation (2.31). Replacing R by $r_2\sin\phi$ and N_θ by the right-hand side of equation (2.41), we obtain

$$\frac{\partial}{\partial\phi}\left[(r_2\sin\phi)N_\phi\right] + r_1\frac{\partial N_{\theta\phi}}{\partial\theta} - r_1r_2p_r\cos\phi + r_2N_\phi\cos\phi = -r_1r_2(\sin\phi)p_\phi \tag{2.42}$$

Now the first term on the left-hand side of equation (2.42) may be evaluated as follows:

$$\frac{\partial}{\partial\phi}\left[(r_2\sin\phi)N_\phi\right] = N_\phi\frac{\partial}{\partial\phi}(r_2\sin\phi) + r_2\sin\phi\frac{\partial N_\phi}{\partial\phi}$$

$$= N_\phi\left\{r_2\cos\phi + \sin\phi\frac{\partial r_2}{\partial\phi}\right\} + r_2\sin\phi\frac{\partial N_\phi}{\partial\phi} \tag{2.43}$$

Consider Figure 2.7, which shows a shell element in meridional section.

The meridional arc length ds is equal to $r_1\,d\phi$. The dimension dR is equal to $ds\cos\phi$. Therefore we may write $dR = (r_1\cos\phi)\,d\phi$, from which

$$\frac{\partial R}{\partial\phi} = r_1\cos\phi$$

Recalling that $R = r_2\sin\phi$, we may write

$$\frac{\partial}{\partial\phi}(r_2\sin\phi) = r_1\cos\phi$$

Figure 2.7 Shell element in meridional section showing geometric differential parameters.

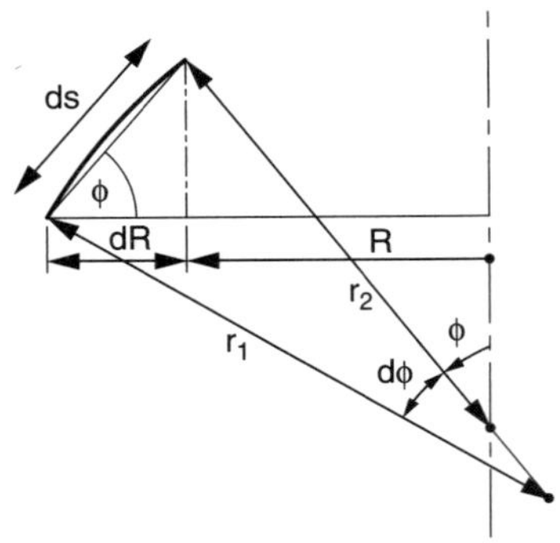

– that is,

$$r_2 \cos\phi + \sin\phi \frac{\partial r_2}{\partial \phi} = r_1 \cos\phi$$

from which

$$\sin\phi \frac{\partial r_2}{\partial \phi} = r_1 \cos\phi - r_2 \cos\phi \tag{2.44}$$

Using this result in the right-hand side of equation (2.43), we may now write

$$\frac{\partial}{\partial \phi}\left[(r_2 \sin\phi) N_\phi\right] = N_\phi \{r_2 \cos\phi + r_1 \cos\phi - r_2 \cos\phi\} + r_2 \sin\phi \frac{\partial N_\phi}{\partial \phi}$$

$$= r_1 N_\phi \cos\phi + r_2 \sin\phi \frac{\partial N_\phi}{\partial \phi} \tag{2.45}$$

Replacing the first term of the left-hand side of equation (2.42) by the right-hand side of equation (2.45), rearranging terms, and using the relationship $N_{\theta\phi} = N_{\phi\theta}$ to eliminate $N_{\theta\phi}$, we finally obtain the equation

$$r_2 \frac{\partial N_\phi}{\partial \phi} \sin\phi + r_1 \frac{\partial N_{\phi\theta}}{\partial \theta} + (r_1 + r_2) N_\phi \cos\phi = r_1 r_2 \left(p_r \cos\phi - p_\phi \sin\phi\right) \tag{2.46}$$

Considering next equation (2.36), we replace R by $r_2 \sin\phi$ and N_θ by the right-hand side of equation (2.41), to obtain

$$\frac{\partial}{\partial \phi}\left[(r_2 \sin\phi) N_{\phi\theta}\right] + r_1 \frac{\partial}{\partial \theta}\left(r_2 p_r - \frac{r_2}{r_1} N_\phi\right) + r_1 N_{\theta\phi} \cos\phi = -r_1 r_2 (\sin\phi) p_\theta \tag{2.47}$$

By analogy with expression (2.45), the first term of the left-hand side of equation (2.47) may be written as

$$r_1 N_{\phi\theta} \cos\phi + r_2 \sin\phi \frac{\partial N_{\phi\theta}}{\partial \phi}$$

so that equation (2.47), when use is made of the relation $N_{\theta\phi} = N_{\phi\theta}$, becomes

$$r_2 \frac{\partial N_{\phi\theta}}{\partial \phi} \sin \phi + 2r_1 N_{\phi\theta} \cos \phi - r_2 \frac{\partial N_{\phi}}{\partial \theta} = -r_1 r_2 \left(\frac{\partial p_r}{\partial \theta} + p_\theta \sin \phi \right) \tag{2.48}$$

Expressions (2.46) and (2.48) are two partial differential equations in N_ϕ and $N_{\phi\theta}$, with variable coefficients in ϕ but not θ. We expand the loading as Fourier series in θ:

$$p_\phi = \sum_{n=0}^{\infty} P_{\phi n} \cos n\theta + \sum_{n=1}^{\infty} P'_{\phi n} \sin n\theta \tag{2.49a}$$

$$p_\theta = \sum_{n=1}^{\infty} P_{\theta n} \sin n\theta + \sum_{n=0}^{\infty} P'_{\theta n} \cos n\theta \tag{2.49b}$$

$$p_r = \sum_{n=0}^{\infty} P_{rn} \cos n\theta + \sum_{n=1}^{\infty} P'_{rn} \sin n\theta \tag{2.49c}$$

Note that the first terms on the right-hand side of each expression represent loadings that are symmetric about the meridian $\theta = 0$, while the second terms represent antisymmetric loadings; $P_{\phi n}$, $P'_{\phi n}$, $P_{\theta n}$, $P'_{\theta n}$, P_{rn} and P'_{rn} are functions of ϕ only.

The general solution may then be assumed to be of the form

$$N_\phi = \sum_{n=0}^{\infty} N_{\phi n} \cos n\theta + \sum_{n=1}^{\infty} N'_{\phi n} \sin n\theta \tag{2.50a}$$

$$N_\theta = \sum_{n=0}^{\infty} N_{\theta n} \cos n\theta + \sum_{n=1}^{\infty} N'_{\theta n} \sin n\theta \tag{2.50b}$$

$$N_{\phi\theta} = \sum_{n=1}^{\infty} N_{\phi\theta n} \sin n\theta + \sum_{n=0}^{\infty} N'_{\phi\theta n} \cos n\theta \tag{2.50c}$$

where the first term on the right-hand side of each expression represents a response of the shell that is symmetric about $\theta = 0$ (caused by the symmetric loading), while the second term represents an antisymmetric response (caused by the antisymmetric loading); $N_{\phi n}$, $N'_{\phi n}$, $N_{\theta n}$, $N'_{\theta n}$, $N_{\phi\theta n}$ and $N'_{\phi\theta n}$ are functions of ϕ only.

Symmetric and antisymmetric loadings are solved for separately, and the results finally added together. Substituting the nth terms of expressions (2.49) and (2.50) into the governing equations (2.46) and (2.48), we obtain

$$\frac{\partial N_{\phi n}}{\partial \phi} + \left(\frac{r_1 + r_2}{r_2} \right) N_{\phi n} \cot \phi + \left(\frac{r_1}{r_2} \right) \left(\frac{n}{\sin \phi} \right) N_{\phi\theta n} = r_1 \left(P_{rn} \cot \phi - P_{\phi n} \right) \tag{2.51a}$$

$$\frac{\partial N_{\phi\theta n}}{\partial \phi} + 2 \left(\frac{r_1}{r_2} \right) N_{\phi\theta n} \cot \phi + \left(\frac{n}{\sin \phi} \right) N_{\phi n} = r_1 \left[\left(\frac{n}{\sin \phi} \right) P_{rn} - P_{\theta n} \right] \tag{2.51b}$$

for the symmetric part, and

$$\frac{\partial N'_{\phi n}}{\partial \phi} + \left(\frac{r_1 + r_2}{r_2}\right) N'_{\phi n} \cot \phi - \left(\frac{r_1}{r_2}\right)\left(\frac{n}{\sin \phi}\right) N'_{\phi\theta n} = r_1 \left(P'_{rn} \cot \phi - P'_{\phi n}\right) \tag{2.52a}$$

$$\frac{\partial N'_{\phi\theta n}}{\partial \phi} + 2\left(\frac{r_1}{r_2}\right) N'_{\phi\theta n} \cot \phi - \left(\frac{n}{\sin \phi}\right) N'_{\phi n} = -r_1 \left[\left(\frac{n}{\sin \phi}\right) P'_{rn} + P'_{\theta n}\right] \tag{2.52b}$$

for the antisymmetric part.

The above equations are usually solved by numerical methods. Closed-form solutions will be developed for the spherical shell and the circular cone in Chapter 4.

2.4. Deformations in axisymmetrically loaded shells of revolution

Displacements in shells are required in their own right (if we need to know by what amount the shell moves or deflects at specified locations), or more importantly, as part of a very useful procedure for evaluating edge effects in axisymmetric shells of revolution (see Chapters 6–8). The latter is relevant at discontinuities of the type discussed in Section 1.5.3 (where the membrane solution on its own is not adequate), and typically involves calculation of edge deformations associated with the membrane solution and with arbitrary edge actions, and enforcement of the relevant compatibility conditions as a means of solving for the actual values of the edge actions.

For axisymmetrically loaded shells of revolution, the basic displacements are v (in the direction of the tangent to the meridian, considered positive when pointing in the direction of increasing ϕ) and w (in the direction of the normal to the shell midsurface, considered positive when pointing away from the meridional centre of curvature) – see Figure 2.8(a).

By reference to Figure 2.8(a), which shows a line element AB of the shell in the vertical meridional section, the elongation of the element, in deforming to the new position $A'B'$, consists of a component e_1 due to the purely radial movement w, and a component e_2 due to the difference in the purely tangential movements v of ends A and B. By reference to Figure 2.8(b), which shows the same line element experiencing only the radial component w, it is clear that e_1 is equal to $w\,d\phi$. Similarly, by reference to Figure 2.8(c), which shows the line element experiencing only the tangential components v, it is evident that e_2 is equal to dv. Thus, the total elongation of line element AB, denoted by e_ϕ, is given by

$$e_\phi = dv + w\,d\phi \tag{2.53}$$

Now the original length of the element (before deformation) was $r_1\,d\phi$. Therefore, as a result of the elongation e_ϕ, the meridional strain ε_ϕ of the element is given by

$$\varepsilon_\phi = \frac{1}{r_1\,d\phi}(dv + w\,d\phi) = \frac{1}{r_1}\left(\frac{dv}{d\phi} + w\right) \tag{2.54}$$

By reference to Figure 2.8(d), the radius of the circle of latitude at point A increases from R to $R + dR$ as a result of the movement of the point to the position A'. Thus, the hoop strain (or circumferential strain) ε_θ is given by

$$\varepsilon_\theta = \frac{2\pi(R + dR) - 2\pi R}{2\pi R} = \frac{dR}{R} \tag{2.55}$$

Figure 2.8 Deformations of a line element of an axisymmetrically loaded shell of revolution.

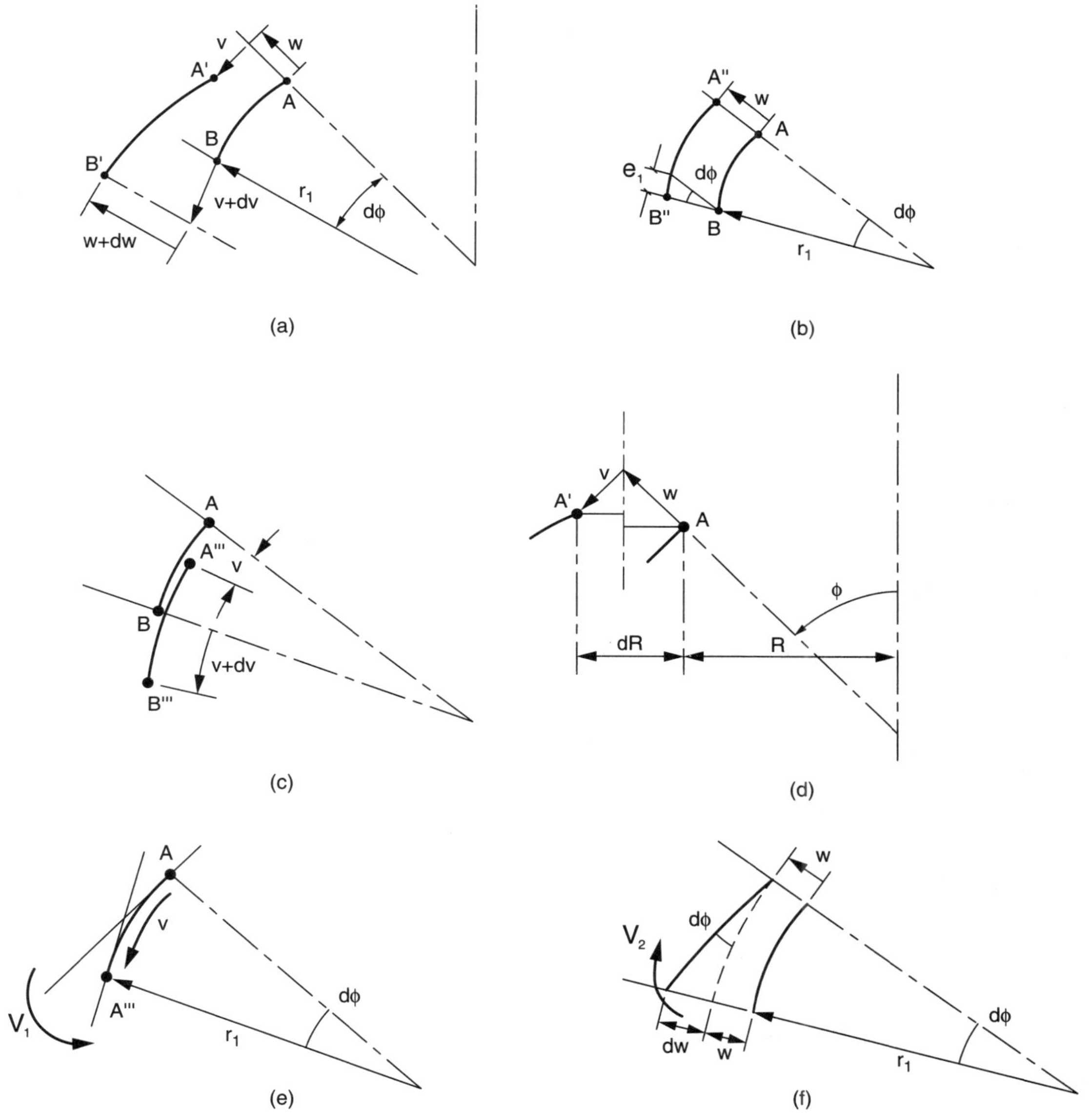

Now, by reference to the same sketch of Figure 2.8(d), the lateral movement dR is equal to the sum of the horizontal components of the basic displacements v and w – that is,

$$dR = v\cos\phi + w\sin\phi \tag{2.56}$$

Therefore,

$$\varepsilon_\theta = \frac{1}{R}\left(v\cos\phi + w\sin\phi\right)$$

$$= \frac{1}{r_2}\left(v \cot \phi + w\right) \tag{2.57}$$

since $R = r_2 \sin \phi$.

From equations (2.54) and (2.57),

$$\frac{dv}{d\phi} + w = r_1 \varepsilon_\phi \tag{2.58}$$

$$v \cot \phi + w = r_2 \varepsilon_6 \tag{2.59}$$

Subtracting equation (2.59) from equation (2.58), we obtain

$$\frac{dv}{d\phi} - v \cot \phi = r_1 \varepsilon_\phi - r_2 \varepsilon_\theta \tag{2.60}$$

By Hooke's law, we have

$$\varepsilon_\phi = \frac{1}{E}\left(\sigma_\phi - \nu \sigma_\theta\right) = \frac{1}{Et}\left(N_\phi - \nu N_\theta\right) \tag{2.61}$$

$$\varepsilon_\theta = \frac{1}{E}\left(\sigma_\theta - \nu \sigma_\phi\right) = \frac{1}{Et}\left(N_\theta - \nu N_\phi\right) \tag{2.62}$$

Using these results in equation (2.60), we obtain

$$\frac{dv}{d\phi} - v \cot \phi = \frac{1}{Et}\left[r_1\left(N_\phi - \nu N_\theta\right) - r_2\left(N_\theta - \nu N_\phi\right)\right]$$

$$= \frac{1}{Et}\left[\left(r_1 + \nu r_2\right)N_\phi - \left(r_2 + \nu r_1\right)N_\theta\right] \tag{2.63}$$

Integrating equation (2.63) by standard procedures, we get

$$v = \sin \phi\left[\int\left(\frac{1}{\sin \phi}\right)\frac{1}{Et}\left[\left(r_1 + \nu r_2\right)N_\phi - \left(r_2 + \nu r_1\right)N_\theta\right]d\phi + k\right] \tag{2.64}$$

where k is a constant of integration to be determined from a boundary condition.

Once v has been obtained, w follows from equations (2.59) and (2.62):

$$w = r_2 \varepsilon_\theta - v \cot \phi = \frac{r_2}{Et}\left(N_\theta - \nu N_\phi\right) - v \cot \phi \tag{2.65}$$

In practical calculations, it is usually the horizontal displacement dR, henceforth denoted by δ (considered positive when away from the axis of revolution of the shell), and the rotation V of the meridian, that need to be known. The flexibility procedure for evaluating bending actions M_e and H_e at an edge (see Section 1.5.4), which will be the method adopted throughout this book, involves superimposing the membrane-solution deformation parameters δ and V with their counterparts associated with arbitrary values of M_e and H_e, such that the ensuing net values of δ and V at a given shell edge are consistent with the actual boundary conditions at that edge, thereby allowing the redundant actions M_e and H_e to be solved for.

Once the edge actions M_e and H_e have been evaluated, the edge effect becomes quantified, and addition of its stresses to those stemming from the membrane solution finally yields the net stresses in the shell. As will be seen in Chapters 6–8, this procedure is analogous to the flexibility method for the structural analysis of indeterminate beams, trusses and frames, with the membrane solution (in shell analysis) corresponding to the effects of the applied loads on the primary (i.e. released) structure, and the bending correction (in shell analysis) corresponding to the effects of the redundant actions on the primary structure.

Since $\varepsilon_\theta = \left(dR / R\right) = \delta / R$ as previously shown, it follows that

$$\delta = R\varepsilon_\theta = (r_2 \sin\phi)\varepsilon_\theta \tag{2.66}$$

Using equation (2.62) to eliminate ε_θ, we finally obtain

$$\delta = \frac{1}{Et}(r_2 \sin\phi)\left(N_\theta - \nu N_\phi\right) \tag{2.67}$$

Result (2.67) coincides with equation (1.3) of Chapter 1, which was then given without derivation.

To obtain the meridional rotation V, consider Figures 2.8(e) and (f). As a result of the movement by an amount v of A to A''' (Figure 2.8(e)), the tangent at A rotates anticlockwise by an amount $V_1 = d\phi = v / r_1$. As a result of the difference dw in the radial movement of the two ends of the line element (Figure 2.8(f)), the line element rotates clockwise by an amount $V_2 = d\phi = dw / (r_1\, d\phi)$. Thus, the net *anticlockwise* rotation V is given by

$$V = \frac{1}{r_1}\left(v - \frac{dw}{d\phi}\right) \tag{2.68}$$

If we differentiate equation (2.59), we obtain

$$\frac{dv}{d\phi}\cot\phi - \frac{v}{\sin^2\phi} + \frac{dw}{d\phi} = \frac{d}{d\phi}\left(r_2\varepsilon_\theta\right) = \frac{d}{d\phi}\left[\frac{r_2}{Et}\left(N_\theta - \nu N_\phi\right)\right] \tag{2.69}$$

If we also multiply equation (2.63) by $\cot\phi$, we obtain

$$\frac{dv}{d\phi}\cot\phi - v\cot^2\phi = \frac{\cot\phi}{Et}\left[\left(r_1 + \nu r_2\right)N_\phi - \left(r_2 + \nu r_1\right)N_\theta\right] \tag{2.70}$$

Subtracting equation (2.69) from (2.70), we obtain

$$v - \frac{dw}{d\phi} = \frac{\cot\phi}{Et}\left[\left(r_1 + \nu r_2\right)N_\phi - \left(r_2 + \nu r_1\right)N_\theta\right] - \frac{d}{d\phi}\left[\frac{r_2}{Et}\left(N_\theta - \nu N_\phi\right)\right] \tag{2.71}$$

Substituting this result into equation (2.68), we may finally write

$$V = \frac{1}{r_1}\left[\frac{\cot\phi}{Et}\left\{\left(r_1 + \nu r_2\right)N_\phi - \left(r_2 + \nu r_1\right)N_\theta\right\} - \frac{d}{d\phi}\left\{\frac{r_2}{Et}\left(N_\theta - \nu N_\phi\right)\right\}\right] \tag{2.72}$$

This expression was given in Chapter 1, without derivation, as equation (1.1).

Thus δ and V at any point on the shell can be computed, once the stress resultants N_ϕ and N_θ of a problem have been obtained. As already stated, it is the value of these deformations at the support and other shell edges/discontinuities that are required in bending-correction calculations.

For the special case of the circular cylinder, we set r_2 equal to a (the radius of the cylinder), r_1 equal to infinity, ϕ equal to 90°, and replace N_ϕ with N_x, $d\phi$ with dx / r_1 (since dx is the limit of $r_1\ d\phi$ as r_1 tends to infinity and $d\phi$ tends to zero), as follows:

$$\delta\,(=w) = \frac{1}{Et}(a\sin 90°)(N_\theta - \nu N_x) = \frac{a}{Et}(N_\theta - \nu N_x) \tag{2.73}$$

$$\begin{aligned} V &= \frac{\cot 90°}{Et}\left\{\left(1+\nu\frac{r_2}{r_1}\right)N_x - \left(\frac{r_2}{r_1}+\nu\right)N_\theta\right\} - \frac{1}{r_1}\frac{d}{(dx/r_1)}\left[\frac{a}{Et}(N_\theta - \nu N_x)\right] \\ &= -a\frac{d}{dx}\left[\frac{1}{Et}(N_\theta - \nu N_x)\right] \end{aligned} \tag{2.74}$$

For the second special case of the circular cone, we set r_2 equal to $s\cot\alpha$, r_1 equal to infinity, ϕ equal to α, and replace N_ϕ with N_s, $d\phi$ with ds / r_1 (since ds is the limit of $r_1\ d\phi$ as r_1 tends to infinity and $d\phi$ tends to zero), as follows:

$$\delta = \frac{1}{Et}(s\cot\alpha\sin\alpha)(N_\theta - \nu N_s) = \frac{1}{Et}(s\cos\alpha)(N_\theta - \nu N_s) \tag{2.75}$$

$$\begin{aligned} V &= \frac{\cot\alpha}{Et}\left\{\left(1+\nu\frac{r_2}{r_1}\right)N_s - \left(\frac{r_2}{r_1}+\nu\right)N_\theta\right\} - \frac{1}{r_1}\frac{d}{(ds/r_1)}\left\{\frac{s\cot\alpha}{Et}(N_\theta - \nu N_1)\right\} \\ &= \cot\alpha\left[\frac{1}{Et}(N_1 - \nu N_\theta) - \frac{d}{ds}\left\{s\frac{1}{Et}(N_\theta - \nu N_1)\right\}\right] \\ &= \cot\alpha\left[\frac{1}{Et}(N_s - \nu N_\theta) - \frac{1}{Et}(N_\theta - \nu N_1) - s\frac{d}{ds}\left\{\frac{1}{Et}(N_\theta - \nu N_1)\right\}\right] \\ &= \cot\alpha\left[\frac{1}{Et}(1+\nu)(N_s - N_\theta) - s\frac{d}{ds}\left\{\frac{1}{Et}(N_\theta - \nu N_1)\right\}\right] \end{aligned} \tag{2.76}$$

For cylindrical, spherical, conical and various other shells of revolution, the above results for δ and V will be employed in Chapters 6–8 to calculate membrane deformations associated with supports, geometric discontinuities and junctions with ring beams, plates or other shells.

Shell Structures in Civil and Mechanical Engineering
ISBN 978-0-7277-6028-9

http://dx.doi.org/10.1680/ssicame.60289.057

Chapter 3
Membrane solutions for various shells of revolution under axisymmetric loading

3.1. Pressure vessels

Pressure vessels are used extensively in industry to contain gases, steam and other vapours under pressures often well in excess of atmospheric pressure. These structures are almost invariably shells of revolution, with the dominant loading condition being uniform internal pressure p. Because the consequences of failure can be very dangerous to both life and property, it is important to ensure that the induced stresses in the walls of pressure vessels do not exceed the rupture stress of the shell material. In this section, the general solutions of Chapter 2 will be applied to derive closed-form results for stress resultants in pressure vessels of various shapes. In all cases, $p_r = p$, $p_\theta = 0$, p_ϕ (or p_x for cylinders, p_s for cones) $= 0$.

3.1.1 Circular cylindrical vessel

Figure 3.1(a) shows a thin circular cylinder of radius a, provided with end closures in the form of flat plates, and subjected to a uniform internal pressure p.

From equation (2.18), the hoop stress resultant is simply

$$N_\theta = pa \tag{3.1}$$

while from equation (2.17), the longitudinal stress resultant is

$$N_x = \int -p_x \, dx + k = k \tag{3.2}$$

since $p_x = 0$. At the closed ends, the total force on the circular end plate due to the internal pressure p is equal to $p\pi a^2$. This is resisted by the N_x forces around the circumference of the cylinder end, which total $2\pi a N_x$. Therefore, as is illustrated in Figure 3.1(b),

$$2\pi a N_x = p\pi a^2$$

implying that

$$N_x = \frac{pa}{2} \tag{3.3}$$

The actual stresses in the cylinder walls are given by

$$\sigma_\theta = \frac{N_\theta}{t} = \frac{pa}{t} \text{ (hoop stress)} \tag{3.4a}$$

Figure 3.1 Pressurised circular cylinder.

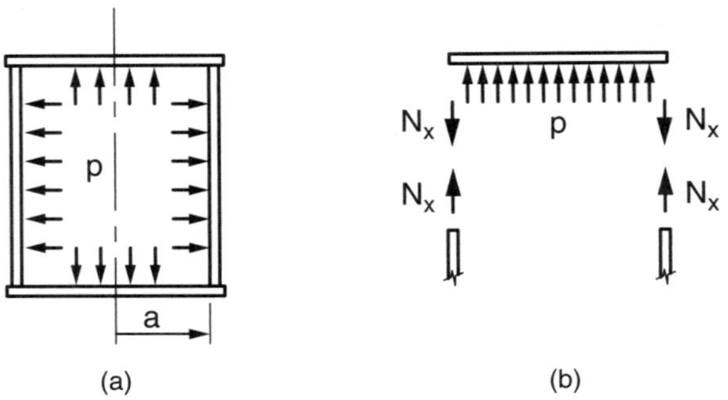

$$\sigma_x = \frac{N_x}{t} = \frac{pa}{2t} \quad \text{(longitudinal stress)} \tag{3.4b}$$

where t is the shell thickness, assumed here to be constant.

From equations (2.73) and (2.74), the membrane deformations of the cylinder are

$$\delta = w = \frac{a}{Et}\left(pa - \nu\frac{pa}{2}\right) = \frac{pa^2}{Et}\left(1 - \frac{\nu}{2}\right) \tag{3.5a}$$

$$V = -a\frac{d}{dx}\left[\frac{1}{Et}\left(pa - \nu\frac{pa}{2}\right)\right] = 0 \tag{3.5b}$$

throughout its length. Thus, the straight meridian does not suffer any rotation, all points on it simply moving outwards by the same amount, as would be expected from the symmetry of the situation. Since the physical connection to the flat plate closures tends to constrain this outward movement of the shell at its edges, bending is induced in the shell in the vicinity of the two ends, in accordance with the generalised indications of Section 1.5.3.

3.1.2 Circular conical vessel

Figure 3.2(a) shows a circular cone of length l from the vertex to the edge, where the sides are inclined at an angle α to the horizontal (assuming the axis of the cone is vertical). The cone is provided with a flat plate closure at its opening, and subjected to a uniform internal pressure p.

From equation (2.27), we have

$$N_\theta = s(\cot\alpha)p = ps\cot\alpha \tag{3.6}$$

and from equation (2.29), we have

$$N_s = \frac{1}{s}\left[\int p(\cot\alpha)\, s\, ds + k\right] = \frac{1}{s}\left[p(\cot\alpha)\frac{s^2}{2} + k\right] \tag{3.7}$$

By reference to Figure 3.2(b), the resultant $p\pi(l\cos\alpha)^2$ of the pressure forces on the end plate, which acts through the vertical axis of revolution of the cone, is resisted by the components in the vertical direction of the N_s shell forces, which amount to $2\pi(l\cos\alpha)N_s \sin\alpha$. Therefore

$$2\pi(l\cos\alpha)N_s\sin\alpha = p\pi(l\cos\alpha)^2$$

implying that

$$N_s = \frac{pl}{2}\cot\alpha \tag{3.8}$$

at $s = l$. This is the boundary condition to be applied to expression (3.7). Thus

$$\frac{1}{l}\left[p(\cot\alpha)\frac{l^2}{2} + k\right] = \frac{pl}{2}\cot\alpha$$

implying that $k = 0$.

The same result could have been obtained by arguing that N_s must remain of finite value at the apex ($s = 0$) of the cone, since engineering materials have finite strength. For this condition to hold, the term in square brackets on the right-hand side of equation (3.7) must vanish when $s = 0$ (implying that $k = 0$), since the term $1/s$ outside the square brackets becomes infinite when $s = 0$.

With k set equal to zero in equation (3.7), we may finally write

$$N_s = \frac{ps}{2}\cot\alpha \tag{3.9}$$

The deformations δ and V are given by equations (2.75) and (2.76), as follows:

$$\delta = \frac{1}{Et}(s\cos\alpha)\, ps\,(\cot\alpha)\left(1 - \frac{v}{2}\right)$$

$$= \frac{ps^2}{Et}\left(\frac{\cos^2\alpha}{\sin\alpha}\right)\left(1 - \frac{v}{2}\right) \tag{3.10}$$

$$V = \frac{\cot\alpha}{Et}\left[(1+v)(-1)\frac{ps}{2}\cot\alpha - sp(\cot\alpha)\left(1 - \frac{v}{2}\right)\right]$$

Figure 3.2 Pressurised circular cone.

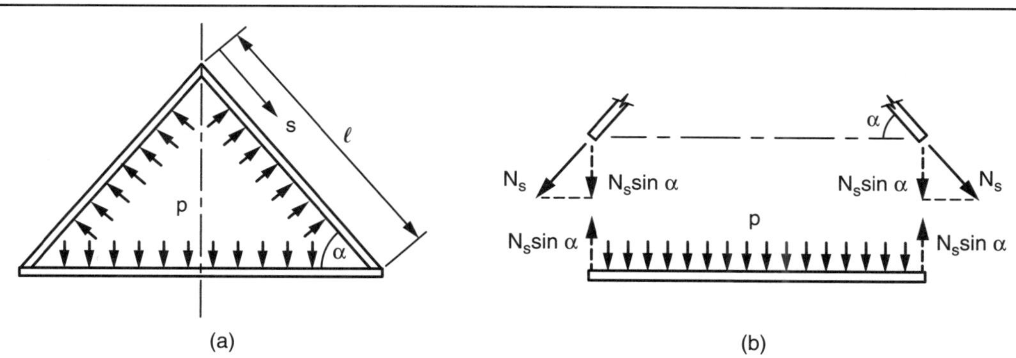

$$= -\frac{3}{2}\frac{ps}{Et}\cot^2\alpha \qquad (3.11)$$

assuming E and t are both constant. Thus N_θ, N_s and V are all directly proportional to the distance s of a point from the vertex of the cone, while the lateral displacement δ is proportional to s^2. As with the circular cylindrical pressure vessel with flat end closures, the circular conical pressure vessel will experience localised bending in the vicinity of its junction with the plate closure, as a result of the constraining effect of the joint.

3.1.3 Spherical vessel

For a spherical pressure vessel of radius a (Figure 3.3), we could proceed on the basis of equation (2.15), with $r_1 = r_2 = a$, to obtain

$$N_\phi = \frac{1}{a\sin^2\phi}\left[a^2 p\int \cos\phi \sin\phi \, d\phi + k\right]$$

$$= \frac{1}{a\sin^2\phi}\left[\frac{pa^2}{2}\sin^2\phi + k\right]$$

At $\phi = 0$ or $\phi = 180°$ (i.e. the *poles* of the closed shell of revolution, through which the axis of revolution passes), the condition that N_ϕ must remain of finite value yields $k = 0$, so that

$$N_\phi = \frac{pa}{2} \qquad (3.12)$$

The hoop stress resultant follows from equation (2.12):

$$N_\theta = pa - \frac{pa}{2} = \frac{pa}{2} \qquad (3.13)$$

The above results can, of course, be deduced more simply from the spherical symmetry of the problem: N_ϕ and N_θ at any point are equal and interchangeable, so that, from equation (2.11), each must be equal to $pa/2$. Also from spherical symmetry, all points on the vessel must move outwards along the normal to the shell mid-surface by the same amount w, so that there are no rotations ($V = 0$) throughout. From equation (2.67), we obtain

$$\delta = \frac{1}{Et}(a\sin\phi)\frac{pa}{2}(1-\nu) = \frac{pa^2}{2Et}(1-\nu)\sin\phi \qquad (3.14)$$

Figure 3.3 Spherical pressure vessel.

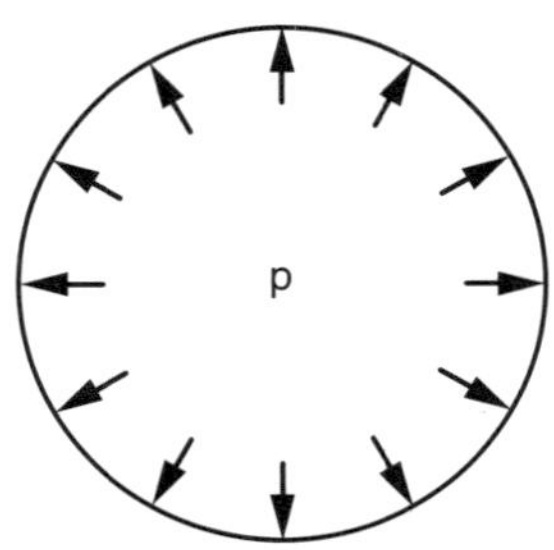

Making use of relation (2.56), with $v = 0$ (owing to spherical symmetry), we obtain the normal displacement as

$$w = \frac{dR}{\sin\phi} = \frac{\delta}{\sin\phi} = \frac{pa^2}{2Et}(1 - v) \tag{3.15}$$

3.1.4 Vessel in the form of an ellipsoid of revolution

For an ellipsoid of revolution, the equation of the meridian is

$$\frac{x^2}{a^2} + \frac{y^2}{b^2} = 1 \tag{3.16}$$

where it is assumed that the y axis coincides with the vertical axis of revolution of the surface, and the x axis is horizontal and coincident with the meridian bisector (perpendicular to the axis of revolution), so that the radius R of a circle of latitude is actually the x coordinate at that location (see Figure 3.4).

For any shell midsurface, the meridional radius of curvature r_1 is given by the well-known relationship of plane-curve geometry

$$r_1 = \frac{\left[1 + \left(\dfrac{dy}{dx}\right)^2\right]^{\frac{3}{2}}}{\dfrac{d^2y}{dx^2}} \tag{3.17}$$

The second principal radius of curvature r_2 is given by the distance between the point in question and the intersection of the axis of revolution of the shell with the normal to the shell midsurface at that point. By reference to Figure 3.4,

Figure 3.4 Pressure vessel in the form of an ellipsoid of revolution.

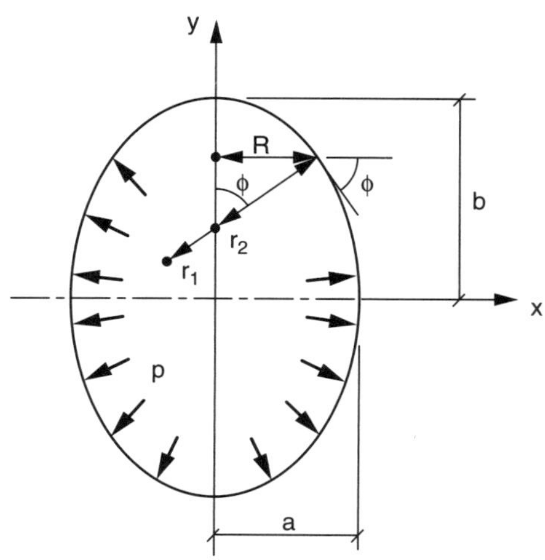

$$r_2 = \frac{R}{\sin\phi} = \frac{x}{\sin\phi} \tag{3.18}$$

From equation (3.16), and the interpretation of the slope at a point as $\phi = \tan^{-1}(dy/dx)$, we can express r_1 and r_2 of the ellipsoid of revolution in terms of the coordinate ϕ. We proceed as follows.

From equation (3.16), we obtain

$$y = \pm\frac{b}{a}\sqrt{a^2 - x^2} \tag{3.19}$$

Differentiating this expression with respect to x, we have

$$\frac{dy}{dx} = \pm\frac{b}{a}\frac{x}{\sqrt{a^2 - x^2}} \tag{3.20}$$

By definition, dy/dx is equal to the slope of the tangent to the meridian at the point in question, which in turn is equal to $\tan\phi$, since ϕ is the angle between the tangent to the meridian and the direction of the x axis. From equation (3.20), we may therefore write

$$\left(\frac{dy}{dx}\right)^2 = \tan^2\phi = \frac{b^2}{a^2}\left(\frac{x^2}{a^2 - x^2}\right) \tag{3.21}$$

Solving equation (3.21) for x^2 in terms of $\tan^2\phi$, we obtain

$$x^2 = \frac{a^4\tan^2\phi}{b^2 + a^2\tan^2\phi}$$

– that is,

$$x = \frac{a^2\tan\phi}{\left(b^2 + a^2\tan^2\phi\right)^{1/2}} = \frac{a^2\sin\phi}{\left(a^2\sin^2\phi + b^2\cos^2\phi\right)^{1/2}} \tag{3.22}$$

Using this result in equation (3.18) then yields r_2:

$$r_2 = \frac{x}{\sin\phi} = \frac{a^2}{\left(a^2\sin^2\phi + b^2\cos^2\phi\right)^{1/2}} \tag{3.23}$$

Substituting result (3.22) into expression (3.19), we obtain

$$y = \pm\frac{b^2\cos\phi}{\left(a^2\sin^2\phi + b^2\cos^2\phi\right)^{1/2}} \tag{3.24}$$

From equation (3.20), the second derivative of y with respect to x is obtained as

$$\frac{d^2y}{dx^2} = \pm\frac{ab}{\left(a^2 - x^2\right)^{3/2}} \tag{3.25}$$

Substituting into equation (3.17) the expression for the first derivative of y (equation (3.20)) and the expression for the second derivative of y (equation (3.25)), we obtain

$$r_1 = \frac{\left[a^2\left(a^2 - x^2\right) + b^2x^2\right]^{3/2}}{a^4b} \tag{3.26}$$

from which we can readily eliminate x, using result (3.22), to obtain

$$r_1 = \frac{a^2b^2}{\left(a^2\sin^2\phi + b^2\cos^2\phi\right)^{3/2}} \tag{3.27}$$

Substituting the expressions for r_1 (equation 3.27) and r_2 (equation 3.23) into the general solution for N_ϕ (equation (2.15) of Chapter 2), we obtain

$$N_\phi = \frac{\left(a^2\sin^2\phi + b^2\cos^2\phi\right)^{1/2}}{a^2\sin^2\phi}\left[pa^4b^2\int\frac{\cos\phi\sin\phi}{\left(a^2\sin^2\phi + b^2\cos^2\phi\right)^2}d\phi + k\right] \tag{3.28}$$

Using standard methods of integration, we obtain

$$\int\frac{\cos\phi\sin\phi}{\left(a^2\sin^2\phi + b^2\cos^2\phi\right)^2}d\phi = -\frac{1}{2\left(a^2 - b^2\right)\left[b^2 + \left(a^2 - b^2\right)\sin^2\phi\right]}$$

so that equation (3.28) becomes

$$N_\phi = \frac{\left(a^2\sin^2\phi + b^2\cos^2\phi\right)^{1/2}}{a^2\sin^2\phi}\left[\frac{-pa^4b^2}{2\left(a^2 - b^2\right)\left[b^2 + \left(a^2 - b^2\right)\sin^2\phi\right]} + k\right] \tag{3.29}$$

At $\phi = 0$ or $\phi = 180°$ (i.e. at the poles), we require that N_ϕ remains of finite value, a condition that yields

$$k = \frac{pa^4}{2\left(a^2 - b^2\right)} \tag{3.30}$$

Equation (3.29) then becomes, after some further simplifications,

$$N_\phi = \frac{pa^2}{2}\frac{1}{\left(a^2\sin^2\phi + b^2\cos^2\phi\right)^{1/2}} \tag{3.31}$$

The hoop stress resultant then follows from equation (2.12); after some simplifications, the result is

$$N_\theta = \frac{pa^2}{2b^2} \frac{b^2 + (b^2 - a^2)\sin^2\phi}{(a^2\sin^2\phi + b^2\cos^2\phi)^{1/2}} \tag{3.32}$$

Note that N_ϕ is always positive (i.e. tensile). On the other hand, N_θ may become compressive over certain regions of the shell, depending on the relative sizes of dimensions a and b. As long as $b > a$, N_θ will be tensile throughout the shell. When $b = a$, the above expressions reduce to $N_\phi = N_\theta = pa/2$, the results for a spherical vessel.

For b smaller than a, we may rewrite N_θ as

$$N_\theta = \frac{pa^2}{2b^2} \frac{b^2 - (a^2 - b^2)\sin^2\phi}{(a^2\sin^2\phi + b^2\cos^2\phi)^{1/2}} \tag{3.33}$$

In a shell of given ratio b/a, where $b < a$, N_θ is tensile from the apex ($\phi = 0$) downwards, until we reach the horizontal circle of latitude corresponding to $\phi = \phi_c$, where it becomes zero. By solving equation (3.33) with the left-hand side set equal to zero, we obtain

$$\sin\phi_c = \frac{b}{\sqrt{a^2 - b^2}} \tag{3.34}$$

Below this level, N_θ becomes negative (i.e. compressive).

Since the equatorial plane ($\phi = \pi/2$) of pressure vessels in the form of complete ellipsoids of revolution (such as the one illustrated in Figure 3.4) is a plane of symmetry for both hoop and meridional effects, we may confine our considerations to the top half of the vessel (i.e. to the region $0 \le \phi \le \pi/2$). Thus, for the domain $0 \le \phi \le \pi/2$, hoop compression occurs in the zone $\phi_c \le \phi \le \pi/2$.

To ensure that no hoop compression occurs in the shell, the relative proportions of b and a must be such that the transition level from tensile N_θ to compressive N_θ is pushed right down to the edge of the domain $0 \le \phi \le \pi/2$ – that is, we require that $\phi_c = \pi/2$. Substituting this value for ϕ_c into equation (3.34), we obtain the relation

$$b = \frac{a}{\sqrt{2}} \tag{3.35}$$

In other words, for all ellipsoidal pressure vessels for which the vertical dimension b exceeds $a/\sqrt{2}$ (i.e. for which $b > a/\sqrt{2}$), the hoop stress resultant N_θ will always be tensile throughout. As the meridional stress resultant N_ϕ is always tensile throughout the vessel regardless of the relative proportions of a and b, it means the entire vessel would then be completely free of any compressive actions, and local buckling need not be checked for vessels satisfying this condition.

3.1.5 Vessel in the form of a torus of circular section

For a circular torus, let ϕ be measured from the vertical axis of the circular cross section, towards the outer surface of the torus, as shown in Figure 3.5. For the outer surface of the torus (over which the Gaussian

Figure 3.5 Pressure vessel in the form of a circular torus.

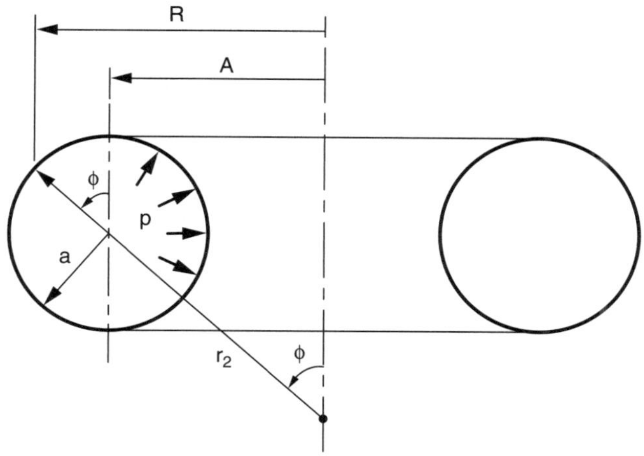

curvature is positive), ϕ is positive ($0 \le \phi \le \pi$), whereas for the inner surface (which is of negative Gaussian curvature), ϕ will be considered to be negative ($0 \ge \phi \ge -\pi$). The principal radii of curvature are as follows (see Figure 3.5):

$$r_1 = a \tag{3.36}$$

$$r_2 = \frac{R}{\sin\phi} = \frac{A + a\sin\phi}{\sin\phi} \tag{3.37}$$

where A is the mean radius of the torus.

Applying equation (2.15), we have

$$N_\phi = \frac{\sin\phi}{(A + a\sin\phi)\sin^2\phi}\left[\int a\left(\frac{A + a\sin\phi}{\sin\phi}\right) p\cos\phi\sin\phi\, d\phi + k\right]$$

$$= \frac{1}{(A + a\sin\phi)\sin\phi}\left[pa\left(A\sin\phi + \frac{a\sin^2\phi}{2}\right) + k\right]$$

At $\phi = 0$ and $\phi = 180°$, N_ϕ must remain finite, implying that $k = 0$. Therefore,

$$N_\phi = pa\left(\frac{A + \frac{a}{2}\sin\phi}{A + a\sin\phi}\right) \tag{3.38}$$

From equation (2.12), we obtain

$$N_\theta = \frac{pa}{2} \tag{3.39}$$

Figure 3.6 Equilibrium in the direction normal to the shell midsurface of a point on the inner surface of a toroid.

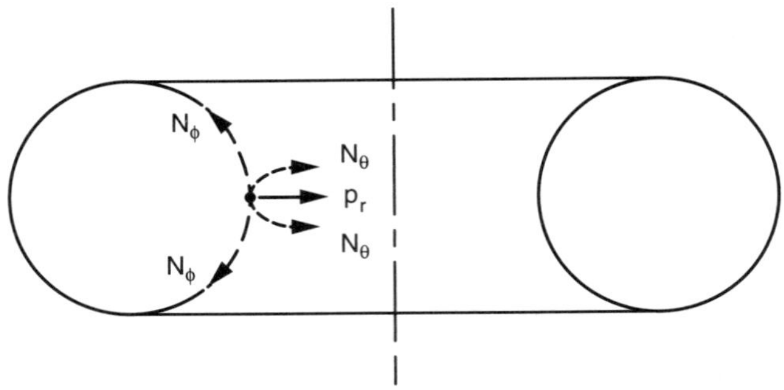

For the inner surface, over which ϕ is negative, we may write $\phi = -|\phi|$ so that

$$N_\phi = pa\left(\frac{A - \frac{a}{2}\sin|\phi|}{A - a\sin|\phi|}\right) \tag{3.40}$$

N_θ would still be given by expression (3.39), a result that may be confirmed by substituting (3.40) in equation (2.11), which, for the inner surface of negative Gaussian curvature, must read (see Figure 3.6)

$$\frac{N_\phi}{|r_1|} - \frac{N_\theta}{|r_2|} = p_r \tag{3.41}$$

At the highest and lowest circles of latitude (i.e. the locations $\phi = 0$ and $\phi = \pm 180°$, where the outer surface meets the inner surface), we expect some localised bending to occur, owing to the vanishing of the curvature $1/r_2$ there, as r_2 jumps from $+\infty$ (just within the outer surface) to $-\infty$ (just within the inner surface). Such *lines of distortion* were discussed in Section 1.5.3. The predicted breakdown of the membrane solution at these two circles may be confirmed by calculating membrane deformations there; incompatibilities will be observed.

3.1.6 Vessel in the form of a torus of elliptical section

For the elliptical cross-section in Figure 3.7, we see that r_1 is the same as for the ellipsoid of revolution – that is,

$$r_1 = \frac{a^2b^2}{\left(a^2\sin^2\phi + b^2\cos^2\phi\right)^{3/2}} \tag{3.42}$$

while r_2 may readily be seen from the same figure to be

$$r_2 = \frac{A}{\sin\phi} + r_2' = \frac{A}{\sin\phi} + \frac{a^2}{\left(a^2\sin^2\phi + b^2\cos^2\phi\right)^{1/2}} \tag{3.43}$$

where r_2' is the second principal radius of an ellipsoid of revolution of the same meridional cross-section as the cross-section of the elliptic torus, and A, as before, is the mean radius of the elliptic torus.

Figure 3.7 Pressure vessel in the form of an elliptic torus.

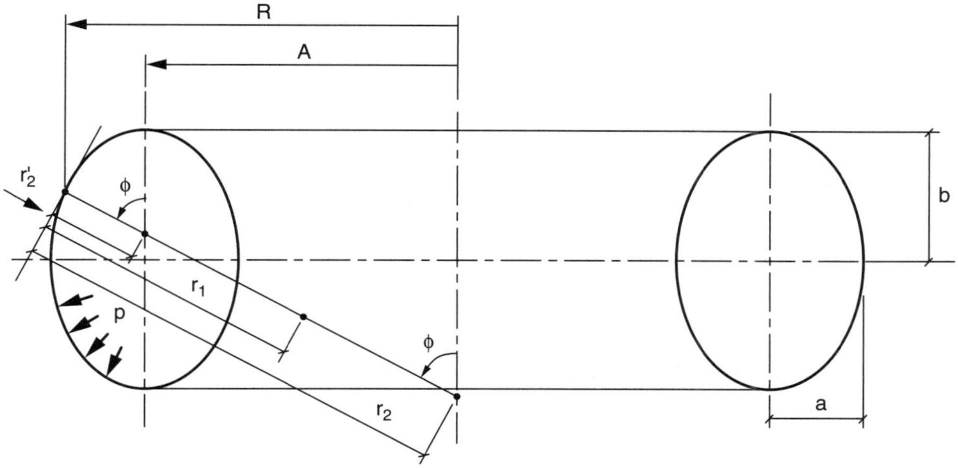

Applying equation (2.15), we obtain

$$N_\phi = \frac{1}{r_2 \sin^2 \phi}\left[p\int \frac{Aa^2b^2 \cos\phi}{\left(a^2 \sin^2 \phi + b^2 \cos^2 \phi\right)^{3/2}}\, d\phi + p\int \frac{a^4b^2 \cos\phi \sin\phi}{\left(a^2 \sin^2 \phi + b^2 \cos^2 \phi\right)^2}\, d\phi + k\right]$$

$$= \frac{1}{r_2 \sin^2 \phi}\left[\frac{pAa^2 \sin\phi}{\left(a^2 \sin^2 \phi + b^2 \cos^2 \phi\right)^{1/2}} - \frac{pa^4b^2}{2\left(a^2 - b^2\right)\left(a^2 \sin^2 \phi + b^2 \cos^2 \phi\right)} + k\right] \tag{3.44}$$

Now

$$r_2 \sin^2 \phi = A \sin\phi + \frac{a^2 \sin^2 \phi}{\left(a^2 \sin^2 \phi + b^2 \cos^2 \phi\right)^{1/2}} \tag{3.45}$$

At $\phi = 0$ and $\phi = 180°$, $r_2 \sin^2 \phi$ vanishes, so for N_ϕ to remain finite, the term in square brackets on the right-hand side of expression (3.44) must also vanish. This gives

$$k = \frac{pa^4}{2\left(a^2 - b^2\right)} \tag{3.46}$$

Substituting (3.45) and (3.46) into (3.44), and simplifying the ensuing expression, we finally obtain

$$N_\phi = \frac{pa^2}{\left(a^2 \sin^2 \phi + b^2 \cos^2 \phi\right)^{1/2}}\left[\frac{A\left(a^2 \sin^2 \phi + b^2 \cos^2 \phi\right)^{1/2} + \dfrac{a^2}{2}\sin\phi}{A\left(a^2 \sin^2 \phi + b^2 \cos^2 \phi\right)^{1/2} + a^2 \sin\phi}\right] \tag{3.47}$$

When $b = a$, this result reduces to that for a circular torus, as expected. For $A = 0$ but $b \neq a$, expression (3.47) reduces to (3.31) for an ellipsoid of revolution, also as expected.

N_θ then follows from equation (2.12). After some simplifications, we eventually obtain

$$N_\theta = \frac{pa^2}{b^2}\left[A\left(\frac{b^2-a^2}{a^2}\right)\sin\phi + \frac{b^2-0.5\left(a^2\sin^2\phi+b^2\cos^2\phi\right)}{\left(a^2\sin^2\phi+b^2\cos^2\phi\right)^{1/2}}\right] \tag{3.48}$$

When $b = a$, the above expression coincides with the corresponding one for the circular torus (expression (3.39)); when $A = 0$ but $b \neq a$, the expression coincides with the N_θ expression for the ellipsoid of revolution (expression (3.32)). The comments that were made for the circular torus regarding the validity of the membrane solution at $\phi = 0$ and $\phi = 180°$ are also applicable for the elliptic toroid.

3.2. Elevated liquid-filled vessels

For elevated liquid-filled vessels, the principal loading condition is hydrostatic pressure, which acts in a direction normal to the shell surface at any given point, and is proportional to the depth of liquid at that point. Thus p_ϕ (or p_x for cylinders, p_s for cones) = 0, $p_\theta = 0$ and $p_r = \gamma h$, where γ is the weight per unit volume of the contained liquid, and h is the depth of liquid at a given point. Throughout this section, the parameters E (Young's modulus) and t (shell thickness) are assumed to be constant for a given shell, unless otherwise indicated.

3.2.1 Circular cylindrical vessel

Figure 3.8 shows a circular cylindrical vessel filled with liquid of unit weight γ, and supported above the ground on vertical columns. From equations (2.17) and (2.18), with $p_r = \gamma x$, we have

$$N_\theta = \gamma a x \tag{3.49}$$

$$N_x = k \tag{3.50}$$

At $x = 0$ (i.e. at the top of the tank), we require that $N_x = 0$, since this edge is free. From equation (3.50), this condition yields $k = 0$ so that

$$N_x = 0 \tag{3.51}$$

The deformations δ and V are obtained from equations (2.73) and (2.74):

$$\delta = \frac{\gamma a^2}{Et}x \tag{3.52}$$

$$V = -\frac{\gamma a^2}{Et} \tag{3.53}$$

At the base, the shell cannot rotate freely, or move freely in the lateral direction, owing to its attachment to a base plate, a concrete slab or some other bottom closure. Thus, the location $x = H$ is a source of bending effects for the shell. If the bottom edge of the cylindrical tank is fully restrained against lateral movement and rotation, then a bending correction (in the form of axisymmetric bending moments M_e and horizontal

Figure 3.8 Elevated circular cylindrical tank.

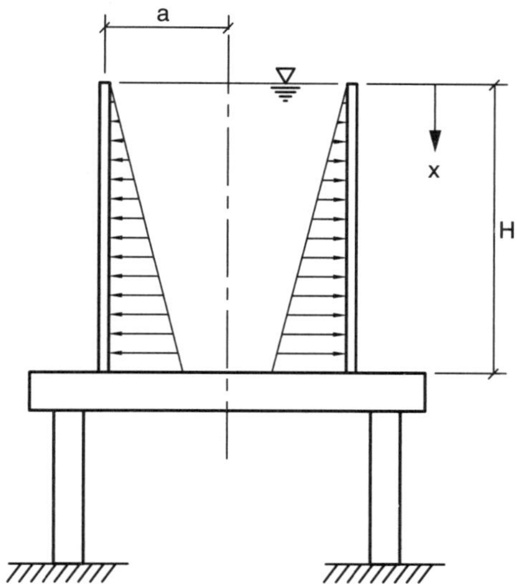

shears H_e) will be necessary. The magnitudes of these edge actions M_e and H_e must be chosen such that the deformations δ and V caused by them, when superimposed with the values of the above membrane-solution deformations at the location $x = H$, yield net values of zero for the lateral movement and the meridional rotation, as is consistent with the actual boundary conditions at the base. This is the flexibility approach to shell analysis, which we will use extensively in Chapters 5, 7 and 8.

3.2.2 Circular conical vessel

Two typical support configurations for circular conical tanks are shown in Figure 3.9. In the first arrangement (Figure 3.9 (a)), the cone is supported at its vertex by a single vertical stem in wine-glass fashion. In the second arrangement (Figure 3.9(b)), the cone is supported along its circular edge by a series of closely spaced vertical columns, usually via a ring beam (not shown); the support conditions may be considered to be essentially axisymmetric.

For the tank completely filled with liquid of unit weight γ, the hydrostatic pressure at the coordinate s (see Figure 3.9) is given by $p_r = \gamma(H - s\sin\alpha)$. Applying expressions (2.27) and (2.29), we obtain

$$N_\theta = \gamma(\cot\alpha)(H - s\sin\alpha)s \tag{3.54}$$

$$N_s = \frac{1}{s}\left[\gamma(\cot\alpha)\int(H - s\sin\alpha)s\;ds \;+\; k\right]$$

$$= \frac{1}{s}\left[\frac{\gamma}{6}\,(\cot\alpha)\left(3Hs^2 - 2s^3\sin\alpha\right) \;+\; k\right] \tag{3.55}$$

For the configuration of Figure 3.9(a), N_s must vanish at the free edge at the top – that is, $N_s = 0$ at $s = H/\sin\alpha$. Applying this condition to equation (3.55), we obtain

$$k = -\frac{\gamma}{6}\cot\alpha\frac{H^3}{\sin^2\alpha} \tag{3.56}$$

so that

$$N_s = \frac{\gamma}{6}(\cot\alpha)\frac{1}{s}\left[3Hs^2 - 2s^3\sin\alpha - \frac{H^3}{\sin^2\alpha}\right] \tag{3.57}$$

At $s = 0$, N_s is infinite, a singularity that we expect of an object of finite weight supported on a point. In practice, the stem has a finite diameter, so that the conical tank is supported on a finite area, and N_s will therefore be finite.

The deformations follow from relations (2.75) and (2.76):

$$\delta = \frac{\gamma}{Et}\left(\frac{\cos^2\alpha}{\sin\alpha}\right)s\left[(H - s\sin\alpha)s - \frac{\nu}{6s}\left\{3Hs^2 - 2s^3\sin\alpha - \frac{H^3}{\sin^2\alpha}\right\}\right] \tag{3.58}$$

$$V = -\frac{\gamma}{Et}(\cot^2\alpha)\left[\frac{1+\nu}{6s}\left\{3Hs^2 - 4s^3\sin\alpha + \frac{H^3}{\sin^2\alpha}\right\}\right.$$
$$\left. + s\left\{H - 2s\sin\alpha - \frac{\nu}{6}\left(3H - 4s\sin\alpha + \frac{H^3}{s^2\sin^2\alpha}\right)\right\}\right] \tag{3.59}$$

For the configuration of Figure 3.9(b), N_s must remain finite at $s = 0$, implying that $k = 0$. This leads to the expression

$$N_s = \frac{\gamma}{6}(\cot\alpha)(3H - 2s\sin\alpha)s \tag{3.60}$$

In this case, the deformations become

Figure 3.9 Elevated conical tank: (a) supported at the vertex; (b) supported along the edge.

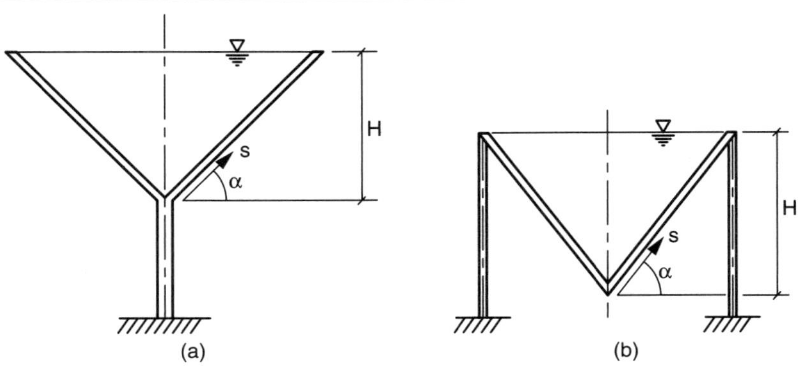

$$\delta = \frac{\gamma}{Et}\left(\frac{\cos^2\alpha}{\sin\alpha}\right)s^2\left[H - s\sin\alpha - \frac{\nu}{6}(3H - 2s\sin\alpha)\right] \tag{3.61}$$

$$V = -\frac{\gamma}{Et}(\cot^2\alpha)\,s\left[\frac{1+\nu}{6}(3H - 4s\sin\alpha) + \left\{H - 2s\sin\alpha - \frac{\nu}{6}(3H - 4s\sin\alpha)\right\}\right] \tag{3.62}$$

Localised bending occurs around the top of the tank, owing to the concentrated line reaction of the vertical supports, and the constraining effect of the ring beam upon the shell.

3.2.3 Spherical vessel

Figure 3.10 shows a spherical vessel of radius a completely filled with liquid of unit weight γ, and supported on a series of closely spaced vertical columns via a ring beam (not shown). For this arrangement, we have $r_1 = r_2 = a$; $p_\phi = p_\theta = 0$; $p_r = \gamma a(1 - \cos\phi)$. Equation (2.15) becomes

$$N_\phi = \frac{1}{a\sin^2\phi}\left[\gamma a^3 \int (1-\cos\phi)\cos\phi\sin\phi\, d\phi + k\right]$$

$$= \frac{1}{a\sin^2\phi}\left[-\gamma a^3\left(\frac{\cos^2\phi}{2} - \frac{\cos^3\phi}{3}\right) + k\right] \tag{3.63}$$

Let ϕ_s be the value of ϕ at the circle of latitude of the supports (see Figure 3.10). The support reactions are assumed to be axisymmetric, given the close spacing of vertical columns and the smoothening effect of the interposing ring beam. In the upper part of the tank ($0 \le \phi \le \phi_s$), N_ϕ must remain finite at the top of the vessel (i.e. as $\phi \to 0$). This condition gives

$$k = \frac{\gamma a^3}{6} \tag{3.64}$$

so that

$$N_\phi = \frac{\gamma a^2}{6}\left(\frac{1-\cos\phi}{1+\cos\phi}\right)(1 + 2\cos\phi) \tag{3.65}$$

From equation (2.12), and using the above result, we obtain

$$N_\theta = \frac{\gamma a^2}{6}\left(\frac{1-\cos\phi}{1+\cos\phi}\right)(5 + 4\cos\phi) \tag{3.66}$$

The deformations in the upper part of the tank are obtained as

$$\delta = \frac{\gamma a^3}{6Et}(\sin\phi)\left(\frac{1-\cos\phi}{1+\cos\phi}\right)\{(5 + 4\cos\phi) - \nu(1 + 2\cos\phi)\} \tag{3.67}$$

$$V = -\frac{\gamma a^2}{Et}\sin\phi \tag{3.68}$$

Figure 3.10 Elevated spherical tank filled with liquid.

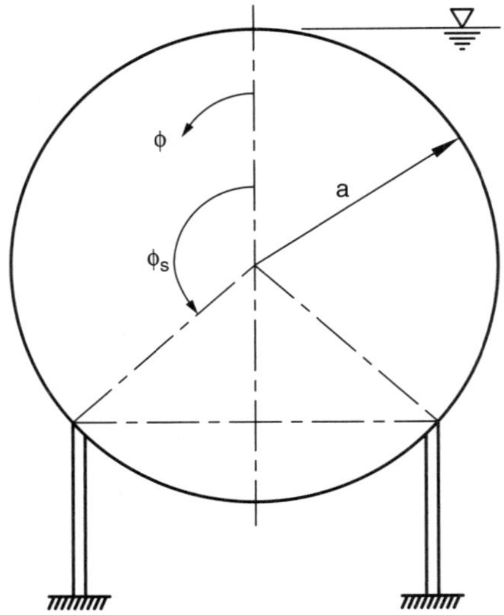

In the lower part of the tank ($\phi_s \leq \phi \leq \pi$), we require that N_ϕ remains finite at the bottom of the vessel (i.e. as $\phi \to \pi$). This leads to the solution

$$k = \frac{5}{6}\gamma a^3 \tag{3.69}$$

so that the meridional stress resultant becomes

$$N_\phi = \frac{\gamma a^2}{6}\left(\frac{5 - 5\cos\phi + 2\cos^2\phi}{1 - \cos\phi}\right) \tag{3.70}$$

The hoop stress resultant then follows from expression (2.12):

$$N_\theta = \frac{\gamma a^2}{6}\left(\frac{1 - 7\cos\phi + 4\cos^2\phi}{1 - \cos\phi}\right) \tag{3.71}$$

Substituting results (3.70) and (3.71) into the general expressions for the deformations (i.e. equations (2.67) and (2.72)), we obtain the following closed-form expressions for the lateral displacement and rotation of the shell in the lower part of the tank:

$$\delta = \frac{\gamma a^3}{6Et}(\sin\phi)\left(\frac{1}{1 - \cos\phi}\right)\left\{(1 - 7\cos\phi + 4\cos^2\phi) - \nu(5 - 5\cos\phi + 2\cos^2\phi)\right\} \tag{3.72}$$

$$V = -\frac{\gamma a^2}{Et}\sin\phi \tag{3.73}$$

Although V at the support level (i.e. at $\phi = \phi_s$) is the same for the upper and lower domains of the vessel, the δ values are different, implying a broken midsurface. As the shell is physically continuous throughout, bending corrective actions are required at the support circle in order to restore continuity, and therefore the support circle is a source of bending disturbances.

As will be seen in Chapter 8, the magnitude of the bending disturbance is a function of the support location (ϕ_s), the angle of inclination of the supports to the shell midsurface, and the stiffness properties of the ring beam. For now, it will suffice to point out that the *equatorial* circle of latitude (i.e. the location $\phi_s = 90°$) is the best location for vertical supports, as the support reactions are then tangential to the shell midsurface, which minimises bending effects.

For spherical metal tanks, stepwise local thickening of the shell is usually adopted in a narrow band around the supports, in order to cater for the very localised bending stresses there. However, in accordance with the considerations of Section 1.5.3, the locations at which the shell thickness is abruptly changed are themselves sources of bending disturbances.

The difference in the values of N_ϕ just below the support and just above the support is

$$N_s = \frac{2}{3} \frac{\gamma a^2}{\sin^2 \phi_s} \tag{3.74}$$

acting tangentially to the shell midsurface, in the direction of increasing ϕ. The component of N_s in the vertical direction is a downward force $N_s \sin \phi_s$ causing compression in the vertical supporting columns; the integral of this component over the support circumference is

$$\frac{2}{3}\left(\frac{\gamma a^2}{\sin^2 \phi_s}\right) \sin \phi_s \times (2\pi a \sin \phi_s) = \frac{4}{3} \gamma \pi a^3 \tag{3.75}$$

which, of course, is equal to the total weight of the contained liquid.

The component of N_s in the horizontal direction, where $\phi_s > 90°$, is an inward force

$$N_s \left|\cos \phi_s\right| = \frac{2}{3} \gamma a^2 \frac{\left|\cot \phi_s\right|}{\sin \phi_s} \tag{3.76}$$

causing compression in the ring beam.

The vertical columns and the horizontal ring beam must be designed for their respective compressive actions.

3.2.4 Intze tank

An Intze tank is a liquid-containing shell of compound geometry, usually comprising a spherical top closure, vertical cylindrical sides, a conical transition in the lower part, and a spherical bottom closure (see Figure 3.11). At the junction of the conical transition and the spherical bottom closure, the compressive meridional actions on either side oppose each other on the horizontal projection, which minimises the net lateral thrust on the ring beam at this level, resulting in an economical design for the ring beam. Let the shell thicknesses of the spherical top, the vertical cylindrical sides, the conical transition and the spherical bottom be denoted by t_1, t_2, t_3 and t_4 respectively, these being constant over their respective domains. The rest of the geometrical parameters of the Intze tank are as shown in Figure 3.11.

Figure 3.11 Geometric parameters of a liquid-filled Intze tank.

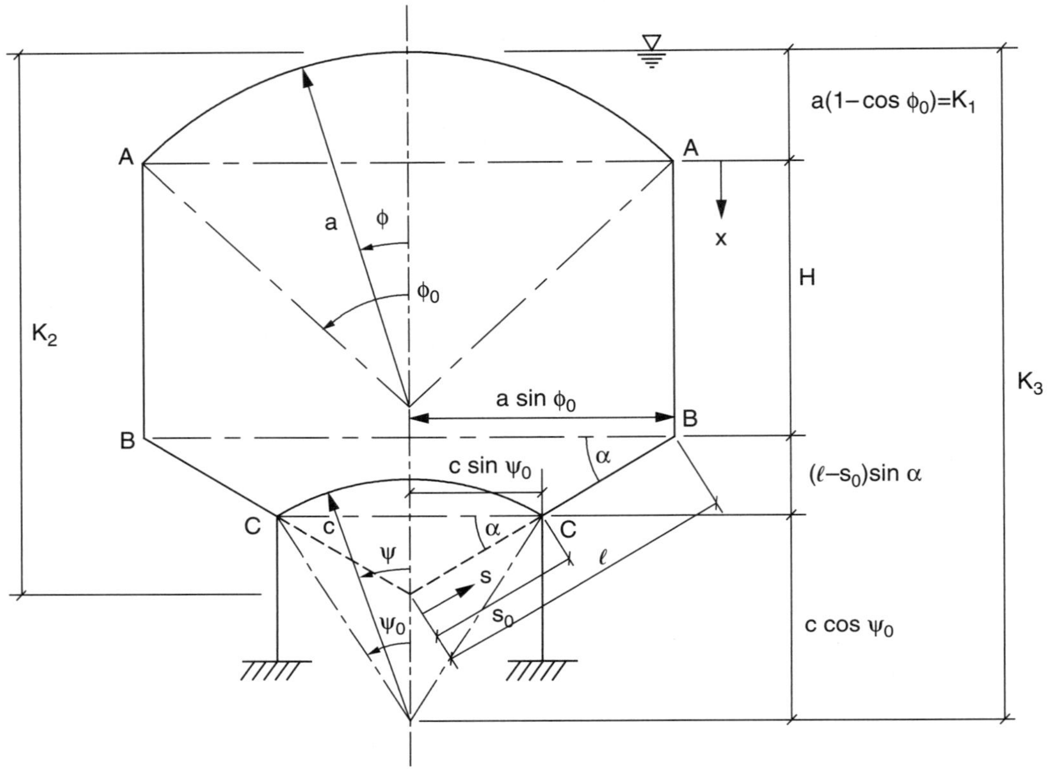

From Figure 3.11, the slope α of the conical portion is given by

$$\cos\alpha = \frac{a\sin\phi_o}{l} = \frac{c\sin\psi_o}{s_o} \tag{3.77}$$

Therefore, the parameters l and s_o of the conical portion may be expressed as

$$l = a\left(\frac{\sin\phi_o}{\cos\alpha}\right) \tag{3.78a}$$

$$s_o = c\left(\frac{\sin\psi_o}{\cos\alpha}\right) \tag{3.78b}$$

Assuming the tank to be filled right up to the top with liquid of weight γ per unit volume, we may express the hydrostatic pressure p_r in the four portions of the tank in terms of the generalised geometric parameters of Figure 3.11, as follows:

(*a*) *Spherical top* ($0 \le \phi \le \phi_o$)

$$p_r = \gamma a\left(1-\cos\phi\right) \tag{3.79}$$

(*b*) *Cylindrical sides* $(0 \le x \le H)$

$$p_r = \gamma\left[a(1-\cos\phi_o)+x\right]$$

$$= \gamma(K_1 + x) \tag{3.80}$$

where

$$K_1 = a(1-\cos\phi_o) \tag{3.81}$$

(*c*) *Conical transition* $(s_o \le s \le l)$

$$p_r = \gamma\left[a(1-\cos\phi_o)+H+(l-s)\sin\alpha\right]$$

$$= \gamma(K_2 - s\sin\alpha) \tag{3.82}$$

where

$$K_2 = a(1-\cos\phi_o)+H+l\sin\alpha = K_1+H+l\sin\alpha \tag{3.83}$$

(*d*) *Spherical bottom* $(0 \le \psi \le \psi_o)$

$$p_r = \gamma\left[a(1-\cos\phi_o)+H+(l-s_o)\sin\alpha - c(\cos\psi-\cos\psi_o)\right]$$

$$= \gamma(K_3 - c\cos\psi) \tag{3.84}$$

where

$$K_3 = a(1-\cos\phi_o)+H+(l-s_o)\sin\alpha + c\cos\psi_o$$

$$= K_1+H+(l-s_o)\sin\alpha + c\cos\psi_o \tag{3.85}$$

In the spherical top, the stress resultants are given by the same expressions as were obtained for the upper domain of the spherical tank – that is,

$$N_\phi = \frac{\gamma a^2}{6}\left(\frac{1-\cos\phi}{1+\cos\phi}\right)(1+2\cos\phi) \tag{3.86a}$$

$$N_\theta = \frac{\gamma a^2}{6}\left(\frac{1-\cos\phi}{1+\cos\phi}\right)(5+4\cos\phi) \tag{3.86b}$$

At the edge of the spherical top (i.e. at $\phi = \phi_o$), let the value of N_ϕ be denoted by N_o – that is,

$$N_o = \frac{\gamma a^2}{6}\left(\frac{1-\cos\phi_o}{1+\cos\phi_o}\right)(1+2\cos\phi_o) \tag{3.87}$$

This has a vertical component $N_o \sin\phi_o$.

The deformations of the upper spherical shell at its edge can be written down from equations (3.67) and (3.68) as follows:

$$\delta_o = \frac{\gamma a^3}{6Et_1}(\sin\phi_o)\left(\frac{1-\cos\phi_o}{1+\cos\phi_o}\right)\{(5+4\cos\phi_o) - v(1+2\cos\phi_o)\} \tag{3.88a}$$

$$V_o = -\frac{\gamma a^2}{Et_1}\sin\phi_o \tag{3.88b}$$

For the cylindrical sides, noting that the radius of the cylinder is equal to $a\sin\phi_o$ (see Figure 3.11), we have, from equations (2.17) and (2.18),

$$N_x = k \tag{3.89a}$$

$$N_\theta = (a\sin\phi_o)\,\gamma(K_1 + x) = \gamma a(\sin\phi_o)(K_1 + x) \tag{3.89b}$$

At the top edge of the cylinder (i.e. at $x = 0$), N_x must equal the vertical component of the meridional stress resultant at the edge of the spherical shell (i.e. at $\phi = \phi_o$), so that expression (3.89a) becomes

$$N_x = N_o \sin\phi_o \tag{3.89c}$$

The deformations in the cylinder are, from equations (2.73) and (2.74), as follows:

$$\delta = \frac{a\sin^2\phi_o}{Et_2}\{\gamma a(K_1 + x) - vN_o\} \tag{3.90a}$$

$$V = -\frac{a\sin\phi_o}{Et_2}\left[\gamma a(\sin\phi_o)\right] = -\frac{\gamma a^2}{Et_2}\sin^2\phi_o \tag{3.90b}$$

At the top of the cylinder ($x = 0$), we have

$$\delta_t = \frac{a\sin^2\phi_o}{Et_2}\{\gamma a K_1 - vN_o\} \tag{3.91a}$$

$$V_t = -\frac{\gamma a^2}{Et_2}\sin^2\phi_o \tag{3.91b}$$

At the bottom ($x = H$) of the cylinder, we have

$$\delta_b = \frac{a\sin^2\phi_o}{Et_2}\{\gamma a(K_1 + H) - vN_o\} \tag{3.92a}$$

$$V_b = -\frac{\gamma a^2}{Et_2}\sin^2\phi_o \tag{3.92b}$$

As in the case of the cylindrical tank (refer to results (3.52) and (3.53)), the lateral deflection δ varies linearly, while the rotation V is constant across the depth of the cylinder.

For the conical portion, we make use of expression (2.29), with p_r given by (3.82), to obtain

$$\begin{aligned} N_s &= \frac{1}{s}\left[\int \gamma(\cot\alpha)(K_2 - s\sin\alpha)\, s\, ds + k\right] \\ &= \frac{1}{s}\left[\gamma(\cot\alpha)\left(\frac{K_2 s^2}{2} - \frac{s^3}{3}\sin\alpha\right) + k\right] \end{aligned} \tag{3.93}$$

At $s = l$, the vertical component of N_s is balanced by the N_x force in the cylindrical portion (the horizontal component of N_s being resisted by a ring beam at the cylinder-cone junction). Thus

$$(N_s)_{s=l}\sin\alpha = (N_x)_{x=H} = N_o\sin\phi_o \tag{3.94}$$

where N_o is as defined by equation (3.87). Evaluating N_s at $s = l$ from equation (3.93), and substituting the result into (3.94), we obtain

$$N_o\left(\frac{\sin\phi_o}{\sin\alpha}\right) = \frac{1}{l}\left[\gamma(\cot\alpha)\left(\frac{K_2 l^2}{2} - \frac{l^3}{3}\sin\alpha\right) + k\right]$$

from which

$$k = lN_o\left(\frac{\sin\phi_o}{\sin\alpha}\right) - \gamma(\cot\alpha)\left(\frac{K_2 l^2}{2} - \frac{l^3}{3}\sin\alpha\right) \tag{3.95}$$

Substituting this result into equation (3.93), we finally obtain N_s as follows:

$$N_s = \frac{1}{s}\left[\gamma(\cot\alpha)\left\{\frac{\sin\alpha}{3}(l^3 - s^3) - \frac{K_2}{2}(l^2 - s^2)\right\} + lN_o\left(\frac{\sin\phi_o}{\sin\alpha}\right)\right] \tag{3.96a}$$

From equations (2.27) and (3.82), we obtain

$$N_\theta = \gamma(\cot\alpha)(K_2 - s\sin\alpha)s \tag{3.96b}$$

Substituting solutions (3.96) into the general expressions (2.75) and (2.76) for deformations, we obtain the closed-form results

$$\delta = \frac{1}{Et_3}(s\cos\alpha)\left[\ \gamma(\cot\alpha)(K_2 - s\sin\alpha)s\right.$$

$$-\frac{v}{s}\left\{\gamma(\cot\alpha)\left(\frac{\sin\alpha}{3}(l^3-s^3)-\frac{K_2}{2}(l^2-s^2)\right)+lN_o\left(\frac{\sin\phi_o}{\sin\alpha}\right)\right\}\Bigg] \tag{3.97a}$$

$$V=\frac{\cot\alpha}{Et_3}\Bigg[\ (1+v)\Bigg\{\ \frac{1}{s}\Bigg[\gamma(\cot\alpha)\left\{\frac{\sin\alpha}{3}(l^3-s^3)-\frac{K_2}{2}(l^2-s^2)\right\}+lN_o\left(\frac{\sin\phi_o}{\sin\alpha}\right)\Bigg]$$

$$-\gamma(\cot\alpha)(K_2-s\sin\alpha)s\ \Bigg\}$$

$$-s\Bigg\{\ \gamma(\cot\alpha)(K_2-2s\sin\alpha)$$

$$-v\Bigg[\gamma(\cot\alpha)\left\{\frac{K_2}{2}\left(\frac{l^2}{s^2}+1\right)-\frac{\sin\alpha}{3}\left(\frac{l^3}{s^2}+2s\right)\right\}-\frac{lN_o}{s^2}\left(\frac{\sin\phi_o}{\sin\alpha}\right)\Bigg]\Bigg\}\Bigg] \tag{3.97b}$$

At the lower edge of the conical shell, the expressions for the deformations δ and V are exactly as given by (3.97), but with s replaced by s_o throughout. At the upper edge of the conical shell, expressions (3.97) with $s=l$ further simplify to

$$\delta_{s=l}=\frac{1}{Et_3}(l\cos\alpha)\left\{\gamma(\cot\alpha)(K_2-l\sin\alpha)l-vN_o\left(\frac{\sin\phi_o}{\sin\alpha}\right)\right\} \tag{3.98a}$$

$$V_{s=l}=\frac{\cot\alpha}{Et_3}\Bigg[\ (1+v)\left\{N_o\left(\frac{\sin\phi_o}{\sin\alpha}\right)-\gamma(\cot\alpha)(K_2-l\sin\alpha)l\right\}$$

$$-l\left\{\gamma(\cot\alpha)(K_2-2l\sin\alpha)-v\left[\gamma(\cot\alpha)(K_2-l\sin\alpha)-\frac{N_o}{l}\left(\frac{\sin\phi_o}{\sin\alpha}\right)\right]\right\}\Bigg] \tag{3.98b}$$

For the spherical bottom, using equation (2.15) with $r_1=r_2=c$, $p_\phi=0$ and p_r as given by equation (3.84), we obtain

$$N_\psi=\frac{1}{c\sin^2\psi}\left[\gamma c^2\int(K_3-c\cos\psi)\cos\psi\sin\psi\,d\psi+k\right]$$

$$=\frac{1}{c\sin^2\psi}\left[-\gamma c^2\left(K_3\frac{\cos^2\psi}{2}-c\frac{\cos^3\psi}{3}\right)+k\right] \tag{3.99}$$

At $\psi=0$, N_ψ must be finite, implying that

$$k=\gamma c^2\left(\frac{K_3}{2}-\frac{c}{3}\right) \tag{3.100}$$

Using this result in expression (3.99), the meridional stress resultant becomes

$$N_\psi = \frac{\gamma c}{6(1+\cos\psi)}\left\{3K_3(1+\cos\psi) - 2c(1+\cos\psi+\cos^2\psi)\right\} \tag{3.101a}$$

From equation (2.12), using the p_r of expression (3.84) and the above result for N_ψ, the hoop stress resultant is obtained as

$$N_\theta = \frac{\gamma c}{6(1+\cos\psi)}\left\{3K_3(1+\cos\psi) + 2c(1-2\cos\psi-2\cos^2\psi)\right\} \tag{3.101b}$$

On the basis of equation (2.67), the lateral displacement is obtained as

$$\delta = \frac{\gamma c^2}{6Et_4}\left(\frac{\sin\psi}{1+\cos\psi}\right)\Big[\left\{3K_3(1+\cos\psi) + 2c(1-2\cos\psi-2\cos^2\psi)\right\}$$
$$- \nu\left\{3K_3(1+\cos\psi) - 2c(1+\cos\psi+\cos^2\psi)\right\}\Big] \tag{3.102a}$$

V is obtained from expression (2.72), with $r_1 = r_2 = c$, and adopting the N_ϕ $(= N_\psi)$ and N_θ expressions of equations (3.101). After some rather lengthy manipulations, the final result is surprisingly simple:

$$V = -\frac{\gamma c^2}{Et_4}\sin\psi \tag{3.102b}$$

Thus V is simply proportional to the sine of the angular coordinate ψ, and is independent of the liquid head $(K_3 - c)$ above the apex of the spherical bottom. The values of δ and V at the edge of the spherical bottom are given by expressions (3.102), with ψ set equal to ψ_o.

The locations marked $A-A$, $B-B$ and $C-C$ in Figure 3.11 are expected to be sources of bending disturbances (i.e. *lines of distortion* of the membrane state of stress) for the following reasons, as per the guidelines of Section 1.5.3:

(*a*) At $A-A$, there is a discontinuity in the slope of the meridian (which jumps from $\tan\phi_o$ to ∞), in the meridional radius of curvature r_1 (which jumps from a to ∞), and in the second principal radius of curvature r_2 (which jumps from a to $a\sin\phi_o$); there may also be a discontinuity in shell thickness if t_2 (for the cylindrical portion) is different from t_1 (for the spherical top).

(*b*) At $B-B$, there is a discontinuity in the slope of the meridian (which jumps from ∞ to $-\tan\alpha$), and in the second principal radius of curvature r_2 (which jumps from $a\sin\phi_o$ to $l/\tan\alpha$, while the meridional radius of curvature remains constant at ∞); there may also be a discontinuity in shell thickness if t_3 (for the conical portion) is different from t_2 (for the cylindrical portion).

(*c*) At $C-C$, there is a discontinuity in the slope of the meridian (which jumps from $-\tan\alpha$ to $+\tan\psi_o$), in the meridional radius of curvature (which jumps from ∞ to c), and in the second principal radius of curvature (which jumps from $s_o/\tan\alpha$ to c); there may also be a discontinuity in shell thickness if t_4 (for the spherical bottom) is different from t_3 (for the conical shell). In addition, concentrated vertical reactions (due to the weight of the tank and its contents) act along the circumference $C-C$.

The membrane solution fails to satisfy the conditions of continuity and/or equilibrium at these locations, necessitating bending corrections in the form of axisymmetric bending moments and horizontal shears (see Chapter 7). Ring beams at $A - A$, $B - B$ and $C - C$ are desirable, as they tend to reduce the magnitude of the bending effects in the shells by absorbing the out-of-balance lateral resultants of membrane meridional shell actions at these locations, which would otherwise have to be resisted by bending of the shell edges. However, ring beams will never eliminate bending, as they are in themselves *lines of distortion* if one considers them as very localised enlargements of shell thickness, whose rigidity (hence deformation response) will therefore differ from that of the adjoining thin shell, necessitating some bending to restore compatibility of deformations.

3.2.5 Membrane solutions for vessels of arbitrary shape

3.2.5.1 General problem description

The present study centres on the use of the membrane hypothesis in exploring the possibility of new shell forms for elevated liquid-storage reservoirs in the form of open-top shells of revolution of arbitrary but smoothly varying meridian profiles of one single geometry.[1] Profiles of interest are those with zero slope at the *pole* (the point on the shell through which the axis of revolution of the shell passes, this also being the lowest point of the vessel), and where the sides of the vessel become increasingly steeper with increasing height above the pole, such a 'bowl-like' shape enhancing storage capacity for reservoirs of a specified height and plan area. The vessels are assumed to be supported at the top (in *hanging* fashion) along their open circular edge by means of vertical columns, with an interposing ring beam being provided to absorb the lateral pull from the shell and to smoothen out the vertical reactions from the columns.

In developing the general expressions for membrane stress resultants, we will adopt rectangular coordinates (x, y) to describe the position of points on the shell meridian. This approach is particularly useful when the profile of the vessel is specified as an equation in Cartesian coordinates, the form that is most convenient from the practical point of view of construction.

Because these profiles are continuous and smoothly varying, and given that the surface loading (i.e. the liquid hydrostatic pressure) is of a gradually varying intensity, it would be expected that the membrane solution gives an excellent estimate of the true state of stress throughout the entire shell, except in the very localised zone adjacent to the line of support. We will first develop a general approach, and then apply this to selected geometric profiles. Of the various cases studied, we will identify, through numerical examples, the most desirable shape of the meridian.

3.2.5.2 Geometric preliminaries

Figure 3.12 shows an arbitrary curve $y = f(x)$, being a typical meridian of the surface of revolution formed by rotating the curve about the y axis. The curve, assumed to be a single-valued function of x over the finite length of the meridian below some upper-bound value $y = c$ (constant), passes through the origin of the system of coordinate axes. The main restriction upon its arbitrariness is that the slope of the curve must be zero at the origin, so that, upon rotation of the curve about the y-axis, the surface of revolution thus formed is smoothly continuous over the pole. Furthermore, only functions whose slope is always increasing with increasing x (at least below the upper-bound $y = c$) will be considered. This restriction rules out curves exhibiting one or more inflexion points, locations at which the membrane hypothesis would certainly break down owing to the infinite rate of change of r_1 there. It is also a restriction on functional grounds, as this ensures a 'bowl-like' shape for the vessel, which is ideal for containment.

The horizontal radial distance R of a point P on the surface, from the vertical axis of rotation of the surface, is actually the coordinate x of the point (i.e. $R = x$). The principal radius of curvature of the surface in the meridional plane, r_1, is given by the usual relationship

$$r_1 = \frac{\left[1+\{f'(x)\}^2\right]^{3/2}}{f''(x)} \tag{3.103}$$

where $f'(x)$ and $f''(x)$ denote first and second derivatives, respectively, of $f(x)$ with respect to x. The second principal radius of curvature of the surface of revolution, r_2, is of course given by the distance between P and the intersection point of the normal to the tangent plane at P and the axis of revolution of the surface, i.e.

$$r_2 = \frac{R}{\sin\phi} = \frac{x}{\sin\phi} \tag{3.104}$$

The angle of inclination (relative to the horizontal) of the tangent at P, denoted by the meridional angle ϕ in Figure 3.12, is of course given by the fundamental definition of slope at a given point of a curve:

$$\tan\phi = f'(x) \tag{3.105}$$

By sketching a right-angled triangle satisfying equation (3.105), as has been done in Figure 3.13(a), it is seen that

$$\sin\phi = \frac{f'(x)}{\left[1+\{f'(x)\}^2\right]^{1/2}} \tag{3.106}$$

so that equation (3.104) becomes

$$r_2 = \frac{x\left[1+\{f'(x)\}^2\right]^{1/2}}{f'(x)} \tag{3.107}$$

Figure 3.12 Geometric parameters of the surface of revolution formed by rotating the curve $y = f(x)$ about the y axis.

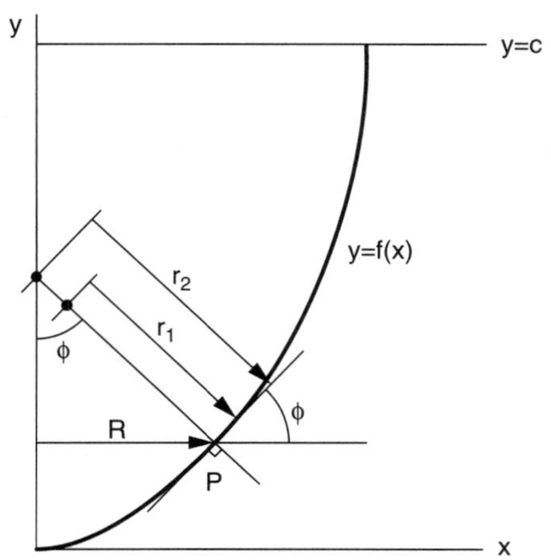

Figure 3.13 Trigonometric interpretations of equation (3.105).

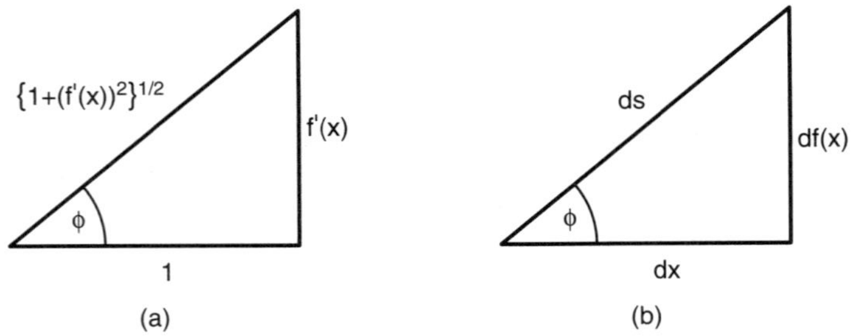

The relationship between an increment in the arc length, ds, and an increment in the x-coordinate, dx, will also be required. From the sketch in Figure 3.13(b), which also satisfies equation (3.105), we may write

$$(ds)^2 = (dx)^2 + \{df(x)\}^2$$
$$= \left\{1 + \left(\frac{df(x)}{dx}\right)^2\right\}(dx)^2 = \left\{1 + (f'(x))^2\right\}(dx)^2$$

so that

$$ds = \left\{1 + (f'(x))^2\right\}^{1/2} dx \tag{3.108}$$

3.2.5.3 Loading

Figure 3.14 depicts a shell of revolution generated by rotating the curve $y = f(x)$ about the y axis, and filled with liquid (of weight γ per unit volume) to a depth $H = f(x_H)$.

The pressure p_r (acting in the direction normal to the shell midsurface) at locations a height $\lambda = f(\xi)$ above the bottom of the tank is given by

$$p_r = \gamma(H - \lambda) = \gamma\{f(x_H) - f(\xi)\} \tag{3.109}$$

Assuming this pressure to act over an elemental annular strip of width ds (and hence of surface area $2\pi\xi ds$), the vertical resultant dW of the elemental forces is a downward force acting along the y axis, and given by

$$dW = 2\pi\xi p_r \, ds \cos\phi$$
$$= 2\pi\xi\gamma\{f(x_H) - f(\xi)\}d\xi$$

noting, from Figure 3.13(b), that $ds\cos\phi = dx$, and from Figure 3.14, that $d\xi = dx$. The variable p_r has also been replaced with the right-hand side of equation (3.109).

Let us consider a shell cap whose edges are defined by the coordinates $\{x, f(x)\}$, as shown in Figure 3.15.

Figure 3.14 Liquid-filled shell of revolution.

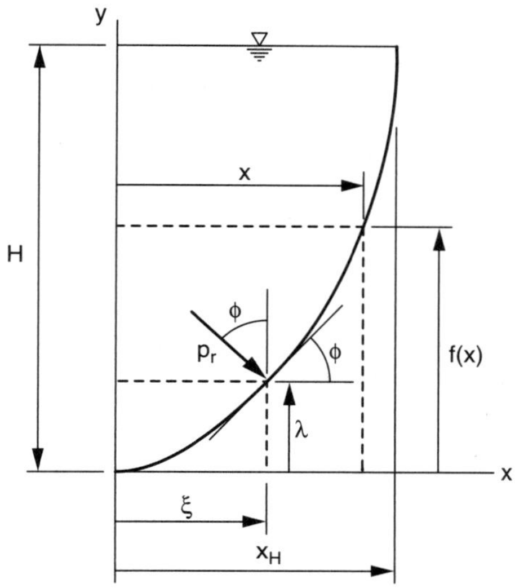

Figure 3.15 Equilibrium of shell cap in the meridional plane: equation (3.113).

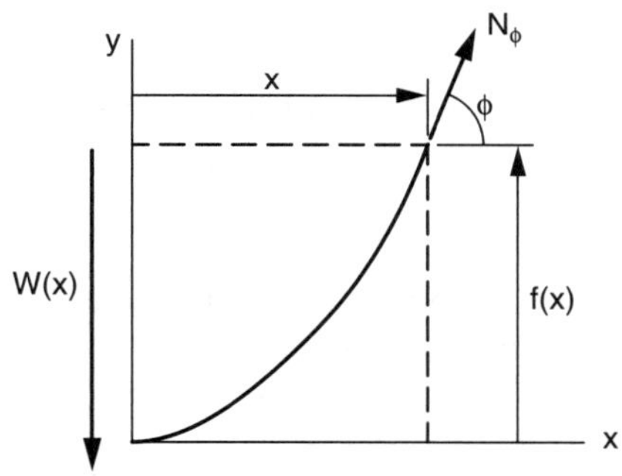

The net vertical resultant $W(x)$ of the hydrostatic pressure acting on the inside surface of this cap is obtained by integrating the dW forces:

$$W(x) = 2\pi\gamma\left\{f(x_H)\int_0^x \xi\, d\xi - \int_0^x \xi f(\xi)\, d\xi\right\}$$

$$= 2\pi\gamma\left\{\frac{x^2}{2} f(x_H) - \int_0^x \xi f(\xi)\, d\xi\right\} \tag{3.110}$$

The total volume V of liquid contained in the reservoir is readily obtained from equation (3.110), by dividing the total weight of liquid held by the tank – this is equal to $W(x_H)$ – by the unit weight γ. Thus

$$V = 2\pi\left\{\frac{x_H^2}{2}f(x_H) - \int_0^{x_H} \xi f(\xi)\, d\xi\right\} \tag{3.111}$$

The total surface area S of the vessel is given by the well-known relation for the surface area of a surface of revolution obtained by rotating the curve $y = f(x)$ about the y axis:[2]

$$S = 2\pi\int_0^{x_H} x\left\{1+\left(f'(x)\right)^2\right\}^{1/2} dx \tag{3.112}$$

3.2.5.4 Equilibrium considerations and stress resultants

The meridional stress resultant N_ϕ may be obtained by considering the equilibrium of a loaded shell cap in the meridional plane (Figure 3.15). The position of the axisymmetric edge of this shell cap is defined by the coordinates $\{x, f(x)\}$ as shown in Figure 3.15. The downward weight $W(x)$ of the liquid in the cap is countered by the upward N_ϕ force along the circular edge of the cap, which is therefore tensile. Thus, resolving forces in the vertical direction, we obtain

$$2\pi x N_\phi \sin\phi = W(x) \tag{3.113}$$

Using the expression for $W(x)$ as given by equation (3.110) and the expression for $\sin\phi$ as given by equation (3.106), we obtain N_ϕ as follows:

$$N_\phi = \gamma\frac{\left\{1+\left(f'(x)\right)^2\right\}^{1/2}}{xf'(x)}\left\{\frac{x^2}{2}f(x_H) - \int_0^{x} \xi f(\xi)\, d\xi\right\} \tag{3.114}$$

The hoop stress resultant N_θ follows from equation (2.11), when we rearrange this to read

$$N_\theta = r_2 p_r - \frac{r_2}{r_1}N_\phi \tag{3.115}$$

Using expressions (3.103), (3.107), (3.109) and (3.114) to eliminate r_1, r_2, p_r and N_ϕ respectively, we obtain a general expression for the stress resultant N_θ, as follows:

$$N_\theta = \gamma\frac{\left\{1+\left(f'(x)\right)^2\right\}^{1/2}}{f'(x)}\left[x\{f(x_H)-f(x)\} - \frac{f''(x)}{f'(x)\left\{1+\left(f'(x)\right)^2\right\}}\left\{\frac{x^2}{2}f(x_H) - \int_0^{x} \xi f(\xi)\, d\xi\right\}\right] \tag{3.116}$$

3.2.5.5 Application to specific profiles of the shell meridian

3.2.5.5.1 STRESS RESULTANTS, VOLUMES AND SURFACE AREAS

Let us explore the feasibility of the following meridian profiles, which all satisfy the geometric specifications of Section 3.2.5.1, and hence are all reasonable containment shapes:

$$(a)\ \ f(x) = kx^2; x \geq 0 \tag{3.117}$$

$$(b)\ \ f(x) = (k\cosh x) - k = k\{(\cosh x - 1)\}; x \geq 0 \tag{3.118}$$

$$(c)\ f(x) = kx^3; x \geq 0 \tag{3.119}$$

$$(d)\ f(x) = b - \frac{b}{a}\left(a^2 - x^2\right)^{1/2}; 0 \leq x \leq a \tag{3.120}$$

These four functions represent, in the order given, *parabolic*, *hyperbolic-cosine* (henceforth referred to simply as *cosh*), *cubic* and *elliptic* variations of the shell meridian. The last function is better recognised as part of an ellipse if it is written in the form

$$\frac{x^2}{a^2} + \frac{(y-b)^2}{b^2} = 1 \tag{3.121}$$

The general results for the stress resultants N_ϕ and N_θ, as represented by equations (3.114) and (3.116), and the general results for the total volume V and total surface area S of the vessel, as represented by equations (3.111) and (3.112), have been applied to the above four profiles. The ensuing profile-specific results for N_ϕ, N_θ, V and S are as follows:

(*a*) *Vessel of parabolic profile*

$$N_\phi = \frac{\gamma}{8}\left(1 + 4k^2x^2\right)^{1/2}\left(2x_H^2 - x^2\right) \tag{3.122}$$

$$N_\theta = \frac{\gamma}{8}\left(1 + 4k^2x^2\right)^{1/2}\left\{4\left(x_H^2 - x^2\right) - \frac{2x_H^2 - x^2}{1 + 4k^2x^2}\right\} \tag{3.123}$$

$$V = \frac{\pi}{2}kx_H^4 \tag{3.124}$$

$$S = \frac{\pi}{6k^2}\left\{\left(1 + 4k^2x_H^2\right)^{3/2} - 1\right\} \tag{3.125}$$

(*b*) *Vessel of cosh profile*

$$N_\phi = \frac{\gamma}{2}\frac{\left(1 + k^2\sinh^2 x\right)^{1/2}}{x\sinh x}\left\{x^2\cosh x - 2x\sinh x + 2(\cosh x - 1)\right\} \tag{3.126}$$

$$N_\theta = \frac{\gamma}{2}\frac{\left(1 + k^2\sinh^2 x\right)^{1/2}}{\sinh x}\left\{2x(\cosh x_H - \cosh x) - \frac{\coth x}{1 + k^2\sinh^2 x} \times \left\{x^2\cosh x_H - 2x\sinh x + 2(\cosh x - 1)\right\}\right\} \tag{3.127}$$

$$V = \pi k\left\{x_H^2\cosh x_H - 2x_H\sinh x_H + 2(\cosh x_H - 1)\right\} \tag{3.128}$$

$$S = 2\pi\int_0^{x_H} x\left(1 + k^2\sinh^2 x\right)^{1/2} dx \tag{3.129}$$

Expression (3.129) is best integrated numerically using Simpson's rule,[2] unless $k = 1$.

(c) *Vessel of cubic profile*

$$N_\phi = \frac{\gamma}{10}\frac{\left(1+9k^2x^4\right)^{1/2}}{3x}\left(5x_H^3 - 2x^3\right) \tag{3.130}$$

$$N_\theta = \frac{\gamma}{10}\frac{\left(1+9k^2x^4\right)^{1/2}}{3x}\left\{10\left(x_H^3 - x^3\right) - \frac{2}{1+9k^2x^4}\left(5x_H^3 - 2x^3\right)\right\} \tag{3.131}$$

Expressions (3.130) and (3.131) are not applicable for x too close to zero – see Section 3.2.5.6.

$$V = \frac{3}{5}\pi k x_H^5 \tag{3.132}$$

$$S = \frac{\pi}{6k}\left[3kx_H^2\left(1+9k^2x_H^4\right)^{1/2} + \ln\left\{3kx_H^2 + \left(1+9k^2x_H^4\right)^{1/2}\right\}\right] \tag{3.133}$$

(d) *Vessel of elliptic profile*

$$N_\phi = \gamma\frac{\left\{\left(b^2-a^2\right)x^2+a^4\right\}^{1/2}}{6ax^2}\left\{2\left[a^3-\left(a^2-x^2\right)^{3/2}\right]-3x^2\left(a^2-x_H^2\right)^{1/2}\right\} \tag{3.134}$$

$$N_\theta = \gamma\frac{\left\{\left(b^2-a^2\right)x^2+a^4\right\}^{1/2}}{6ax^2}\left\{ 6x^2\left\{\left(a^2-x^2\right)^{1/2}-\left(a^2-x_H^2\right)^{1/2}\right\}\right.$$

$$\left. - \frac{a^4}{\left(b^2-a^2\right)x^2+a^4}\left[2\left\{a^3-\left(a^2-x^2\right)^{3/2}\right\}-3x^2\left(a^2-x_H^2\right)^{1/2}\right]\right\} \tag{3.135}$$

$$V = \frac{\pi}{3}\frac{b}{a}\left[2\left\{a^3-\left(a^2-x_H^2\right)^{3/2}\right\}-3x_H^2\left(a^2-x_H^2\right)^{1/2}\right] \tag{3.136}$$

$$S = \frac{\pi}{a}\left(b^2-a^2\right)^{1/2}\left[a\left\{\left(\frac{a^2b^2}{b^2-a^2}\right)-a^2\right\}^{1/2} + \left(\frac{a^2b^2}{b^2-a^2}\right)\sin^{-1}\left\{\frac{\left(b^2-a^2\right)^{1/2}}{b}\right\}\right.$$

$$-\left(a^2-x_H^2\right)^{1/2}\left\{\left(\frac{a^2b^2}{b^2-a^2}\right)-\left(a^2-x_H^2\right)\right\}^{1/2}$$

$$\left. +\left(\frac{a^2b^2}{b^2-a^2}\right)\sin^{-1}\left\{\frac{\left(a^2-x_H^2\right)^{1/2}\left(b^2-a^2\right)^{1/2}}{ab}\right\}\right] \tag{3.137}$$

Special case: $x_H = a,\ f(x_H) = b$

$$N_\phi = \gamma \frac{\left\{\left(b^2 - a^2\right)x^2 + a^4\right\}^{1/2}}{3ax^2}\left\{a^3 - \left(a^2 - x^2\right)^{3/2}\right\} \tag{3.138}$$

$$N_\theta = \gamma \frac{\left\{\left(b^2 - a^2\right)x^2 + a^4\right\}^{1/2}}{3ax^2}\left[3x^2\left(a^2 - x^2\right)^{1/2} - \frac{a^4\left\{a^3 - \left(a^2 - x^2\right)^{3/2}\right\}}{\left(b^2 - a^2\right)x^2 + a^4}\right] \tag{3.139}$$

$$V = \frac{2}{3}\pi b a^2 \tag{3.140}$$

$$S = \frac{\pi}{a}\left(b^2 - a^2\right)^{1/2}\left[a\left\{\left(\frac{a^2b^2}{b^2 - a^2}\right) - a^2\right\}^{1/2} + \left(\frac{a^2b^2}{b^2 - a^2}\right)\sin^{-1}\left\{\frac{\left(b^2 - a^2\right)^{1/2}}{b}\right\}\right] \tag{3.141}$$

3.2.5.5.2 VALUES AT THE POLE LOCATION

The above expressions for N_ϕ and N_θ may be indeterminate at the pole (i.e. at the location $x = 0$). To overcome this problem, use is made of symmetry – that is, $N_\phi = N_\theta$ and $r_1 = r_2$ at the pole – so that from equation (3.115),

$$N_\phi(0) = N_\theta(0) = \frac{1}{2}\, p_r(0)\, r_1(0) \tag{3.142}$$

From the original condition that $f'(0) = 0$, equation (3.103) reduces to

$$r_1(0) = \frac{1}{f''(0)} \tag{3.143}$$

at the pole, while equation (3.109) becomes

$$p_r(0) = \gamma\{f(x_H) - f(0)\} = \gamma\, f(x_H) \tag{3.144}$$

since $f(0) = 0$ (recalling that the origin of the $\{x, y\}$ coordinate system is at the pole). Thus, equation (3.142) may simply be written as

$$N_\phi(0) = N_\theta(0) = \frac{\gamma}{2}\frac{f(x_H)}{f''(0)} \tag{3.145}$$

Evaluating expression (3.145) for the four vessel profiles in turn, we obtain

(*a*) *Vessel of parabolic profile*

$$N_\phi(0) = N_\theta(0) = \frac{\gamma}{4}x_H^2 \tag{3.146}$$

(*b*) *Vessel of cosh profile*

$$N_\phi(0) = N_\theta(0) = \frac{\gamma}{2}(\cosh x_H - 1) \tag{3.147}$$

(*c*) *Vessel of cubic profile*

$$N_\phi(0) = N_\theta(0) = \frac{\gamma}{12}\frac{x_H^3}{x} = \infty \tag{3.148}$$

This result is not valid, as explained below.

(*d*) *Vessel of elliptic profile*

$$N_\phi(0) = N_\theta(0) = \frac{\gamma}{2}a\left\{a - \left(a^2 - x_H^2\right)^{1/2}\right\} \tag{3.149}$$

Special case: $x_H = a,\ f(x_H) = b$

$$N_\phi(0) = N_\theta(0) = \frac{\gamma}{2}a^2 \tag{3.150}$$

The result obtained for the cubic profile (expression (3.148)) is, quite obviously, not valid. The stress resultants cannot be infinite, in reality. We conclude from this result that the membrane hypothesis completely breaks down at the pole. In fact, for this profile, the values of stress resultants calculated for the general locations (equations (3.130) and (3.131)) do not give a correct estimate of the actual state of stress in the shell, if one approaches the pole too closely. Considerable bending has to occur around the pole, as, quite clearly, the membrane hypothesis fails to satisfy equilibrium at the pole.

The underlying reason for this breakdown of the membrane hypothesis is the rate of change of the principal radii of curvature r_1 and r_2. From equation (3.115), r_1 and r_2 must not *both* tend to infinity at any point on the shell, if the membrane actions are to satisfy force equilibrium in the direction normal to the shell midsurface. The tending to infinity of *both* r_1 and r_2 would represent a total loss of curvature of the shell at the point, so that the shell assumes the characteristics of a flat plate at that point. Since the rates of change of r_1 and r_2 have a considerable bearing on the validity of the results, we will examine these in greater depth.

3.2.5.6 Reliability of the solution in the vicinity of the pole

In view of the singularity in the cubic-shell solution noted at the polar location, the question arises as to how much reliance can be placed on the membrane-solution results for locations around the pole, even in cases where finite values of the stress resultants can be calculated at the pole. Recalling from Chapter 1 that one of the conditions for the validity of the membrane solution is that the principal radii of curvature r_1 and r_2 must not vary too rapidly in a given region, we may answer this question by simply examining how rapidly r_1 and r_2 change with respect to arc length s over the domain of the vessel. For locations around the pole, the rates of change of r_1 and r_2 with respect to arc length s are nearly equal to dr_1 / dx and dr_2 / dx respectively, since in these regions, $ds \approx dx$ for vessels of present interest (i.e. those satisfying the condition $f'(0) = 0$). However, we will investigate the exact relationships (i.e. dr_1 / ds and dr_2 / ds), so that our conclusions remain valid at locations that are distant from the pole.

For any function of x, say $g(x)$,

$$\frac{dg(x)}{ds} = \frac{dg(x)}{dx}\frac{dx}{ds} = g'(x)\frac{dx}{ds} \tag{3.151}$$

From equation (3.108),

$$\frac{dx}{ds} = \frac{1}{\left\{1+\left(f'(x)\right)^2\right\}^{1/2}} \tag{3.152}$$

so that equation (3.151) becomes

$$\frac{dg(x)}{ds} = \frac{g'(x)}{\left\{1+\left(f'(x)\right)^2\right\}^{1/2}} \tag{3.153}$$

The principal radii of curvature r_1 and r_2 may be evaluated by means of equations (3.103) and (3.107) respectively. Their rates of change along the meridian may then be deduced on the basis of equation (3.153). For the four profiles of interest, general results for r_1, r_2, dr_1 / ds and dr_2 / ds, as well as specific values at (or approaching) the pole, are given below. Comments and conclusions on the reliability of the membrane solution around the pole are also included:

(a) Parabolic variation

$$r_1 = \frac{\left(1+4k^2x^2\right)^{3/2}}{2k} \tag{3.154}$$

$$r_2 = \frac{\left(1+4k^2x^2\right)^{1/2}}{2k} \tag{3.155}$$

$$\frac{dr_1}{ds} = 6kx \tag{3.156}$$

$$\frac{dr_2}{ds} = \frac{2kx}{1+4k^2x^2} \tag{3.157}$$

At $x = 0$, $r_1 = r_2 = 1/(2k)$ and $dr_1 / ds = dr_2 / ds = 0$. Thus, the principal radii of curvature tend towards a finite value at the pole, and their rates of change are practically zero around the pole. For this profile, the membrane solution is valid throughout the practical ranges of x.

(b) Cosh variation

$$r_1 = \frac{\left(1+k^2\sinh^2 x\right)^{3/2}}{k\cosh x} \tag{3.158}$$

$$r_2 = \frac{x\left(1+k^2\sinh^2 x\right)^{1/2}}{k\sinh x} \tag{3.159}$$

$$\frac{dr_1}{ds} = \frac{\sinh x\left(3k^2\cosh^2 x - k^2\sinh^2 x - 1\right)}{k\cosh^2 x} \tag{3.160}$$

$$\frac{dr_2}{ds} = \frac{\left(1+k^2\sinh^2 x\right)\sinh x - x\left(1-k^2\sinh^2 x\right)\cosh x}{k\left(1+k^2\sinh^2 x\right)\sinh^2 x} \tag{3.161}$$

As $x \to 0$,

$$r_1 \to \frac{1}{k}$$

$$r_2 \to \frac{1}{k}\left(\frac{x}{\sinh x}\right) \to \frac{1}{k}$$

$$\frac{dr_1}{ds} \to \left(\frac{3k^2-1}{k}\right)\sinh x \to 0$$

$$\frac{dr_2}{ds} \to \left(\frac{\sinh x - x}{k\sinh^2 x}\right) \to \frac{x}{6k} \to 0$$

Again, r_1 and r_2 tend towards a finite value at the pole, being practically invariant in the proximity of the pole. The membrane solution is valid throughout the practical ranges of x.

(*c*) *Cubic variation*

$$r_1 = \frac{\left(1+9k^2x^4\right)^{3/2}}{6kx} \tag{3.162}$$

$$r_2 = \frac{\left(1+9k^2x^4\right)^{1/2}}{3kx} \tag{3.163}$$

$$\frac{dr_1}{ds} = \frac{45k^2x^4-1}{6kx^2} \tag{3.164}$$

$$\frac{dr_2}{ds} = \frac{1}{3kx^2}\left(\frac{9k^2x^4-1}{9k^2x^4+1}\right) \tag{3.165}$$

As $x \to 0$, $r_1 \to \infty$, $r_2 \to \infty$, $dr_1/ds \to -\infty$ and $dr_2/ds \to -\infty$. Thus, both principal radii of curvature change very rapidly around the pole, the rates of change tending towards infinity as the infinite values of r_1 and r_2 at the pole are approached. This is a clear indication of the very unreliable nature of the membrane solution in the polar vicinity, and of the complete breakdown of the membrane hypothesis at the pole itself. Consideration of equations (3.103) and (3.107) shows that the membrane hypothesis breaks down at the poles of all profiles of the form $y = kx^n$, where n is a positive number greater than 2, since for such shells, both principal *curvatures* vanish at the pole (i.e. the pole is a 'flat' region).

(*d*) *Elliptic variation*

$$r_1 = \frac{1}{ba^4}\left\{\left(b^2 - a^2\right)x^2 + a^4\right\}^{3/2} \tag{3.166}$$

$$r_2 = \frac{1}{b}\left\{\left(b^2 - a^2\right)x^2 + a^4\right\}^{1/2} \tag{3.167}$$

$$\frac{dr_1}{ds} = 3\frac{b^2 - a^2}{ba^3}x\left(a^2 - x^2\right)^{1/2} \tag{3.168}$$

$$\frac{dr_2}{ds} = \frac{a}{b}\left(b^2 - a^2\right)\frac{x\left(a^2 - x^2\right)^{1/2}}{\left(b^2 - a^2\right)x^2 + a^4} \tag{3.169}$$

At $x = 0$, $r_1 = r_2 = a^2 / b$; as $x \to 0$, $dr_1 / ds \to 0$ and $dr_2 / ds \to 0$. Hence, as with the parabolic and the cosh profiles, r_1 and r_2 tend towards a finite value at the pole, their rates of change along the meridian gradually diminishing to zero at the pole. Thus, for the elliptic profile, the membrane solution is valid throughout the entire length of the meridian (edge effects aside, of course).

3.2.5.7 Numerical examples and conclusions

It is required to construct a shell-of-revolution liquid-storage vessel of overall height H and opening radius (at the top) x_H, where these dimensions are specified as $H = 15$m and $x_H = 5$m (i.e. the diameter of the circular top is 10 m). Fitting the parabolic, cosh, cubic and elliptic curves – see equations (3.117) to (3.120) – through the point $\{5, 15\}$ of the xy coordinate plane gives

(*a*) $k = 0.6$ for the parabolic variation
(*b*) $k = 0.2049$ for the cosh variation
(*c*) $k = 0.12$ for the cubic variation
(*d*) $a = 5$ and $b = 15$ for the elliptic variation, adopting the special case when $x_H = a$

Plots of these profiles are shown in Figure 3.16. For better visualisation of the full cross-sections of the vessels, mirror images of the curves in the vertical plane $x = 0$ are also shown.

Using the closed-form results of Section 3.2.5.5, with the applicable values of k now known, the stress resultants N_ϕ and N_θ have been worked out for each shell of revolution generated by these profiles. The variations of these membrane actions with vertical distance y above the pole are shown in Figure 3.17; for small values of y, it is clearer to plot variations with respect to horizontal distance x from the axis of symmetry of the shells, as shown in the inset figure. It can be seen that all stresses are tensile throughout, except in a very narrow band adjacent to the supported circular edge at the top, where a small amount of hoop compression occurs in all cases. Hence buckling is unlikely to be a problem in tanks of this type.

Disregarding the variations in the vicinity of the pole for the cubic case (since the membrane hypothesis is not valid around the pole for this profile), the peak stress-resultant values are observed to be hoop quantities in all cases, occurring at a height above the pole of about a third of the full height for

Figure 3.16 Plots of meridian profiles for the four numerical cases.

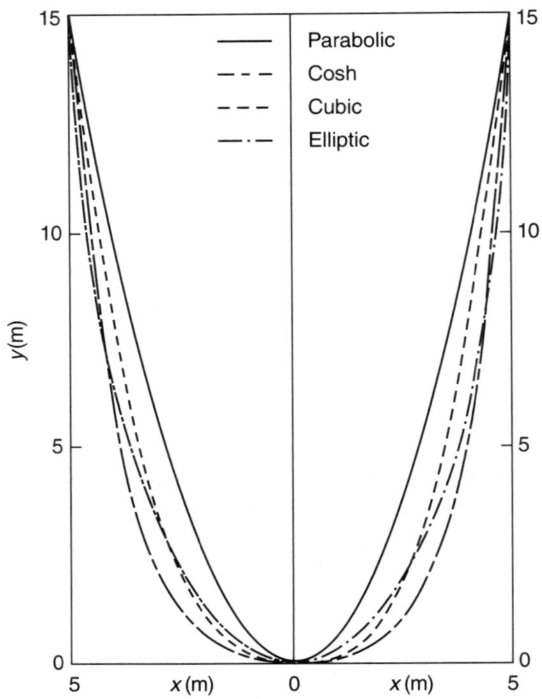

the parabolic profile, and somewhat less than a third of the full height in the other three cases. These peak values of the membrane-stress resultants – which will be denoted simply as N_p – are read off as follows:

(*a*) $N_p / \gamma = 28.6$ m² for the parabolic profile
(*b*) $N_p / \gamma = 36.8$ m² for the cosh profile
(*c*) $N_p / \gamma = 33.6$ m² for the cubic profile
(*d*) $N_p / \gamma = 34.7$ m² for the elliptic profile

The volume capacities V and shell surface areas S of the four vessels are also obtained from the closed-form results of Section 3.2.5.5 as follows:

(*a*) $V = 589$ m³; $S = 326$ m² for the parabolic vessel
(*b*) $V = 811$ m³; $S = 401$ m² for the cosh vessel
(*c*) $V = 707$ m³; $S = 368$ m² for the cubic vessel
(*d*) $V = 785$ m³; $S = 386$ m² for the elliptic vessel

Defining a *shape efficiency* parameter η as the volume V of liquid held by the vessel per unit volume of shell material used in the construction of the vessel, we have

$$\eta = \frac{V}{tS} \tag{3.170}$$

where S is the total surface area of the vessel and t is the thickness of the shell, which we will assume to be constant throughout the depth of the vessel.

Since compressive stresses are insignificant and buckling is not an issue, the required minimum shell thickness t may be determined from the peak value of the tensile-stress variations – that is, $t = N_p / \sigma_m$ where σ_m is the tensile strength of the material of construction. Let us consider a vessel that is provided with this minimum value of shell thickness. Substituting this value of t in equation (3.170), we obtain

Figure 3.17 Variations of membrane stress resultants with the coordinate y for the four numerical cases. The inset shows variations with respect to the coordinate x for the lower portions of the tanks.

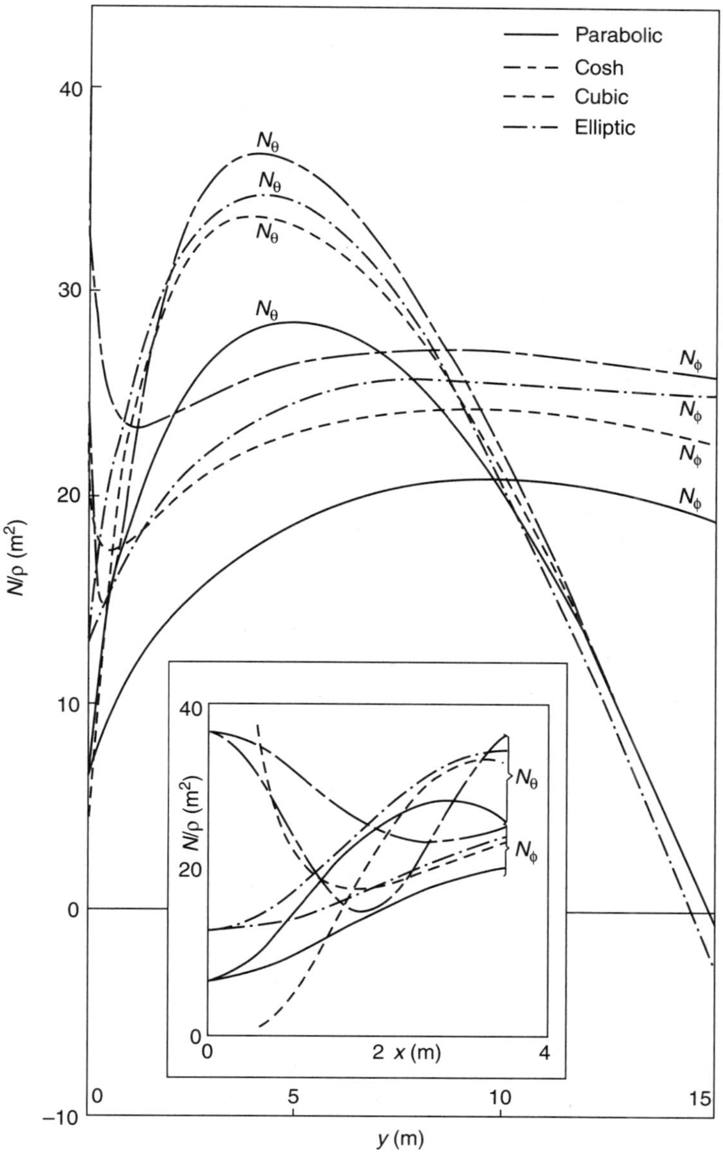

$$\eta = \frac{V\sigma_m}{SN_p} = \frac{V(\sigma_m / \gamma)}{S(N_p / \gamma)} = \beta \frac{V}{S(N_p / \gamma)} \tag{3.171}$$

where $\beta = \sigma_m / \gamma$ (this is constant for a given material of construction and a given retained liquid).

Substituting into equation (3.171) the values of V, S and N_p / γ as given above for each of the four numerical examples, we obtain the shape efficiency parameters as

(*a*) $\eta = \{589/(326 \times 28.6)\}\ \text{m}^{-1} \times \beta = 0.063\ \text{m}^{-1} \times \beta$ for the parabolic profile
(*b*) $\eta = \{811/(401 \times 36.8)\}\ \text{m}^{-1} \times \beta = 0.055\ \text{m}^{-1} \times \beta$ for the cosh profile
(*c*) $\eta = \{707/(368 \times 33.6)\}\ \text{m}^{-1} \times \beta = 0.057\ \text{m}^{-1} \times \beta$ for the cubic profile
(*d*) $\eta = \{785/(386 \times 34.7)\}\ \text{m}^{-1} \times \beta = 0.059\ \text{m}^{-1} \times \beta$ for the elliptic profile

Note that β has units of metres, and the parameter η is a dimensionless number.

Thus, disregarding the effects of the bending behaviour around the pole predicted for the case of the cubic profile (such effects tend to be very localised in any case), it may be concluded that, for typical vessel sizes of the proportions exemplified in the present study, the most efficient profile (i.e. the one with the highest ratio of retained volume of liquid to volume of construction material used), of the four types that have been studied, is the parabola, followed by the half-ellipse, then the cubic profile, with the cosh profile being the least efficient. Obviously, this conclusion is only valid for the particular numerical proportions that have been considered. For more general conclusions covering all likely combinations of geometric parameters of the vessel, a more comprehensive parametric study is necessary.

The present study has demonstrated the usefulness of the membrane hypothesis in searching for alternative shell forms for elevated liquid-containment vessels. The concept of a shape efficiency parameter η has facilitated direct comparisons among the various alternatives. We have also gained some insights into shell regions in which the membrane solution is not valid, despite both the shell geometry and the applied loading not exhibiting any obvious discontinuities.

3.2.6 Egg-shaped digester of parabolic ogival form

3.2.6.1 Functional benefits of the egg shape

Egg-shaped sludge digesters have become popular in relatively recent times owing to their superior functional performance and lower maintenance costs in comparison with conventional cylindrical digesters.[3,4] The smooth curved profile of the egg shape permits better mixing of the sludge, while the greater volume-to-surface ratio reduces heat losses. Better circulation of sludge in turn results in reduced accumulation of deposits at the bottom of the digester, and consequent reductions in maintenance costs. Furthermore, the deposits that do settle to the bottom of the egg shell are easy to remove as they all collect in one relatively small area at the bottom of the digester, and the removal of the deposits may be carried out on a continuous basis. Similarly, the tapered shape of the top reduces the free surface area of the sludge, allowing the crust that forms on the surface of the sludge to be removed more conveniently. Admittedly the more complex geometry of the egg shell implies higher initial costs of construction, but these are offset in the long term by the lower maintenance costs.

These innovative structures are usually constructed as thin shells of revolution in concrete, and must be designed to withstand the hydrostatic pressure from the contained liquid, as well as other relevant effects. As regards the precise shape of the egg shell, a number of mathematical shell surfaces may be envisaged, and the stress distribution will very much depend on the chosen form. Here, we will study the structural feasibility of the parabolic ogival shell as a sludge digester.[5] From a functional point of view, this is a good candidate, given the pointed ends (top and bottom), the bulging middle, and a profile that varies smoothly and continuously from the top all the way to the bottom. Other possible shapes for egg-shaped digesters will be discussed in Chapter 8, taking into account discontinuity effects.

We will assume that the vessel is anchored to the ground at its bottom end, with most of its height projecting above the ground unsupported (like a giant egg sitting on the ground); there are no visible supporting columns above the ground. Usually the tapered end of the bottom of the vessel is built into a thick base slab at or below ground level, which in turn is supported on a system of vertical piles designed to resist both vertical and lateral loads.

3.2.6.2 Geometry of the parabolic ogival shell

The parabolic ogival shell of revolution is formed by rotating through 360° a parabola that is symmetrical about the horizontal x axis, about the vertical y axis (which therefore is the axis of revolution of the shell), giving a meridional cross-section as depicted in Figure 3.18. The shell is therefore symmetrical not only about the vertical y axis of rotation, but also about the horizontal 'equatorial' plane containing the x axis. Let the overall height of the shell be H, and the equatorial diameter be D, as shown in Figure 3.18.

With the origin O taken at the intersection of the axis of revolution and the equatorial plane, the equation of the generating meridian of the shell of revolution is

$$\frac{D}{2} - x = ky^2 \tag{3.172}$$

where k is a constant.

Figure 3.18 Geometrical parameters of the parabolic ogival shell of revolution.

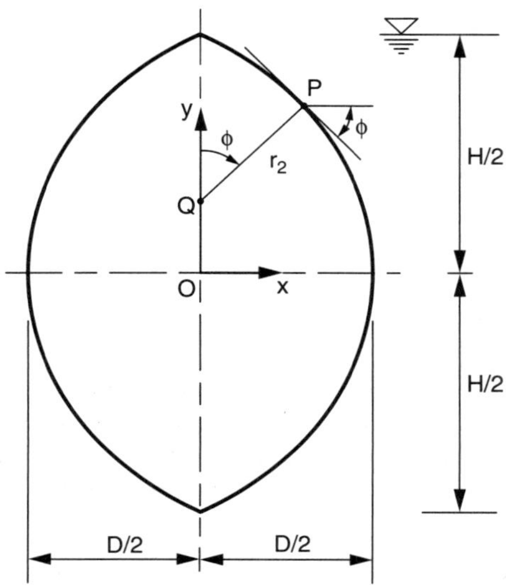

When $x = 0$, $y = \pm H/2$. From this condition, it follows that

$$k = \frac{2D}{H^2} \tag{3.173}$$

From equation (3.172), and making use of result (3.173), we may write

$$y = \pm \frac{1}{\sqrt{k}}\left(\frac{D}{2} - x\right)^{1/2} = \pm \frac{H}{\sqrt{2D}}\left(\frac{D}{2} - x\right)^{1/2} \tag{3.174}$$

Let us consider an arbitrary point P on the shell midsurface (Figure 3.18). As in previous considerations, the angular coordinate ϕ is the angle measured from the upward direction of the axis of revolution of the shell, to the normal at point P. From the diagram, it is evident that

$$\tan\phi = -\frac{dy}{dx} = \pm \frac{H}{2\sqrt{2D}}\left(\frac{D}{2} - x\right)^{-1/2} \tag{3.175}$$

At the upper pole ($x = 0,\ y = +H/2$), if $\phi = \phi_o$, then

$$\phi_o = \tan^{-1}\left(\frac{+H}{2D}\right) \tag{3.176a}$$

Similarly, at the lower pole ($x = 0, y = -H/2$), if $\phi = \phi_o'$, then

$$\phi_o' = \tan^{-1}\left(\frac{-H}{2D}\right) = \pi - \phi_o \tag{3.176b}$$

From equation (3.175), we have

$$\tan^2\phi = \frac{H^2}{8D}\left(\frac{D}{2} - x\right)^{-1} \tag{3.177}$$

leading to the result

$$x = \frac{4D^2\sin^2\phi - H^2\cos^2\phi}{8D\sin^2\phi} \tag{3.178}$$

When $\phi = 90°$, equation (3.178) yields $x = D/2$, as expected.

At the point P, let the two principal radii of curvature be denoted by r_1 and r_2 as before. The first (r_1) is, of course, the actual radius of curvature of the parabolic meridian at the point P, while the second (r_2) is equal to the distance between P and Q, where Q is the point of intersection of the axis of revolution of the shell and the normal at P (Figure 3.18). Thus, by reference to Figure 3.18, we may write

$$r_2 = \frac{x}{\sin\phi} = \frac{4D^2\sin^2\phi - H^2\cos^2\phi}{8D\sin^3\phi} \tag{3.179}$$

The other radius of curvature is given by the usual relationship

$$r_1 = \frac{\left[1+\left(\dfrac{dy}{dx}\right)^2\right]^{3/2}}{\dfrac{d^2y}{dx^2}} \tag{3.180}$$

Now, from equation (3.175),

$$\frac{d^2y}{dx^2} = \mp\frac{H}{4\sqrt{2D}}\left(\frac{D}{2}-x\right)^{-3/2} \tag{3.181}$$

Substituting expressions (3.175), (3.178) and (3.181) into equation (3.180), we obtain r_1 as follows

$$r_1 = \frac{\left[4D(D-2x)+H^2\right]^{3/2}}{4DH} = \frac{H^2}{4D\sin^3\phi} \tag{3.182}$$

(The negative symbol has been dropped, since r_1 is positive throughout.)

3.2.6.3 Loading components and shell stress resultants

Assuming the shell is completely filled with liquid of weight γ per unit volume, the depth of liquid at the vertical coordinate y is given by

$$d = \frac{H}{2} - y \tag{3.183a}$$

From equations (3.174) and (3.175),

$$y = \pm\frac{H}{\sqrt{2D}}\left(\frac{D}{2}-x\right)^{1/2} = \left(\pm\frac{H}{\sqrt{2D}}\right)\left(\pm\frac{H}{2\sqrt{2D}\tan\phi}\right) = \frac{H^2}{4D\tan\phi} \tag{3.183b}$$

Making use of this result, equation (3.183a) becomes

$$d = \frac{H}{2} - \frac{H^2}{4D\tan\phi} = \frac{2HD\sin\phi - H^2\cos\phi}{4D\sin\phi} \tag{3.183c}$$

Thus, the loading component p_r normal to the shell midsurface is given by

$$p_r = \gamma d = \gamma\left(\frac{2HD\sin\phi - H^2\cos\phi}{4D\sin\phi}\right) \tag{3.184}$$

The loading component p_ϕ in the direction of the tangent to the shell meridian is, of course, zero, since hydrostatic pressure acts purely perpendicular to the shell midsurface.

Using expressions (3.179), (3.182) and (3.184) for r_2, r_1 and p_r respectively, we may evaluate the integral of the general solution for the meridional stress resultant N_ϕ (equation (2.15)), with p_ϕ set equal to zero. After some simplifications, we obtain

$$\begin{aligned} N_\phi = \frac{\gamma H^3}{16D^2}\left(\frac{\sin\phi}{4D^2\sin^2\phi - H^2\cos^2\phi}\right)\Bigg[&-\frac{D(4D^2+H^2)}{\sin^2\phi}+\frac{DH^2}{2\sin^4\phi} \\ &-H(4D^2+H^2)\left(\frac{\cos\phi}{\sin\phi}\right)+\frac{H}{3}(4D^2+2H^2)\left(\frac{\cos\phi}{\sin^3\phi}\right)(1+2\sin^2\phi) \\ &-\frac{H^3}{15}\left(\frac{\cos\phi}{\sin^5\phi}\right)(3+4\sin^2\phi+8\sin^4\phi)+k\Bigg] \end{aligned} \tag{3.185}$$

where k is a constant of integration. Since the entire shell is unsupported except at the very bottom, we only need one boundary condition for the whole vessel. If the top opening of the vessel is a free edge, then $N_\phi = 0$ at $\phi = \phi_o$. This condition gives

$$\begin{aligned} k = &\frac{D(4D^2+H^2)}{\sin^2\phi_o}-\frac{DH^2}{2\sin^4\phi_o}+H(4D^2+H^2)\left(\frac{\cos\phi_o}{\sin\phi_o}\right) \\ &-\frac{H}{3}(4D^2+2H^2)\left(\frac{\cos\phi_o}{\sin^3\phi_o}\right)(1+2\sin^2\phi_o) \\ &+\frac{H^3}{15}\left(\frac{\cos\phi_o}{\sin^5\phi_o}\right)(3+4\sin^2\phi_o+8\sin^4\phi_o) \end{aligned} \tag{3.186}$$

With N_ϕ now known, the hoop stress resultant follows from equation (2.12). After eliminating r_1, r_2 and p_r, we obtain

$$N_\theta = (4D^2\sin^2\phi - H^2\cos^2\phi)\left[\frac{\gamma(2HD\sin\phi - H^2\cos\phi)}{32D^2\sin^4\phi}-\frac{N_\phi}{2H^2}\right] \tag{3.187}$$

3.2.6.4 Volume capacity of tank

Considering elemental horizontal discs of radius x and thickness dy, the volume of the tank is twice the integral summation of such discs between the equatorial plane and the apex – that is,

$$V = 2\int_0^{\frac{H}{2}} \pi x^2 dy \tag{3.188}$$

From equations (3.172) and (3.173),

$$x = \frac{D}{2} - ky^2 = \frac{D}{2} - \frac{2D}{H^2}y^2 \tag{3.189}$$

Therefore

$$V = 2\pi\int_0^{\frac{H}{2}}\left(\frac{D^2}{4} - 2\frac{D^2}{H^2}y^2 + 4\frac{D^2}{H^4}y^4\right)dy$$
$$= 2\pi\left[\frac{D^2}{4}y - \frac{2}{3}\frac{D^2}{H^2}y^3 + \frac{4}{5}\frac{D^2}{H^4}y^5\right]_0^{\frac{H}{2}} = \frac{2\pi}{15}D^2H \tag{3.190}$$

3.2.6.5 Parametric study

Making use of equation (3.176a) to eliminate $\sin\phi_o$ and $\cos\phi_o$ from expression (3.186), we obtain, after simplifications

$$k = \frac{D}{30H^2}\left(112D^4 + 120D^2H^2 + 15H^4\right) \tag{3.191}$$

If we define a non-dimensional parameter of the ogival shell as follows

$$\xi = \frac{H}{D} \tag{3.192}$$

(i.e. ξ is the height-to-diameter ratio of the tank), we may recast equation (3.191) in the form

$$k = \frac{D^3}{30\xi^2}\left(112 + 120\xi^2 + 15\xi^4\right) \tag{3.193}$$

We use this result to eliminate the constant k from expression (3.185), and use relation (3.192) to eliminate D from expressions (3.185) and (3.187). This yields the following non-dimensional form of the expressions for the stress resultants:

$$\frac{N_\phi}{\gamma H^2} = \frac{\xi}{16}\left(\frac{\sin\phi}{4\sin^2\phi - \xi^2\cos^2\phi}\right)\left[-\left(\frac{4+\xi^2}{\sin^2\phi}\right) + \left(\frac{\xi^2}{2\sin^4\phi}\right) - \xi(4+\xi^2)\left(\frac{\cos\phi}{\sin\phi}\right)\right.$$
$$+\frac{\xi}{3}(4+2\xi^2)\left(\frac{\cos\phi}{\sin^3\phi}\right)(1+2\sin^2\phi) - \frac{\xi^3}{15}\left(\frac{\cos\phi}{\sin^5\phi}\right)(3+4\sin^2\phi+8\sin^4\phi)$$
$$\left.+\frac{1}{30\xi^2}\left(112+120\xi^2+15\xi^4\right)\right] \tag{3.194a}$$

$$\frac{N_\theta}{\gamma H^2} = \frac{1}{32\xi^2}(4\sin^2\phi - \xi^2\cos^2\phi)\left[\left(\frac{2\xi\sin\phi - \xi^2\cos\phi}{\sin^4\phi}\right) - \left(\frac{\xi\sin\phi}{4\sin^2\phi - \xi^2\cos^2\phi}\right)\right.$$
$$\times\left[-\left(\frac{4+\xi^2}{\sin^2\phi}\right) + \left(\frac{\xi^2}{2\sin^4\phi}\right) - \xi(4+\xi^2)\left(\frac{\cos\phi}{\sin\phi}\right) + \frac{\xi}{3}(4+2\xi^2)\left(\frac{\cos\phi}{\sin^3\phi}\right)(1+2\sin^2\phi)\right.$$
$$\left.\left.-\frac{\xi^3}{15}\left(\frac{\cos\phi}{\sin^5\phi}\right)(3+4\sin^2\phi+8\sin^4\phi) + \frac{1}{30\xi^2}\left(112+120\xi^2+15\xi^4\right)\right]\right] \tag{3.194b}$$

From these final expressions, it is evident that for tanks with the same shape (i.e. tanks of the same height-to-diameter ratio ξ), stress resultants in the shell are directly proportional to H^2 (or to D^2, since $D \propto H$). For instance, doubling the height H or diameter D of the tank, while maintaining the parameter ξ constant, will quadruple the stress resultants N_ϕ and N_θ in the shell. This is how the *scale* of the structure will affect its design.

Non-dimensional stress variations $N_\phi / \gamma H^2$ and $N_\theta / \gamma H^2$ have been plotted versus the meridional angle ϕ, for various values of ξ ranging from 1.0 to 3.0, which covers the most practical proportions for egg-shaped sludge digesters. The results are shown in Figure 3.19. From equations (3.176), it is clear that the computation of stress variations is only relevant in the interval $\phi_o \leq \phi \leq \phi_o'$, i.e. $\tan^{-1}(\xi/2) \leq \phi \leq \pi - \tan^{-1}(\xi/2)$.

Having identified ξ as the key parameter that governs the relative variation of stress resultants over the surface of the shell, let us express the volume capacity of the tank (equation (3.190)) also in terms of this parameter, as follows:

$$V = \left(\frac{2\pi\xi}{15}\right) D^3 = \left(\frac{2\pi}{15\xi^2}\right) H^3 \qquad (3.195)$$

Table 3.1 summarises the ranges of ϕ (ϕ_o to ϕ_o') and non-dimensional tank capacities (V / H^3) for the ξ values appearing in Figure 3.19.

Figure 3.19 Non-dimensional stress variations for the liquid-filled parabolic ogival shell: (a) meridional stresses; (b) hoop stresses.

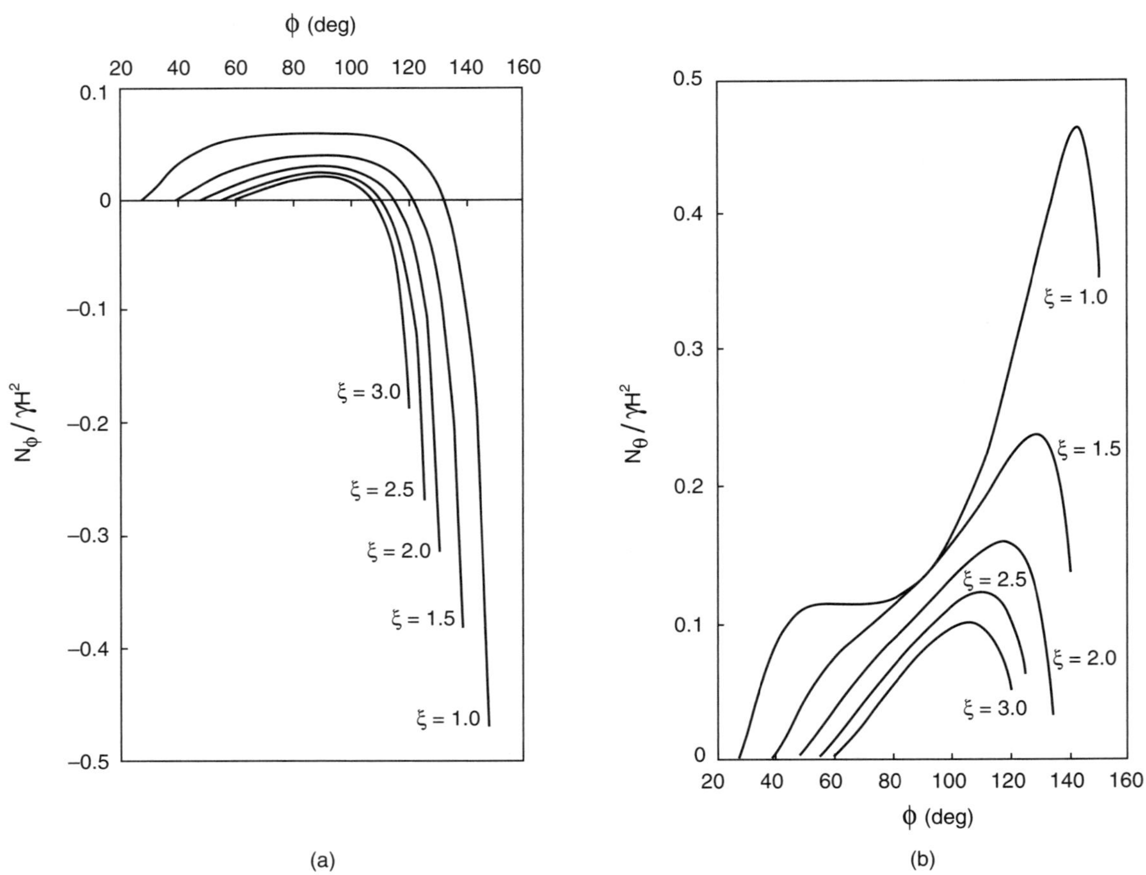

Table 3.1 Ranges of ϕ and non-dimensional tank volumes for various ξ.

ξ	Range of ϕ	V/H^3
1.0	26.6° to 153.4°	0.41888
1.5	36.9° to 143.1°	0.18617
2.0	45.0° to 135.0°	0.10472
2.5	51.3° to 128.7°	0.06702
3.0	56.3° to 123.7°	0.04654

3.2.6.6 Discussion of results

For all values of the parameter ξ, hoop stress resultants N_θ remain positive (tensile) throughout the parabolic digester, rising from zero at the apex ($\phi = \phi_o$), to some peak value below the equatorial level, before beginning to drop in magnitude with further increase in ϕ. In terms of the non-dimensional stresses $N_\theta / \gamma H^2$, the magnitude and location of this peak value are approximately {0.46;143°}, {0.24;129°}, {0.16;118°}, {0.125;110°} and {0.100;107°} for $\xi = 1.0$, 1.5, 2.0, 2.5 and 3.0, respectively (Figure 3.19b).

Thus, for a given height H of the digester, the peak value of the hoop stress resultant N_θ increases rapidly as ξ is reduced from 3.0, through 2.5, 2.0 and 1.5, to 1.0 (i.e. as the diameter D of the tank is increased). However, by reference to Table 3.1, the capacity gain as ξ is reduced is more rapid than the increase in the peak hoop stress, implying that structural efficiency increases with reduction in ξ. Here, structural efficiency η is defined as the ratio of non-dimensional tank volume to non-dimensional peak hoop-stress resultant – that is,

$$\eta = \frac{V}{H^3} \times \frac{\gamma H^2}{\left(N_\theta\right)_{peak}} = \frac{V\gamma}{H\left(N_\theta\right)_{peak}} \tag{3.196}$$

For $\xi = 1.0$, 1.5, 2.0, 2.5 and 3.0, the parameter η works out at approximately 0.91, 0.78, 0.65, 0.54 and 0.47, respectively.

A peculiarity of the hoop-stress variation for $\xi = 1.0$ is the insensitivity of the hoop stress resultant to changes in ϕ over the region $\phi = 54°$ to $\phi = 76°$, a behaviour that might be expected of the region of the parabolic meridian in the neighbourhood of the equator ($\phi = 90°$, around which ϕ is changing rapidly with relatively little change in the vertical coordinate y), but not away from the equator ($\phi = 54°$ to 76°). Also, for a tank of given height H, the equatorial value of the hoop stress resultant appears to be insensitive to ξ in the range $\xi = 1.0$ to $\xi = 1.5$ (note the touching curves).

Turning now to the meridional stress variations (Figure 3.19a), it is noted that the stress resultants N_ϕ rise from zero at the apex ($\phi = \phi_o$), to a peak tensile value around the equator ($\phi = 90°$), before beginning to decrease and becoming negative (compressive) in the regions of the tank below the equatorial level. The peak tensile meridional values obtained at the equatorial level are all considerably lower than the peak tensile hoop values that were noted earlier, being approximately (in non-dimensional terms) 0.058, 0.039, 0.029, 0.023 and 0.019 for $\xi = 1.0$, 1.5, 2.0, 2.5 and 3.0, respectively.

The case $\xi = 1.0$ exhibits the greatest insensitivity of the meridional stress-resultant variation with meridional angle ϕ in the neighbourhood of the equatorial plane ($\phi = 90°$), partly because, in this region, the meridional angle ϕ changes more rapidly with respect to the vertical coordinate y the smaller ξ becomes: when $\xi = 1.0$, ϕ changes significantly over the neighbourhood of the equator, while the depth coordinate y and the meridional stress resultant N_ϕ (which is roughly inversely proportional to $\sin\phi$ in the neighbourhood of the equatorial plane) change relatively little.

The crossover from tension to compression occurs at $\phi \approx 132°$, 120°, 114°, 110° and 106° for $\xi = 1.0$, 1.5, 2.0, 2.5 and 3.0, respectively. Beyond these changeover values of ϕ, compression rapidly increases with ϕ (i.e. with depth of liquid), so that the supports of the tank must be positioned not too far below the changeover values of ϕ in order to cut off the excessive meridional compression that would otherwise occur in the shell were the shell to continue unsupported over its lower regions. This would reduce the likelihood of meridional buckling of the shell in the lower regions. The advantage of locating the supports exactly at the tension-to-compression changeover values of ϕ would be not only the total elimination of zones of compression (and hence of zones of possible local instability), but also the cut-off of the higher hoop-tension peaks (corresponding to $\xi = 1.0$, 1.5 and 2.0) noted earlier.

Assuming, then, that compression in the tank can be eliminated or minimised by careful choice of support location as indicated above, the design of the concrete shell may be based on the noted hoop and meridional tensile actions. For instance, to cater for the peak hoop tensile non-dimensional stress resultant of 0.46 noted for the case $\xi = 1.0$, one should design the shell to withstand a tensile force of $0.46\gamma H^2$ (per metre width) in the hoop direction. For a very large tank with $H = 50$ m, this force amounts to 0.46(9.81)(2500) = 11 281.5 kN/m, which, over a shell thickness of say 0.5 m, would result in a tensile stress of 22.6 N/mm^2 in the material of the shell. Clearly, there would be a need to provide hoop steel reinforcement and/or prestressing to withstand this stress while minimising cracking, and this can easily be achieved following the usual design procedures.

Although the structural efficiency η of the tank (equation (3.196)) is highest at the lower end of the range of ξ (i.e. as ξ approaches 1.0), the range $1.5 \le \xi \le 2.0$ is recommended for practical egg-shaped digesters of parabolic ogival profile, since the slope of the shell is sufficiently steep at the poles ($37° \le \phi_o \le 45°$) to allow effective prestressing.

3.3. Roofs and domes

For this type of shell structures, the main form of axisymmetric loading is self-weight and superimposed dead loads. For either of these, let q denote the intensity of the distributed loading as a vertical force per unit area of the shell midsurface. If we want to make a distinction between superimposed dead load and self-weight, we can write q_i for the former and q_w for self-weight; the latter is equal to γt, where γ now denotes the weight per unit volume of the shell material, and t, as before, denotes shell thickness. The loading q_i may vary along the meridian, as may shell thickness t, hence q_w. For now, unless otherwise stated, q will be assumed to be constant, and no distinction will be made between q_i and q_w ($= \gamma t$). When we resolve q in the direction of the normal to the shell midsurface and the direction of the tangent to the meridian, we obtain the components p_r and p_ϕ (p_s in the case of conical shells) respectively, required for the evaluation of the membrane stress resultants.

3.3.1 Conical roof

For the conical shell structures shown in Figure 3.20, the loading components are

$$p_r = -q\cos\alpha \tag{3.197a}$$

Figure 3.20 Conical shell roofs: (a) edge supported; (b) centrally supported.

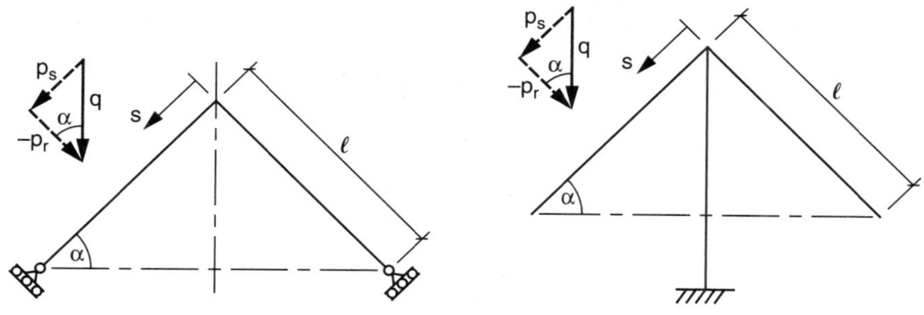

$$p_s = q\sin\alpha \tag{3.197b}$$

Substituting the p_r loading component into the general expression for N_θ (equation (2.27)), we obtain

$$N_\theta = -qs\cot\alpha\cos\alpha \tag{3.198}$$

which, quite clearly, is compressive throughout.

Substituting the two loading components into expression (2.29), we obtain

$$N_s = \frac{1}{s}\left[q\int(-\cos\alpha\cot\alpha - \sin\alpha)s\,ds + k\right]$$

$$= \frac{1}{s}\left[-q(\cos\alpha\cot\alpha + \sin\alpha)\frac{s^2}{2} + k\right]$$

$$= \frac{1}{s}\left[-\frac{qs^2}{2\sin\alpha} + k\right] \tag{3.199}$$

For the edge-supported shell (Figure 3.20a), we require that N_s remains finite at the apex ($s = 0$), which implies that

$$k = 0 \tag{3.200}$$

so that

$$N_s = -\frac{qs}{2\sin\alpha} \tag{3.201}$$

which is also compressive throughout.

Figure 3.21 Free body diagram of one half of the ring beam, showing the external forces H_l on the ring beam in equilibrium with the internal hoop force T.

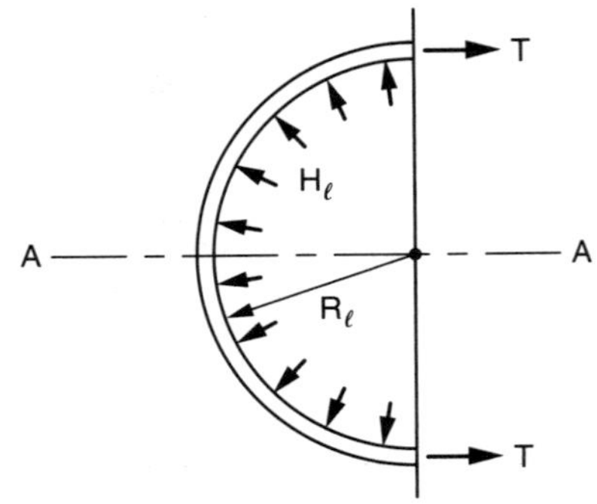

The membrane deformations then follow from expressions (2.75) and (2.76) as

$$\delta^m = -\frac{qs^2}{2Et}(\cot\alpha)\left(2\cos^2\alpha - \nu\right) \tag{3.202a}$$

$$V^m = -\frac{qs}{2Et}\left(\frac{\cos\alpha}{\sin^2\alpha}\right)\left(1 + 2\nu\sin^2\alpha - 4\cos^2\alpha\right) \tag{3.202b}$$

In the above, we have introduced the superscript m (for *membrane*) to distinguish these deformations from those stemming from other effects to be considered below.

In practice, a ring beam is usually provided at the base, instead of the somewhat ideal membrane support conditions depicted in Figure 3.20(a).

At the base, N_s attains its maximum compressive value:

$$\left(N_s\right)_{s=l} = -\frac{ql}{2\sin\alpha} \tag{3.203}$$

This has a horizontal component H_l of magnitude given by

$$H_l = \left|\left(N_s\right)_{s=l}\right|\cos\alpha = \frac{ql}{2}\cot\alpha \tag{3.204}$$

H_l is a uniformly distributed horizontal radial force per unit length of the shell-edge circumference, acting outwards upon the ring beam as shown in Figure 3.21. It is resisted by a tensile hoop force T_r in the ring beam.

From a consideration of horizontal equilibrium in the direction AA of the free-body diagram of half of the ring beam shown in Figure 3.21, we may write

$$H_l\left(2R_l\right) = 2T_r \tag{3.205}$$

from which we obtain

$$T_r = H_l R_l$$

$$= \frac{ql^2}{2}\left(\frac{\cos^2\alpha}{\sin\alpha}\right) \tag{3.206}$$

since the radius R_l of the shell at $s = l$ (which is the same as the radius of the ring beam) is equal to $l\cos\alpha$.

This tensile hoop force produces a direct strain ε_r in the ring beam given by

$$\varepsilon_r = \frac{\sigma_r}{E_r} = \frac{T_r}{A_r E_r} = \frac{ql^2}{2E_r b_r d_r}\left(\frac{\cos^2\alpha}{\sin\alpha}\right) \tag{3.207a}$$

where σ_r is the direct stress in the ring beam, E_r is Young's modulus of the ring beam, and $A_r = b_r d_r$ is the cross-sectional area of the ring beam (assumed to be of rectangular shape with breadth b_r in the horizontal direction and depth d_r in the vertical direction).

The hoop strain in the ring beam is also given by the relation

$$\varepsilon_r = \frac{\delta_r^m}{R_l} \tag{3.207b}$$

where δ_r^m is the outward lateral movement of the ring beam (subscript r) due to membrane thrust (superscript m) of the shell, and R_l, as already explained, is the radius of the circle of latitude of the conical shell at $s = l$, which of course is equal to the radius of the ring beam.

From equations (3.207a) and (3.207b), we obtain

$$\delta_r^m = \frac{ql^3}{2E_r b_r d_r}\left(\frac{\cos^3\alpha}{\sin\alpha}\right) \tag{3.208}$$

Denoting the vertical component of N_s at the edge of the shell by G_l, we have

$$G_l = \left|(N_s)_{s=l}\right|\sin\alpha = \frac{ql}{2} \tag{3.209}$$

This is resisted by vertical reactions of the support, which in practice are provided by a vertical cylindrical wall or closely spaced vertical columns beneath the ring beam.

Since the deformations δ and V of the ring beam and the shell edge are, in general, not compatible, corrective bending-moment and shearing-force actions, applied axisymmetrically at the shell-ring junction as shown in Figure 3.22, will be required such that the net lateral displacements δ and net rotations V of the shell edge and the ring beam match:

$$\delta_{sh}^m + \delta_{sh}^H + \delta_{sh}^M = \delta_r^m + \delta_r^H + \delta_r^M \tag{3.210a}$$

$$V_{sh}^m + V_{sh}^H + V_{sh}^M = V_r^m + V_r^H + V_r^M \tag{3.210b}$$

Figure 3.22 Corrective actions at the interface of the conical roof and the ring beam.

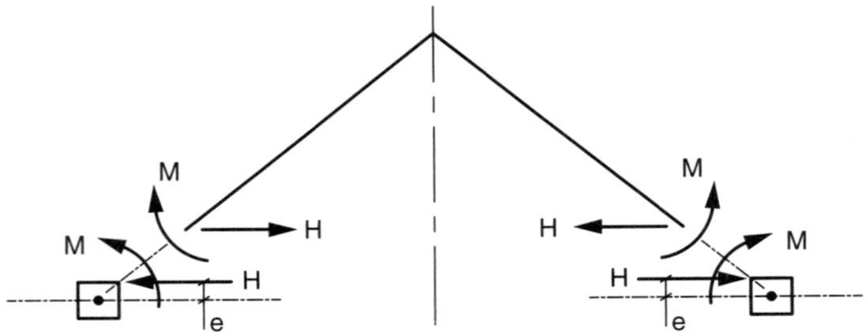

In the above expressions, superscripts m, H and M refer to the membrane contribution, the contribution due to the forces H, and the contribution due to the moments M, respectively; subscripts sh and r refer to the shell edge and the ring beam, respectively. The parameter V_r^m is often taken as zero since the line of action of the shell is usually designed to pass through the centroid of the ring-beam cross-section, so that membrane thrust has no eccentricity and therefore no rotating effect on the ring-beam cross-section. On the other hand, V_r^H, though usually relatively small, is generally non-zero because of the eccentricity e of H (see Figure 3.22).

The ring beam must be designed for the combined effects of T_r (expression (3.206)), M and H; similarly, the shell must be designed for the superimposed effects of the membrane solution and the edge actions M and H. The effects of M and H upon the shell, which tend to be localised to the vicinity of the shell edge, can only be determined on the basis of the bending theory (of which they constitute the homogeneous solution), which will be considered in later chapters. Further considerations of ring-beam design appear in Chapter 7.

For the conical shelter of Figure 3.20(b), where the shell is supported by a central column in mushroom fashion, we require that N_s vanishes at the edge of the shell, since the edge is free. The condition $N_s = 0$ at $s = l$, when substituted into equation (3.199), leads to the solution

$$k = \frac{ql^2}{2\sin\alpha} \tag{3.211}$$

so that

$$N_s = \frac{q}{2s\sin\alpha}\left(l^2 - s^2\right) \tag{3.212}$$

N_θ is, of course, still given by expression (3.198).

As expected, N_s has a singularity at $s = 0$, because the roof is here assumed to be supported on a point. However, a real column support will have a finite diameter, and N_s at the interface of the column and the shell will therefore have a finite value.

3.3.2 Spherical dome

Figure 3.23 shows a spherical dome of radius a and angle of opening ϕ_s. The loading components are

$$p_r = -q\cos\phi \tag{3.213a}$$

Figure 3.23 Spherical dome with no opening at the top.

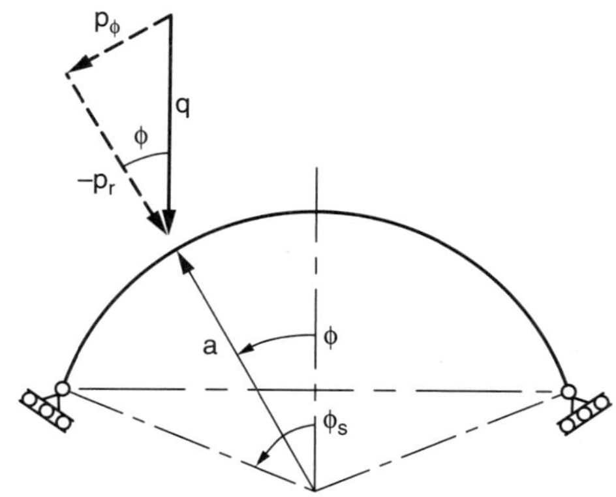

$$p_\phi = q \sin \phi \tag{3.213b}$$

When we put $r_1 = r_2 = a$, and substitute the above expressions for p_ϕ and p_r into equation (2.15), we readily obtain

$$N_\phi = \frac{1}{a \sin^2 \phi}\left[qa^2 \cos \phi + k\right] \tag{3.214}$$

At $\phi = 0$, N_ϕ must be finite, which gives us

$$k = -qa^2 \tag{3.215}$$

Substituting this result into expression (3.214), we obtain

$$N_\phi = -qa\left(\frac{1}{1+\cos \phi}\right) \tag{3.216a}$$

and, from equation (2.12),

$$N_\theta = qa\left(\frac{1}{1+\cos \phi} - \cos \phi\right) \tag{3.216b}$$

Using these results in equations (2.67) and (2.72), we obtain the deformations of the shell as

$$\delta = \frac{qa^2}{Et}\left(\frac{1+\nu}{1+\cos \phi} - \cos \phi\right)\sin \phi \tag{3.217a}$$

$$V = -\frac{qa}{Et}(2+\nu)\sin \phi \tag{3.217b}$$

Figure 3.24 Spherical dome with opening at the top: (a) with no concentrated edge load and no ring beam around the opening; (b) with concentrated vertical edge load *P* but no ring beam around the opening; (c) with concentrated vertical edge load *P* and a ring beam around the opening.

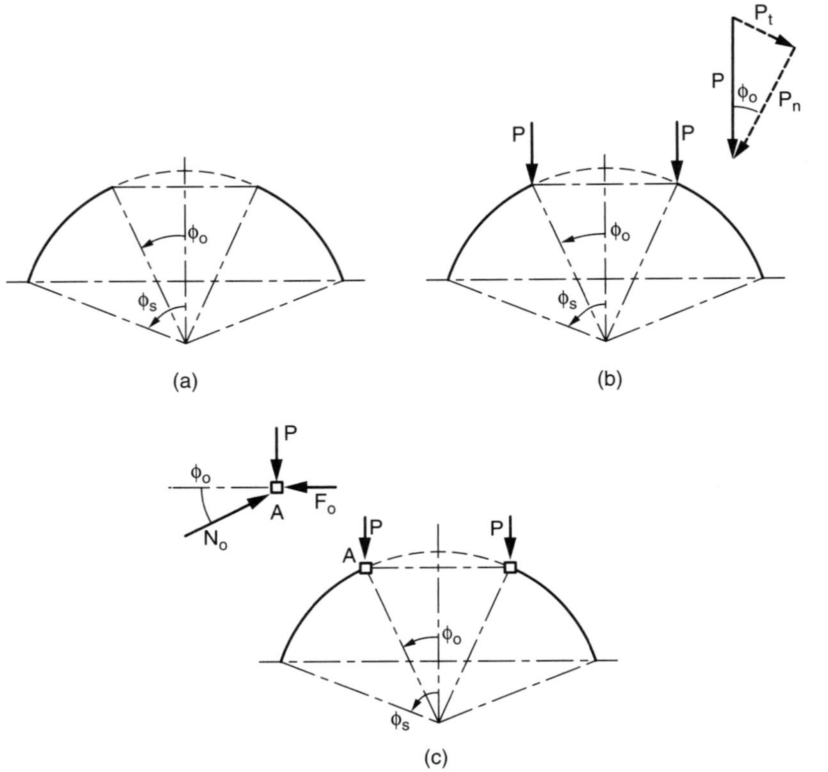

If the dome has an axisymmetric circular opening around the top at the circle of latitude defined by ϕ_o (see Figure 3.24(a)), we require that N_ϕ vanishes at the free edge ($\phi = \phi_o$). From equation (3.214), this implies that

$$k = -qa^2 \cos \phi_o \tag{3.218}$$

and the stress resultants become

$$N_\phi = qa\left(\frac{\cos \phi - \cos \phi_o}{\sin^2 \phi}\right) \tag{3.219a}$$

$$N_\theta = -qa\left(\cos \phi + \frac{\cos \phi - \cos \phi_o}{\sin^2 \phi}\right) \tag{3.219b}$$

If a vertical line load *P* per unit length acts along the circumference of the opening, and a ring beam is not provided around the opening (Figure 3.24(b)), the shell can resist by membrane action only the component $P_t = P \sin \phi_o$, which acts in the plane of the shell midsurface. The component $P_n = P \cos \phi_o$ is perpendicular to the shell midsurface, and therefore cannot be resisted by membrane action; it can only be resisted by considerable bending of the shell at the edge, which may be quite severe. The net effects in the shell will then be the sum of the membrane effects (evaluated on the basis that N_ϕ at $\phi = \phi_o$ must equal the component P_t) and the flexural effects of a transverse line loading P_n applied at the edge $\phi = \phi_o$ (which are obtained from the homogeneous solution of the bending-theory equations).

On the other hand, if a ring beam is provided around the opening (as shown in Figure 3.24(c)), a point A at the upper edge of the shell can now be more or less in equilibrium under a system of three non-colinear forces: P (applied load), N_o (membrane thrust from the shell) and F_o (outward thrust from the ring beam). It becomes unnecessary for the shell to mobilise large bending actions in order to resist P. However, some bending will still occur at the shell edge due to deformation incompatibilities between the shell and the ring beam (as explained in Section 3.3.1), but this is relatively small in comparison with the bending associated with Figure 3.24(b).

For the arrangement shown in Figure 3.24(c), resolving vertically, we have

$$N_o \sin\phi_o = P \quad \Rightarrow \quad N_o = \frac{P}{\sin\phi_o} \tag{3.220}$$

N_o, being an upward force upon point A, represents a compression within the shell. To obtain the value of k in expression (3.214), we therefore impose the boundary condition that $N_\phi = -N_o$ at $\phi = \phi_o$. This gives

$$k = -\left(qa^2 \cos\phi_o + Pa\sin\phi_o\right) \tag{3.221}$$

so that

$$N_\phi = qa\left(\frac{\cos\phi - \cos\phi_o}{\sin^2\phi}\right) - P\frac{\sin\phi_o}{\sin^2\phi} \tag{3.222a}$$

$$N_\theta = -qa\left[\cos\phi + \left(\frac{\cos\phi - \cos\phi_o}{\sin^2\phi}\right)\right] + P\frac{\sin\phi_o}{\sin^2\phi} \tag{3.222b}$$

Resolving the forces at point A horizontally, we have

$$N_o \cos\phi_o = F_o \quad \Rightarrow \quad F_o = N_o \cos\phi_o = \left(\frac{P}{\sin\phi_o}\right)\cos\phi_o = P\cot\phi_o \tag{3.223}$$

The outward thrust F_o is a uniformly distributed horizontal radial force per unit length of the circumference of the shell opening, supplied by the compressed ring beam. Based on the same equilibrium considerations as for the ring beam of Figure 3.21, the hoop compression C_r in the ring beam at the top of the dome, in the presence of the vertical line load P, is given by

$$C_r = F_o a\sin\phi_o = Pa\cos\phi_o \tag{3.224}$$

If at the base of the spherical domes of Figures 3.23 and 3.24 a ring beam is provided, all the ring-beam considerations of Section 3.3.1 apply, with the modification that N_s at the edge ($s = l$) of the conical shell of Figure 3.20(a) is now replaced with the value of N_ϕ at the supported lower edge ($\phi = \phi_s$) of the spherical dome, while the angle α of the cone is replaced with the angle ϕ_s throughout.

3.3.3 Spherical dome of gradually varied thickness

The thickness of concrete domes is often increased in the approaches to the support, in order to accommodate any edge effects there. Since concrete can easily be moulded into a variety of shapes, this local thickening is usually effected in a gradual manner, so as to avoid thickness-discontinuity bending disturbances.

Starting from the apex of the dome, where $\phi = 0$, and following a meridian, the shell thickness is assumed to remain constant at $t = t_o$ as ϕ is increased, down to a certain location defined by the coordinate ϕ_e. Beyond this location, the thickness is gradually increased (symmetrically about the midsurface of the shell) to a maximum value of t_s at the base of the dome, where $\phi = \phi_s$. Figure 3.25 depicts a dome with such a variation of thickness. In the range $\phi_e \leq \phi \leq \phi_s$, the thickness is assumed to vary in accordance with a quadratic law of the form

$$t = \mu(\beta + \phi)^2 \tag{3.225}$$

where μ and β are constants.

The above variation has been adopted for two reasons. First, and as will be shown in Section 8.3.3, this particular law of thickness variation is amenable to closed-form solution of the associated approximate bending-theory equation. Therefore for a shell with such a thickness variation, the net effects (membrane + bending) can be obtained in closed form, for the benefit of the shell designer. In this section, we will consider the calculation of membrane effects only; in Section 8.3.3, we will develop a procedure for estimating edge-related bending effects.

The second reason for adopting the thickness variation of equation (3.225) is that, even if the actual variation does not conform to this equation, values for μ and β can always be chosen such that the ensuing thickness variation, as calculated on the basis of equation (3.225), closely approximates the actual variation.

Figure 3.25 Spherical dome with gradually varied thickness in the lower part.

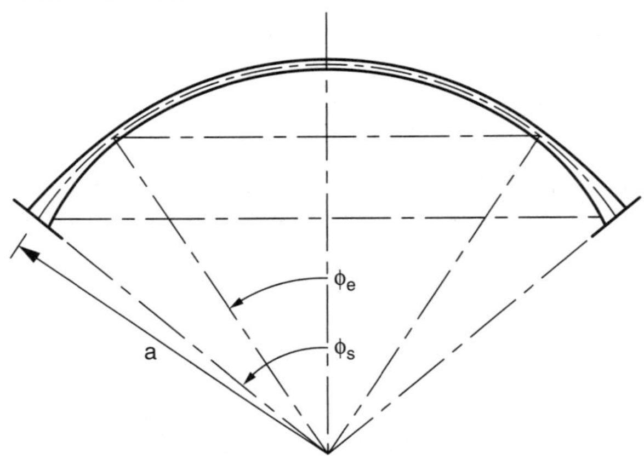

Let us write the loading components as

$$p_r = -q(\phi)\cos\phi \tag{3.226a}$$

$$p_\phi = q(\phi)\sin\phi \tag{3.226b}$$

where q is now shown as a function of ϕ.

A distinction between (i) superimposed dead load $q(\phi) = q_i$, and (ii) self-weight loading $q(\phi) = q_w = \gamma t(\phi)$, both of which were mentioned at the beginning of Section 3.3, is now desirable. For the latter loading, and as explained earlier, γ simply denotes the weight per unit volume of the shell material, while $t(\phi)$ denotes the shell thickness at location ϕ.

While q_i will be assumed to be constant over the entire surface of the dome, q_w will only be constant over the constant-thickness portion of the shell ($0 \le \phi \le \phi_e$), where its value is given by $q_w = \gamma t_o$. Thus, using relations (3.226) in equations (2.15) and (2.12), with $r_1 = r_2 = a$, and imposing the usual condition that N_ϕ must remain finite at the apex of the shell to evaluate k, we obtain the following simple results

$$N_\phi = -q_i a\left(\frac{1}{1+\cos\phi}\right) \tag{3.227a}$$

$$N_\theta = q_i a\left(\frac{1}{1+\cos\phi} - \cos\phi\right) \tag{3.227b}$$

for superimposed dead load $q(\phi) = q_i$ = constant, valid throughout the full range of ϕ (i.e. $0 \le \phi \le \phi_s$), and

$$N_\phi = -\gamma a t_o\left(\frac{1}{1+\cos\phi}\right) \tag{3.228a}$$

$$N_\theta = \gamma a t_o\left(\frac{1}{1+\cos\phi} - \cos\phi\right) \tag{3.228b}$$

for self-weight $q(\phi) = \gamma t_o$ = constant, valid only in the range $0 \le \phi \le \phi_e$ in which the shell thickness is constant.

At the edge of the constant-thickness portion of the dome, where $\phi = \phi_e$, the stress resultants in equations (3.228) assume the boundary values

$$\left(N_\phi\right)_{\phi=\phi_e} = -\gamma a t_o\left(\frac{1}{1+\cos\phi_e}\right) \tag{3.229a}$$

$$\left(N_\theta\right)_{\phi=\phi_e} = \gamma a t_o\left(\frac{1}{1+\cos\phi_e} - \cos\phi_e\right) \tag{3.229b}$$

In the gradually thickened lower portion of the dome ($\phi_e \leq \phi \leq \phi_s$), self-weight is no longer constant: $q(\phi) = \gamma t(\phi)$, where $t(\phi) = \mu(\beta + \phi)^2$ (as given by equation (3.225)). To obtain the membrane stress resultants for this region, we first substitute the expression for $t(\phi)$ into equations (3.226), which become

$$p_r = -\gamma\mu(\beta + \phi)^2 \cos\phi \tag{3.230a}$$

$$p_\phi = \gamma\mu(\beta + \phi)^2 \sin\phi \tag{3.230b}$$

Substituting the above expressions for p_r and p_ϕ into equation (2.15), we obtain

$$N_\phi = \frac{1}{a\sin^2\phi}\left[-\gamma a^2 \mu \int (\beta + \phi)^2 \sin\phi \, d\phi + k\right] \tag{3.231}$$

The integral in the above equation may be evaluated by parts:

$$\int (\beta + \phi)^2 \sin\phi \, d\phi = -(\beta + \phi)^2 \cos\phi + 2(\beta + \phi)\sin\phi + 2\cos\phi \tag{3.232}$$

Putting this result into equation (3.231), we obtain

$$N_\phi = \frac{1}{a\sin^2\phi}\left[-\gamma a^2 \mu\left\{-(\beta + \phi)^2 \cos\phi + 2(\beta + \phi)\sin\phi + 2\cos\phi\right\} + k\right] \tag{3.233}$$

Equating the value of N_ϕ just below the location $\phi = \phi_e$ (as given by expression (3.233) with ϕ set equal to ϕ_e) to the value of N_ϕ just above the same location (as given by expression (3.229a)), we obtain the solution for k as

$$k = -\gamma a^2 \mu\left\{(\beta + \phi_e)^2 \cos\phi_e - 2(\beta + \phi_e)\sin\phi_e - 2\cos\phi_e + \frac{t_o}{\mu}(1 - \cos\phi_e)\right\} \tag{3.234}$$

Eliminating k from expression (3.233), we may finally write down the meridional stress resultant in the range $\phi_e \leq \phi \leq \phi_s$ in the explicit form

$$N_\phi = \frac{\gamma a \mu}{\sin^2\phi}\Bigg[\left\{(\beta + \phi)^2 \cos\phi - 2(\beta + \phi)\sin\phi - 2\cos\phi\right\} \\ -\left\{(\beta + \phi_e)^2 \cos\phi_e - 2(\beta + \phi_e)\sin\phi_e - 2\cos\phi_e\right\} - \frac{t_o}{\mu}(1 - \cos\phi_e)\Bigg] \tag{3.235a}$$

The other stress resultant follows from relation (2.12) as

$$N_\theta = -\frac{\gamma a \mu}{\sin^2\phi}\Bigg[\left\{(\beta + \phi)^2 (\cos\phi)(1 + \sin^2\phi) - 2(\beta + \phi)\sin\phi - 2\cos\phi\right\}$$

$$-\left\{(\beta+\phi_e)^2 \cos\phi_e - 2(\beta+\phi_e)\sin\phi_e - 2\cos\phi_e\right\} - \frac{t_o}{\mu}(1-\cos\phi_e) \Bigg] \tag{3.235b}$$

At the location $\phi = \phi_e$, equation (3.235b) reduces to

$$(N_\theta)_{\phi=\phi_e} = \gamma a t_o\left(\frac{1}{1+\cos\phi_e} - \cos\phi_e\right) \tag{3.236}$$

This expression is precisely equation (3.229b). Thus, at the location $\phi = \phi_e$, the stress resultant N_ϕ just within the (upper) constant-thickness portion of the shell is equal to N_ϕ just within the (lower) portion of gradually varied thickness (a condition that we imposed for the evaluation of the constant k), and the N_θ values on either side of this location have also turned out to be exactly equal to each other. Now from equation (2.67), the lateral displacement δ is a function of the variables $\{N_\phi, N_\theta, t, a, E, \nu\}$ only. As these parameters are the same on either side of the junction $\phi = \phi_e$, it follows that the δ values are the same, and there is no hoop-strain incompatibility between the two portions of the shell. Therefore, as with uniform superimposed dead load, self-weight does not give rise to a significant bending disturbance at $\phi = \phi_e$.

Note, however, that a bending disturbance still occurs at $\phi = \phi_e$, owing to a difference in the values of the meridional rotation V at the locations immediately on either side of the circle $\phi = \phi_e$. Although the variables $\{N_\phi, N_\theta, t, a, E, \nu\}$ for the two adjoining sides of the location $\phi = \phi_e$ are equal to each other, the rate of change of the quantity $(N_\theta - \nu N_\phi)/t$ with respect to ϕ is different, which, by reference to expression (2.72), implies that the V values differ. However, since the flexural effects implied by membrane-solution rotations are generally relatively small in most shells, such a discrepancy in the V values would not be expected to give rise to any significant bending effects at $\phi = \phi_e$. This is in keeping with the general rules of Section 1.5.3, which predict that a discontinuity in the variation of the *first derivative* of shell thickness with respect to ϕ, rather than in the variation of the shell thickness itself, exhibits relatively weak bending effects.

For the evaluation of shell-edge redundants at the support (an aspect of the bending considerations of Chapter 8), we will need to know the value of N_ϕ^m at the support ($\phi = \phi_s$), as well as δ_s^m and V_s^m, which are the values of lateral displacement and meridional rotation at the support as calculated from the membrane solution. For uniform superimposed dead load, evaluation of expressions (3.227), (2.67) and (2.72) at the location $\phi = \phi_s$ yields

$$(N_\phi^m)_{\phi=\phi_s} = -q_i a\left(\frac{1}{1+\cos\phi_s}\right) \tag{3.237}$$

$$\delta_s^m = \frac{q_i a^2}{E t_s}\left(\frac{1+\nu}{1+\cos\phi_s} - \cos\phi_s\right)\sin\phi_s \tag{3.238a}$$

$$V_s^m = -\frac{q_i a}{E t_s}\left\{(2+\nu)\sin\phi_s - \left(\frac{2}{\beta+\phi_s}\right)\left(\frac{1+\nu}{1+\cos\phi_s} - \cos\phi_s\right)\right\} \tag{3.238b}$$

For self-weight loading, evaluation of expressions (3.235), (2.67) and (2.72) at the support yields

$$\left(N_\phi^m\right)_{\phi=\phi_s} = \gamma a t_s \left(\frac{F_s}{\sin^2\phi_s}\right) \tag{3.239}$$

$$\delta_s^m = -\frac{\gamma a^2}{E}\left\{\left(\frac{1+\nu}{\sin^2\phi_s}\right)F_s + \cos\phi_s\right\}\sin\phi_s \tag{3.240a}$$

$$V_s^m = -\frac{\gamma a}{E}\left\{(2+\nu)\sin\phi_s + \left(\frac{2}{\beta+\phi_s}\right)\left(\frac{1+\nu}{\sin^2\phi_s}\right)F_s\right\} \tag{3.240b}$$

where

$$F_s = \left(\cos\phi_s - \frac{t_o}{t_s}\right) - \frac{2}{\beta+\phi_s}\left(\sin\phi_s - \left(\frac{t_o}{t_s}\right)^{1/2}\sin\phi_e\right) - \frac{2}{(\beta+\phi_s)^2}(\cos\phi_s - \cos\phi_e) \tag{3.241}$$

3.3.4 Domes of arbitrary shape

3.3.4.1 General considerations

Figure 3.26 shows a dome in the form of a surface of revolution obtained by rotating the plane curve $y = f(x)$ about the y axis. The x axis is horizontal, the origin is at the apex of the dome, and $f'(0) = 0$ (i.e. the slope of the curve is zero at the apex).

The surface area $S(x)$ of the cap in the domain $0 \le \xi \le x$ is given by an expression of the same form as equation (3.112):

$$S(x) = 2\pi\int_0^x \xi\left[1+\{f'(\xi)\}^2\right]^{1/2} d\xi \tag{3.242}$$

(As in Section 3.2.5, we use a single prime to denote the first derivative of a function with respect to ξ or x, and a double prime to denote the second derivative.) Assuming the dome to be of constant thickness t, with the weight per unit volume of the shell material being denoted by γ as before, the self-weight $W(x)$ of this cap is then

$$W(x) = \gamma\, t\, S(x) \tag{3.243}$$

Figure 3.26 Dome in the form of a surface of revolution of arbitrary meridian shape.

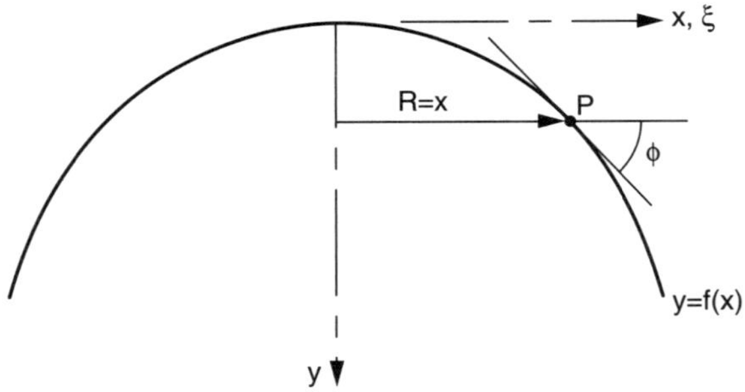

Figure 3.27 Vertical equilibrium of a shell cap: equation (3.245).

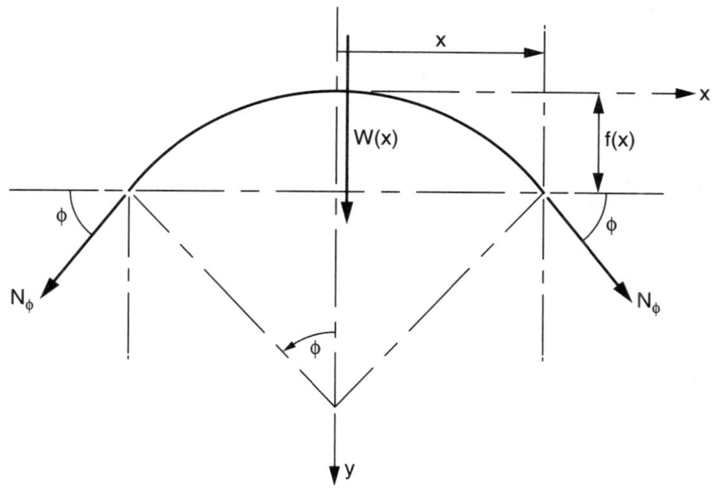

In general, if the vertical loading per unit area of the shell midsurface is constant and equal to q (i.e. $q = q_w = \gamma t$ for self-weight; $q = q_i$ for superimposed dead load), then

$$W(x) = q\,S(x) \tag{3.244}$$

Considering the vertical equilibrium of such a cap (see Figure 3.27), we have

$$2\pi x\,N_\phi \sin\phi + W(x) = 0 \tag{3.245}$$

so that

$$N_\phi = -\frac{W(x)}{2\pi x \sin\phi} = -\frac{q\,S(x)}{2\pi x}\frac{\left[1+\{f'(x)\}^2\right]^{1/2}}{f'(x)} \tag{3.246}$$

where use has been made of relation (3.106) to eliminate $\sin\phi$, and of relation (3.244) to eliminate $W(x)$. Using equation (3.242) to eliminate $S(x)$, we obtain

$$N_\phi = -q\frac{\left[1+\{f'(x)\}^2\right]^{1/2}}{x\,f'(x)}\int_0^x \xi\left[1+\{f'(\xi)\}^2\right]^{1/2} d\xi \tag{3.247}$$

To obtain N_θ, we first write down the expression for p_r in terms of x:

$$\begin{aligned} p_r &= -q\cos\phi \\ &= -q\frac{1}{\left[1+\{f'(x)\}^2\right]^{1/2}} \end{aligned} \tag{3.248}$$

where use has been made of the trigonometric relationship in Figure 3.13(a) to eliminate $\cos\phi$.

The hoop stress resultant N_θ then follows from equation (2.12), when we substitute the results for r_1 and r_2 (equations (3.103) and (3.107)), and the results for N_ϕ and p_r (equations (3.247) and (3.248)). The final expression for N_θ is obtained as follows:

$$N_\theta = q\left[-\frac{x}{f'(x)} + \frac{f''(x)}{\{f'(x)\}^2}\frac{\int_0^x \xi\left[1+\{f'(\xi)\}^2\right]^{1/2} d\xi}{\left[1+\{f'(x)\}^2\right]^{1/2}}\right] \tag{3.249}$$

3.3.4.2 Dome in the form of a paraboloid of revolution

In this case, the equation of the meridian is

$$f(x) = kx^2, \qquad x \geq 0 \tag{3.250}$$

From expression (3.242), we obtain

$$S(x) = \frac{\pi}{6k^2}\left\{\left(1+4k^2x^2\right)^{3/2} - 1\right\} \tag{3.251}$$

From expressions (3.247) and (3.249), we obtain the closed-form results

$$N_\phi = -\frac{q}{24k^3x^2}\left(1+4k^2x^2\right)^{1/2}\left\{\left(1+4k^2x^2\right)^{3/2} - 1\right\} \tag{3.252a}$$

$$N_\theta = -\frac{q}{2k}\left[1 - \frac{1}{12k^2x^2}\left\{\left(1+4k^2x^2\right) - \frac{1}{\left(1+4k^2x^2\right)^{1/2}}\right\}\right] \tag{3.252b}$$

3.3.4.3 Dome in the form of an ellipsoid of revolution

Writing the equation of the meridian as

$$f(x) = b - \frac{b}{a}\left(a^2 - x^2\right)^{1/2}, \qquad 0 \leq x \leq a \tag{3.253}$$

we obtain, from equation (3.242), and assuming $b \geq a$,

$$S(x) = \frac{\pi}{a}\left(b^2 - a^2\right)^{1/2}\left[a\left\{\left(\frac{a^2b^2}{b^2-a^2}\right) - a^2\right\}^{1/2} + \left(\frac{a^2b^2}{b^2-a^2}\right)\sin^{-1}\left\{\frac{\left(b^2-a^2\right)^{1/2}}{b}\right\}\right.$$

$$\left. -\left(a^2 - x^2\right)^{1/2}\left\{\left(\frac{a^2b^2}{b^2-a^2}\right) - \left(a^2 - x^2\right)\right\}^{1/2}\right.$$

$$+\left(\frac{a^2b^2}{b^2-a^2}\right)\sin^{-1}\left\{\frac{\left(a^2-x^2\right)^{1/2}\left(b^2-a^2\right)^{1/2}}{ab}\right\}\Bigg]\Bigg] \tag{3.254}$$

Denoting the right-hand side of the above expression by S, we obtain, from expressions (3.247) and (3.249), the closed-form results

$$N_\phi = -\frac{q}{2\pi b}\frac{S}{x^2}\left\{a^4+\left(b^2-a^2\right)x^2\right\}^{1/2} \tag{3.255a}$$

$$N_\theta = -\frac{qa}{b}\left[\left(a^2-x^2\right)^{1/2}-\frac{Sa^3}{2\pi x^2}\frac{1}{\left\{a^4+\left(b^2-a^2\right)x^2\right\}^{1/2}}\right] \tag{3.255b}$$

3.4. Cooling towers

Cooling towers at thermal power stations are usually designed as hyperboloids of revolution of one sheet. With reference to Figure 3.28, the equation of the meridian is

$$\frac{x^2}{a^2}-\frac{y^2}{b^2}=1, \qquad x \geq a \tag{3.256}$$

where the y axis coincides with the vertical axis of revolution of the shell midsurface, the x axis is horizontal, and the radius R of a circle of latitude is equal to the coordinate x.

Figure 3.28 Cooling tower in the form of a hyperboloid of revolution of one sheet.

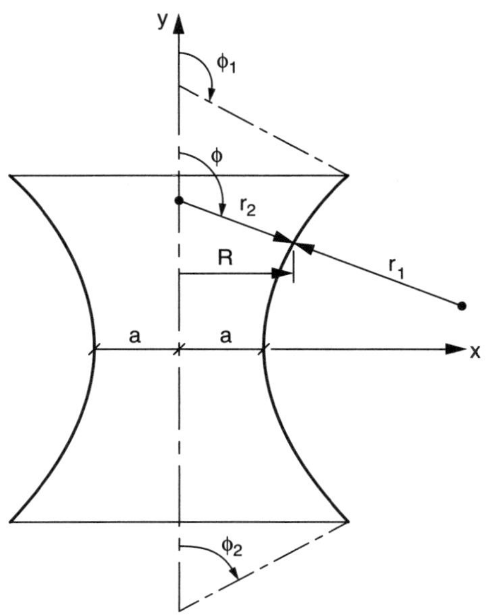

From equation (3.256), we have

$$y = \pm \frac{b}{a} \sqrt{x^2 - a^2} \tag{3.257}$$

Differentiating this expression with respect to x, we obtain

$$\frac{dy}{dx} = \pm \frac{b}{a} \frac{x}{\sqrt{x^2 - a^2}} = \tan \phi \tag{3.258}$$

Therefore,

$$\tan^2 \phi = \frac{b^2}{a^2} \left(\frac{x^2}{x^2 - a^2} \right) \tag{3.259}$$

which on rearranging yields

$$x^2 = \frac{a^4 \tan^2 \phi}{a^2 \tan^2 \phi - b^2} \tag{3.260}$$

– that is, for the meridian in the positive x domain of Figure 3.28,

$$x = \frac{a^2 \tan \phi}{\left(a^2 \tan^2 \phi - b^2\right)^{1/2}} = \frac{a^2 \sin \phi}{\left(a^2 \sin^2 \phi - b^2 \cos^2 \phi\right)^{1/2}} \tag{3.261}$$

This solution is real if $a^2 \sin^2 \phi > b^2 \cos^2 \phi$ – that is, if

$$\tan^2 \phi > \frac{b^2}{a^2} \tag{3.262}$$

which, by reference to equation (3.259), is always so for a hyperbolic meridian.

The second principal radius of curvature is then given by

$$r_2 = \frac{x}{\sin \phi} = \frac{a^2}{\left(a^2 \sin^2 \phi - b^2 \cos^2 \phi\right)^{1/2}} \tag{3.263}$$

Substituting result (3.261) into expression (3.257), we obtain

$$y = \pm \frac{b^2 \cos \phi}{\left(a^2 \sin^2 \phi - b^2 \cos^2 \phi\right)^{1/2}} \tag{3.264}$$

From equation (3.258), the second derivative of y with respect to x is obtained as

$$\frac{d^2 y}{dx^2} = \pm \frac{ab}{\left(x^2 - a^2\right)^{3/2}} \tag{3.265}$$

Substituting expressions (3.258) and (3.265) into equation (3.17), we obtain

$$|r_1| = \frac{\left[a^2\left(x^2 - a^2\right) + b^2 x^2\right]^{3/2}}{a^4 b} \tag{3.266a}$$

from which we can readily eliminate x, using result (3.261), to obtain

$$|r_1| = \frac{a^2 b^2}{\left(a^2 \sin^2 \phi - b^2 \cos^2 \phi\right)^{3/2}} \tag{3.266b}$$

Now r_1, being on the side of the shell midsurface opposite to that on which the axis of revolution lies, must be taken as negative – that is,

$$r_1 = \frac{-a^2 b^2}{\left(a^2 \sin^2 \phi - b^2 \cos^2 \phi\right)^{3/2}} \tag{3.267}$$

Let q be the downward vertical loading per unit area of the shell midsurface. In a cooling tower, this comprises mostly self-weight $q_w = \gamma t$, where γ is the weight per unit volume of the shell material and t is the shell thickness. Let us assume q is constant for simplicity. Then the loading components p_ϕ (tangential to the meridian, and always pointing towards the base of the cooling tower) and p_r (normal to the shell midsurface, and always pointing away from the axis of revolution) are given, as in the case of domes, by the relations

$$p_\phi = q \sin \phi \tag{3.268a}$$

$$p_r = -q \cos \phi \tag{3.268b}$$

where, as shown in Figure 3.28, ϕ is the angle between the positive direction of the y axis and the normal to the shell midsurface at the point in question (i.e. ϕ *decreases* as one goes down the meridian from the top of the cooling tower towards the bottom).

Substituting the expressions for r_1 (equation (3.267)) and r_2 (equation (3.263)), and those for p_ϕ and p_r as given by equations (3.268), into the general solution for N_ϕ (equation (2.15)), we obtain

$$N_\phi = \frac{\left(a^2 \sin^2 \phi - b^2 \cos^2 \phi\right)^{1/2}}{a^2 \sin^2 \phi}\left[q a^4 b^2 \int \frac{\sin \phi}{\left(a^2 \sin^2 \phi - b^2 \cos^2 \phi\right)^2} d\phi + k\right] \tag{3.269}$$

Now

$$\int \frac{\sin \phi}{\left(a^2 \sin^2 \phi - b^2 \cos^2 \phi\right)^2} d\phi = -\frac{1}{4a^3\left(a^2 + b^2\right)^{1/2}}\left[\left(\frac{2a\left(a^2+b^2\right)^{1/2}\cos\phi}{a^2 - \left(a^2+b^2\right)\cos^2\phi}\right) + \ln\left(\frac{a + \left(a^2+b^2\right)^{1/2}\cos\phi}{a - \left(a^2+b^2\right)^{1/2}\cos\phi}\right)\right] \tag{3.270}$$

Therefore,

$$N_\phi = \frac{\left(a^2 \sin^2 \phi - b^2 \cos^2 \phi\right)^{1/2}}{a^2 \sin^2 \phi}\left[-\frac{qab^2}{4\left(a^2+b^2\right)^{1/2}}\left\{\left(\frac{2a\left(a^2+b^2\right)^{1/2}\cos\phi}{a^2-\left(a^2+b^2\right)\cos^2\phi}\right)+\ln\left(\frac{a+\left(a^2+b^2\right)^{1/2}\cos\phi}{a-\left(a^2+b^2\right)^{1/2}\cos\phi}\right)\right\}+k\right] \tag{3.271}$$

At the top edge, where $\phi = \phi_1$, N_ϕ must vanish since this edge is free. This gives

$$k = \frac{qab^2}{4\left(a^2+b^2\right)^{1/2}}\left\{\left(\frac{2a\left(a^2+b^2\right)^{1/2}\cos\phi_1}{a^2-\left(a^2+b^2\right)\cos^2\phi_1}\right)+\ln\left(\frac{a+\left(a^2+b^2\right)^{1/2}\cos\phi_1}{a-\left(a^2+b^2\right)^{1/2}\cos\phi_1}\right)\right\} \tag{3.272}$$

so that N_ϕ may finally be written down as

$$\begin{aligned} N_\phi = {} & \frac{qb^2}{4a\left(a^2+b^2\right)^{1/2}}\frac{\left(a^2 \sin^2 \phi - b^2 \cos^2 \phi\right)^{1/2}}{\sin^2 \phi} \times \\ & \left[\left\{\left(\frac{2a\left(a^2+b^2\right)^{1/2}\cos\phi_1}{a^2-\left(a^2+b^2\right)\cos^2\phi_1}\right)+\ln\left(\frac{a+\left(a^2+b^2\right)^{1/2}\cos\phi_1}{a-\left(a^2+b^2\right)^{1/2}\cos\phi_1}\right)\right\}\right. \\ & \left.-\left\{\left(\frac{2a\left(a^2+b^2\right)^{1/2}\cos\phi}{a^2-\left(a^2+b^2\right)\cos^2\phi}\right)+\ln\left(\frac{a+\left(a^2+b^2\right)^{1/2}\cos\phi}{a-\left(a^2+b^2\right)^{1/2}\cos\phi}\right)\right\}\right] \end{aligned} \tag{3.273}$$

Figure 3.29 Equilibrium of a point on a hyperbolic surface of revolution in the direction normal to the shell midsurface: equation (3.274).

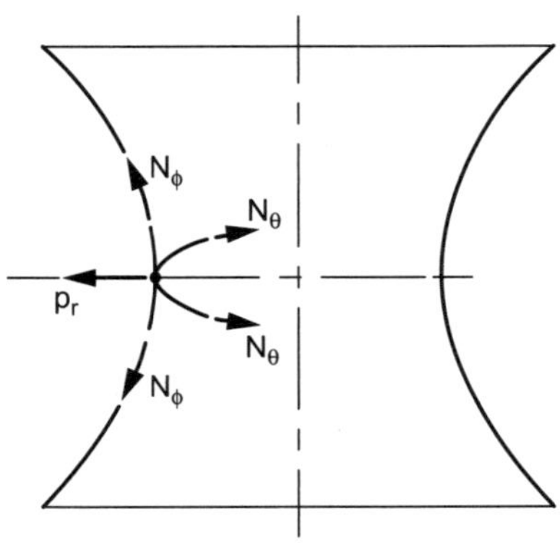

By reference to Figure 3.29, equilibrium of a shell element in the direction normal to the shell midsurface yields

$$\frac{N_\phi}{|r_1|} + p_r = \frac{N_\theta}{|r_2|} \tag{3.274}$$

from which

$$N_\theta = r_2\left(p_r + \frac{N_\phi}{|r_1|}\right) \tag{3.275}$$

since $|r_2| = r_2$. Like the final expression for the stress resultant N_ϕ, this expression for N_θ is a closed-form result, since p_r, r_2, $|r_1|$ and N_ϕ are all known from expressions (3.268b), (3.263), (3.266) and (3.273), respectively.

References

1. Zingoni A. On membrane solutions for elevated shell-of-revolution tanks of certain meridional profiles. *Thin-Walled Structures*, 1995, **22** (2), 121–142.
2. Ayres F. and Mendelson E. *Calculus*. McGraw-Hill, London, 1992.
3. Sutter G. and Jager P. Egg-shaped anaerobic digesters, Taiwan. *Structural Engineering International*, 1994, **4** (3), 160–163.
4. Jager P. Two egg-shaped digesters, Australia. *Structural Engineering International*, 1997, **7** (4), 271–274.
5. Zingoni A. Parametric stress distribution in shell-of-revolution sludge digesters of parabolic ogival form. *Thin-Walled Structures*, 2002, **40** (7, 8), 691–702.

Shell Structures in Civil and Mechanical Engineering
ISBN 978-0-7277-6028-9

http://dx.doi.org/10.1680/ssicame.60289.123

Chapter 4
Membrane solutions for shells of revolution under non-axisymmetric loading

The membrane theory for general shells of revolution subjected to non-axisymmetric loading was presented in Section 2.3.2, and a solution strategy was indicated in Section 2.3.3. In this chapter, we will develop general solutions for two common types of shells of revolution, and apply these solutions to the non-axisymmetric loading condition of wind pressure.

4.1. Spherical shell

4.1.1 General solution

Consider the nth terms of equations (2.49) and (2.50). For a spherical shell, we set $r_1 = r_2 = a$ in the governing equations for the nth terms (i.e. equations (2.51) and (2.52)). These become

$$\frac{dN_{\phi n}}{d\phi} + 2N_{\phi n}\cot\phi + \left(\frac{n}{\sin\phi}\right)N_{\phi\theta n} = a\left(P_{rn}\cot\phi - P_{\phi n}\right) \tag{4.1a}$$

$$\frac{dN_{\phi\theta n}}{d\phi} + 2N_{\phi\theta n}\cot\phi + \left(\frac{n}{\sin\phi}\right)N_{\phi n} = a\left[\left(\frac{n}{\sin\phi}\right)P_{rn} - P_{\theta n}\right] \tag{4.1b}$$

for the symmetric part of the loading, and

$$\frac{dN'_{\phi n}}{d\phi} + 2N'_{\phi n}\cot\phi - \left(\frac{n}{\sin\phi}\right)N'_{\phi\theta n} = a\left(P'_{rn}\cot\phi - P'_{\phi n}\right) \tag{4.2a}$$

$$\frac{dN'_{\phi\theta n}}{d\phi} + 2N'_{\phi\theta n}\cot\phi - \left(\frac{n}{\sin\phi}\right)N'_{\phi n} = -a\left[\left(\frac{n}{\sin\phi}\right)P'_{rn} + P'_{\theta n}\right] \tag{4.2b}$$

for the antisymmetric part of the loading.

For the symmetric response, and taking any nth term of the loading, let

$$F = N_{\phi n} + N_{\phi\theta n} \tag{4.3a}$$

$$G = N_{\phi n} - N_{\phi\theta n} \tag{4.3b}$$

so that

$$N_{\phi n} = \tfrac{1}{2}(F + G) \tag{4.4a}$$

$$N_{\phi\theta n} = \tfrac{1}{2}(F - G) \tag{4.4b}$$

Similarly, for the nth component of the antisymmetric response, we introduce the parameters

$$F' = N'_{\phi n} + N'_{\phi\theta n} \tag{4.5a}$$

$$G' = N'_{\phi n} - N'_{\phi\theta n} \tag{4.5b}$$

so that

$$N'_{\phi n} = \tfrac{1}{2}(F' + G') \tag{4.6a}$$

$$N'_{\phi\theta n} = \tfrac{1}{2}(F' - G') \tag{4.6b}$$

Substituting expressions (4.4) and (4.6) into equations (4.1) and (4.2) respectively, we obtain two pairs of equations:

$$\frac{1}{2}\frac{d}{d\phi}(F+G) + (F+G)\cot\phi + \frac{1}{2}\left(\frac{n}{\sin\phi}\right)(F-G) = a\left(P_{rn}\cot\phi - P_{\phi n}\right) \tag{4.7a}$$

$$\frac{1}{2}\frac{d}{d\phi}(F-G) + (F-G)\cot\phi + \frac{1}{2}\left(\frac{n}{\sin\phi}\right)(F+G) = a\left[\left(\frac{n}{\sin\phi}\right)P_{rn} - P_{\theta n}\right] \tag{4.7b}$$

for the symmetric response, and

$$\frac{1}{2}\frac{d}{d\phi}(F'+G') + (F'+G')\cot\phi - \frac{1}{2}\left(\frac{n}{\sin\phi}\right)(F'-G') = a\left(P'_{rn}\cot\phi - P'_{\phi n}\right) \tag{4.8a}$$

$$\frac{1}{2}\frac{d}{d\phi}(F'-G') + (F'-G')\cot\phi - \frac{1}{2}\left(\frac{n}{\sin\phi}\right)(F'+G') = -a\left[\left(\frac{n}{\sin\phi}\right)P'_{rn} + P'_{\theta n}\right] \tag{4.8b}$$

for the antisymmetric response.

Addition and subtraction of equations (4.7) yields

$$\frac{dF}{d\phi} + F\left(\frac{2\cos\phi + n}{\sin\phi}\right) = a\left\{\left(\frac{\cos\phi + n}{\sin\phi}\right)P_{rn} - P_{\phi n} - P_{\theta n}\right\} \tag{4.9a}$$

$$\frac{dG}{d\phi} + G\left(\frac{2\cos\phi - n}{\sin\phi}\right) = a\left\{\left(\frac{\cos\phi - n}{\sin\phi}\right)P_{rn} - P_{\phi n} + P_{\theta n}\right\} \tag{4.9b}$$

while addition and subtraction of equations (4.8) gives

$$\frac{dF'}{d\phi} + F'\left(\frac{2\cos\phi - n}{\sin\phi}\right) = a\left\{\left(\frac{\cos\phi - n}{\sin\phi}\right)P'_{rn} - P'_{\phi n} - P'_{\theta n}\right\} \tag{4.10a}$$

$$\frac{dG'}{d\phi} + G'\left(\frac{2\cos\phi + n}{\sin\phi}\right) = a\left\{\left(\frac{\cos\phi + n}{\sin\phi}\right)P'_{rn} - P'_{\phi n} + P'_{\theta n}\right\} \tag{4.10b}$$

Now the general solution for any differential equation of the form

$$\frac{dy}{d\phi} + y\, f(\phi) = g(\phi) \tag{4.11a}$$

may easily be shown, through the introduction of an integrating factor, to be

$$y = e^{-\int f(\phi)d\phi}\left[\int g(\phi)\, e^{\int f(\phi)d\phi}\, d\phi + C\right] \tag{4.11b}$$

where C is a constant of integration.

For $f(\phi) = \left(\dfrac{2\cos\phi + n}{\sin\phi}\right)$, we have

$$\int f(\phi)\, d\phi = \ln\left(\sin^2\phi\, \tan^n\frac{\phi}{2}\right)$$

$$e^{\int f(\phi)d\phi} = \sin^2\phi\, \tan^n\frac{\phi}{2}$$

$$e^{-\int f(\phi)d\phi} = \frac{1}{\sin^2\phi\, \tan^n\dfrac{\phi}{2}} = \frac{\cot^n\dfrac{\phi}{2}}{\sin^2\phi} \tag{4.12a}$$

For $f(\phi) = \left(\dfrac{2\cos\phi - n}{\sin\phi}\right)$, we have

$$\int f(\phi)\, d\phi = \ln\left(\sin^2\phi\, \cot^n\frac{\phi}{2}\right)$$

$$e^{\int f(\phi)d\phi} = \sin^2\phi \cot^n \frac{\phi}{2}$$

$$e^{-\int f(\phi)d\phi} = \frac{1}{\sin^2\phi \cot^n \frac{\phi}{2}} = \frac{\tan^n \frac{\phi}{2}}{\sin^2\phi} \tag{4.12b}$$

Applying these results to equations (4.9) and (4.10), we obtain the general solutions for F, G, F' and G' as follows:

$$F = \frac{\cot^n \frac{\phi}{2}}{\sin^2\phi}\left[C_n + a\int\left\{\left(\frac{\cos\phi + n}{\sin\phi}\right)P_{rn} - P_{\phi n} - P_{\theta n}\right\}\sin^2\phi \tan^n \frac{\phi}{2}\, d\phi\right] \tag{4.13a}$$

$$G = \frac{\tan^n \frac{\phi}{2}}{\sin^2\phi}\left[D_n + a\int\left\{\left(\frac{\cos\phi - n}{\sin\phi}\right)P_{rn} - P_{\phi n} + P_{\theta n}\right\}\sin^2\phi \cot^n \frac{\phi}{2}\, d\phi\right] \tag{4.13b}$$

$$F' = \frac{\tan^n \frac{\phi}{2}}{\sin^2\phi}\left[C'_n + a\int\left\{\left(\frac{\cos\phi - n}{\sin\phi}\right)P'_{rn} - P'_{\phi n} - P'_{\theta n}\right\}\sin^2\phi \cot^n \frac{\phi}{2}\, d\phi\right] \tag{4.14a}$$

$$G' = \frac{\cot^n \frac{\phi}{2}}{\sin^2\phi}\left[D'_n + a\int\left\{\left(\frac{\cos\phi + n}{\sin\phi}\right)P'_{rn} - P'_{\phi n} + P'_{\theta n}\right\}\sin^2\phi \tan^n \frac{\phi}{2}\, d\phi\right] \tag{4.14b}$$

4.1.2 Dome subjected to wind loading

Consider the hemispherical dome shown in Figure 4.1. Wind is assumed to act upon the dome in a constant horizontal direction, with the pressure on a vertical plane perpendicular to the direction of the wind being constant and equal to p. The direction from which the wind is blowing will be taken as the origin plane for θ (i.e. the plane $\theta = 0$). For simplicity, and consistent with the approach adopted in the analysis of conventional structures, the resultant effect of the wind action on the shell is assumed to be a pressure normal to the shell surface.

With reference to Figure 4.1, this pressure is greatest at point A (where $\{\phi = 90°, \theta = 0\}$), and zero at points B $\{\phi = 90°, \theta = \pm 90°\}$ and C $\{\phi = 0\}$. A reasonable approximation for the loading component p_r may therefore be taken as

$$p_r = -p \sin\phi \cos\theta \tag{4.15}$$

This distribution implies a positive pressure on the windward side of the dome and a suction on the leeward side, a pattern that is consistent with experimental observation.

The other loading components on the shell, p_ϕ and p_θ, are both taken as zero, as the effects of tangential friction are negligible in comparison with the effects of normal pressure.

Figure 4.1 Hemispherical dome subjected to wind loading.

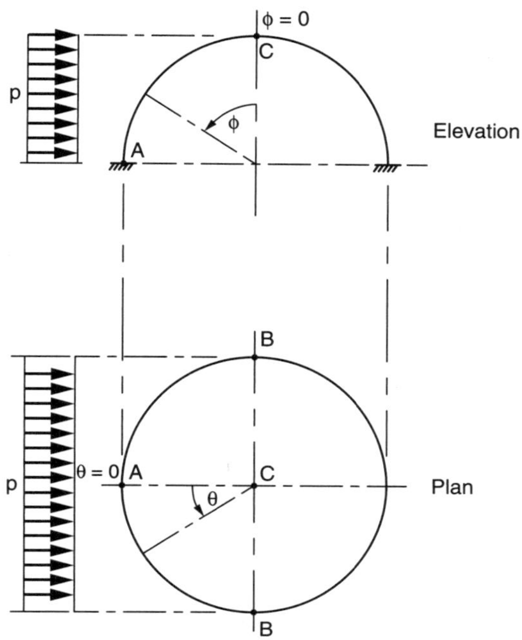

The right-hand side of equation (4.15) corresponds to the second term ($n = 1$) of the first series on the right-hand side of expression (2.49c). Thus,

$$p_r = -p \sin\phi \cos\theta = P_{r1} \cos\theta \tag{4.16}$$

so that

$$P_{r1} = -p \sin\phi \tag{4.17}$$

As the loading is symmetric about the vertical plane $\theta = 0$, we need to evaluate expressions (4.13) with n set equal to 1; $P_{\phi 1}$ and $P_{\theta 1}$ are zero, while P_{r1} is given by equation (4.17). Thus,

$$F = \frac{\cot\dfrac{\phi}{2}}{\sin^2\phi}\left[C_1 - pa\int(1+\cos\phi)\ \sin^2\phi \tan\frac{\phi}{2}\ d\phi\right] \tag{4.18a}$$

$$G = \frac{\tan\dfrac{\phi}{2}}{\sin^2\phi}\left[D_1 + pa\int(1-\cos\phi)\ \sin^2\phi\ \cot\frac{\phi}{2}\ d\phi\right] \tag{4.18b}$$

Now,

$$\tan\frac{\phi}{2} = \left(\frac{1-\cos\phi}{1+\cos\phi}\right)^{1/2} = \frac{1}{\cot\dfrac{\phi}{2}} \tag{4.19a}$$

so that

$$(1+\cos\phi)\left(\tan\frac{\phi}{2}\right) = (1-\cos\phi)\left(\cot\frac{\phi}{2}\right) = \sin\phi \tag{4.19b}$$

Both integrals in equations (4.18a) and (4.18b) therefore reduce to

$$\int \sin^3\phi \; d\phi = -\int (1-\cos^2\phi)\; d(\cos\phi) = -\left(\cos\phi - \frac{\cos^3\phi}{3}\right) \tag{4.19c}$$

Expressions (4.18) thus become

$$\begin{aligned} F &= \frac{\cot\dfrac{\phi}{2}}{\sin^2\phi}\left[C_1 + pa\left(\cos\phi - \frac{\cos^3\phi}{3}\right)\right] \\ &= \frac{1}{(\sin\phi)(1-\cos\phi)}\left[C_1 + pa\left(\cos\phi - \frac{\cos^3\phi}{3}\right)\right] \end{aligned} \tag{4.20a}$$

$$\begin{aligned} G &= \frac{\tan\dfrac{\phi}{2}}{\sin^2\phi}\left[D_1 - pa\left(\cos\phi - \frac{\cos^3\phi}{3}\right)\right] \\ &= \frac{1}{(\sin\phi)(1+\cos\phi)}\left[D_1 - pa\left(\cos\phi - \frac{\cos^3\phi}{3}\right)\right] \end{aligned} \tag{4.20b}$$

From relations (4.4), we obtain

$$\begin{aligned} N_{\phi 1} &= \tfrac{1}{2}(F+G) \\ &= \frac{1}{2}\,\frac{1}{\sin^3\phi}\left\{(C_1+D_1)+(C_1-D_1)\cos\phi + 2pa\cos^2\phi\left(1-\frac{\cos^2\phi}{3}\right)\right\} \end{aligned} \tag{4.21a}$$

$$\begin{aligned} N_{\phi 1} &= \tfrac{1}{2}(F-G) \\ &= \frac{1}{2}\,\frac{1}{\sin^3\phi}\left\{(C_1-D_1)+(C_1+D_1)\cos\phi + 2pa\cos\phi\left(1-\frac{\cos^2\phi}{3}\right)\right\} \end{aligned} \tag{4.21b}$$

From relations (2.50a) and (2.50c), we further obtain

$$N_\phi = N_{\phi 1}\cos\theta$$

$$= \frac{1}{2}\left(\frac{\cos\theta}{\sin^3\phi}\right)\left\{(C_1 + D_1) + (C_1 - D_1)\cos\phi \ + \ 2pa\cos^2\phi\left(1 - \frac{\cos^2\phi}{3}\right)\right\} \tag{4.22a}$$

$$N_{\phi\theta} = N_{\phi\theta 1}\sin\theta$$

$$= \frac{1}{2}\left(\frac{\sin\theta}{\sin^3\phi}\right)\left\{(C_1 - D_1) + (C_1 + D_1)\cos\phi \ + \ 2pa\cos\phi\left(1 - \frac{\cos^2\phi}{3}\right)\right\} \tag{4.22b}$$

To determine the constants C_1 and D_1, we impose the conditions that both N_ϕ and $N_{\phi\theta}$ must remain finite at the apex of the dome ($\phi = 0$). That is, for each right-hand side of the above expressions, the numerator term in braces must vanish when $\phi = 0$, since the denominator term ($\sin^3\phi$) vanishes when $\phi = 0$. These two conditions yield the same result:

$$C_1 = -\tfrac{2}{3}pa \tag{4.23a}$$

with D_1 remaining unknown. We note that the finiteness condition for N_ϕ and $N_{\phi\theta}$ at $\phi = 0$ also implies that the first and second derivatives of the terms in braces of the right-hand sides of expressions (4.22) must vanish at $\phi = 0$, since the first and second derivatives of the denominator term ($\sin^3\phi$) vanish at $\phi = 0$. The vanishing condition for the first derivative of the term in braces leads to the identity $0 = 0$ for both expressions (4.22a) and (4.22b), while the vanishing condition for the second derivative leads to the solution

$$D_1 = \tfrac{2}{3}pa \tag{4.23b}$$

for both expressions.

Substituting the results for C_1 and D_1 into expressions (4.22), we finally obtain

$$N_\phi = \frac{pa}{3}\left(\frac{\cos\phi}{\sin^3\phi}\right)\left\{(3 - \cos^2\phi)\cos\phi \ - \ 2\right\}\cos\theta \tag{4.24a}$$

$$N_{\phi\theta} = \frac{pa}{3}\left(\frac{1}{\sin^3\phi}\right)\left\{(3 - \cos^2\phi)\cos\phi \ - \ 2\right\}\sin\theta \tag{4.24b}$$

The hoop stress resultant N_θ follows from equation (2.41), with $r_1 = r_2 = a$, p_r given by equation (4.15), and N_ϕ as given by expression (4.24a) above:

$$N_\theta = -\frac{pa}{3}\left(\frac{1}{\sin^3\phi}\right)\left[\left\{(3 - \cos^2\phi)\cos\phi \ - \ 2\right\}\cos\phi \ + \ 3\sin^4\phi\right]\cos\theta \tag{4.24c}$$

N_ϕ and N_θ attain their greatest magnitudes with respect to θ in the vertical meridional planes $\theta = 0$ and $\theta = 180°$ (which are parallel to the direction of the wind). In these planes, the shear stress resultant $N_{\phi\theta}$ is zero, which is what we would expect from the symmetry of the loading about the line AC in the plan view of Figure 4.1. Along the meridians defined by $\theta = \pm 90°$ (which lie in the vertical meridional planes perpendicular to the direction of the wind), both N_ϕ and N_θ vanish, while the shearing force $N_{\phi\theta}$ attains its maximum value for a given circle of latitude. Figure 4.2 depicts the variations of N_ϕ and N_θ along the meridian $\theta = 0$, and the variation of $N_{\phi\theta}$ along the meridian $\theta = 90°$.

From Figure 4.2, we note that the maximum stress resultant in the shell is a hoop compression $N_\theta = pa$, occurring at the base of the dome ($\phi = 90°$), at point A (see Figure 4.1). (A hoop tension of the same magnitude occurs on the leeward side at the point diametrically opposite point A.) The magnitude of the stress associated with this value of N_θ is $p(a/t)$.

To obtain a feel for the magnitude of wind-induced stresses, let us compare the maximum wind-induced stress in a spherical concrete dome of radius $a = 30$m and a/t ratio of 300, with the maximum values of self-weight stresses. For the purposes of this comparison, we will take the wind pressure as 1 kN/m^2. In wind-loading calculations, the design pressure due to a 40 m/s wind acting on a plane perpendicular to the direction of the wind is usually of this order. The largest value of the wind-induced stresses in the dome is therefore $1000 \times 300 \times 10^{-6}$ N/mm^2 = 0.30 N/mm^2. For a hemispherical dome of constant thickness, the largest value of self-weight stresses is qa/t (see equations (3.216)) – that is, γa, where γ is the unit weight of concrete. Assuming $\gamma = 24$ kN/m^3, we obtain $\gamma a = 24 \times 10^3 \times 10^{-9} \times 30 \times 10^3$ N/mm^2 = 0.72 N/mm^2.

Thus, while both wind-induced and self-weight stresses in the above example are of relatively low magnitudes, wind-induced stresses are considerably smaller than self-weight stresses. This will generally be the case for the sizes of domes usually encountered in practice, except in instances of very slender domes (i.e. with an a/t ratio much higher than 300), or in environments experiencing strong winds, where wind effects then assume a greater relative importance. Although our study has been based on the spherical dome, the same observation holds for other geometries. Wind-induced stresses may be ignored in the preliminary design of the dome, unless the shell is very slender, or the winds are particularly strong.

Figure 4.2 Results for simplified wind loading.

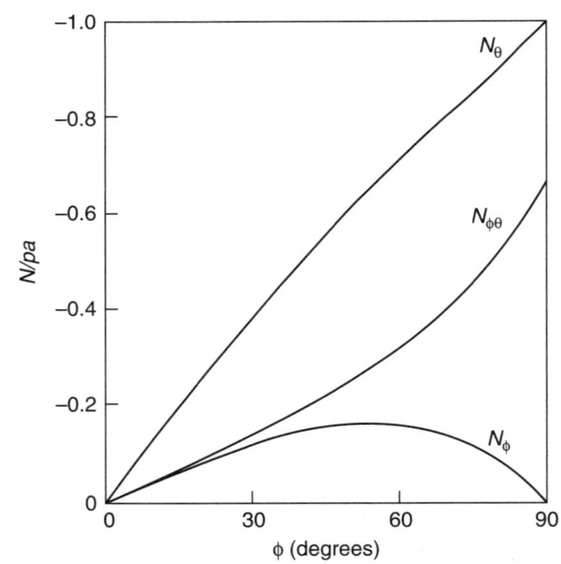

4.2. Conical shell
4.2.1 General solution

For a circular conical shell, and considering only the symmetric response (the antisymmetric response is treated in a similar way, as was shown for the spherical shell), we may express the loading components as sums of harmonic terms

$$p_s = \sum_{n=0}^{\infty} P_{sn} \cos n\theta \tag{4.25a}$$

$$p_\theta = \sum_{n=1}^{\infty} P_{\theta n} \sin n\theta \tag{4.25b}$$

$$p_r = \sum_{n=0}^{\infty} P_{rn} \cos n\theta \tag{4.25c}$$

and the stress resultants similarly as

$$N_s = \sum_{n=0}^{\infty} N_{sn} \cos n\theta \tag{4.26a}$$

$$N_\theta = \sum_{n=0}^{\infty} N_{\theta n} \cos n\theta \tag{4.26b}$$

$$N_{s\theta} = \sum_{n=1}^{\infty} N_{s\theta n} \sin n\theta \tag{4.26c}$$

where P_{sn}, $P_{\theta n}$, P_{rn}, N_{sn}, $N_{\theta n}$ and $N_{s\theta n}$ are functions of s only.

From relation (2.40), we have

$$N_{\theta n} = P_{rn}\, s \cot\alpha \tag{4.27}$$

Substituting the nth terms of series (4.25) and (4.26) into the governing equations (2.38) and (2.39), with $N_{\theta n}$ given by equation (4.27), we obtain the two equations

$$(\cos n\theta)\frac{d}{ds}(sN_{sn}) + \frac{1}{\cos\alpha}N_{s\theta n}(n\cos n\theta) = P_{rn}(s\cot\alpha)(\cos n\theta) - sP_{sn}(\cos n\theta)$$

$$(\sin n\theta)\frac{d}{ds}(sN_{s\theta n}) - \frac{1}{\cos\alpha}P_{rn}(s\cot\alpha)(n\sin n\theta) + N_{s\theta n}(\sin n\theta) = -\; sP_{\theta n}(\sin n\theta)$$

which simplify to

$$s\frac{dN_{sn}}{ds} + N_{sn} + \left(\frac{n}{\cos\alpha}\right)N_{s\theta n} = s\left(P_{rn}\cot\alpha - P_{sn}\right)$$

$$s \frac{dN_{s\theta n}}{ds} + N_{s\theta n} + N_{s\theta n} = s\left[\left(\frac{n}{\sin\alpha}\right)P_{rn} - P_{\theta n}\right]$$

Dividing throughout by s, we obtain

$$\frac{dN_{sn}}{ds} + \frac{1}{s}N_{sn} + \frac{n}{s\cos\alpha}N_{s\theta n} = P_{rn}\cot\alpha - P_{sn} \tag{4.28a}$$

$$\frac{dN_{s\theta n}}{ds} + \frac{2}{s}N_{s\theta n} = \frac{n}{\sin\alpha}P_{rn} - P_{\theta n} \tag{4.28b}$$

The first of the above equations involves both N_{sn} and $N_{s\theta n}$. The second equation involves $N_{s\theta n}$ only, so let us tackle this first. Applying the formulae of equations (4.11) with the variable s of present considerations replacing the ϕ in those expressions, and with $f(s) = 2/s$, we obtain the solution of equation (4.28b) as follows:

$$N_{s\theta n} = \frac{1}{s^2}\left[C_1 + \int\left\{\left(\frac{n}{\sin\alpha}\right)P_{rn} - P_{\theta n}\right\}s^2\,ds\right] \tag{4.29}$$

Applying the same formulae to equation (4.28a), again with the variable s replacing the ϕ in those expressions, and with $f(s) = 1/s$, we obtain the solution for N_{sn} as

$$N_{sn} = \frac{1}{s}\left[D_1 + \int\left\{P_{rn}\cot\alpha - P_{sn} - \left(\frac{n}{s\cos\alpha}\right)N_{s\theta n}\right\}s\,ds\right] \tag{4.30}$$

where $N_{s\theta n}$ is now known (equation (4.29)).

4.2.2 Conical roof subjected to wind loading

As before, we assume $p_s = p_\theta = 0$, and

$$p_r = -p\sin\alpha\cos\theta \tag{4.31}$$

so that only the term corresponding to $n = 1$ is relevant, and

$$P_{r1} = -p\sin\alpha \tag{4.32}$$

Substituting this value for P_{r1} in equation (4.29), with $n = 1$ and $P_{\theta 1} = 0$, we obtain

$$N_{s\theta 1} = \frac{1}{s^2}\left(C_1 - \frac{ps^3}{3}\right) \tag{4.33}$$

so that

$$N_{s\theta} = N_{s\theta 1}\sin\theta$$

$$= \frac{1}{s^2}\left(C_1 - \frac{ps^3}{3}\right)\sin\theta \tag{4.34}$$

Using this result in equation (4.30), with $n = 1$ and $P_{s1} = 0$, we obtain

$$N_{s1} = \frac{1}{s}\left[D_1 + \int\left\{-p\cos\alpha - \left(\frac{1}{s\cos\alpha}\right)\left(\frac{1}{s^2}\right)\left(C_1 - \frac{ps^3}{3}\right)\right\} s\ ds\right]$$

$$= \frac{1}{s}\left[D_1 - \frac{ps^2}{2}\cos\alpha + \frac{1}{\cos\alpha}\left(\frac{C_1}{s} + \frac{ps^2}{6}\right)\right] \tag{4.35}$$

so that

$$N_s = N_{s1}\cos\theta$$

$$= \frac{1}{s}\left[D_1 - \frac{ps^2}{2}\cos\alpha + \frac{1}{\cos\alpha}\left(\frac{C_1}{s} + \frac{ps^2}{6}\right)\right]\cos\theta \tag{4.36}$$

For the roof supported at the edge (see Figure 3.18(a)), $N_{s\theta}$ must be finite at the apex ($s = 0$). Imposing this condition on equation (4.34) yields $C_1 = 0$. Applying the finiteness condition on N_s at the apex, we require the term in square brackets on the right-hand side of equation (4.36) to vanish when $s = 0$. With C_1 already shown to be zero, this term now reads

$$D_1 - \frac{ps^2}{2}\cos\alpha + \frac{1}{\cos\alpha}\frac{ps^2}{6}$$

Setting this to zero at $s = 0$ yields $D_1 = 0$. The stress resultants $N_{s\theta}$ and N_s thus finally become

$$N_{s\theta} = -\frac{ps}{3}\sin\theta \tag{4.37a}$$

$$N_s = \frac{ps}{6}\left(\frac{1}{\cos\alpha} - 3\cos\alpha\right)\cos\theta \tag{4.37b}$$

From equations (4.26b), (4.27) and (4.32), we find

$$N_\theta = -p\sin\alpha\left(s\cot\alpha\right)\cos\theta$$

$$= -ps\cos\alpha\cos\theta \tag{4.37c}$$

For the roof supported at the vertex (see Figure 3.18(b)), we first apply the boundary condition that $N_{s\theta}$ must vanish at the edge of the shell ($s = l$). Equation (4.34) then yields

$$C_1 = \frac{pl^3}{3} \tag{4.38a}$$

Applying the vanishing condition for N_s at $s = l$, equation (4.36) yields

$$D_1 = -\left(\frac{\sin^2\alpha}{\cos\alpha}\right)\frac{pl^2}{2} \tag{4.38b}$$

Substituting the values of the constants C_1 and D_1 into equations (4.34) and (4.36), we finally obtain the stress resultants $N_{s\theta}$ and N_s as follows

$$N_{s\theta} = \frac{p}{3s^2}\left(l^3 - s^3\right)\sin\theta \tag{4.39a}$$

$$N_s = \frac{p}{\cos\alpha}\left\{\left(\frac{l^3 - s^3}{3s^2}\right) - \frac{\left(l^2 - s^2\right)\sin^2\alpha}{2s}\right\}\cos\theta \tag{4.39b}$$

N_θ is, of course, the same as before (equation 4.37c), being independent of boundary conditions. As expected, both solutions for $N_{s\theta}$ and N_s exhibit singularities at $s = 0$ due to the concentrated point reaction of the central support. In a real structure, the central support would have a finite diameter, and the stress resultants $N_{s\theta}$ and N_s would have finite values at the interface of the support and the conical shell.

Shell Structures in Civil and Mechanical Engineering
ISBN 978-0-7277-6028-9

http://dx.doi.org/10.1680/ssicame.60289.135

Chapter 5
Axisymmetric bending of cylindrical shells

5.1. Introduction

Circumferentially closed circular cylinders are widely used in the pressure-vessel industry, and for liquid containment. In pressure vessels and vertical cylindrical tanks, the loading distribution is predominantly axisymmetric. Membrane solutions for cylindrical pressure vessels and tanks were developed in Chapter 3. In this chapter, we will investigate the bending behaviour of cylindrical shells that are subjected to axisymmetric loading conditions.

For loadings that are constant over the shell surface (e.g. uniform gas or vapour pressure), or which vary continuously over the shell surface in a gradual and smooth manner (e.g. hydrostatic pressure), the membrane solutions of Chapter 3 are adequate over most of the shell surface, except near shell edges, in the vicinity of abrupt (stepwise) changes of shell thickness, around junctions of the cylinder with other shells (such as spherical, conical or toroidal segments) or other structural elements (like ring beams, stiffeners or plates), and in the vicinity of any other geometric and loading discontinuities of the type discussed in Section 1.5.3.

The membrane solution fails to satisfy boundary conditions at such discontinuities, so that we have to resort to the more general bending theory, which allows for flexural and transverse-shear effects. However, and as discussed in Section 1.5.4, we do not discard the membrane solution in the proximity of these discontinuities, but instead we adopt it as a particular solution of the full set of differential equations. The homogeneous solution is the response of the shell when only edge loads are applied and no surface loads are present.

By combining the membrane solution with the true homogeneous solution of the differential equations, we get a total solution that is adaptable to the actual boundary conditions of the problem. Strictly speaking, the membrane solution is not a true particular solution of the general bending-theory equations, except in a few cases. Fortunately, and as pointed out in Section 1.5.4, it turns out that this approximation is generally very accurate.

After obtaining the solution of the relevant differential equations for the cylindrical shell, we will fully understand why the effects of the homogeneous solution are localised to the vicinity of the shell edge or interior discontinuity in question. It will also be clear how the curvature property of the shell plays a role in confining the effects of the homogeneous solution to the proximity of the associated edge or discontinuity. This property of the homogeneous solution also explains why the membrane solution on its own is adequate in the *interior regions* of the shell (i.e. locations remote from edges and discontinuities); for such locations, the flexural effects associated with the homogeneous solution (and stemming from shell edges and discontinuities) would have decayed sufficiently to leave the membrane effects dominating.

In the treatment that follows, we will first derive the governing equation for the axisymmetric bending of circular cylinders, and obtain a general solution. We will then apply the general solution to selected problems of practical importance.

5.2. Derivation of the governing equation

5.2.1 Equilibrium considerations

Figure 5.1 shows an element of an axisymmetrically loaded circular cylindrical shell, with dimensions $a\,d\theta$ in the θ (circumferential) direction and dx in the x (axial) direction. The cylinder (of which the element is a part) is of radius a. In addition to the in-plane stress resultants N_x and N_θ, we now have bending moments M_x (acting over the edges of length $a\,d\theta$) and M_θ (acting over the edges of length dx); these are considered positive when tending to curl the edges of the shell element away from the axis of the cylinder. We also have a transverse shearing force Q_x (acting over the edges of length $a\,d\theta$). Because of the axisymmetry of the loading, the transverse shearing force Q_θ (which would have acted over the edges of length dx) is zero, as are the in-plane shear stress resultants ($N_{x\theta}$ and $N_{\theta x}$) and twisting moments ($M_{x\theta}$ and $M_{\theta x}$). All actions are per unit length of the relevant shell edge.

Also owing to axisymmetry, stress and deformation quantities, while generally varying across the element in the x direction, remain constant in the θ (circumferential) direction. The surface-loading components that are consistent with the conditions of axisymmetry are p_x (in the positive x direction) and p_r (in the direction normal to the shell midsurface), but these are not shown in the figure since we will only be concerned with finding a solution to the homogeneous form of the bending-theory equations, for reasons already explained.

Resolving the elemental forces in the x direction, we have

$$\left(N_x + dN_x\right) a\,d\theta \;-\; N_x a\; d\theta \;=\; 0$$

– that is,

$$dN_x = 0 \tag{5.1}$$

Figure 5.1 Bending element of an axisymmetrically loaded circular cylinder.

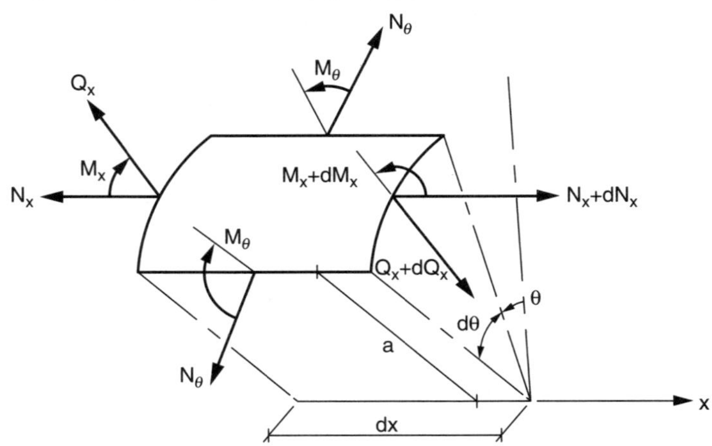

On integrating this equation, we obtain

$$N_x = k \quad \text{(constant)} \tag{5.2}$$

Thus, the value of N_x within the shell is constant, being simply equal to the known value of the applied axial force (per unit circumferential length) at one end of the cylinder; if one end of the cylinder is free of any such axial force, then $N_x = 0$ at that end, and hence throughout the shell. Henceforth, we will take the N_x of bending considerations as zero, since any applied axial force can be allowed for within the membrane solution.

Resolving the elemental forces in the direction normal to the shell midsurface, we have

$$2N_\theta \; dx \; \frac{d\theta}{2} + \left(Q_x + dQ_x\right) \; a \; d\theta - Q_x \; a \; d\theta = 0$$

– that is,

$$N_\theta \, dx + a \; dQ_x = 0$$

which gives

$$\frac{dQ_x}{dx} + \frac{N_\theta}{a} = 0 \tag{5.3}$$

Taking moments in the plane of the M_x actions about the centre of the element, we have

$$\left(M_x + dM_x\right) \; a \; d\theta - M_x \, a \; d\theta - Q_x \; a \; d\theta \; \frac{dx}{2} - \left(Q_x + dQ_x\right) \; a \; d\theta \; \frac{dx}{2} = 0$$

which, on ignoring the relatively small term $\left(dQ_x \;\; d\theta \;\; dx\right) \; a/2$ and simplifying, becomes

$$\frac{dM_x}{dx} - Q_x = 0 \tag{5.4}$$

Equations (5.1), (5.3) and (5.4) are the only ones we can obtain from the static equilibrium of the element. Equation (5.1), in one variable N_x, was solved independently of the other two equations, and will not be considered further. On the other hand, equations (5.3) and (5.4) contain three unknowns Q_x, N_θ and M_x. Unlike problems of the membrane theory, which in general are statically determinate, problems of the bending theory of shells are statically indeterminate, and require consideration of strain-displacement and stress-strain relations for their solution.

5.2.2 Strain-displacement relations and Hooke's law

Let u be the extension of the shell in the x direction and w the displacement normal to the shell midsurface, positive when away from the axis of the cylinder. Consider a meridional line element of the cylinder, of length dx as shown in Figure 5.2. As a result of the actions within the shell, end A of the element displaces to A', while end B moves to B'.

Figure 5.2 Displacements of a meridional element of the cylinder.

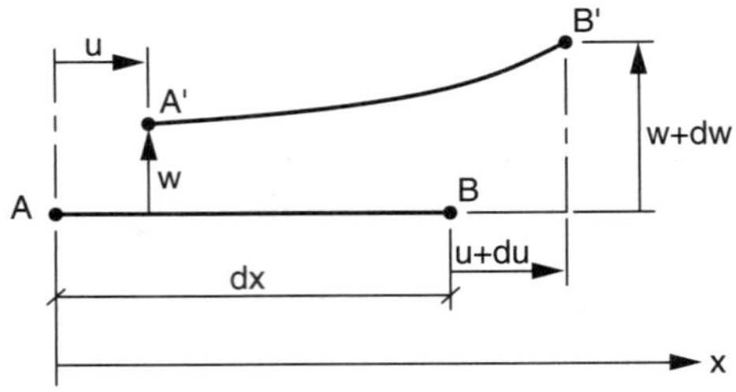

The direct strain ε_x in the x direction is given by

$$\varepsilon_x = \frac{(u+du)-u}{dx} = \frac{du}{dx} \tag{5.5a}$$

The direct strain ε_θ in the circumferential direction is, of course, the same as the strain of the radius. As dw is small in comparison with w, ε_θ is essentially given by

$$\varepsilon_\theta = \frac{w}{a} \tag{5.5b}$$

across the entire length dx of the element, where a is the radius of the cylinder.

The change in curvature χ_x along the meridian is approximately given by the rate of change of slope along the meridian – that is,

$$\chi_x \approx \frac{d}{dx}\left(\frac{dw}{dx}\right) = \frac{d^2w}{dx^2} \tag{5.6a}$$

Strictly speaking, χ_x is given by the relation

$$\chi_x = \frac{\dfrac{d^2w}{dx^2}}{\left\{1+\left(\dfrac{dw}{dx}\right)^2\right\}^{3/2}}$$

of curve geometry, but since deformations are small and $dw / dx << 1$, the approximation of equation (5.6a) is valid.

In general situations where deformations are not necessarily axisymmetric, the change in curvature along the circumferential direction, χ_θ, is similarly defined in terms of arc length s_θ along the circumferential direction:

$$\chi_\theta \approx \frac{\partial}{\partial s_\theta}\left(\frac{\partial w}{\partial s_\theta}\right) = \frac{\partial^2 w}{\partial s_\theta^2} = \frac{1}{a^2}\frac{\partial^2 w}{\partial \theta^2} \tag{5.6b}$$

However, in axisymmetric problems of present interest, $\chi_\theta = 0$ since the deformation w does not vary with respect to θ.

Now the relationship between direct strains and direct stresses is given by Hooke's law:

$$\varepsilon_x = \frac{1}{E}(\sigma_x - \nu\sigma_\theta) \tag{5.7a}$$

$$\varepsilon_\theta = \frac{1}{E}(\sigma_\theta - \nu\sigma_x) \tag{5.7b}$$

Rearrangement of these expressions gives the stresses as

$$\sigma_x = \frac{E}{1-\nu^2}(\varepsilon_x + \nu\varepsilon_\theta) \tag{5.8a}$$

$$\sigma_\theta = \frac{E}{1-\nu^2}(\varepsilon_\theta + \nu\varepsilon_x) \tag{5.8b}$$

The stresses may in turn be expressed in terms of the displacements u and w by using relations (5.5) to eliminate the strains. In this way, we obtain the stress resultants as follows:

$$N_x = \sigma_x t = \frac{Et}{1-\nu^2}\left(\frac{du}{dx} + \nu\frac{w}{a}\right) \tag{5.9a}$$

$$N_\theta = \sigma_\theta t = \frac{Et}{1-\nu^2}\left(\frac{w}{a} + \nu\frac{du}{dx}\right) \tag{5.9b}$$

It has already been pointed out that N_x of bending considerations may be taken as zero. Setting the right-hand side of equation (5.9a) to zero, we obtain

$$\frac{du}{dx} = -\nu\frac{w}{a} \tag{5.10}$$

We then use this result to eliminate du / dx from equation (5.9b), as follows

$$N_\theta = \frac{Et}{1-\nu^2}\left(\frac{w}{a} - \nu^2\frac{w}{a}\right) = \frac{Et}{1-\nu^2}\left(\frac{w}{a}\right)(1-\nu^2) = Et\,\frac{w}{a} \tag{5.11}$$

Let us consider the direct strains in the shell caused by a bending moment M_x. For our element of length dx, Figure 5.3 shows a portion of the shell thickness in longitudinal section, between the midsurface NS (for *neutral surface*, by analogy with the neutral axis of a beam) and the surface AB lying at a depth z ($z \le t/2$) below the midsurface. The fibre along NS is initially straight, but assumes a curvature $1/r_x$ due to the bending effect of M_x. Let the rotation of the right-hand end of a longitudinal fibre relative to the left-hand end be denoted by $d\phi$.

Before deformation, the fibre on the neutral surface is of length dx. This fibre remains of length dx during deformation (i.e. $ds = dx$), but as a result of this rotation of the shell element through an angle $d\phi$, the fibre at depth z below the neutral surface stretches by an amount e as illustrated in Figure 5.3. At level AB, we therefore have

Figure 5.3 Flexural strain in the shell thickness.

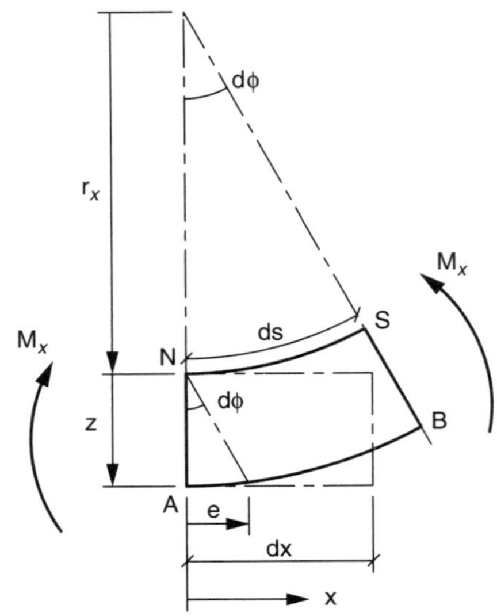

$$\varepsilon_x = \frac{e}{dx} = \frac{e}{ds} = \frac{e}{r_x\ d\phi}$$

From Figure 5.3, the following relationship is evident:

$$e = z \tan d\phi \approx z\ d\phi$$

Therefore,

$$\varepsilon_x = \frac{z\ d\phi}{r_x\ d\phi} = \frac{z}{r_x} = z\frac{1}{r_x} = z\,\chi_x \tag{5.12a}$$

where χ_x is the change of curvature of the shell element along the axial direction. Similarly, for bending in the hoop direction, the following relationship generally holds

$$\varepsilon_\theta = z\,\chi_\theta \tag{5.12b}$$

where χ_θ is the change of curvature of the shell element along the circumferential direction. In the present problem, $\chi_\theta = 0$ owing to axisymmetry, which means that

$$\varepsilon_\theta = 0 \tag{5.13}$$

Let us next consider the direct stresses across the shell thickness caused by a bending moment M_x (Figure 5.4). The element shown in longitudinal view on the left has unit width in the circumferential direction, as shown in the diagram on the right.

The elemental force at a depth z below the neutral surface, where the stress intensity is σ_x, is given by

$$dF_x = \sigma_x\, dA = \sigma_x\,(1\times dz)$$

Figure 5.4 Flexural stresses across the shell thickness.

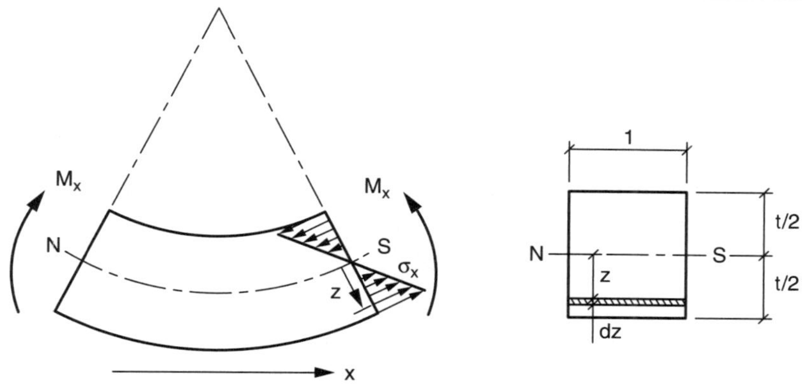

where dA is the elemental area shown shaded in the diagram on the right. The elemental moment dM_x of the elemental force about the midsurface is therefore

$$dM_x = z\ dF_x = z\ \sigma_x\ dz$$

The total moment M_x across the shell thickness is obtained by integration:

$$M_x = \int_{-t/2}^{+t/2} \sigma_x\ z\ dz \tag{5.14}$$

Replacing σ_x by the expression on the right-hand side of equation (5.8a), with ε_θ set equal to zero (result (5.13)), we may write

$$M_x = \frac{E}{1-\nu^2} \int_{-t/2}^{+t/2} \varepsilon_x\ z\ dz \tag{5.15}$$

Eliminating ε_x from the above equation, using relation (5.12a), we obtain

$$M_x = \frac{E}{1-\nu^2}\ \chi_x \int_{-t/2}^{+t/2} z^2\ dz = \frac{Et^3}{12(1-\nu^2)}\ \chi_x = D\,\chi_x \tag{5.16}$$

where

$$D = \frac{Et^3}{12(1-\nu^2)} \tag{5.17}$$

is called the flexural rigidity of the shell.

Now χ_x is given in terms of displacement w by expression (5.6a). We may therefore finally write

$$M_x = D\frac{d^2w}{dx^2} \tag{5.18}$$

Proceeding in a similar manner for M_θ, we arrive at an expression analogous to equation (5.14):

$$M_\theta = \int_{-t/2}^{+t/2} \sigma_\theta\ z\ dz \tag{5.19}$$

Substituting for σ_θ, using the right-hand side of equation (5.8b) with ε_θ set equal to zero as already explained, we obtain

$$M_\theta = \frac{E\nu}{1-\nu^2}\int_{-t/2}^{+t/2} \varepsilon_x z dz = \nu D \chi_x = \nu M_x \tag{5.20}$$

5.2.3 Bending equation in terms of *w*

From equation (5.4), we have

$$Q_x = \frac{dM_x}{dx} \tag{5.21}$$

Substituting this result into equation (5.3), we obtain

$$\frac{d^2 M_x}{dx^2} + \frac{N_\theta}{a} = 0 \tag{5.22}$$

Eliminating M_x (using result (5.18)) and N_θ (using result (5.11)) from the above equation, and taking the shell thickness t as constant so that the flexural rigidity D is also constant, we get a differential equation in w:

$$D\frac{d^4 w}{dx^4} + \frac{Et}{a^2} w = 0 \tag{5.23}$$

or

$$\frac{d^4 w}{dx^4} + \frac{12\left(1-\nu^2\right)}{a^2 t^2} w = 0 \tag{5.24}$$

or

$$\frac{d^4 w}{dx^4} + 4\beta^4 w = 0 \tag{5.25}$$

where

$$\beta^4 = \frac{3\left(1-\nu^2\right)}{a^2 t^2} \tag{5.26}$$

Equation (5.25) is a fourth-order ordinary differential equation in w, with constant coefficients. It has the same appearance as the equation for a beam on an elastic foundation, whose homogeneous form is usually given as follows:[1]

$$\frac{d^4 w}{dx^4} + \left(\frac{k}{EI}\right) w = 0 \tag{5.27}$$

where EI is the flexural rigidity of the beam and k is the modulus of the foundation. It can be seen that the deflection of the cylinder is the same as that of a beam of unit width and depth t (so that $I = t^3/12$), resting on an elastic foundation whose k value is equal to $Et(1-\nu^2)/a^2$.

5.3. General solution

Following the standard procedure (refer to any text on ordinary differential equations), we assume a solution to equation (5.25) of the form

$$w = A\,e^{\lambda x}$$

so that

$$\frac{d^4 w}{dx^4} = A\,\lambda^4\,e^{\lambda x}$$

Substituting these expressions into the governing equation (5.25), we obtain

$$A\lambda^4 e^{\lambda x} + 4\beta^4 A e^{\lambda x} = 0$$

leading to the characteristic equation

$$\lambda^4 + 4\beta^4 = 0$$

with roots

$$\lambda = (-1)^{1/4}\,\sqrt{2}\beta = \pm\frac{1}{\sqrt{2}}(1 \pm i)\sqrt{2}\beta = \pm\,\beta(1 \pm i)$$

Thus, the general solution to equation (5.25) is the linear combination

$$w = A_1' e^{\beta(1+i)x} + A_2' e^{\beta(1-i)x} + A_3' e^{-\beta(1+i)x} + A_4' e^{-\beta(1-i)x}$$

where A_1', A_2', A_3' and A_4' are constants. This solution may readily be expressed in terms of exponential and trigonometric terms, as follows:

$$w = e^{\beta x}\left(A_1 \cos\beta x + A_2 \sin\beta x\right) + e^{-\beta x}\left(B_1 \cos\beta x + B_2 \sin\beta x\right) \tag{5.28}$$

where A_1, A_2, B_1 and B_2 are new constants to be determined from the boundary conditions at the ends of the cylinder. It must be emphasized that this solution is only valid in the case of cylindrical shells of constant thickness, for which the parameter β is constant. If the shell thickness varies only moderately (by not more than 25%) over the portion of the shell experiencing bending effects, an average value for β may be adopted with good accuracy.

5.4. The case of a long cylinder

Let us assume that an edge bending moment M_e and a transverse shearing force H_e, both per unit length of the edge circumference, are applied to the edge $x = 0$ of the cylinder, as depicted in Figure 5.5. The other edge is assumed to be at a sufficiently large distance that it does not have an appreciable influence on the response of the shell at $x = 0$.

The above assumption allows us to adapt the general solution for w so that it has only two constants of integration corresponding to the boundary conditions at $x = 0$. Now Saint-Venant's principle[2] states that a system of forces and moments statically equivalent to zero force and zero couple, applied to a small part of a body or structure, gives rise to stresses and strains that are localised to the vicinity of the applied loads. The self-equilibrating moments and forces in Figure 5.5, applied at one end of the shell, constitute such a

Figure 5.5 Long cylinder subjected to axisymmetric edge actions: (a) bending moment M_e per unit length of shell edge; (b) transverse shear force H_e per unit length of shell edge.

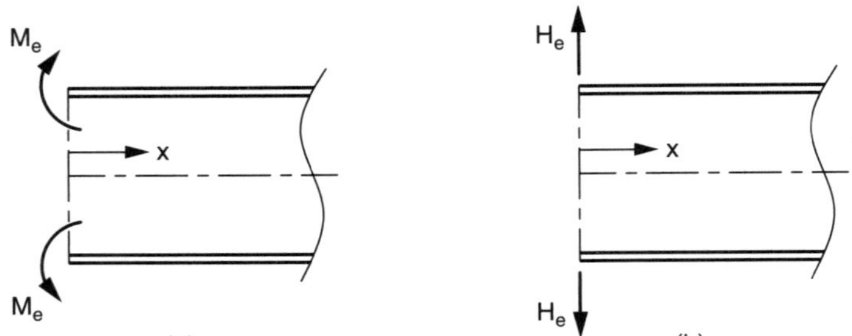

system. In expression (5.28), the part with $e^{\beta x}$ increases with increasing x, which is contrary to Saint-Venant's principle. On the other hand, the part with $e^{-\beta x}$ decays with increasing x, and is therefore consistent with Saint-Venant's principle. Thus for our solution, we take

$$w = e^{-\beta x}\left(B_1 \cos\beta x + B_2 \sin\beta x\right) \tag{5.29}$$

To obtain the values of the two constants B_1 and B_2, we need to consider the two boundary conditions at $x = 0$, for each of the cases of Figure 5.5. For the case shown in Figure 5.5(a), we have

$$\left(M_x\right)_{x=0} = D\left(\frac{d^2w}{dx^2}\right)_{x=0} = M_e \tag{5.30a}$$

$$\left(Q_x\right)_{x=0} = \left(\frac{dM_x}{dx}\right)_{x=0} = D\left(\frac{d^3w}{dx^3}\right)_{x=0} = 0 \tag{5.30b}$$

Now, from equation (5.29),

$$\frac{dw}{dx} = \beta e^{-\beta x}\left\{\left(B_2 - B_1\right)\cos\beta x - \left(B_2 + B_1\right)\sin\beta x\right\} \tag{5.31a}$$

$$\frac{d^2w}{dx^2} = -\,2\beta^2 e^{-\beta x}\left(B_2 \cos\beta x - B_1 \sin\beta x\right) \tag{5.31b}$$

$$\frac{d^3w}{dx^3} = 2\beta^3 e^{-\beta x}\left\{\left(B_1 + B_2\right)\cos\beta x + \left(B_2 - B_1\right)\sin\beta x\right\} \tag{5.31c}$$

Implementing boundary condition (5.30b), we obtain

$$B_1 + B_2 = 0 \quad\Rightarrow\quad B_1 = -B_2$$

Implementing boundary condition (5.30a), we obtain

$$B_2 = -\frac{M_e}{2\beta^2 D} \quad \left(= -B_1\right) \tag{5.32}$$

Substituting for B_1 and B_2 in equation (5.29), we obtain the deflection w_{M_e} due to the axisymmetric edge moment M_e as follows:

$$w_{M_e} = \frac{e^{-\beta x}}{2\beta^2 D}(\cos\beta x - \sin\beta x) M_e \tag{5.33}$$

For the case shown in Figure 5.5(b), the boundary conditions are

$$(M_x)_{x=0} = D\left(\frac{d^2 w}{dx^2}\right)_{x=0} = 0 \tag{5.34a}$$

$$(Q_x)_{x=0} = \left(\frac{dM_x}{dx}\right)_{x=0} = D\left(\frac{d^3 w}{dx^3}\right)_{x=0} = H_e \tag{5.34b}$$

Implementing boundary condition (5.34a) with the help of expression (5.31b), we obtain

$$B_2 = 0 \tag{5.35a}$$

Implementing boundary condition (5.34b) with the help of expression (5.31c), we get

$$B_1 = \frac{H_e}{2\beta^3 D} \tag{5.35b}$$

Thus, the deflection w_{H_e} due to the axisymmetric transverse force H_e is given by

$$w_{H_e} = \frac{e^{-\beta x}}{2\beta^3 D}(\cos\beta x)\, H_e \tag{5.36}$$

In the general case of M_e and H_e acting simultaneously over the edge $x = 0$, we superimpose their individual results, to obtain:

$$w = w_{M_e} + w_{H_e} = \frac{e^{-\beta x}}{2\beta^2 D}\left\{(\cos\beta x - \sin\beta x)\, M_e + \left(\frac{1}{\beta}\cos\beta x\right) H_e\right\} \tag{5.37}$$

The interior actions in the shell then follow as

$$N_x^b = 0 \tag{5.38a}$$

$$N_\theta^b = \frac{Etw}{a}$$

$$= \frac{Et}{a}\,\frac{e^{-\beta x}}{2\beta^2 D}\left\{(\cos\beta x - \sin\beta x)\, M_e + \left(\frac{1}{\beta}\cos\beta x\right) H_e\right\} \tag{5.38b}$$

$$M_x = D\frac{d^2 w}{dx^2}$$

$$= e^{-\beta x}\left\{(\cos\beta x + \sin\beta x)\, M_e + \left(\frac{1}{\beta}\sin\beta x\right) H_e\right\} \tag{5.39a}$$

$$M_\theta = \nu M_x$$

$$= \nu e^{-\beta x}\left\{(\cos\beta x + \sin\beta x)\, M_e + \left(\frac{1}{\beta}\sin\beta x\right) H_e\right\} \tag{5.39b}$$

with the superscript b in N_x^b and N_θ^b serving as a reminder that these stress resultants are associated with the bending edge effects (rather than the membrane solution).

Now the direct stress σ associated with a stress resultant N is simply N/t, where t is, of course, the shell thickness. This stress is uniform across the shell thickness. From simple bending theory, the maximum direct stress σ_{max} associated with a bending moment M is given by $\pm(M/I)z_{max}$, where z_{max} is the maximum value of z (the distance of a fibre from the shell midsurface). With z_{max} equal to $t/2$ and I (the second moment of area) for unit width of the shell being $t^3/12$, the maximum direct stress due to bending, which occurs on the surfaces of the shell, becomes $\pm 6M/t^2$. The net stresses on the inner and outer surfaces of the shell, due to the system of edge actions $\{M_e,\ H_e\}$ and interior membrane actions $\{N_x^m,\ N_\theta^m\}$, may therefore be written down as follows:

$$\sigma_x^T = \frac{N_x^m}{t} + \frac{N_x^b}{t} \pm \frac{6M_x}{t^2} = \frac{N_x^m}{t} \pm \frac{6M_x}{t^2} \tag{5.40a}$$

$$\sigma_\theta^T = \frac{N_\theta^m}{t} + \frac{N_\theta^b}{t} \pm \frac{6M_\theta}{t^2} = \frac{N_\theta^m}{t} + \frac{N_\theta^b}{t} \pm \frac{6\nu M_x}{t^2} \tag{5.40b}$$

recalling that $N_x^b = 0$ and $M_\theta = \nu M_x$.

The deformation δ has been defined in earlier chapters as the displacement perpendicular to the axis of revolution of the shell, positive when away from the axis of revolution. In the special case of the circular cylinder, δ coincides with w. At the edge $x = 0$ of the shell, the deformation δ_e^b due to the bending-related edge actions M_e and H_e is therefore given by

$$\delta_e^b = \left(w^b\right)_{x=0} = \frac{1}{2\beta^2 D}\left(M_e + \frac{1}{\beta}H_e\right) \tag{5.41a}$$

where w^b is given by expression (5.37), the superscript b now being used to distinguish the w caused by the bending-related actions M_e and H_e from the w due to membrane action (denoted by w^m).

The deformation V was defined in earlier chapters as the rotation of the meridian of a shell of revolution, the positive direction being as indicated in Section 2.4. In the special case of circular cylinders, V coincides with dw/dx. The edge rotation V_e^b due to the bending-related edge actions M_e and H_e is thus

$$V_e^b = \left(\frac{dw^b}{dx}\right)_{x=0} = -\frac{1}{2\beta D}\left(2M_e + \frac{1}{\beta}H_e\right) \tag{5.41b}$$

Superimposing membrane and bending-related effects, the net deformations δ_e^T and V_e^T at the edge of the shell are thus

$$\delta_e^T = \delta_e^m + \delta_e^b = \delta_e^m + \frac{1}{2\beta^2 D}\left(M_e + \frac{1}{\beta}H_e\right) \tag{5.42a}$$

$$V_e^T = V_e^m + V_e^b = V_e^m - \frac{1}{2\beta D}\left(2M_e + \frac{1}{\beta}H_e\right) \tag{5.42b}$$

where δ_e^m and V_e^m are the calculated membrane deformations at the shell edge ($x = 0$), corresponding to a specific loading over the surface of the shell. For a given problem, we have to impose appropriate boundary conditions in order to evaluate the two unknowns M_e and H_e. For example, if the edge is rigidly clamped, the set of conditions $\{\delta_e^T = 0,\ \ V_e^T = 0\}$ applies. Once M_e and H_e are known, all stresses and deformations in the shell become completely known through the closed-form results that have been developed. This will become clearer in Section 5.6, when we consider specific loading arrangements.

From the results that have been obtained in this section, we observe that edge effects, while oscillating rapidly between positive and negative, generally diminish exponentially with increasing distance from the edge of the shell. The amplitude of the oscillations is proportional to $e^{-\beta x}$. The higher the value of β, the more rapid is the decay of the amplitude with increasing distance x. Now $e^{-\beta x}$ falls to 5% of its initial value of unity (at $x = 0$) when $\beta x = 3$. For example, in the case of a concrete shell with $\nu = 0.15$, $a = 3$m and $t = 0.1$m (giving $\beta = 2.389\text{m}^{-1}$), the parameter $e^{-\beta x}$ diminishes from 1.00 (at $x = 0$) to 0.05 over a distance of $x = 1.26$m, a relatively short distance compared with the diameter of the cylinder.

We will arbitrarily define long shells as those for which the length l equals or exceeds $3/\beta$, and short shells as those for which $l < 3/\beta$. Short shells require the simultaneous consideration of all four boundary conditions at the two edges. For long shells, the two edges may be treated independently of each other without too much loss of accuracy. In the axisymmetric bending analysis of circular cylindrical shells, the maximum error of the long-shell approximation does not exceed the order of 5% provided that the shell is longer than $3/\beta$.

5.5. The case of a short cylinder

For short shells in general, all four constants of integration in expression (5.28) must be retained, since all four boundary conditions at the two edges ($x = 0$ and $x = l$) of the shell now significantly influence the stresses at any point on the shell. Consider a short cylinder subjected to axisymmetric bending moments and shear forces as shown in Figure 5.6.

The boundary conditions for the evaluation of the constants A_1, A_2, B_1 and B_2 are as follows:

$$\left(M_x\right)_{x=0} = D\left(\frac{d^2 w}{dx^2}\right)_{x=0} = M_1 \tag{5.43a}$$

Figure 5.6 Short cylinder subjected to axisymmetric bending moments and transverse shear forces at the two ends.

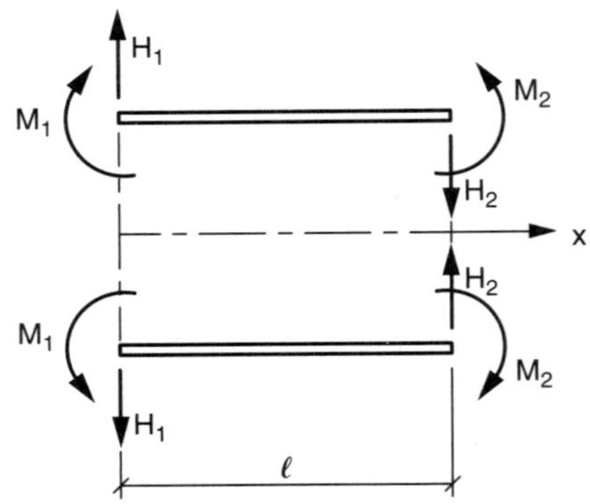

Figure 5.7 Decomposition of the loading shown in Figure 5.6 into symmetric and antisymmetric components.

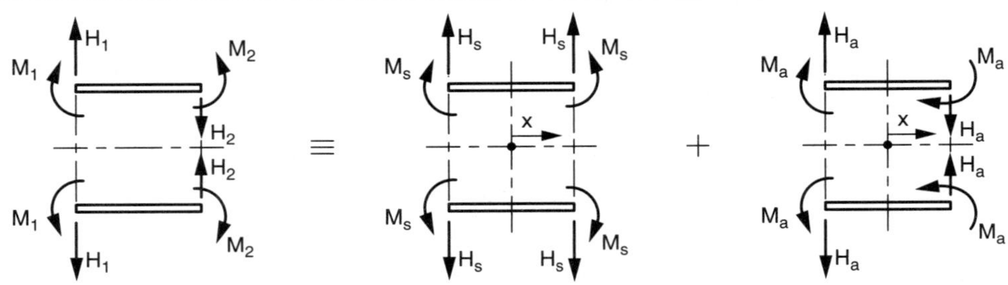

$$(Q_x)_{x=0} = D\left(\frac{d^3w}{dx^3}\right)_{x=0} = H_1 \qquad (5.43b)$$

$$(M_x)_{x=l} = D\left(\frac{d^2w}{dx^2}\right)_{x=l} = M_2 \qquad (5.43c)$$

$$(Q_x)_{x=l} = D\left(\frac{d^3w}{dx^3}\right)_{x=l} = H_2 \qquad (5.43d)$$

We obtain four simultaneous equations in the four unknowns $\{A_1,\ A_2,\ B_1,\ B_2\}$. The solution procedure is rather cumbersome. To simplify this, consider the scheme depicted in Figure 5.7, in which the edge loading $\{M_1,\ H_1,\ M_2,\ H_2\}$ has been decomposed into a symmetric system $\{M_s,\ H_s\}$ and an antisymmetric system $\{M_a,\ H_a\}$, such that

$$M_s = \frac{M_1 + M_2}{2} \qquad (5.44a)$$

$$H_s = \frac{H_1 - H_2}{2} \tag{5.44b}$$

$$M_a = \frac{M_1 - M_2}{2} \tag{5.45a}$$

$$H_a = \frac{H_1 + H_2}{2} \tag{5.45b}$$

This type of load decomposition takes advantage of the mirror symmetry of a cylindrical shell about the transverse plane through the middle of the cylinder, and may be regarded as a particular application of the more general group-theoretic techniques developed by the author for the analyses of structural configurations exhibiting symmetry properties.[3–5] We take the origin of the x axis to be the transverse plane of symmetry through the middle of the cylinder, as shown in Figure 5.7. Instead of expression (5.28), let us adopt the following alternative form for the general solution, where the coordinate x is now measured from the centre:

$$w = C_1 \sin\beta x \sinh\beta x + C_2 \sin\beta x \cosh\beta x + C_3 \cos\beta x \sinh\beta x + C_4 \cos\beta x \cosh\beta x \tag{5.46}$$

For the symmetric loading, the response w will also be symmetric about the plane $x = 0$, hence the expression for w must be an even function of x. Hence $C_2 = C_3 = 0$ and

$$w_s = C_1 \sin\beta x \sinh\beta x + C_4 \cos\beta x \cosh\beta x \tag{5.47}$$

The constants C_1 and C_4 are obtained from the conditions

$$(M_x)_{x=l/2} = D\left(\frac{d^2 w_s}{dx^2}\right)_{x=l/2} = M_s \tag{5.48a}$$

$$(Q_x)_{x=l/2} = D\left(\frac{d^3 w_s}{dx^3}\right)_{x=l/2} = -\ H_s \tag{5.48b}$$

For the antisymmetric loading, the response w will be antisymmetric about the centre plane $x = 0$, so that the expression for w must be an odd function of x. Hence $C_1 = C_4 = 0$ and

$$w_a = C_2 \sin\beta x \cosh\beta x + C_3 \cos\beta x \sinh\beta x \tag{5.49}$$

The constants C_2 and C_3 are evaluated from the boundary conditions

$$(M_x)_{x=l/2} = D\left(\frac{d^2 w_a}{dx^2}\right)_{x=l/2} = -\ M_a \tag{5.50a}$$

$$(Q_x)_{x=l/2} = D\left(\frac{d^3 w_a}{dx^3}\right)_{x=l/2} = H_a \tag{5.50b}$$

Note that, instead of solving a set of four simultaneous equations as in the original problem, the present scheme requires the solution of two independent sets of simultaneous equations, each in only two unknowns. This represents a considerable simplification of the problem.

Having solved the two independent systems for their respective constants, we may then write w_s in terms of M_s and H_s, and w_a in terms of M_a and H_a. The net deflection w of the combined symmetric and antisymmetric loading systems finally follows by superposition:

$$w = w_s + w_a \tag{5.51}$$

The symmetry-adapted actions $\{M_s, \ H_s\}$ and $\{M_a, \ H_a\}$ occurring in this last expression are eliminated by using the relations in equations (5.44) and (5.45), thereby casting w in terms of the original end variables $\{M_1, \ H_1, \ M_2, \ H_2\}$ of the cylindrical-shell problem.

Fortunately for many practical problems, the two ends of the cylinder are always sufficiently far apart to permit the decoupling of edge effects between the two ends, allowing each end to be treated independently (like the edge of a long cylinder).

5.6. Practical applications

5.6.1 Long cylinder with a radial line load

A radial ring load P (per unit length of circumference) acts on a long cylinder at a location that is remote from either end of the cylinder, as shown in Figure 5.8(a). Clearly, the deformation of the cylinder is symmetric about the transverse plane of the ring load, so that we may conceptually separate the cylinder into two portions each with an end load of $P/2$, and simply impose the condition of zero slope at the adjacent ends, as depicted in Figure 5.8(b).

Let us consider the portion on the right-hand side of the plane of symmetry SS. The H_e of previous generalised considerations, positive when pointing away from the axis of the cylinder for the left-hand end of the cylinder, now takes on the value $-P/2$. In order to maintain a configuration of zero slope at the edge, it is clear that a moment M_e, acting in the sense shown in Figure 5.8(b), is necessary. Taking the origin of the x axis as the plane of separation SS, the required value of M_e is determined by imposing the boundary condition of zero slope at $x = 0$. From equation (5.41b), with V_e^b set equal to zero, we obtain

Figure 5.8 Long cylinder with an axisymmetric radial line load P per unit length of circumference: (a) basic problem; (b) free-body diagram.

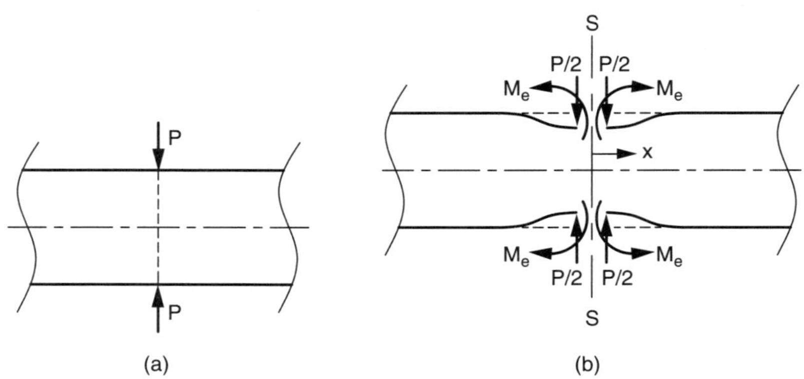

$$M_e = -\frac{1}{2\beta} H_e$$

$$= -\frac{1}{2\beta}\left(-\frac{P}{2}\right) = \frac{P}{4\beta} \qquad (5.52)$$

The deflection then follows from expression (5.37), with M_e as given by equation (5.52) and H_e set equal to $-P/2$:

$$w = -\frac{Pe^{-\beta x}}{8\beta^3 D}\left(\cos\beta x + \sin\beta x\right) \qquad (5.53)$$

The maximum deflection occurs at $x = 0$, and is given by

$$w_{\max} = \left(w\right)_{x=0} = -\frac{P}{8\beta^3 D} \qquad (5.54)$$

The maximum bending moment also occurs at $x = 0$, and is given by expression (5.39a) with x set equal to zero:

$$M_{\max} = \left(M_x\right)_{x=0} = M_e = \frac{P}{4\beta} \qquad (5.55)$$

5.6.2 Long cylinder with internal pressure

For a pressurised cylinder with built-in ends (i.e. rigidly clamped against rotation, transverse displacement and axial displacement), bending stresses are induced in the vicinity of the ends, owing to the constraining effect of the supports on the free deformation of the shell.

Figure 5.9(a) shows a pressurised cylinder with both ends rigidly built into the supports. The distance between the rigid faces of the supports is assumed to be fixed. For the membrane analysis of the vessel, we assume that the two circular edges of the cylinder can rotate freely and translate freely in the direction transverse to the axis of the cylinder (as if the two edges were pinned on rollers). Under these conditions, the resultants of the pressure forces on the circular ends of the enclosure are directly resisted by the rigid faces of the support; there is no membrane axial force transmitted into the shell walls because there is no relative

Figure 5.9 Long cylinder with built-in ends and subjected to uniform internal pressure p: (a) basic arrangement; (b) bending and shear actions at left end of cylinder.

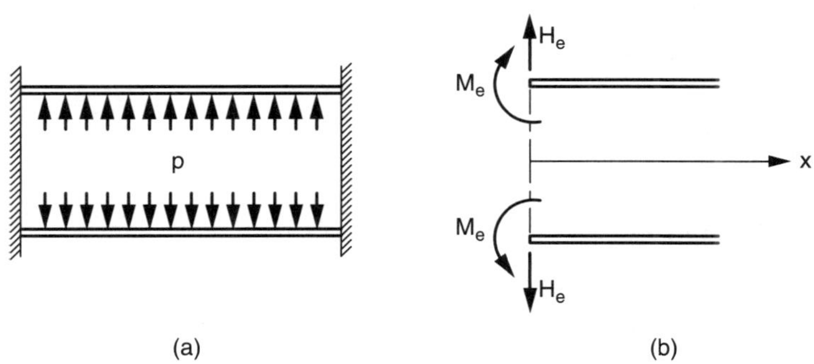

movement between the two opposite ends of the cylinder. From equation (3.2), it follows that $N_x^m = k = 0$. The membrane hoop stress resultant N_θ^m is given by expression (3.1).

The membrane deformations δ^m and V^m then follow from expressions (2.73) and (2.74):

$$\delta^m = w^m = \frac{pa^2}{Et} \tag{5.56a}$$

$$V^m = 0 \tag{5.56b}$$

These are independent of the coordinate x, as would be expected from the symmetry of the situation in Figure 5.9(a) when the shell edges are permitted to translate freely outwards.

Since in reality the shell can neither rotate nor move transversally at the built-in ends of the cylinder (Figure 5.9(a)), we have to supplement the membrane solution in the vicinity of the ends, by introducing axisymmetric bending actions M_e and H_e at the edges of the shell such that the net transverse displacements δ_e^T and net meridional rotations V_e^T are corrected to zero at the ends. Given the symmetry of the configuration, we need only consider one end of the cylinder ($x = 0$). From expressions (5.42), and using the results in equations (5.56), we have

$$\delta_e^T = \delta_e^m + \delta_e^b = \frac{pa^2}{Et} + \frac{1}{2\beta^2 D}\left(M_e + \frac{1}{\beta} H_e\right) = 0 \tag{5.57a}$$

$$V_e^T = V_e^m + V_e^b = 0 - \frac{1}{2\beta D}\left(2M_e + \frac{1}{\beta} H_e\right) = 0 \tag{5.57b}$$

Solving these two simultaneous equations, we obtain

$$M_e = \frac{2pa^2}{Et}\beta^2 D = \frac{p}{2\beta^2} \tag{5.58a}$$

$$H_e = -\frac{4pa^2}{Et}\beta^3 D = -\frac{p}{\beta} \tag{5.58b}$$

Substituting these values of M_e and H_e into expression (5.37) then gives the bending-related deflection w^b as

$$w^b = -\frac{p}{4\beta^4 D} e^{-\beta x}\left(\cos\beta x + \sin\beta x\right) \tag{5.59}$$

and the corresponding meridional bending moment as

$$M_x = \frac{p}{2\beta^2} e^{-\beta x}\left(\cos\beta x - \sin\beta x\right) \tag{5.60}$$

The analysis of pressurised cylinders joined to other shell geometries to form compound-shell pressure vessels is dealt with in Chapter 7.

5.6.3 Vertical water tank

As a last example, consider the tank of Section 3.2.1 (Figure 3.8). The membrane deformations at the bottom of the tank are given by expressions (3.52) and (3.53), with $x = H$ (height of tank):

$$\delta_e^m = \frac{\gamma a^2}{Et} H \tag{5.61a}$$

$$V_e^m = -\frac{\gamma a^2}{Et} \tag{5.61b}$$

For the purposes of computing the edge effects at the bottom of the tank, it is convenient to change the coordinate system of Figure 3.8, and now take the origin to be the bottom of the tank.

If the walls of the tank are rigidly built into a rigid base, we impose the conditions that the net transverse displacement and the net meridional rotation at the bottom of the tank must be zero. Using the results of equations (5.61) in expressions (5.42) and equating the net deformations to zero, we obtain

$$\delta_e^T = \delta_e^m + \delta_e^b = \frac{\gamma a^2 H}{Et} + \frac{1}{2\beta^2 D}\left(M_e + \frac{1}{\beta}H_e\right) = 0 \tag{5.62a}$$

$$V_e^T = V_e^m + V_e^b = -\frac{\gamma a^2}{Et} - \frac{1}{2\beta D}\left(2M_e + \frac{1}{\beta}H_e\right) = 0 \tag{5.62b}$$

Solving these two equations simultaneously, we obtain the results

$$M_e = 2D\beta(\beta H - 1)\frac{\gamma a^2}{Et} = \frac{\gamma H}{2\beta^2}\left(1 - \frac{1}{\beta H}\right) \tag{5.63a}$$

$$H_e = -2D\beta^2(2\beta H - 1)\frac{\gamma a^2}{Et} = -\frac{\gamma H}{2\beta^2}\left(2\beta - \frac{1}{H}\right) \tag{5.63b}$$

from which w^b and the internal shell actions associated with the bending edge effects then follow from expressions (5.37) and (5.38, 5.39). The total stresses within the shell are finally obtained by superposition of membrane and bending-related stresses (equations (5.40)) .

Other practical applications of the theory developed in this chapter may be seen in the text of Timoshenko and Woinowsky-Krieger.[6] Tabulated formulae for a wide range of problems of the axisymmetric bending of circular cylinders are contained in the reference work of Young,[7] while Ghali[8] presents comprehensive tables specifically for circular cylindrical tanks and silos.

References

1. Hetényi M. *Beams on Elastic Foundation*. The University of Michigan Press, Ann Arbor (Michigan), 1946.
2. Timoshenko S.P. and Goodier J.N. *Theory of Elasticity*. McGraw-Hill, New York, 1982.
3. Zingoni A., Pavlovic M.N. and Zlokovic G.M. A symmetry-adapted flexibility approach for multi-storey space frames: Symmetry-adapted loads. *Structural Engineering Review*, 1995, **7** (2), 121–130.
4. Zingoni A. A group-theoretic formulation for symmetric finite elements. *Finite Elements in Analysis and Design*, 2005, **41** (6), 615–635.

5. Zingoni A. *Vibration Analysis and Structural Dynamics for Civil Engineers: Essentials and Group-Theoretic Formulations*. CRC Press/Taylor & Francis, London, 2015.
6. Timoshenko S.P. and Woinowsky-Krieger S. *Theory of Plates and Shells*. McGraw-Hill, New York, 1959.
7. Young W.C. *Roark's Formulas for Stress and Strain*. McGraw-Hill, New York, 1989.
8. Ghali A. *Circular Storage Tanks and Silos*. E. & F.N. Spon, London, 1979.

Shell Structures in Civil and Mechanical Engineering
ISBN 978-0-7277-6028-9

http://dx.doi.org/10.1680/ssicame.60289.155

Chapter 6
Axisymmetric bending of general shells of revolution

6.1. Introduction

The membrane theory of general shells of revolution was developed in Chapter 2, and applied to specific shell geometries and load types in Chapter 3. To allow the assessment of edge effects around the locations at which the membrane hypothesis breaks down, a theory for the axisymmetric bending of circular cylinders was presented in Chapter 5. Owing to the geometric simplicity of the cylinder (curvature in one principal direction only), the obtained governing differential equation was of relatively simple form, permitting an exact solution to be written down for shells of constant thickness, and closed-form results presented for specific problems.

In the present chapter, we will develop a theory for the axisymmetric bending of general shells of revolution (i.e. those with double curvature). Unlike in the case of the circular cylinder, the governing differential equations will be relatively complex, making it difficult to obtain exact mathematical solutions for the general shell of revolution. However, by specialising the theory to simpler shell geometries, it is possible to obtain closed-form analytical solutions without too much difficulty, if certain approximations are introduced. Such solutions will be developed for the spherical shell and the conical shell. As will be explained at the end, the solution for the spherical shell may be used as a basis for estimating edge effects in general shells of revolution.

As in the case of the cylindrical shell, the approach that will be adopted is to consider the membrane solution as the particular integral of the complete bending-theory equations. The membrane solution is already known from the considerations of Chapters 2 and 3, so now we only need to find solutions for the homogeneous form of the bending-theory equations. The latter, with an adequate number of constants of integration, makes the total solution adaptable to the actual (as opposed to membrane) boundary conditions at the shell edge(s), and also accounts for the localised edge effects. Consistent with this approach, we will take the surface-loading components p_ϕ and p_r as zero (p_θ is already zero owing to axisymmetry), and proceed to find the solution for the edge-loaded shell (i.e. the homogeneous solution).

6.2. Derivation of the governing differential equations

6.2.1 Equilibrium considerations

Figure 6.1 shows an element of a shell of revolution experiencing axisymmetric bending. In addition to the direct stress resultants N_ϕ and N_θ, we now have a transverse shear force Q_ϕ acting in the meridional section (Q_θ being zero owing to axisymmetry), and bending moments M_ϕ and M_θ acting across opposite edges as shown in the figure, with the symbols denoting quantities per unit length of the associated edge.

Let us first consider equilibrium of the shell element in the direction of the tangent to a meridian. The contributions of the stress resultants N_ϕ and N_θ to the net force in the direction of increasing ϕ were obtained in

Figure 6.1 Element of a shell of revolution experiencing axisymmetric bending.

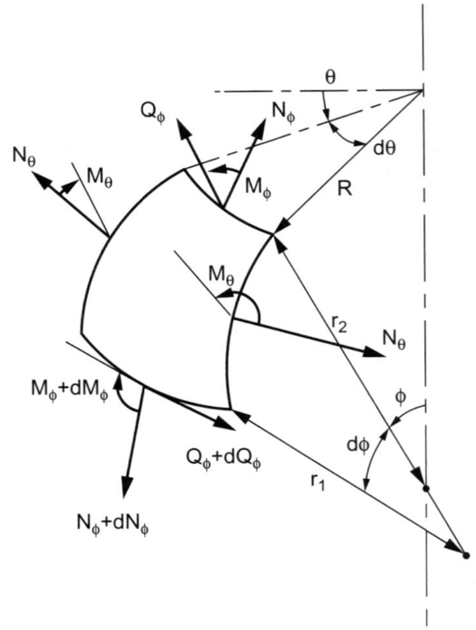

Section 2.1.3. All that is required now is to include the contribution of the transverse shear force Q_ϕ, which was absent in the membrane element of Figure 2.1.

From Figure 6.2, which shows a view of the element in the vertical (meridional) section, we may write down the resultant of the transverse shear forces in the direction *OT* as follows:

$$-Q_\phi R\, d\theta \sin\frac{d\phi}{2} - \left(Q_\phi + dQ_\phi\right)\left(R + dR\right) d\theta \sin\frac{d\phi}{2} \approx -RQ_\phi\, d\theta\, d\phi$$

Summing up the contributions of the N_ϕ, N_θ and Q_ϕ actions in the direction of the tangent to a meridian (see Section 2.1.3 for the contributions of the N_ϕ and the N_θ actions), we have

$$d\left(RN_\phi\right) d\theta - N_\theta r_1 \cos\phi\, d\phi\, d\theta - RQ_\phi\, d\phi\, d\theta = 0$$

Dividing throughout by $d\phi\, d\theta$, we obtain the equation

$$\frac{d\left(RN_\phi\right)}{d\phi} - r_1 N_\theta \cos\phi - RQ_\phi = 0 \tag{6.1}$$

Let us next consider equilibrium of the shell element in the direction of the normal to the shell midsurface. From Figure 6.2, the resultant of the transverse shear forces in the direction *ON* is

$$Q_\phi R\, d\theta \cos\frac{d\phi}{2} - \left(Q_\phi + dQ_\phi\right)\left(R + dR\right) d\theta \cos\frac{d\phi}{2} \approx -\left(Q_\phi\, dR + R\, dQ_\phi\right) d\theta = -d\left(RQ_\phi\right) d\theta$$

Figure 6.2 Contributions of Q_ϕ to net forces in the directions OT and ON: section of the element in the vertical plane of a meridian, showing transverse shears at upper and lower ends.

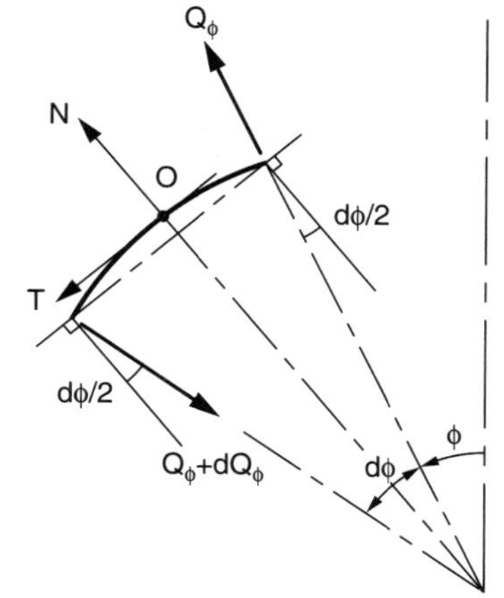

Summing up the contributions of the N_ϕ, N_θ and Q_ϕ actions in the direction of the normal to the shell midsurface (see Section 2.1.3 for the contributions of the N_ϕ and the N_θ actions), we have

$$-RN_\phi\, d\phi\, d\theta - r_1 \sin\phi\, N_\theta\, d\phi\, d\theta - d\left(RQ_\phi\right) d\theta = 0$$

which, upon dividing throughout by $d\phi\, d\theta$, becomes

$$RN_\phi + r_1 N_\theta \sin\phi + \frac{d\left(RQ_\phi\right)}{d\phi} = 0 \tag{6.2}$$

Finally, we consider moment equilibrium of the shell element in the vertical meridional section. By reference to both Figures 6.1 and 6.2, and taking clockwise moments as positive, we may write down the net moment, due to the M_ϕ and Q_ϕ actions, approximately as follows:

$$\left(M_\phi + dM_\phi\right)\left(R + dR\right) d\theta - M_\phi\, R\, d\theta - Q_\phi\, R\, d\theta\, r_1\, d\phi$$

$$\approx \left(M_\phi\, dR + R\, dM_\phi\right) d\theta - r_1 RQ_\phi\, d\phi\, d\theta$$

$$= d\left(RM_\phi\right) d\theta - r_1 RQ_\phi\, d\phi\, d\theta$$

The contribution of the M_θ actions (to moment equilibrium of the shell element in the meridional section) is less obvious. To visualise this, consider a view of the element along the normal to the shell midsurface, looking towards the axis of revolution of the shell (Figure 6.3). Bending actions over the four edges of the

Figure 6.3 Direct view of the shell element along the normal and looking towards the axis of revolution of the shell, showing bending actions as vectors.

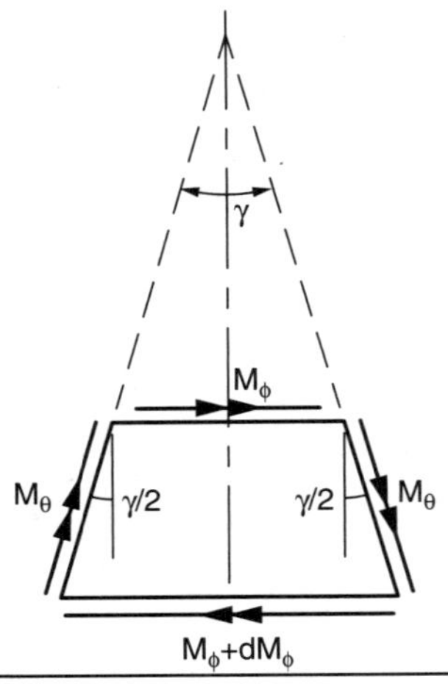

element are drawn as vectors, where the line of the vector represents the axis about which the associated moment acts, and the arrows of the vector indicate the direction of viewing (along this axis) for which the moment appears clockwise.

The resultant of the M_θ vectors in the direction of the $\left(M_\phi + dM_\phi\right)$ vector is clearly

$$-2\, M_\theta r_1\, d\phi \sin\frac{\gamma}{2} \approx -\, r_1 M_\theta\, \gamma\, d\phi$$

recalling that the length of the edges over which the M_θ actions act is $r_1\, d\phi$. Now, the angle γ is the same as appears in Figure 2.6(a), and is therefore equal to $(\cos\phi)\, d\theta$ (see Section 2.3.2). Therefore, the above expression for the contribution of M_θ becomes

$$-\, r_1 M_\theta\, (\cos\phi)\, d\theta\, d\phi$$

Summing up the contributions of the M_ϕ, Q_ϕ and M_θ actions in the meridional section, we have

$$d\left(RM_\phi\right) d\theta - r_1 RQ_\phi\, d\phi\, d\theta - r_1 M_\theta\, (\cos\phi)\, d\phi\, d\theta = 0$$

Dividing throughout by $d\phi\, d\theta$, we obtain

$$\frac{d\left(RM_\phi\right)}{d\phi} - r_1 M_\theta(\cos\phi) - r_1 RQ_\phi = 0 \tag{6.3}$$

The above equilibrium considerations have yielded three equations ((6.1)–(6.3)) in five unknowns N_ϕ, N_θ, M_ϕ, M_θ and Q_ϕ. We conclude from this that the problem of the axisymmetric bending of shells of revolution, like that for the axisymmetric bending of circular cylinders, is statically indeterminate. To obtain a solution, we need to invoke strain-displacement relations and Hooke's law.

6.2.2 Strain-displacement relations and Hooke's law

Axisymmetric displacement components of the shell of revolution, namely v in the direction of the tangent to the meridian and w in the direction of the normal to the shell midsurface, were discussed in Section 2.4. In that section, the following strain-displacement relations were established:

$$\varepsilon_\phi = \frac{1}{r_1}\left(\frac{dv}{d\phi} + w\right) \tag{6.4a}$$

$$\varepsilon_\theta = \frac{1}{r_2}\left(v\cot\phi + w\right) \tag{6.4b}$$

Direct strains are related to direct stresses through Hooke's law:

$$\varepsilon_\phi = \frac{1}{E}\left(\sigma_\phi - \nu\sigma_\theta\right) \tag{6.5a}$$

$$\varepsilon_\theta = \frac{1}{E}\left(\sigma_\theta - \nu\sigma_\phi\right) \tag{6.5b}$$

Rearranging these expressions, we obtain

$$\sigma_\phi = \frac{E}{1-\nu^2}\left(\varepsilon_\phi + \nu\varepsilon_\theta\right) \tag{6.6a}$$

$$\sigma_\theta = \frac{E}{1-\nu^2}\left(\varepsilon_\theta + \nu\varepsilon_\phi\right) \tag{6.6b}$$

and the stress resultants follow as

$$N_\phi = \sigma_\phi t = \frac{Et}{1-\nu^2}\left\{\frac{1}{r_1}\left(\frac{dv}{d\phi} + w\right) + \frac{\nu}{r_2}\left(v\cot\phi + w\right)\right\} \tag{6.7a}$$

$$N_\theta = \sigma_\theta t = \frac{Et}{1-\nu^2}\left\{\frac{1}{r_2}\left(v\cot\phi + w\right) + \frac{\nu}{r_1}\left(\frac{dv}{d\phi} + w\right)\right\} \tag{6.7b}$$

when use is made of relations (6.4) to eliminate ε_ϕ and ε_θ.

The bending moments M_ϕ and M_θ can also be expressed in terms of the basic displacements v and w. In Section 2.4, we obtained the following expression for the meridional rotation V:

$$V = \frac{1}{r_1}\left(v - \frac{dw}{d\phi}\right) \tag{6.8}$$

Figure 6.4 Effect of V on the meridional curvature χ_ϕ.

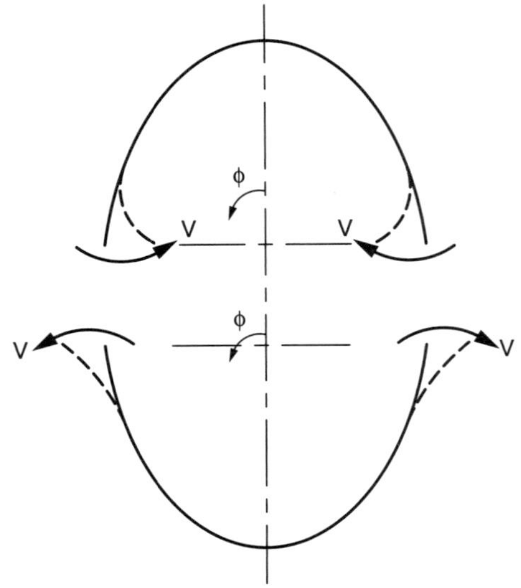

Now the change in meridional curvature χ_ϕ is approximately given by the rate of change of meridional rotation with distance along the meridian – that is, $|\chi_\phi| = |dV/ds_\phi|$. To deduce the correct sign of χ_ϕ relative to dV/ds_ϕ, onsider Figure 6.4, which shows a positive rotation V prescribed at the edges of shell caps closed over the top ($\phi = 0$) and over the bottom ($\phi = 180°$). As depicted in the figure, V (which of course is an axisymmetric deformation) is taken as positive when anticlockwise looking on the left side of the axis of revolution.

For the upper cap, and in accordance with Saint Venant's principle,[1] V tends to diminish in magnitude with increasing distance from the shell edge – that is, with decreasing ϕ, as shown. Therefore, $dV / d\phi$ is positive. However, the associated change of curvature χ_ϕ, which by definition has the same sign as M_ϕ, is clearly negative in the upper part. For the lower cap, V also tends to diminish with increasing distance from the shell edge (i.e. with increasing ϕ), so in this case, $dV / d\phi$ is negative. The associated change of curvature χ_ϕ is clearly positive (i.e. in the same sense as that due to a positive bending moment) in the lower part.

For both cases in Figure 6.4 (downward- and upward-facing shell edges), we have established that the sign of $dV / d\phi$ in the vicinity of the shell edge is opposite to that of the associated χ_ϕ. Thus, the relationship between χ_ϕ and V, taking proper account of the positive sign conventions, must read (since s_ϕ, by definition, increases in the same direction as ϕ)

$$\chi_\phi \approx -\frac{dV}{ds_\phi} \tag{6.9}$$

But $ds_\phi = r_1\, d\phi$. Therefore,

$$\chi_\phi \approx -\frac{dV}{r_1\, d\phi} = -\frac{1}{r_1}\frac{dV}{d\phi} = -\frac{1}{r_1}\frac{d}{d\phi}\left\{\frac{1}{r_1}\left(v - \frac{dw}{d\phi}\right)\right\} \tag{6.10}$$

when use is made of relation (6.8) to eliminate V.

Figure 6.5 Line element of shell in the meridional section, showing change of second principal radius of curvature r_2 due to meridional rotation V.

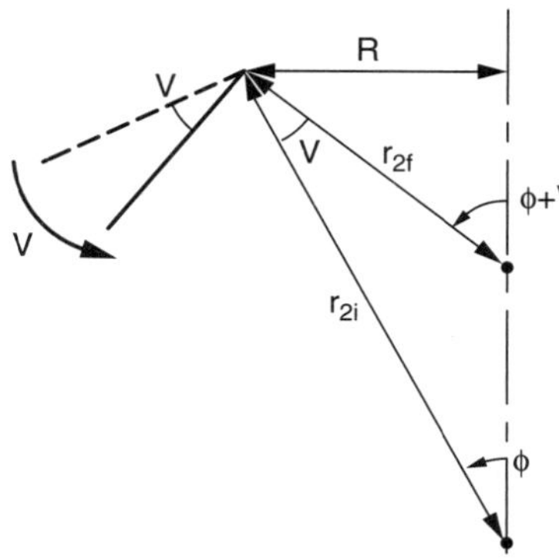

As a result of the rotation V, the second principal radius of curvature r_2 changes from an initial value of r_{2i} to a final value of r_{2f}, as illustrated in Figure 6.5. From the triangles in the diagram, we may write

$$r_{2i} \sin \phi = r_{2f} \sin(\phi + V) \quad (= R)$$

Now

$$\sin(\phi + V) = \sin \phi \cos V + \cos \phi \sin V \approx \sin \phi + V \cos \phi$$

(using the approximations $\cos V \approx 1$ and $\sin V \approx V$ for small rotations V). Therefore,

$$r_{2i} \sin \phi \approx r_{2f} (\sin \phi + V \cos \phi)$$

from which

$$r_{2f} = \frac{r_{2i} \sin \phi}{\sin \phi + V \cos \phi}$$

and hence

$$\frac{1}{r_{2f}} = \frac{\sin \phi + V \cos \phi}{r_{2i} \sin \phi} = \frac{1}{r_{2i}} + \frac{1}{r_{2i}} V \cot \phi$$

The change in the second principal curvature χ_θ is given by

$$\chi_\theta = -\left(\frac{1}{r_{2f}} - \frac{1}{r_{2i}} \right)$$

$$= -\left(\frac{1}{r_{2i}} + \frac{1}{r_{2i}} V \cot\phi - \frac{1}{r_{2i}}\right)$$

$$= -\frac{1}{r_{2i}} V \cot\phi \qquad (6.11)$$

the minus sign outside the brackets being required for consistency of positive sign conventions.

Eliminating V from the above expression (using relation (6.8)), and dropping the i in the subscript of r_{2i} in order to return to the original notation, we may now write

$$\chi_\theta = -\frac{1}{r_1 r_2}\left(v - \frac{dw}{d\phi}\right)\cot\phi \qquad (6.12)$$

Note that for the circular cylinder, $\phi = 90°$ throughout so that $\cot\phi = 0$. Therefore $\chi_\theta = 0$ throughout, a result consistent with the assumption that we made in Chapter 5.

For direct strains caused by bending, the following relations (similar to relations (5.12) for the cylindrical shell) hold:

$$\varepsilon_\phi = z\chi_\phi \qquad (6.13a)$$

$$\varepsilon_\theta = z\chi_\theta \qquad (6.13b)$$

Substituting the above relations into expressions (6.6), we obtain

$$\sigma_\phi = \frac{Ez}{1-\nu^2}\left(\chi_\phi + \nu\chi_\theta\right) \qquad (6.14a)$$

$$\sigma_\theta = \frac{Ez}{1-\nu^2}\left(\chi_\theta + \nu\chi_\phi\right) \qquad (6.14b)$$

The bending moments M_ϕ and M_θ are given by the following expressions, similar to relation (5.14) for the cylindrical shell:

$$M_\phi = \int_{-t/2}^{+t/2} \sigma_\phi \, z \, dz \qquad (6.15a)$$

$$M_\theta = \int_{-t/2}^{+t/2} \sigma_\theta \, z \, dz \qquad (6.15b)$$

where t is, of course, the shell thickness. Substituting into the above relations the results for σ_ϕ and σ_θ as given by equations (6.14), and performing the integrations, we obtain

$$M_\phi = D\left(\chi_\phi + \nu\chi_\theta\right) \qquad (6.16a)$$

$$M_\theta = D(\chi_\theta + \nu\chi_\phi) \tag{6.16b}$$

where D is as defined by equation (5.17).

Substituting the results for χ_ϕ and χ_θ (as given by equations (6.10) and (6.12) respectively) into the above relations, we finally obtain expressions for the bending moments in terms of the basic displacements:

$$M_\phi = -\frac{D}{r_1}\left[\frac{d}{d\phi}\left\{\frac{1}{r_1}\left(v - \frac{dw}{d\phi}\right)\right\} + \frac{\nu}{r_2}\left(v - \frac{dw}{d\phi}\right)\cot\phi\right] \tag{6.17a}$$

$$M_\theta = -\frac{D}{r_1}\left[\frac{1}{r_2}\left(v - \frac{dw}{d\phi}\right)\cot\phi + \nu\frac{d}{d\phi}\left\{\frac{1}{r_1}\left(v - \frac{dw}{d\phi}\right)\right\}\right] \tag{6.17b}$$

6.2.3 Reduction of the equilibrium equations

By substituting the expressions for N_ϕ and N_θ (equations 6.7) and those for M_ϕ and M_θ (equations 6.17) into the equilibrium equations ((6.1)–(6.3)), we can obtain three equations in three unknowns v, w and Q_ϕ, which, in principal, may then be solved in terms of these three variables. However, a simpler approach involves reducing equations (6.1)–(6.3) to two ordinary differential equations of second order, in the variables V and Q_ϕ, as follows:

We begin by writing expressions (6.17) as

$$M_\phi = -D\left\{\frac{1}{r_1}\frac{dV}{d\phi} + \frac{\nu}{r_2}V\cot\phi\right\} \tag{6.18a}$$

$$M_\theta = -D\left\{\frac{1}{r_2}V\cot\phi + \frac{\nu}{r_1}\frac{dV}{d\phi}\right\} \tag{6.18b}$$

where V is given by expression (6.8). Substituting these expressions for M_ϕ and M_θ into equation (6.3), we obtain, with R replaced by $r_2 \sin\phi$,

$$\frac{d}{d\phi}\left[(r_2\sin\phi)D\left\{\frac{1}{r_1}\frac{dV}{d\phi} + \frac{\nu}{r_2}V\cot\phi\right\}\right] - r_1 D\left\{\frac{1}{r_2}V\cot\phi + \frac{\nu}{r_1}\frac{dV}{d\phi}\right\}\cos\phi = -r_1 r_2(\sin\phi)Q_\phi$$

which on simplification becomes

$$\left[D\left(\frac{r_2}{r_1}\right)\sin\phi\right]\frac{d^2V}{d\phi^2} + \left[D\left(\frac{r_2}{r_1}\right)\cos\phi + \sin\phi\frac{d}{d\phi}\left(D\frac{r_2}{r_1}\right)\right]\frac{dV}{d\phi}$$

$$+\left[\nu\left(\cos\phi\frac{dD}{d\phi} - D\sin\phi\right) - D\left(\frac{r_1}{r_2}\right)\frac{\cos^2\phi}{\sin\phi}\right]V = -r_1 r_2(\sin\phi)Q_\phi \tag{6.19}$$

This second-order differential equation in V, with Q_ϕ appearing on the right-hand side only as a non-derivative term, is the first of the required governing equations. To obtain the second, we begin by establishing a relationship between N_ϕ and Q_ϕ. A comparison of equations (6.1) and (6.2) shows that we may eliminate N_θ between the two expressions by multiplying the first equation by $\sin\phi$ and the second equation by $\cos\phi$, and then adding the ensuing expressions. The result is

$$(\sin\phi)\frac{d(RN_\phi)}{d\phi}+(\cos\phi)RN_\phi-(\sin\phi)RQ_\phi+(\cos\phi)\frac{d(RQ_\phi)}{d\phi}=0$$

which may be written as

$$\frac{d}{d\phi}(RN_\phi\sin\phi)+\frac{d}{d\phi}(RQ_\phi\cos\phi)=0$$

implying that

$$RN_\phi\sin\phi=-RQ_\phi\cos\phi$$

– that is,

$$N_\phi=-Q_\phi\cot\phi \tag{6.20}$$

Substituting this result into equation (6.2) and rearranging, we obtain

$$\begin{aligned}N_\theta&=\frac{1}{r_1\sin\phi}\left[RQ_\phi\cot\phi-\frac{d(RQ_\phi)}{d\phi}\right]\\&=-\left(\frac{r_2}{r_1}\frac{dQ_\phi}{d\phi}+\frac{1}{r_1}\frac{dr_2}{d\phi}Q_\phi\right)\end{aligned} \tag{6.21}$$

The rotation V can be expressed in terms of N_ϕ and N_θ, as was shown in Section 2.4. We then substitute the above expressions for N_ϕ and N_θ (i.e. equations (6.20) and (6.21)) into the expression for V in terms of N_ϕ and N_θ (i.e. equation (2.72)), to obtain an expression for V in terms of Q_ϕ. The result is as follows:

$$\begin{aligned}&\frac{r_2^2}{tr_1}\frac{d^2Q_\phi}{d\phi^2}+\left[\frac{r_2^2}{tr_1}\cot\phi+\frac{r_2}{tr_1}\frac{dr_2}{d\phi}+\frac{d}{d\phi}\left(\frac{r_2^2}{tr_1}\right)\right]\frac{dQ_\phi}{d\phi}\\&\quad+\left[\frac{r_2}{tr_1}\frac{d^2r_2}{d\phi^2}+\left\{\frac{d}{d\phi}\left(\frac{r_2}{tr_1}\right)\right\}\frac{dr_2}{d\phi}+\left\{\left(\frac{r_2+\nu r_1}{r_1}\right)\frac{dr_2}{d\phi}-(r_1+\nu r_2)\cot\phi\right\}\frac{\cot\phi}{t}\right.\\&\quad\left.-\nu\left\{(\cot\phi)\frac{d}{d\phi}\left(\frac{r_2}{t}\right)-\left(\frac{r_2}{t}\right)\frac{1}{\sin^2\phi}\right\}\right]Q_\phi=r_1EV\end{aligned} \tag{6.22}$$

where the material properties E and ν are assumed to be constant throughout the shell.

Equations (6.19) and (6.22) are the required governing differential equations for the axisymmetric bending of general shells of revolution. They are often known as the Reissner–Meissner equations after the investigators who first developed these (see Chapter 1). Owing to the complexity of the coefficients of the differential terms, it would be extremely difficult to find a general solution for these equations. Only for special geometries – namely, the sphere, the circular torus, the circular cylinder and the circular cone – can solutions be obtained. In the next sections, we will develop practical solutions for the spherical shell and the conical shell, and extend these to more general situations.

6.3. Practical solutions for the spherical shell

6.3.1 Introduction

For the spherical shell, we set both r_1 and r_2 equal to a (radius of the sphere). Assuming, for the time being, that the shell thickness t is constant over the shell region of interest (so that the flexural rigidity D is also constant), we find that equations (6.19) and (6.22) simplify to the following:

$$\frac{d^2V}{d\phi^2} + (\cot\phi)\frac{dV}{d\phi} - \left(\nu + \cot^2\phi\right)V = -\frac{a^2}{D}Q_\phi \tag{6.23a}$$

$$\frac{d^2Q_\phi}{d\phi^2} + (\cot\phi)\frac{dQ_\phi}{d\phi} + \left(\nu - \cot^2\phi\right)Q_\phi = EtV \tag{6.23b}$$

A general solution for the above pair of equations can be obtained in the form of a hypergeometric series. The development of such a solution may be seen in earlier texts on shell theory.[2,3] In the present treatment, we will not pursue the rigorous solution of the above equations. As pointed out in Section 1.4.5.1, the hypergeometric-series solution is too cumbersome for practical calculations, and has the major disadvantage of poor convergence, particularly for shells of relatively high a/t ratios. We will develop two approximate solutions for the spherical shell, which are convenient for practical computations.

6.3.2 Geckeler's approximation

For thin spherical shells for which the angle ϕ is not small, it can be shown, through a study of the rigorous hypergeometric-series solution, that bending-related effects applied or prescribed at the shell edge, such as M_ϕ, Q_ϕ or V, decay rapidly with increasing distance from the edge. It follows, therefore, that in equations (6.23), the second derivatives (with respect to ϕ) of Q_ϕ and V are much larger than their corresponding first derivatives, and the first derivatives of Q_ϕ and V are, in turn, much larger than Q_ϕ and V themselves.

Thus, equation (6.23a) may be simplified by neglecting the V' and V terms on the left-hand side, and retaining only the V'' term. Here, a single prime denotes a first derivative with respect to ϕ, and a double prime denotes a second derivative. Similarly, equation (6.23b) may be simplified by neglecting the Q_ϕ' and Q_ϕ terms on the left-hand side, in relation to the Q_ϕ'' term. In this way, the following approximate relations are obtained:

$$\frac{d^2V}{d\phi^2} \approx -\frac{a^2}{D}Q_\phi \tag{6.24a}$$

$$\frac{d^2Q_\phi}{d\phi^2} \approx EtV \tag{6.24b}$$

Eliminating V from the above equations, we obtain the fourth-order ordinary differential equation

$$\frac{d^4 Q_\phi}{d\phi^4} + 4\lambda^4 Q_\phi = 0 \tag{6.25}$$

where

$$\lambda^4 = 3\left(1 - \nu^2\right)\left(\frac{a}{t}\right)^2 \tag{6.26}$$

The dimensionless quantity λ is called the *shell slenderness parameter*. Equation (6.25) is of the same form as that for the axisymmetric bending of the circular cylinder (equation 5.25), and hence its general solution may be taken as follows:

$$Q_\phi = e^{\lambda\phi}\left(C_1 \cos\lambda\phi + C_2 \sin\lambda\phi\right) + e^{-\lambda\phi}\left(C_3 \cos\lambda\phi + C_4 \sin\lambda\phi\right) \tag{6.27}$$

In the cases where the shell has a single edge rather than two edges, only two terms of the right-hand side of equation (6.27) are necessary. Considering a shell cap that is closed over the apex (i.e. around $\phi = 0$), we see that only the first two terms are relevant, since these are the ones that decay with increasing distance from the edge, as is required by Saint Venant's principle.[1] For such a shell cap, and letting $\psi = \phi_e - \phi$ (where, as before, ϕ_e is the meridional angle corresponding to the shell edge), the solution can be rewritten as follows:

$$Q_\phi = Ce^{-\lambda\psi} \sin\left(\lambda\psi + \beta\right) \tag{6.28}$$

where C and β are constants to be determined from the static boundary conditions at the edge.

Knowing Q_ϕ, the stress resultant N_ϕ follows from equation (6.20). Setting r_1 and r_2 in equation (6.21) equal to a, and noting that $dr_2 / d\phi$ is now zero, we obtain

$$N_\theta = -\frac{dQ_\phi}{d\phi} \tag{6.29}$$

The rotation V follows from the second of equations (6.24):

$$V \approx \frac{1}{Et}\frac{d^2 Q_\phi}{d\phi^2} \tag{6.30}$$

The bending moments M_ϕ and M_θ are obtained from relations (6.18) when terms in V are neglected in comparison with terms in $dV / d\phi$. With r_1 and r_2 set equal to a, we obtain

$$M_\phi \approx -\frac{D}{a}\frac{dV}{d\phi} \tag{6.31a}$$

$$M_\theta \approx -\nu\frac{D}{a}\frac{dV}{d\phi} = \nu M_\phi \tag{6.31b}$$

where V is as given by equation (6.30).

We will also require the lateral displacement δ. From equation (2.67), with $r_2 = a$, we have

$$\delta = \frac{a}{Et}(\sin\phi)\left(N_\theta - \nu N_\phi\right) \tag{6.32}$$

Comparing expressions (6.20) and (6.29), we observe that N_θ is proportional to the first derivative of Q_ϕ, while N_ϕ is proportional to Q_ϕ. Consistent with the accuracy of the present approach, we should, therefore, neglect N_ϕ in expression (6.32) in relation to N_θ (especially as the magnitude of the N_ϕ term is further reduced by Poisson's ratio ν). However, the approximation $N_\phi << N_\theta$ is not valid as $\phi \to 0$ or $\phi \to \pi$, since $\cot\phi$ then becomes large, so that $|Q_\phi \cot\phi|$ (magnitude of N_ϕ) becomes comparable with $|dQ_\phi / d\phi|$ (magnitude of N_θ).

That ϕ must not approach 0 or π too closely is also clearly evident from expressions (6.23), where $(\cot\phi)V'$ and $(\cot^2\phi)V$ become comparable with V'' (and therefore cannot be neglected), while $(\cot\phi)Q'_\phi$ and $(\cot^2\phi)Q_\phi$ become comparable with Q''_ϕ, when either ϕ itself or its supplement $\pi - \phi$ approaches zero.

Assuming, for the time being, that the shell is sufficiently *non-shallow* (exactly how non-shallow the shell ought to be will be discussed later), we may therefore write

$$\delta \approx \frac{a}{Et}(\sin\phi)N_\theta \tag{6.33}$$

where N_θ is given by equation (6.29).

The expressions for the various internal actions and deformations of interest are summarised below. In arriving at some of these, we make use of the differential relationship $d\phi = -\,d\psi$, which stems from the relationship between the coordinates: $\phi = \phi_s - \psi$. The superscript b in N^b_ϕ, N^b_θ, V^b and δ^b serves as a reminder that these quantities relate to the bending solution.

$$N^b_\phi = -\,Q_\phi \cot\phi$$

$$= -\left\{\cot\left(\phi_e - \psi\right)\right\} Ce^{-\lambda\psi}\sin\left(\lambda\psi + \beta\right) \tag{6.34}$$

$$N^b_\theta = -\frac{dQ_\phi}{d\phi}$$

$$= -\,\lambda\sqrt{2}\,Ce^{-\lambda\psi}\sin\left(\lambda\psi + \beta - \frac{\pi}{4}\right) \tag{6.35}$$

$$V^b \approx \frac{1}{Et}\frac{d^2Q_\phi}{d\phi^2}$$

$$= -\frac{2\lambda^2}{Et}Ce^{-\lambda\psi}\cos\left(\lambda\psi + \beta\right) \tag{6.36}$$

$$M_\phi \approx -\frac{D}{a}\frac{dV}{d\phi}$$

$$= \frac{a}{\lambda\sqrt{2}} Ce^{-\lambda\psi} \sin\left(\lambda\psi + \beta + \frac{\pi}{4}\right) \tag{6.37}$$

$$M_\theta \approx \nu M_\phi$$

$$= \frac{a\,\nu}{\lambda\sqrt{2}} Ce^{-\lambda\psi} \sin\left(\lambda\psi + \beta + \frac{\pi}{4}\right) \tag{6.38}$$

$$\delta^b \approx \frac{a}{Et}(\sin\phi)\, N_\theta^b$$

$$= -\frac{a}{Et}\lambda\sqrt{2}\left\{\sin(\phi_e - \psi)\right\} Ce^{-\lambda\psi} \sin\left(\lambda\psi + \beta - \frac{\pi}{4}\right) \tag{6.39}$$

6.3.3 Hetényi's improved approximation

We begin by introducing the following new variables:

$$\bar{Q} = Q_\phi\sqrt{\sin\phi} \tag{6.40a}$$

$$\bar{V} = V\sqrt{\sin\phi} \tag{6.40b}$$

Upon substituting these new variables into equations (6.23), it turns out that the terms containing the first derivatives of $\bar{Q}$ and $\bar{V}$ automatically vanish, so that only the terms containing $\bar{Q}$ and $\bar{V}$ need to be neglected in comparison with the terms containing the second derivatives of $\bar{Q}$ and $\bar{V}$. Equations (6.23) thus reduce to

$$\frac{d^2\bar{Q}}{d\phi^2} = Et\bar{V} \tag{6.41a}$$

$$\frac{d^2\bar{V}}{d\phi^2} = -\frac{a^2}{D}\bar{Q} \tag{6.41b}$$

By proceeding in the same manner as with Geckeler's approximation, and going back to the original notation, we finally obtain the following results for internal actions and deformations:

$$Q_\phi = \frac{e^{-\lambda\psi}}{\sqrt{\sin(\phi_e - \psi)}} C\sin(\lambda\psi + \beta) \tag{6.42}$$

$$N_\phi^b = -Q_\phi \cot\phi$$

$$= -\left\{\cot(\phi_e - \psi)\right\}\frac{e^{-\lambda\psi}}{\sqrt{\sin(\phi_e - \psi)}} C\sin(\lambda\psi + \beta) \tag{6.43}$$

$$N_\theta^b = -\frac{dQ_\phi}{d\phi}$$

$$= \frac{\lambda}{2}\frac{e^{-\lambda\psi}}{\sqrt{\sin(\phi_e - \psi)}} C\left\{2\cos(\lambda\psi+\beta) - (k_1 + k_2)\sin(\lambda\psi+\beta)\right\} \tag{6.44}$$

$$V^b \approx \frac{1}{Et}\left\{\frac{d^2Q_\phi}{d\phi^2} + (\cot\phi)\frac{dQ_\phi}{d\phi}\right\}$$

$$\approx -\frac{2\lambda^2}{Et}\frac{e^{-\lambda\psi}}{\sqrt{\sin(\phi_e - \psi)}} C\cos(\lambda\psi+\beta) \tag{6.45}$$

$$M_\phi = -\frac{D}{a}\left\{\frac{dV}{d\phi} + \nu V\cot\phi\right\}$$

$$= \frac{a}{2\lambda}\frac{e^{-\lambda\psi}}{\sqrt{\sin(\phi_e - \psi)}} C\left\{k_1\cos(\lambda\psi+\beta) + \sin(\lambda\psi+\beta)\right\} \tag{6.46}$$

$$M_\theta = -\frac{D}{a}\left\{V(\cot\phi) + \nu\frac{dV}{d\phi}\right\}$$

$$= \frac{a}{4\nu\lambda}\frac{e^{-\lambda\psi}}{\sqrt{\sin(\phi_e - \psi)}} C\left\{\left[(1+\nu^2)(k_1 + k_2) - 2k_2\right]\cos(\lambda\psi+\beta) + 2\nu^2\sin(\lambda\psi+\beta)\right\} \tag{6.47}$$

$$\delta^b = \frac{a}{Et}(\sin\phi)\left(N_\theta^b - \nu N_\phi^b\right)$$

$$= \frac{\lambda a}{Et}\left\{\sin(\phi_e - \psi)\right\}\frac{e^{-\lambda\psi}}{\sqrt{\sin(\phi_e - \psi)}} C\left\{\cos(\lambda\psi+\beta) - k_2\sin(\lambda\psi+\beta)\right\} \tag{6.48}$$

where

$$k_1 = 1 - \left(\frac{1-2\nu}{2\lambda}\right)\cot(\phi_e - \psi) \tag{6.49a}$$

$$k_2 = 1 - \left(\frac{1+2\nu}{2\lambda}\right)\cot(\phi_e - \psi) \tag{6.49b}$$

Note that the basic relations on the first lines of the N_ϕ^b and N_θ^b expressions are the same for the Geckeler and Hetényi approximations. However, for V^b, M_ϕ, M_θ and δ^b, the assumed relations differ between the two approximations (compare the first lines of equations (6.36)–(6.39) for the Geckeler approximation, versus the first lines of equations (6.45)–(6.48) for the Hetényi approximation), each relation being consistent with the accuracy of the associated approximation.

6.3.4 Evaluation of constants and edge deformations

Figure 6.6 shows a spherical shell subjected to an arbitrary bending moment M_e per unit length and a horizontal force H_e per unit length, both acting along the circular edge of the shell. By reference to this figure, the values of the constants C and β occurring in equations (6.28) and (6.34)–(6.39), or (6.42)–(6.48), may readily be established in terms of these actions, and the edge deformations resulting from the application of M_e or H_e written down.

For the application of M_e only (Figure 6.6a), the relevant static boundary conditions are

$$\left(M_\phi\right)_{\phi=\phi_e} = M_e \tag{6.50a}$$

$$\left(N_\phi^b\right)_{\phi=\phi_e} = 0 \tag{6.50b}$$

Considering the Geckeler approximation, the above conditions lead to the solutions

$$C = \frac{2\lambda}{a} M_e \tag{6.51a}$$

$$\beta = 0 \tag{6.51b}$$

so that the deformations caused at the edge by M_e become

Figure 6.6 Spherical shell subjected to axisymmetric edge actions: (a) bending moment M_e per unit length; (b) horizontal shear force H_e per unit length (axis of revolution of shell is vertical).

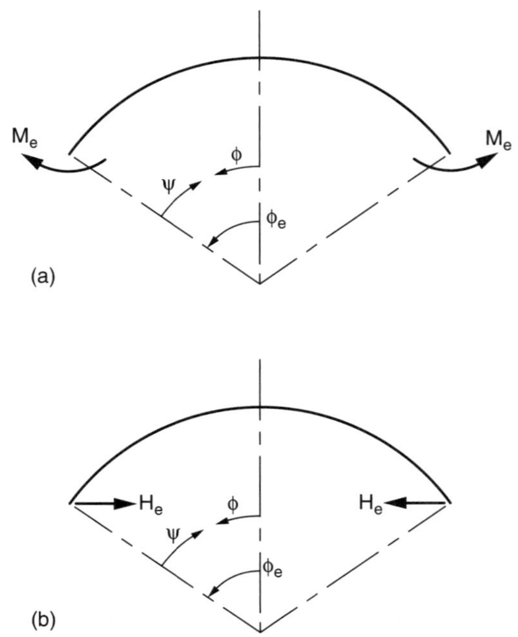

$$\left(V^b\right)_{\phi=\phi_e} = -\frac{4\lambda^3}{Eat} M_e \tag{6.52a}$$

$$\left(\delta^b\right)_{\phi=\phi_e} = \frac{2\lambda^2}{Et} \left(\sin\phi_e\right) M_e \tag{6.52b}$$

In the case of the Hetényi approximation, the same conditions lead to the solutions

$$C = \frac{2\lambda}{aK_1} \sqrt{\sin\phi_e}\; M_e \tag{6.53a}$$

$$\beta = 0 \tag{6.53b}$$

so that the deformations caused at the edge by M_e become

$$\left(V^b\right)_{\phi=\phi_e} = -\frac{4\lambda^3}{EatK_1} M_e \tag{6.54a}$$

$$\left(\delta^b\right)_{\phi=\phi_e} = \frac{2\lambda^2}{EtK_1} \left(\sin\phi_e\right) M_e \tag{6.54b}$$

where

$$K_1 = 1 - \left(\frac{1-2v}{2\lambda}\right) \cot\phi_e \tag{6.55}$$

When H_e is applied instead of M_e (Figure 6.6b), the boundary conditions change to

$$\left(M_\phi\right)_{\phi=\phi_e} = 0 \tag{6.56a}$$

$$\left(N_\phi^b\right)_{\phi=\phi_e} = -H_e \cos\phi_e \tag{6.56b}$$

For the Geckeler approximation, the above conditions lead to the solutions

$$C = -\sqrt{2}\; H_e \sin\phi_e \tag{6.57a}$$

$$\beta = -\frac{\pi}{4} \tag{6.57b}$$

so that the deformations caused at the edge by H_e become

$$\left(V^b\right)_{\phi=\phi_e} = \frac{2\lambda^2}{Et} \left(\sin\phi_e\right) H_e \tag{6.58a}$$

$$\left(\delta^b\right)_{\phi=\phi_e} = -\frac{2\lambda a}{Et}\left(\sin^2\phi_e\right)H_e \tag{6.58b}$$

For the Hetényi approximation, the same boundary conditions lead to the solutions

$$C = -\left\{\frac{\left(1+K_1^2\right)^{1/2}}{K_1}\right\}\left(\sin\phi_e\right)^{3/2} H_e \tag{6.59a}$$

$$\beta = -\tan^{-1}\left(K_1\right) \tag{6.59b}$$

so that the deformations caused at the edge by H_e become

$$\left(V^b\right)_{\phi=\phi_e} = \frac{2\lambda^2}{EtK_1}\left(\sin\phi_e\right) H_e \tag{6.60a}$$

$$\left(\delta^b\right)_{\phi=\phi_e} = -\frac{\lambda a}{Et}\left(K_2+\frac{1}{K_1}\right)\left(\sin^2\phi_e\right)H_e \tag{6.60b}$$

where

$$K_2 = 1-\left(\frac{1+2\nu}{2\lambda}\right)\cot\phi_e \tag{6.61}$$

For the simultaneous application of M_e and H_e, the net deformations V_e^b and δ_e^b at the edge of the shell are obtained by superimposing the effects of M_e and H_e. The results are:

Geckeler approximation

$$V_e^b = -\frac{4\lambda^3}{Eat}M_e + \frac{2\lambda^2}{Et}\left(\sin\phi_e\right)H_e \tag{6.62a}$$

$$\delta_e^b = \frac{2\lambda^2}{Et}\left(\sin\phi_e\right)M_e - \frac{2\lambda a}{Et}\left(\sin^2\phi_e\right)H_e \tag{6.62b}$$

which may be written in matrix form as

$$\begin{bmatrix} V_e^b \\ \delta_e^b \end{bmatrix} = \begin{bmatrix} -\dfrac{4\lambda^3}{Eat} & \dfrac{2\lambda^2}{Et}\left(\sin\phi_e\right) \\ \dfrac{2\lambda^2}{Et}\left(\sin\phi_e\right) & -\dfrac{2\lambda a}{Et}\left(\sin^2\phi_e\right) \end{bmatrix} \begin{bmatrix} M_e \\ H_e \end{bmatrix} \tag{6.63}$$

Hetényi approximation

$$V_e^b = -\frac{4\lambda^3}{EatK_1}M_e + \frac{2\lambda^2}{EtK_1}\left(\sin\phi_e\right)H_e \tag{6.64a}$$

$$\delta_e^b = \frac{2\lambda^2}{EtK_1}(\sin\phi_e)M_e - \frac{\lambda a}{Et}\left(K_2 + \frac{1}{K_1}\right)(\sin^2\phi_e)H_e \tag{6.64b}$$

which may be written in matrix form as

$$\begin{bmatrix} V_e^b \\ \delta_e^b \end{bmatrix} = \begin{bmatrix} -\dfrac{4\lambda^3}{EatK_1} & \dfrac{2\lambda^2}{EtK_1}(\sin\phi_e) \\ \dfrac{2\lambda^2}{EtK_1}(\sin\phi_e) & -\dfrac{\lambda a}{Et}\left(K_2 + \dfrac{1}{K_1}\right)(\sin^2\phi_e) \end{bmatrix} \begin{bmatrix} M_e \\ H_e \end{bmatrix} \tag{6.65}$$

The 2 × 2 square matrices on the right-hand sides of equations (6.63) and (6.65) may be regarded as *flexibility matrices*, since they link edge deformations $\{V_e^b, \delta_e^b\}$ to the edge actions $\{M_e, H_e\}$ causing them. Notice that these flexibility matrices are symmetrical about the top-left to bottom-right diagonal, consistent with the reciprocal theorem of elasticity.[1] The elements of the flexibility matrix of a given shell are often referred to in the literature as *influence coefficients*.

6.3.5 Evaluation of edge redundants and interior shell actions

When we superimpose membrane-solution and bending-related effects, the net deformations V_e^T and δ_e^T at the edge of the shell may be written as

$$V_e^T = V_e^b + V_e^m \tag{6.66a}$$

$$\delta_e^T = \delta_e^b + \delta_e^m \tag{6.66b}$$

where V_e^m and δ_e^m are the membrane-solution deformations at the edge of the shell, calculated for a specific loading over the surface of the shell, as illustrated in Chapter 3.

In matrix form, we may therefore write

$$\begin{bmatrix} V_e^T \\ \delta_e^T \end{bmatrix} = \begin{bmatrix} I_{11} & I_{12} \\ I_{21} & I_{22} \end{bmatrix} \begin{bmatrix} M_e \\ H_e \end{bmatrix} + \begin{bmatrix} V_e^m \\ \delta_e^m \end{bmatrix} \tag{6.67}$$

where the influence coefficients $\{I_{11}, I_{12}, I_{21}, I_{22}\}$ are the elements of the flexibility matrix of equation (6.63) or equation (6.65), depending on whether one is using the Geckeler or the Hetényi approximation.

For a given problem, in order to evaluate M_e and H_e, we need to impose appropriate boundary conditions at the shell edge. For instance, at a clamped edge of the shell, we must have $V_e^T = 0$ and $\delta_e^T = 0$; at the junction of two shells, for continuity, the V_e^T and δ_e^T of one shell must match the V_e^T and δ_e^T of the other shell – that is, $V_{e1}^T = V_{e2}^T$ and $\delta_{e1}^T = \delta_{e2}^T$, where subscripts 1 and 2 are used here to distinguish the two adjacent shells. In some instances, one or both of M_e and H_e may be statically determinate, making it unnecessary to set up expressions for V_e^T and/or δ_e^T. However, in the most general cases, we require expressions for both V_e^T and δ_e^T in order to solve for M_e and H_e via the relevant compatibility conditions.

Once M_e and H_e are known, the bending-related interior actions N_ϕ^b, N_θ^b, M_ϕ and M_θ are obtained by superimposing the effects of M_e and H_e. To achieve this superposition, we go back to expressions (6.34, 6.35, 6.37, 6.38) in the case of the Geckeler approximation or expressions (6.43, 6.44, 6.46, 6.47) in the case of the

Hetényi approximation, substitute the values of C and β for M_e or H_e as appropriate (see Section 6.3.4), and then add the ensuing expressions for the application of M_e to the corresponding expressions for the application of H_e. For the two approximations of interest, the results are as follows:

Geckeler approximation

$$N_\phi^b = -\left\{\cot(\phi_e - \psi)\right\} e^{-\lambda\psi} \left\{\frac{2\lambda}{a}(\sin\lambda\psi) M_e - (\sin\phi_e)(\sin\lambda\psi - \cos\lambda\psi) H_e\right\} \tag{6.68a}$$

$$N_\theta^b = -\, e^{-\lambda\psi} \left\{\frac{2\lambda^2}{a}(\sin\lambda\psi - \cos\lambda\psi) M_e + 2\lambda(\sin\phi_e)(\cos\lambda\psi) H_e\right\} \tag{6.68b}$$

$$M_\phi = e^{-\lambda\psi} \left\{(\sin\lambda\psi + \cos\lambda\psi) M_e - \frac{a}{\lambda}(\sin\phi_e)(\sin\lambda\psi) H_e\right\} \tag{6.68c}$$

$$M_\theta = \nu M_\phi \tag{6.68d}$$

Hetényi approximation

$$N_\phi^b = -\left\{\cot(\phi_e - \psi)\right\} \frac{e^{-\lambda\psi}}{\sqrt{\sin(\phi_e - \psi)}} \left[\frac{2\lambda}{aK_1}\sqrt{\sin\phi_e}\,(\sin\lambda\psi)\, M_e - \left(\frac{\sqrt{1+K_1^2}}{K_1}\right)(\sin\phi_e)^{3/2} \sin\left(\lambda\psi - \tan^{-1} K_1\right) H_e \right] \tag{6.69a}$$

$$N_\theta^b = \frac{\lambda}{K_1} \frac{\sqrt{\sin\phi_e}}{\sqrt{\sin(\phi_e - \psi)}} e^{-\lambda\psi} \left[\left\{2\cos\lambda\psi - (k_1 + k_2)\sin\lambda\psi\right\} \frac{\lambda}{a} M_e - \left(\frac{\sqrt{1+K_1^2}}{2}\right)(\sin\phi_e) \times \left\{2\cos\left(\lambda\psi - \tan^{-1} K_1\right) - (k_1 + k_2)\sin\left(\lambda\psi - \tan^{-1} K_1\right)\right\} H_e \right] \tag{6.69b}$$

$$M_\phi = \frac{1}{K_1} \frac{\sqrt{\sin\phi_e}}{\sqrt{\sin(\phi_e - \psi)}} e^{-\lambda\psi} \left[\left\{k_1\cos\lambda\psi + \sin\lambda\psi\right\} M_e - \frac{a}{\lambda}\left(\frac{\sqrt{1+K_1^2}}{2}\right)(\sin\phi_e) \times \left\{k_1\cos\left(\lambda\psi - \tan^{-1} K_1\right) + \sin\left(\lambda\psi - \tan^{-1} K_1\right)\right\} H_e \right] \tag{6.69c}$$

$$M_\theta = \frac{a}{4\nu\lambda}\frac{e^{-\lambda\psi}}{\sqrt{\sin(\phi_e-\psi)}}\left[\frac{2\lambda}{aK_1}\sqrt{\sin\phi_e}\times\left\{\left[(1+\nu^2)(k_1+k_2)-2k_2\right]\cos\lambda\psi+2\nu^2\sin\lambda\psi\right\}M_e\right.$$
$$-\left(\frac{\sqrt{1+K_1^2}}{K_1}\right)(\sin\phi_e)^{3/2}\times$$
$$\left.\left\{\left[(1+\nu^2)(k_1+k_2)-2k_2\right]\cos(\lambda\psi-\tan^{-1}K_1)+2\nu^2\sin(\lambda\psi-\tan^{-1}K_1)\right\}H_e\right] \quad (6.69d)$$

The total stresses in the shell, maximum on the inner and outer surfaces of the shell (where bending stresses are maximum), are then simply given by superimposing the membrane stresses with those stemming from the edge effects, as follows:

$$\sigma_\phi^T = \frac{N_\phi^m}{t}+\frac{N_\phi^b}{t}\pm\frac{6M_\phi}{t^2} \quad (6.70a)$$

$$\sigma_\theta^T = \frac{N_\theta^m}{t}+\frac{N_\theta^b}{t}\pm\frac{6M_\theta}{t^2} \quad (6.70b)$$

The upper and lower signs associated with the M_ϕ and M_θ terms refer to the inner and outer surfaces of the shell, respectively. In these final expressions, the actual numerical values of N_ϕ^b/t, N_θ^b/t, $6M_\phi/t^2$ and $6M_\theta/t^2$ depend on which approximation – Geckeler or Hetényi – is being used to assess the edge effects, while the membrane stresses N_ϕ^m/t and N_θ^m/t are, of course, independent of the method used to assess the edge disturbances.

As may be verified by inspection of the derived closed-form expressions, the terms N_θ^b/t and $6M_\phi/t^2$ are each larger than the term N_ϕ^b/t by a factor of the order of λ. Thus, when using the Geckeler approximation to assess edge effects, whether we neglect the contribution of N_ϕ^b to net meridional stresses, or take it into account, is of not much consequence to the accuracy of the results. However, when using the more accurate approach of Hetényi, the term N_ϕ^b/t must always be retained, as its neglect would affect the overall accuracy of the σ_ϕ^T stresses.

6.3.6 Results for various loadings and boundary conditions

6.3.6.1 Spherical shell with built-in edge

Figure 6.7 shows a spherical shell of radius a, provided with various support conditions at the edge $\phi=\phi_s$ (the subscript s denotes 'support'). When the edge of the shell is fixed, as shown in Figure 6.7(a), the boundary conditions are:

$$V_s^T = 0 \quad (6.71a)$$

$$\delta_s^T = 0 \quad (6.71b)$$

Expressions for membrane-solution deformations V_s^m and δ_s^m at the edge of the shell, for various surface loadings, were presented in Chapter 3. Expressions for deformations V_s^b and δ_s^b due to arbitrary edge actions M_s and H_s have been presented as equations (6.62) for the Geckeler approximation and equations (6.64) for the Hetényi approximation.

Combining membrane-solution and bending-related deformations, we obtain expressions for total deformations V_s^T and δ_s^T, in which M_s and H_s still appear as unknowns. Setting these expressions to zero (in accordance with the conditions of equations (6.71)) enables the unknown actions M_s and H_s to be solved for. Various load cases will now be considered.

Figure 6.7 Spherical shell with various support conditions: (a) built-in edge; (b) pinned edge on tangentially reacting rollers; (c) pinned edge on vertically reacting rollers; (d) pinned edge fully restrained against translation.

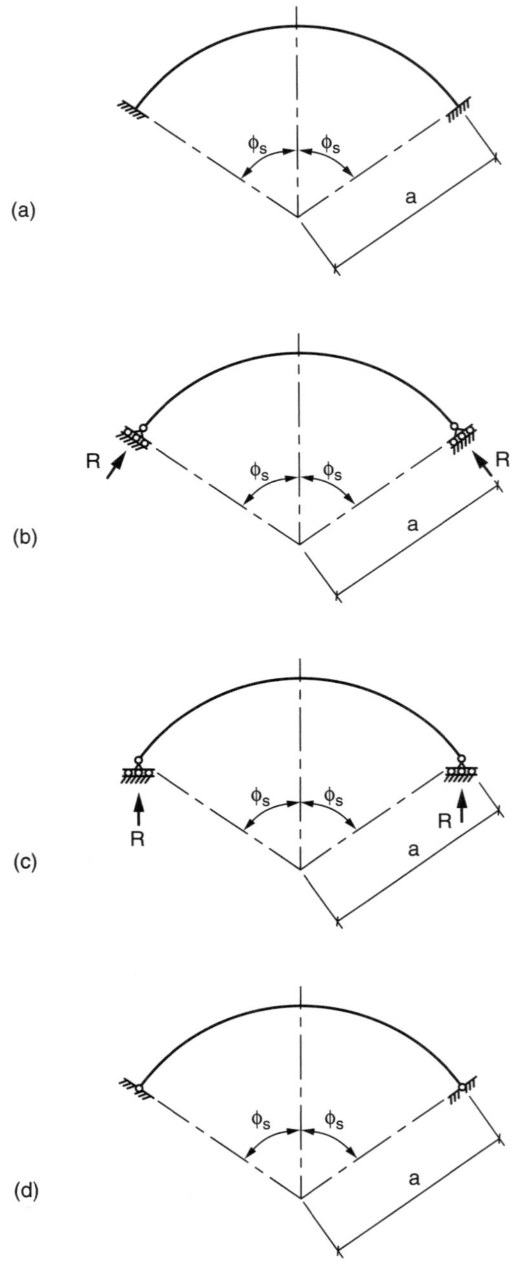

6.3.6.1.1 UNIFORM INTERNAL PRESSURE

As seen in Section 3.1.3, this loading gives rise to the following membrane stress resultants:

$$N_\phi^m = N_\theta^m = \frac{pa}{2} \tag{6.72}$$

Membrane-solution deformations V_s^m and δ_s^m at the edge of the shell are known from Section 3.1.3. Using these in the expanded expressions for V_s^T and δ_s^T, the corrective bending-related actions required at the support are obtained as follows:

Geckeler approximation

$$M_s = \frac{pa^2}{4\lambda^2}(1-\nu) \tag{6.73a}$$

$$H_s = \frac{pa}{2\lambda}\left(\frac{1-\nu}{\sin\phi_s}\right) \tag{6.73b}$$

Hetényi approximation

$$M_s = \frac{pa^2}{4\lambda^2}(1-\nu)\frac{1}{K_2} \tag{6.74a}$$

$$H_s = \frac{pa}{2\lambda}\left(\frac{1-\nu}{\sin\phi_s}\right)\frac{1}{K_2} \tag{6.74b}$$

6.3.6.1.2 HYDROSTATIC PRESSURE

If the shell cap is subjected to an internal hydrostatic pressure due to a retained liquid of weight γ per unit volume, such a pressure increasing linearly with depth of liquid from a value of zero at the apex ($\phi = 0$) to a maximum value at the base ($\phi = \phi_s$), then the membrane stress resultants are the same as were obtained in Section 3.2.3 for the upper domain of the liquid-filled spherical vessel – that is,

$$N_\phi^m = \frac{\gamma a^2}{6}\left(\frac{1-\cos\phi}{1+\cos\phi}\right)(1+2\cos\phi) \tag{6.75a}$$

$$N_\theta^m = \frac{\gamma a^2}{6}\left(\frac{1-\cos\phi}{1+\cos\phi}\right)(5+4\cos\phi) \tag{6.75b}$$

Membrane-solution deformations V_s^m and δ_s^m at the edge of the shell are given by expressions (3.67) and (3.68) with ϕ set equal to ϕ_s. Using these membrane-solution deformations in the expanded expressions for V_s^T and δ_s^T, and evaluating the unknown actions M_s and H_s (i.e. edge redundants) on the basis of conditions (6.71), we obtain

Geckeler approximation

$$M_s = \frac{\gamma a^3}{\lambda^2}\left[\frac{1}{12}\left(\frac{1-\cos\phi_s}{1+\cos\phi_s}\right)\{5+4\cos\phi_s - \nu(1+2\cos\phi_s)\} - \left(\frac{\sin\phi_s}{2\lambda}\right)\right] \tag{6.76a}$$

$$H_s = \frac{\gamma a^2}{\lambda}\left[\frac{1}{6}\left(\frac{1}{\sin\phi_s}\right)\left(\frac{1-\cos\phi_s}{1+\cos\phi_s}\right)\{5+4\cos\phi_s - \nu(1+2\cos\phi_s)\} - \left(\frac{1}{2\lambda}\right)\right] \tag{6.76b}$$

Hetényi approximation

$$M_s = \frac{\gamma a^3}{\lambda^2 K_2}\left[\frac{1}{12}\left(\frac{1-\cos\phi_s}{1+\cos\phi_s}\right)\{5+4\cos\phi_s - \nu(1+2\cos\phi_s)\} - \left(\frac{\sin\phi_s}{2\lambda}\right)\left(\frac{1+K_1K_2}{2}\right)\right] \tag{6.77a}$$

$$H_s = \frac{\gamma a^2}{\lambda K_2}\left[\frac{1}{6}\left(\frac{1}{\sin\phi_s}\right)\left(\frac{1-\cos\phi_s}{1+\cos\phi_s}\right)\{5+4\cos\phi_s - \nu(1+2\cos\phi_s)\} - \left(\frac{1}{2\lambda}\right)\right] \tag{6.77b}$$

6.3.6.1.3 VERTICAL LOADING UNIFORMLY DISTRIBUTED OVER SHELL SURFACE

This type of loading was dealt with in Section 3.3.2. The solutions obtained for the membrane stress resultants are

$$N_\phi^m = -qa\left(\frac{1}{1+\cos\phi}\right) \tag{6.78a}$$

$$N_\theta^m = qa\left(\frac{1}{1+\cos\phi} - \cos\phi\right) \tag{6.78b}$$

Membrane-solution deformations V_s^m and δ_s^m at the edge of the shell are given by expressions (3.217) with ϕ set equal to ϕ_s. Inserting these membrane-solution deformations in the expanded expressions for V_s^T and δ_s^T, and imposing conditions (6.71) to evaluate the edge redundants, we obtain the results

Geckeler approximation

$$M_s = \frac{qa^2}{\lambda^2}\left[\frac{1}{2}\left(\frac{1+\nu}{1+\cos\phi_s} - \cos\phi_s\right) - \left(\frac{2+\nu}{2\lambda}\right)\sin\phi_s\right] \tag{6.79a}$$

$$H_s = \frac{qa}{\lambda}\left(\frac{1}{\sin\phi_s}\right)\left[\left(\frac{1+\nu}{1+\cos\phi_s} - \cos\phi_s\right) - \left(\frac{2+\nu}{2\lambda}\right)\sin\phi_s\right] \tag{6.79b}$$

Hetényi approximation

$$M_s = \frac{qa^2}{\lambda^2 K_2}\left[\frac{1}{2}\left(\frac{1+\nu}{1+\cos\phi_s} - \cos\phi_s\right) - \left(\frac{1+K_1K_2}{2}\right)\left(\frac{2+\nu}{2\lambda}\right)\sin\phi_s\right] \tag{6.80a}$$

$$H_s = \frac{qa}{\lambda K_2}\left(\frac{1}{\sin\phi_s}\right)\left[\left(\frac{1+\nu}{1+\cos\phi_s} - \cos\phi_s\right) - \left(\frac{2+\nu}{2\lambda}\right)\sin\phi_s\right] \tag{6.80b}$$

6.3.6.2 Spherical shell with other support conditions

6.3.6.2.1 PINNED EDGE ON TANGENTIALLY REACTING ROLLERS

For this type of support, illustrated in Figure 6.7(b), the shell can move freely in the direction normal to its midsurface, and it can also rotate freely in the meridional plane, at its edge. The force reaction R (per unit length) at the support acts in a direction tangential to the shell midsurface, as shown in Figure 6.7(b). Assuming redundants M_s and H_s at the edge of the shell, and applying the static conditions that (*i*) the net moment must be zero at the edge, and (*ii*) the net force in the direction normal to the shell midsurface must be zero at the edge, we obtain

$$M_s = 0 \tag{6.81a}$$

$$H_s \sin\phi_s = 0 \tag{6.81b}$$

Hence, both M_s and H_s are zero, so no bending disturbance is present at, and in the vicinity of, the edge. In fact, the above equilibrium considerations show that, with this type of support, M_s and H_s will always be zero for any given surface loading. It follows, therefore, that a homogeneous solution or edge correction (whether based on the Geckeler approximation, the Hetényi approximation, or any other method) is not required for all values of ϕ_s, the stress situation within the shell being given solely by the particular solution, which in the present type of problems, is essentially the same as the membrane solution.

6.3.6.2.2 PINNED EDGE ON VERTICALLY REACTING ROLLERS

In this case, the supports are assumed to permit free rotation of the shell at its edge, and to permit free translation of the shell on the horizontal plane of support (Figure 6.7c). Thus, the rollers provide a force reaction R (per unit length) at the support that acts in the vertical direction. Assuming edge redundants M_s and H_s as before, and imposing the static conditions of (*i*) zero net moment at the shell edge, and (*ii*) zero net force in the horizontal direction at the shell edge, we have

$$M_s = 0 \tag{6.82a}$$

$$H_s - \left(N_\phi^m\right)_s \cos\phi_s = 0 \tag{6.82b}$$

where $\left(N_\phi^m\right)_s$ is the value of the membrane stress resultant N_ϕ^m at the shell edge. The moment redundant M_s is, of course, always zero for pinned support conditions, regardless of the specific loading applied over the surface of the shell. On the other hand, from the second of equations (6.82) with $\left(N_\phi^m\right)_s$ given by equation (6.78a) for $\phi = \phi_s$, the force redundant H_s is obtained as

$$H_s = -qa\left(\frac{\cos\phi_s}{1+\cos\phi_s}\right) \tag{6.83}$$

Quite clearly, this expression for H_s holds for both the Geckeler and the Hetényi approximation, since H_s has been deduced from purely equilibrium conditions at the edge of the shell.

6.3.6.2.3 PINNED EDGE FULLY RESTRAINED AGAINST TRANSLATION

The shell edge in Figure 6.7(d) is hinged to the support in such a way that it can rotate freely in the meridional plane, but is fully restrained against translation in any direction. The boundary conditions therefore comprise a static and a kinematic one:

$$M_s = 0 \tag{6.84a}$$

$$\delta_s^T = 0 \tag{6.84b}$$

From the first condition, which is an expression of moment equilibrium at the edge of the shell, we automatically see that no moment corrective edge action is required for this type of support, regardless of the type of applied surface loading. When the second condition is written in expanded form, with δ_s^m being given by equation (3.217a) when ϕ is set equal to ϕ_s, and M_s set equal to zero as already established, we readily solve for H_s to obtain

Geckeler approximation

$$H_s = \frac{qa}{2\lambda}\left(\frac{1}{\sin\phi_s}\right)\left(\frac{1+\nu}{1+\cos\phi_s} - \cos\phi_s\right) \tag{6.85}$$

Hetényi approximation

$$H_s = \frac{qa}{\lambda}\left(\frac{K_1}{1+K_1K_2}\right)\left(\frac{1}{\sin\phi_s}\right)\left(\frac{1+\nu}{1+\cos\phi_s} - \cos\phi_s\right) \tag{6.86}$$

6.3.7 Accuracy of the approximations of Hetényi and Geckeler

The Geckeler approximation provides a simple and highly convenient means for estimating the bending-disturbance stresses in thin spherical shells of constant thickness that are subjected to axisymmetric loading conditions. As this approximation forms the basis of many practical calculations involving spherical shells, it is desirable to know the degree of confidence that can be placed on the accuracy of the ensuing results, by assessing the magnitude of the errors inherent in Geckeler's simplification. While it constitutes only an approximation in the case of spherical shells and other doubly curved shells of revolution of positive Gaussian curvature, the Geckeler approximation happens to be exact for circular cylindrical shells. Thus, its accuracy for spherical shells (and other synclastic shells of revolution) would be expected to be good if the sides of the shell in the vicinity of the edge are sufficiently steep (i.e. near cylindrical).

Calladine and Paskaran[4] have noted that there is no appreciable difference between the solution of the exact equations and the approximation of Hetényi for $\lambda\phi > 2.5$. This criterion (for regarding the latter method as exact for practical purposes) yields a minimum value for ϕ of approximately 20° for λ approaching the value 7, which may be taken as the lower limit for thin shells for the purposes of the present study. In the same paper, it is observed that there is, in turn, little difference between the results of the Geckeler approximation and those of the approximation of Hetényi if $\lambda(\tan\phi) >> 1$. However, since it was not the aim of that work to consider the accuracy of these approximations in detail, the authors do not explicitly state by what factor $\lambda(\tan\phi)$ must be greater than unity for the difference between the two approximations to be considered negligible or, better still, as corresponding to a specified error.

Hoff[5] also observed that the Geckeler approximation is sufficiently accurate (no error estimate is given in explicit terms) in the range of half-angle of opening of the shell $30° \le \phi_e \le 90°$, provided $a/t > 100$. This limit upon a/t, suggested by Hoff in the light of the accuracy levels demanded in aerospace calculations, appears to be unnecessarily restrictive for civil-engineering applications.

The shells that are of present concern are all assumed to be relatively thin (with the lower limit for thinness arbitrarily taken as corresponding to $a/t \approx 30$ or $\lambda \approx 7$) and relatively non-shallow (i.e. $30° \le \phi_e \le 150°$). Now, within this range of shell thinness and non-shallowness, the approximation of Hetényi practically coincides with the exact hypergeometric-series solution. The excellent agreement between Hetényi's approximation and the exact series solution, even for unfavourable combinations of relatively low a/t ratios and relatively shallow shells, has been demonstrated[2,6] by consideration of a shell with $a/t = 30$ and $\phi_e = 35°$. Leckie[7] has put forward a very strict criterion for the acceptability of Hetényi's approximation, namely that $\lambda\phi > 3\sqrt{2}$. If we take the smallest value of λ consistent with our definition of a thin shell (i.e. $\lambda = 7$), we obtain $\phi \approx 35°$ as the minimum angle (in the range $0 \le \phi \le 90°$) beyond which the method of Hetényi becomes very accurate, according to Leckie's stringent criterion.

As regards the actual accuracy of Hetényi's approximation relative to the exact hypergeometric-series solution, Nielsen[8] gives the error of the approximation as $\rho^2/(1+\rho)$ relative to unity, where $\rho = (\cot\phi)/(\lambda\sqrt{2})$, which indicates that the error is only 2.6% for the most unfavourable combination of $\lambda = 7$ and $\phi = 30°$. According to Nielsen's estimate, the error would always be less than 1% if, for all values of λ in the range of interest (i.e. for $\lambda \ge 7$), the angle ϕ is at least equal to 44° (i.e. $44° \le \phi \le 90°$), or, for all values of ϕ in the range of interest (i.e. $30° \le \phi \le 90°$), the slenderness parameter λ exceeds 12. We conclude that Hetényi's approximation may be considered as practically exact, and hence a suitable basis for assessing the accuracy of the Geckeler approximation, provided that $a/t \ge 30$ and $30° \le \phi_e \le 150°$, which are the ranges of present interest. The inconvenience of the hypergeometric-series solution, especially with regard to convergence of the solution, has been noted in earlier sections.

Regardless of the actual loading over the surface of the shell and the specific boundary conditions at the edge, an order-of-magnitude estimate of the accuracy of the Geckeler approximation relative to the Hetényi approximation may be made on the basis of comparisons between the respective sets of the homogeneous-solution expressions.[9,10]

It is noted that if K_1 (equation 6.55) and K_2 (equation 6.61) are both equal to unity, then the coefficients for M_e in the Hetényi-approximation equations (6.54) become equal to those in the Geckeler-approximation equations (6.52), while the coefficients of H_e in the Hetényi-approximation equations (6.60) become equal to those in the Geckeler-approximation equations (6.58). From Section 6.3.4, these coefficients are, in fact, the influence coefficients for the respective edge actions. Thus, if K_1 and K_2 are both equal to 1.0, the influence coefficients of the two methods are identical, but if K_1 or K_2 or both are not equal to 1.0, then the influence coefficients of the two methods will differ.

However, equality of influence coefficients between the two approximations is not sufficient to make them coincide in every respect. A comparison of the Geckeler-approximation equations (6.28) and (6.34–6.39) versus their Hetényi-approximation counterparts (equations (6.42)–(6.48)) reveals that, even if their influence coefficients are identical, the two methods would only coincide at the shell edge, deviating, albeit to a very small extent, elsewhere. To see this, we compare equation (6.53a) versus (6.51a), and equation (6.59a) versus (6.57a), and note that the constant C for the Hetényi approximation exceeds its Geckeler-approximation counterpart by a factor of $(\sin\phi_e)^{1/2}$ if $K_1 = 1$. Using this result in the comparison of the equation set (6.28, 6.34–6.39) versus the equation set (6.42–6.48), we see that, away from the edge, all Hetényi-approximation quantities exceed their Geckeler-approximation counterparts by a factor of $\{\sin\phi_e / \sin(\phi_e - \psi)\}^{1/2}$ when K_2, like K_1, is assumed to be equal to unity. The difference is very small, being of the order of $\left|1 - \{\sin\phi_e / \sin(\phi_e - \psi)\}^{1/2}\right|$ in comparison with unity. For instance, the order of the difference is less than 1% within 2° of the edge, if within the range of interest ($30° \le \phi_e \le 150°$), the additional restriction $60° \le \phi_e \le 120°$ is applied. This difference is generally of minor importance in a comparison of the performance of the Geckeler approximation versus the Hetényi approximation, and may therefore be ignored.

Thus, we need only consider the deviations from unity of the parameters K_1 and K_2, as these provide a reasonable estimate of the discrepancy between the Geckeler approximation and the Hetényi approximation. This means that we may simply compare the influence coefficients of the two approximations. From this, we conclude that the error of the Geckeler approximation relative to the Hetényi approximation is of the order of $|1-K_1|$ or $|1-K_2|$ in comparison with unity – that is, $(1-2\nu)(\cot\phi_e)/(2\lambda)$ or $(1+2\nu)(\cot\phi_e)/(2\lambda)$. The larger error of $(1+2\nu)(\cot\phi_e)/(2\lambda)$ is the one that governs accuracy.

These simple considerations indicate that the error vanishes, for all practical purposes, as $\phi_e \to 90°$, or as $\lambda \to \infty$ (infinity), which is in keeping with the initial assumptions regarding the validity of the Geckeler formulation. If $\nu << 1$ (i.e. if the value of Poisson's ratio is nearly zero), then the above error expression becomes simply $(\cot\phi_e)/(2\lambda)$.

6.3.8 Edge interaction and decoupling errors in spherical shell frusta

6.3.8.1 Introduction

In the course of analysis of symmetrically loaded spherical shells of constant thickness, we sometimes encounter situations where the homogeneous part of the total solution has to be obtained by applying uniformly distributed actions along two edges that seem to be too close to each other to allow the set of boundary conditions at one edge to be treated independently of that at the other.[9,11]

One example is the case of a liquid-retaining spherical vessel (such as the tank shown in Figure 3.10) supported on closely spaced vertical columns through an interposing circular ring beam, such that the support conditions are essentially axisymmetric. As the vessel is filled, or, conversely, when the liquid contained in the vessel falls from full capacity, the free surface of the liquid (a source of bending effects owing to the first-derivative discontinuity in pressure-intensity variation with respect to ϕ) gradually approaches the plane of the supports (another source of bending disturbances, due to the concentrated vertical column reactions there, coupled with any lateral or rotational constraints imposed on the shell by reason of its connection to the ring beam).

Another instance where we might need to know the relative importance of interactive effects between two adjacent shell edges (before deciding whether or not to adopt the simplified analysis, which treats the two edges separately) is that of a spherical shell constructed as a series of segments of different thicknesses. Liquid-retaining steel vessels are typical examples, with the portion of the tank around the supports often being somewhat thicker than the remainder. In fact, it is not uncommon for the shell thickness in such containers to be changed in stepwise fashion at specific levels, since, where material savings outweigh increased fabrication costs (as in the case of multi-installations of large vessels of the type built by the Chicago Bridge and Iron Company[12]), such changes of shell thickness may result in smaller overall costs. Between adjacent thickness steps, which constitute sources of bending disturbances owing to the abrupt shell-slenderness change there, the thickness of the steel shell is constant. Equally common are large thin-walled steel vessels of constant thickness throughout, which, to control hoop deformations and/or the tendency for local buckling, are stiffened circumferentially along horizontal parallel circles of latitude, the locations of the stiffeners constituting sources of bending disturbances.

In all the above situations, a simple method of analysis such as the Geckeler approximation can still be used, with corrective edge actions in the form of axisymmetric bending moments and shearing forces being necessary at the edge(s) of each shell region (a region being any portion of the shell bound by one or two *lines of distortion* – see Section 1.5.3), in order to achieve continuity of shell deformations across the edges of the regions. However, if two edges happen to be too close to each other (i.e. if the angle subtended at the centre of curvature of the shell by the width of a region, in a meridional section of the shell, is relatively small), a knowledge of the magnitude of edge-interaction effects is desirable.

Edge-disturbance interaction may also be encountered in concrete domes. Although the thickness in these is either constant or made to vary smoothly, the provision of a skylight opening around the top produces an additional edge, thus raising the question of how significant interaction effects are between the upper edge (opening) and the lower edge (support line) of the dome. In these structures, any axisymmetric line loads also constitute sources of bending effects, and these sources may significantly interact between themselves or with adjacent external edges of the shell.

Within a region bounded by two parallel circles of latitude constituting sources of bending disturbances (such a region will henceforth be termed a *frustum* owing to its shape), the results of any interaction of bending effects between the two edges of the region cannot, in general, be deduced from a mere superposition of the distributions that are derived by considering each of the two sources separately; instead, all the four boundary conditions at the two edges of the region must be considered simultaneously from the outset. However, it is possible and, indeed, desirable to *decouple* the respective effects (i.e. to neglect interaction by considering the pair of boundary conditions at one edge independently of that at the other), if the loss in accuracy incurred by so doing is insignificant. The computational effort associated with the rigorous analysis that allows for edge interaction is generally several times greater than that associated with the simplified approach of decoupling the edge effects.

As regards previous studies of interaction phenomena in spherical-shell frusta, Stern and Tsui[13] have evaluated influence coefficients giving the deformations that occur at the edges of a thin spherical frustum, as a result of axisymmetric lateral forces and bending moments applied along these edges. Their results are based on both the exact hypergeometric-series solution, and an approximate but highly accurate asymptotic-series solution employing Bessel–Kelvin functions. On the basis of extensive plots of these coefficients, they observed the conditions under which the influence coefficients corresponding to the deformations at one (unloaded) edge due to actions applied at the other edge become negligible in relation to the influence coefficients corresponding to the deformations at the loaded edge. This enabled them to give certain guidelines on decoupling the bending-disturbance effects stemming from the two edges.

The guidelines of Stern and Tsui are essentially equivalent to stating that coupling is insignificant when $\lambda(\Delta\phi) \geq 4.24$, where $\Delta\phi$ is the angular separation of the bounding edges of the frustum. It must be pointed out, however, that they do not actually state the criterion in this simple form. Here, we have interpreted their criterion in terms of the parameter $\lambda(\Delta\phi)$. Thus, according to the guidance of Stern and Tsui, slenderness-parameter values of $\lambda = 7$ and $\lambda = 30$ (which we may consider as the lower and upper limits, respectively, of shell slenderness for most practical applications) require angular separations of at least $\Delta\phi \approx 35°$ and $\Delta\phi = 8°$, respectively, for decoupling of edge effects to be a valid approximation.

Figure 6.8 Spherical shell frustum under the combined action of M_1, H_1, M_2 and H_2.

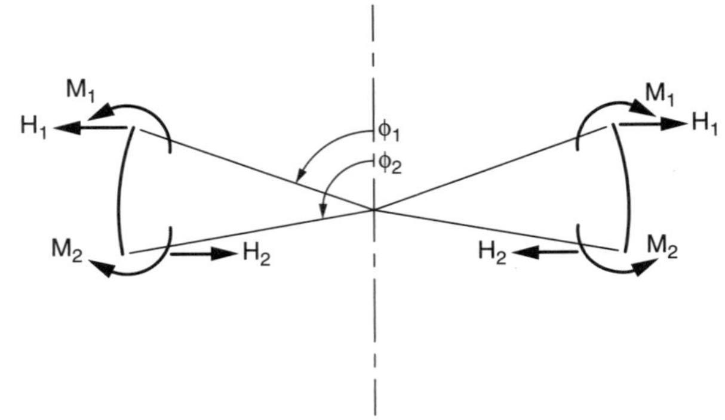

However, the authors do not appear to state the error associated with the criterion $\lambda(\Delta\phi) \geq 4.24$, and the errors associated with arbitrary values of the parameter $\lambda(\Delta\phi)$.

Kocsis[14] has observed that, for $a/t \geq 40$ and $\Delta\phi \geq 20°$, the lateral displacement δ and the meridional rotation V of the loaded edge of a two-edged spherical segment do not differ by more than 5% from those corresponding to the same edge if the shell does not have an opening at the other end. Such conclusions were deduced from parametric-form plots of influence coefficients derived on the basis of an approximate but relatively elaborate method employing Legendre functions. While $\Delta\phi = 20°$ would thus be the minimum separation of edges that permits the neglecting of edge interaction without exceeding an error of the order of 5% for a shell with $a/t = 40$, no generalised criterion giving the minimum $\Delta\phi$ consistent with the same error of 5% for *other* values of a/t appears to have been given, let alone minimum $\Delta\phi$ specifications for not only arbitrary a/t values, but also arbitrary permissible errors.

The present treatment of edge interaction in non-shallow spherical shell frusta is based on previous work of the author on this particular problem.[9,11] The shell edges are assumed to be sufficiently close to each other to require the simultaneous consideration of all the four static boundary conditions at the two edges. The meridional angles of the shell edges will be denoted by ϕ_1 (for the upper edge) and ϕ_2 (for the lower edge), so that the angular separation of the edges is given by $\Delta\phi = (\phi_2 - \phi_1)$. This arrangement is illustrated in Figure 6.8.

The approach is based on the Geckeler theory, whose accuracy (in terms of the shell slenderness parameter λ and meridional angle ϕ_e of the shell edge) has been discussed in Section 6.3.7. In the present cases of shell frusta, the smaller of ϕ_1 and $\pi - \phi_2$ is the ϕ_e value that governs the accuracy of the basic theory, together with the shell slenderness parameter λ (assumed to be constant). In the considerations of the present section, the accuracy of the *basic theory* (i.e. the Geckeler approximation) will be assumed to be adequate in itself, so that we need only concern ourselves with the errors of *neglecting edge interaction*, an approach that in itself constitutes an approximation even if the basic theory is exact (i.e. even if we base our theory on the hypergeometric-series solution).

We will derive influence-coefficient expressions that show the interaction component very clearly. We will quantify decoupling errors as *continuous* functions of a single fundamental parameter β $\left(= \lambda(\Delta\phi)\right)$, and establish decoupling criteria for various error tolerances.

6.3.8.2 Generalised interior actions and deformations

Starting with the fourth-order ordinary differential equation of Geckeler for the bending of a thin non-shallow spherical shell of constant thickness subjected to axisymmetric edge loads (i.e. equation (6.25)), and adopting the complete version of the general solution (i.e. retaining all the four constants of integration) as given by equation (6.27), the interior actions and deformations of interest are readily obtained as follows:

$$N_\phi = -Q_\phi \cot\phi$$

$$= -(\cot\phi)\left\{e^{\lambda\phi}\left(C_1\cos\lambda\phi + C_2\sin\lambda\phi\right) + e^{-\lambda\phi}\left(C_3\cos\lambda\phi + C_4\sin\lambda\phi\right)\right\} \tag{6.87a}$$

$$N_\theta = -\frac{dQ_\phi}{d\phi}$$

$$= -\lambda\left[e^{\lambda\phi}\left\{(C_1+C_2)\cos\lambda\phi - (C_1-C_2)\sin\lambda\phi\right\} - e^{-\lambda\phi}\left\{(C_3-C_4)\cos\lambda\phi + (C_3+C_4)\sin\lambda\phi\right\}\right] \tag{6.87b}$$

$$M_\phi \approx -\frac{a}{4\lambda^4}\frac{d^3Q_\phi}{d\phi^3}$$

$$= -\frac{a}{2\lambda}\left[e^{\lambda\phi}\left\{(C_2-C_1)\cos\lambda\phi - (C_1+C_2)\sin\lambda\phi\right\} + e^{-\lambda\phi}\left\{(C_3+C_4)\cos\lambda\phi + (C_4-C_3)\sin\lambda\phi\right\}\right] \tag{6.87c}$$

$$M_\theta \approx \nu M_\phi \tag{6.87d}$$

$$V = \frac{1}{Et}\frac{d^2Q_\phi}{d\phi^2}$$

$$= \frac{2\lambda^2}{Et}\left\{e^{\lambda\phi}\left(C_2\cos\lambda\phi - C_1\sin\lambda\phi\right) - e^{-\lambda\phi}\left(C_4\cos\lambda\phi - C_3\sin\lambda\phi\right)\right\} \tag{6.87e}$$

$$\delta \approx -\frac{a}{Et}(\sin\phi)\frac{dQ_\phi}{d\phi}$$

$$= -\frac{\lambda a}{Et}(\sin\phi)\left[e^{\lambda\phi}\left\{(C_1+C_2)\cos\lambda\phi - (C_1-C_2)\sin\lambda\phi\right\} - e^{-\lambda\phi}\left\{(C_3-C_4)\cos\lambda\phi + (C_3+C_4)\sin\lambda\phi\right\}\right] \tag{6.87f}$$

The constants C_1, C_2, C_3 and C_4 are obtained by considering the static boundary conditions at edges 1 and 2 when arbitrary edge actions M_1, H_1, M_2 and H_2 are applied simultaneously as shown in Figure 6.8. These boundary conditions are as follows:

$$\left(M_\phi\right)_{\phi=\phi_1} = M_1 \tag{6.88a}$$

$$\left(N_{\phi}\right)_{\phi=\phi_1} = -H_1 \cos\phi_1 \tag{6.88b}$$

$$\left(M_{\phi}\right)_{\phi=\phi_2} = M_2 \tag{6.88c}$$

$$\left(N_{\phi}\right)_{\phi=\phi_2} = -H_2 \cos\phi_2 \tag{6.88d}$$

After a rather lengthy solution procedure, these conditions lead to the following results for the constants $C_1,\ldots, C_4$:

$$\begin{aligned} C_1 = \frac{1}{K} e^{-\frac{\lambda}{2}(\phi_1+\phi_2)} \times \Big[& \left\{ -e^{\frac{1}{2}\beta}\left(2\sin\beta\sin\lambda\phi_2 + \sin\lambda\phi_1\right) + e^{-\frac{3}{2}\beta}\sin\lambda\phi_1 \right\} \frac{2\lambda}{a} M_1 \\ & + \left\{ e^{\frac{1}{2}\beta}\left\{\left(\sin\lambda\phi_1 - \cos\lambda\phi_1\right) - 2\left(\sin\beta\right)\left(\sin\lambda\phi_2 - \cos\lambda\phi_2\right)\right\} - e^{-\frac{3}{2}\beta}\left(\sin\lambda\phi_1 - \cos\lambda\phi_1\right)\right\}\left(\sin\phi_1\right) H_1 \\ & + \left\{ e^{-\frac{1}{2}\beta}\left(2\sin\beta\sin\lambda\phi_1 - \sin\lambda\phi_2\right) + e^{\frac{3}{2}\beta}\sin\lambda\theta_2 \right\} \frac{2\lambda}{a} M_2 \\ & + \left\{ e^{-\frac{1}{2}\beta}\left\{\left(\sin\lambda\phi_2 - \cos\lambda\phi_2\right) + 2\sin\beta\left(\sin\lambda\phi_1 - \cos\lambda\phi_1\right)\right\} - e^{\frac{3}{2}\beta}\left(\sin\lambda\phi_2 - \cos\lambda\phi_2\right)\right\}\left(\sin\phi_2\right) H_2 \Big] \end{aligned} \tag{6.89a}$$

$$\begin{aligned} C_2 = -\frac{1}{K} e^{-\frac{\lambda}{2}(\phi_1+\phi_2)} \times \Big[& \left\{ -e^{\frac{1}{2}\beta}\left(2\sin\beta\cos\lambda\phi_2 + \cos\lambda\phi_1\right) + e^{-\frac{3}{2}\beta}\cos\lambda\phi_1 \right\} \frac{2\lambda}{a} M_1 \\ & + \left\{ e^{\frac{1}{2}\beta}\left\{\left(\sin\lambda\phi_1 + \cos\lambda\phi_1\right) - 2\sin\beta\left(\sin\lambda\phi_2 + \cos\lambda\phi_2\right)\right\} - e^{-\frac{3}{2}\beta}\left(\sin\lambda\phi_1 + \cos\lambda\phi_1\right)\right\}\left(\sin\phi_1\right) H_1 \\ & + \left\{ e^{-\frac{1}{2}\beta}\left(2\sin\beta\cos\lambda\phi_1 - \cos\lambda\phi_2\right) + e^{\frac{3}{2}\beta}\cos\lambda\phi_2 \right\} \frac{2\lambda}{a} M_2 \\ & + \left\{ e^{-\frac{1}{2}\beta}\left\{\left(\sin\lambda\phi_2 + \cos\lambda\phi_2\right) + 2\left(\sin\beta\right)\left(\sin\lambda\phi_1 + \cos\lambda\phi_1\right)\right\} - e^{\frac{3}{2}\beta}\left(\sin\lambda\phi_2 + \cos\lambda\phi_2\right)\right\}\left(\sin\phi_2\right) H_2 \Big] \end{aligned} \tag{6.89b}$$

$$\begin{aligned} C_3 = \frac{1}{K} e^{\frac{\lambda}{2}(\phi_1+\phi_2)} \times \Big[& \left\{ -e^{-\frac{1}{2}\beta}\left(-2\sin\beta\sin\lambda\phi_2 + \sin\lambda\phi_1\right) + e^{\frac{3}{2}\beta}\sin\lambda\phi_1 \right\} \frac{2\lambda}{a} M_1 \\ & + \left\{ -e^{-\frac{1}{2}\beta}\left\{\left(\sin\lambda\phi_1 + \cos\lambda\phi_1\right) + 2\left(\sin\beta\right)\left(\sin\lambda\phi_2 + \cos\lambda\phi_2\right)\right\} + e^{\frac{3}{2}\beta}\left(\sin\lambda\phi_1 + \cos\lambda\phi_1\right)\right\}\left(\sin\phi_1\right) H_1 \end{aligned}$$

$$+\left\{-e^{\frac{1}{2}\beta}\left(2\sin\beta\sin\lambda\phi_1+\sin\lambda\phi_2\right)+e^{-\frac{3}{2}\beta}\sin\lambda\phi_2\right\}\frac{2\lambda}{a}M_2$$

$$+\left\{-e^{\frac{1}{2}\beta}\left\{\left(\sin\lambda\phi_2+\cos\lambda\phi_2\right)-2\sin\beta\left(\sin\lambda\phi_1+\cos\lambda\phi_1\right)\right\}+e^{-\frac{3}{2}\beta}\left(\sin\lambda\phi_2+\cos\lambda\phi_2\right)\right\}\left(\sin\phi_2\right)H_2\Bigg] \tag{6.89c}$$

$$C_4=\frac{1}{K}e^{\frac{\lambda}{2}(\phi_1+\phi_2)}\times\Bigg[\left\{e^{-\frac{1}{2}\beta}\left(-2\sin\beta\cos\lambda\phi_2+\cos\lambda\phi_1\right)-e^{\frac{3}{2}\beta}\cos\lambda\phi_1\right\}\frac{2\lambda}{a}M_1$$

$$+\left\{-e^{-\frac{1}{2}\beta}\left\{\left(\sin\lambda\phi_1-\cos\lambda\phi_1\right)+2(\sin\beta)\left(\sin\lambda\phi_2-\cos\lambda\phi_2\right)\right\}+e^{\frac{3}{2}\beta}\left(\sin\lambda\phi_1-\cos\lambda\phi_1\right)\right\}\left(\sin\phi_1\right)H_1$$

$$+\left\{e^{\frac{1}{2}\beta}\left(2\sin\beta\cos\lambda\phi_1+\cos\lambda\phi_2\right)-e^{-\frac{3}{2}\beta}\cos\lambda\phi_2\right\}\frac{2\lambda}{a}M_2$$

$$+\left\{-e^{\frac{1}{2}\beta}\left\{\left(\sin\lambda\phi_2-\cos\lambda\phi_2\right)-2\,(\sin\beta)\left(\sin\lambda\phi_1-\cos\lambda\phi_1\right)\right\}+e^{-\frac{3}{2}\beta}\left(\sin\lambda\phi_2-\cos\lambda\phi_2\right)\right\}\left(\sin\phi_2\right)H_2\Bigg] \tag{6.89d}$$

where

$$\beta=\lambda\left(\phi_2-\phi_1\right) \tag{6.90}$$

$$K=e^{2\beta}+e^{-2\beta}-2\left(1+2\sin^2\beta\right) \tag{6.91}$$

6.3.8.3 Influence coefficients and decoupling errors

The deformations at the shell edges (shown in Figure 6.9) are related to the edge actions causing them (shown in Figure 6.8) through a flexibility matrix of influence coefficients as follows:

Figure 6.9 Deformations V_1, δ_1, V_2 and δ_2 at the edges of the spherical shell frustum.

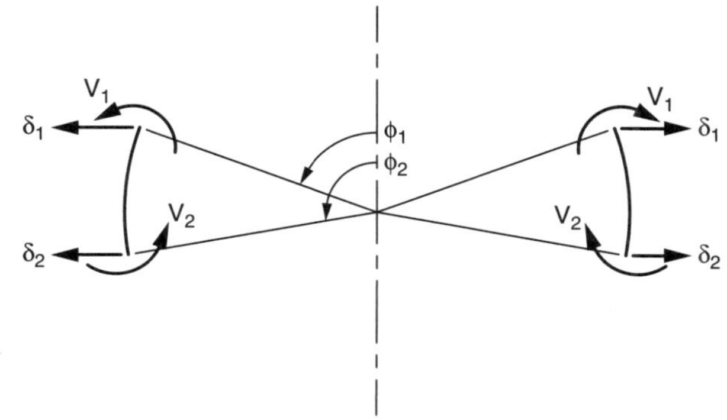

$$\begin{bmatrix} V_1 \\ \delta_1 \\ V_2 \\ \delta_2 \end{bmatrix} = \begin{bmatrix} I_{11} & I_{12} & I_{13} & I_{14} \\ I_{21} & I_{22} & I_{23} & I_{24} \\ I_{31} & I_{32} & I_{33} & I_{34} \\ I_{41} & I_{42} & I_{43} & I_{44} \end{bmatrix} \begin{bmatrix} M_1 \\ H_1 \\ M_2 \\ H_2 \end{bmatrix} \tag{6.92}$$

(The concept of influence coefficients was introduced in Section 6.3.4.) The four influence coefficients for each of the four actions M_1, H_1, M_2 or H_2 are generated by substituting the obtained expressions for the constants C_1, C_2, C_3 and C_4 (equations (6.89)), with the value of the action in question set equal to unity while the other three actions are set equal to zero, into equations (6.87e) and (6.87f), and evaluating the resulting deformation expressions for $\phi = \phi_1$ or $\phi = \phi_2$ as appropriate. The procedure is rather lengthy. However, the end results are surprisingly simple: the influence coefficients are essentially functions of only one fundamental variable β (the product of the shell slenderness parameter and the angular separation of the edges – see equation (6.90)). This greatly facilitates a study of the influence coefficients and, in particular, of the errors inherent in the simplified approach where edge interaction is neglected.

With this aim in view, the influence-coefficient expressions have been written in a form in which the basic component (corresponding to a total absence of interaction effects) and the additional correction (accounting for interaction effects) are clearly distinguishable. These two components correspond to the first and second terms, respectively, of the portion in the square brackets of the expressions below, the first being either 1 or 0:

$$I_{11} = \frac{4\lambda^3}{Eat}\left[1 + \frac{2}{K}\left\{1 + 2\sin^2\beta + \sin 2\beta - e^{-2\beta}\right\}\right] \tag{6.93a}$$

$$I_{12} = \frac{2\lambda^2}{Et}(\sin\phi_1)\left[1 + \frac{2}{K}\left\{1 + 2\sin^2\beta - \cos 2\beta\right\}\right] \tag{6.93b}$$

$$I_{13} = -\frac{4\lambda^3}{Eat}\left[0 + \frac{2}{K}\left\{e^{\beta}(\sin\beta + \cos\beta) + e^{-\beta}(\sin\beta - \cos\beta)\right\}\right] \tag{6.93c}$$

$$I_{14} = \frac{2\lambda^2}{Et}(\sin\phi_2)\left[0 + \frac{4}{K}(\sin\beta)\left(e^{\beta} - e^{-\beta}\right)\right] \tag{6.93d}$$

$$I_{21} = I_{12} \tag{6.93e}$$

$$I_{22} = \frac{2\lambda a}{Et}(\sin^2\phi_1)\left[1 + \frac{2}{K}\left\{1 + 2\sin^2\beta - \sin 2\beta - e^{-2\beta}\right\}\right] \tag{6.93f}$$

$$I_{23} = -\left(\frac{\sin\phi_1}{\sin\phi_2}\right) I_{14} \tag{6.93g}$$

$$I_{24} = \frac{2\lambda a}{Et}(\sin\phi_1)(\sin\phi_2)\left[0 + \frac{2}{K}\left\{e^{\beta}(\sin\beta - \cos\beta) + e^{-\beta}(\sin\beta + \cos\beta)\right\}\right] \tag{6.93h}$$

$$I_{31} = -I_{13} \tag{6.93i}$$

$$I_{32} = -I_{23} = \left(\frac{\sin\phi_1}{\sin\phi_2}\right) I_{14} \tag{6.93j}$$

$$I_{33} = -I_{11} \tag{6.93k}$$

$$I_{34} = \left(\frac{\sin\phi_2}{\sin\phi_1}\right) I_{12} \tag{6.93l}$$

$$I_{41} = -I_{14} \tag{6.93m}$$

$$I_{42} = -I_{24} \tag{6.93n}$$

$$I_{43} = I_{34} = \left(\frac{\sin\phi_2}{\sin\phi_1}\right) I_{12} \tag{6.93o}$$

$$I_{44} = -\left(\frac{\sin^2\phi_2}{\sin^2\phi_1}\right) I_{22} \tag{6.93p}$$

The percentage errors by which the deformations at the loaded edges are underestimated (relative to the true deformations) by using the influence coefficients for a single edge (i.e. by ignoring edge interaction, which is equivalent to assuming that the deformation response at the loaded edge of the frustum is the same as would occur there were the shell without the other opening), instead of the exact loaded-edge influence coefficients taking into account the boundary conditions of the other (unloaded) edge, will be denoted by e_1, e_2 and e_3. These errors correspond to the assessment of (*i*) V_i due to M_i, (*ii*) V_i due to H_i or δ_i due to M_i, and (*iii*) δ_i due to H_i, respectively, where $i = 1$ or 2 (i.e. errors e_1, e_2 and e_3 stem from the neglect of the interaction component in the expressions for $\{I_{11}, I_{33}\}$, $\{I_{12}, I_{34}, I_{21}, I_{43}\}$ and $\{I_{22}, I_{44}\}$, respectively).

When edge interaction is neglected, it is also inherently assumed, for the purposes of evaluating the redundant corrective actions, that actions applied at one edge do not contribute any deformations at the other edge. However, if the two edges are sufficiently close to each other, relatively substantial deformations occur at the second edge as a result of actions applied at the first. The ratios of the actual deformations at the far (i.e. unloaded) edge to the corresponding deformations at the edge at which the loading is applied (i.e. the near edge) will be denoted by e_4, e_5 and e_6. Thus, these ratios are an assessment of (*i*) V_j due to M_i relative to V_i due to M_i, (*ii*) V_j due to H_i relative to V_i due to H_i, or δ_j due to M_i relative to δ_i due to M_i, and (*iii*) δ_j due to H_i relative to δ_i due to H_i, respectively, where $i = 1$ or 2, $j = 1$ or 2, but $i \neq j$ (i.e. the ratios e_4, e_5 and e_6 are a relative measure of the negligibility of $\{I_{13}, I_{31}\}$, $\{I_{14}, I_{32}, I_{23}, I_{41}\}$ and $\{I_{24}, I_{42}\}$, respectively).

Thus,

$$e_1 = \frac{f_1}{1+f_1} \times 100\ \% \tag{6.94a}$$

$$e_2 = \frac{f_2}{1+f_2} \times 100\ \% \tag{6.94b}$$

$$e_3 = \frac{f_3}{1+f_3} \times 100\ \% \qquad (6.94c)$$

$$e_4 = \frac{f_4}{1+f_1} \qquad (6.95a)$$

$$e_5 = \frac{f_5}{1+f_2} \qquad (6.95b)$$

$$e_6 = \frac{f_6}{1+f_3} \qquad (6.95c)$$

where

$$f_1 = \frac{2}{K}\left(1 + 2\sin^2\beta + \sin 2\beta - e^{-2\beta}\right) \qquad (6.96a)$$

$$f_2 = \frac{2}{K}\left(1 + 2\sin^2\beta - \cos 2\beta\right) \qquad (6.96b)$$

$$f_3 = \frac{2}{K}\left(1 + 2\sin^2\beta - \sin 2\beta - e^{-2\beta}\right) \qquad (6.96c)$$

$$f_4 = \frac{2}{K}\left\{e^{\beta}\left(\sin\beta + \cos\beta\right) + e^{-\beta}\left(\sin\beta - \cos\beta\right)\right\} \qquad (6.97a)$$

$$f_5 = \frac{4}{K}\sin\beta\left(e^{\beta} - e^{-\beta}\right) \qquad (6.97b)$$

$$f_6 = \frac{2}{K}\left\{e^{\beta}\left(\sin\beta - \cos\beta\right) + e^{-\beta}\left(\sin\beta + \cos\beta\right)\right\} \qquad (6.97c)$$

The decoupling errors $e_1,\ldots,e_6$ are plotted in Figure 6.10.

With reference to Figure 6.10, it may be stated that in general, when $\beta < 2$, edge interaction ought to be considered under all circumstances, as the errors incurred by not doing so are always substantial (i.e. more than 10%). At the other extreme, when $\beta > 6$, there is no point in allowing for edge interaction under any conceivable practical circumstances, since the additional accuracy earned by such rigour would amount to less than 1%. However, when $2 < \beta < 6$, edge interaction may be neglected, depending on the boundary conditions and, of course, the magnitude of the errors that can be tolerated.

In the case of a spherical-shell frustum that does not require corrective actions along one of the two edges (i.e. one of the edges is free), then the curves e_1, e_2 and e_3 alone determine the decoupling criteria that allow the stresses in the vicinity of the loaded edge to be assessed independently of the boundary conditions at the free edge. Under these circumstances, and considering just the three plots e_1, e_2 and e_3, the decoupling criteria for various error tolerances are $\beta = 2.1$ (10% error), $\beta = 2.4$ (5% error), $\beta = 2.7$ (2% error) and $\beta = 2.9$ (1% error).

Figure 6.10 Variations of decoupling errors with the parameter β.

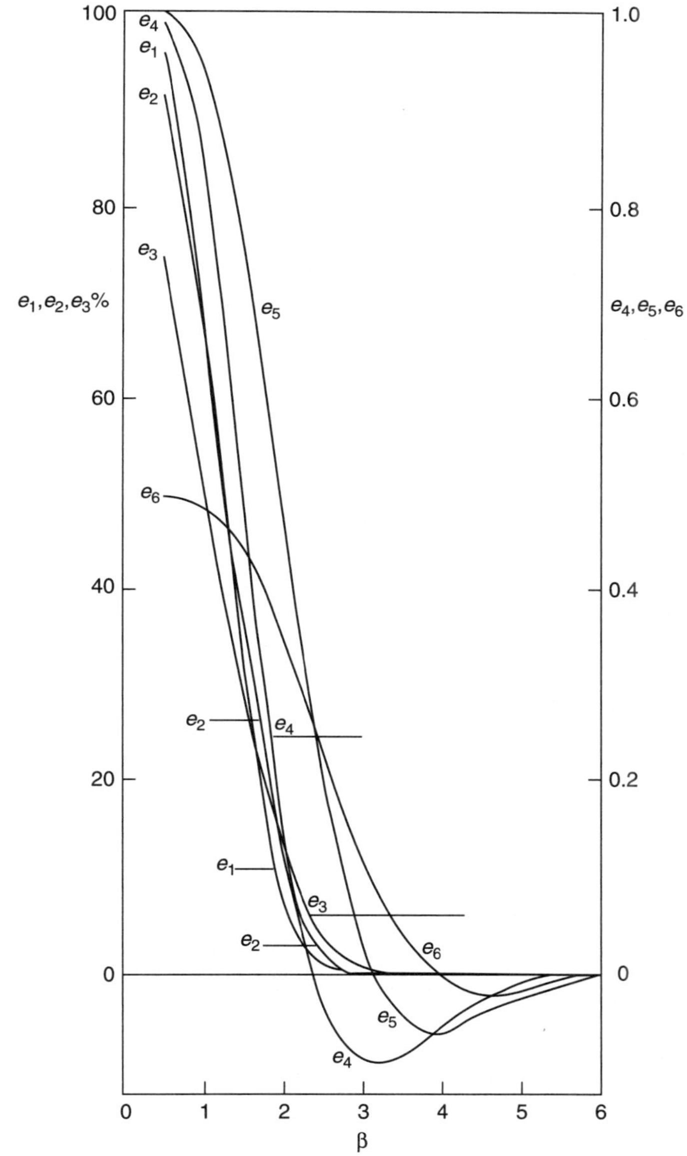

In the more general circumstances of a spherical-shell frustum requiring corrective actions at both edges, all six plots ($e_1,\ldots,e_6$) must be taken into account. Beyond certain values of β in excess of $\beta = 2$, the magnitudes of the unloaded edge deformations relative to their loaded-edge counterparts (i.e. the ratios e_4, e_5 and e_6) actually *increase* with increasing β (i.e. with increasing angular edge separation $\Delta\phi$, for a given λ), which is an unexpected trend; they attain local peaks with respect to β (which occur in the negative domain of the plots, all beyond $\beta = 3$), before gradually diminishing essentially monotonically with further increase in β.

Beyond $\beta = 3$, e_1, e_2 and e_3 would all have diminished to less than 1%, while ratios e_4, e_5 and e_6 would still be of appreciable magnitudes. Thus e_4, e_5 and e_6 dictate the decoupling criteria, which are $\beta = 3.10$ (10% error), $\beta = 4.25$ (5% error), $\beta = 5.25$ (2% error) and $\beta = 5.50$ (1% error).

6.3.8.4 Concluding remarks

Within the framework of the Geckeler approximation, the results presented in this study enable the stresses and deformations in a spherical-shell frustum to be obtained rigorously through the derived closed-form solutions for the generalised integration constants that take into account all the possible boundary conditions. Through considerations of influence coefficients, the present approach has permitted errors inherent in the assumption of decoupled edge effects to be split into two sets, each with a clear physical interpretation. More importantly, such errors have been quantified in terms of a single parameter β.

The study has provided a fairly general, yet simple and concise treatment of the problem of edge interaction and decoupling errors in spherical shells that are thin ($a/t \geq 30$) and non-shallow ($\phi_1 \geq 30°$ and $\phi_2 \leq 150°$). As the basic theory itself (i.e. the Geckeler approximation) is only accurate if the shell is thin and both edges lie in the non-shallow range, the results for decoupling errors are most reliable for shell frusta for which $a/t \geq 100$ and $\{\phi_1 \geq 60°, \phi_2 \leq 120°\}$.

Some remarks on practical analysis seem to be in order. Although the present study has made available all the formulae necessary for a rigorous analysis of spherical frusta and multi-region shells, the computational effort involved is, in general, several times greater than were edge effects assumed to be decoupled. This is not so much due to the relatively lengthy nature of the expressions for the constants C_1, C_2, C_3 and C_4 for a typical two-edged shell region (as given by equations (6.89)), but rather, to the fact that the solution procedure for the full set of redundants associated with a typical multi-region shell structure would generally require considerable computational effort, since it would involve a relatively large set of simultaneous equations.

For each frustum-shaped region (a region being any portion of the shell demarcated by Gol'denveizer's *lines of distortion*), four unknowns $\{M_1, H_1, M_2, H_2\}$ would require to be assumed at the edges, while for a cap-shaped end region, two unknowns $\{M, H\}$ would require to be assumed at the one edge of such a region. In accordance with the procedure adopted in most practical shell calculations, the assumed redundants at the edges of the various regions of a multi-region shell are evaluated by enforcing appropriate boundary conditions (of continuity of deformations between adjacent regions, equilibrium at certain locations, or kinematic constraints at supported edges) on the superimposed effects of the redundants and the membrane solution.

Owing to interactive effects across the widths of all frustum-shaped shell regions, a multi-region shell structure would therefore generally yield a system of coupled boundary-condition equations in the unknown redundants assumed at the edges of the various regions, necessitating the use of computer subroutines for the solution of sets of simultaneous equations. In the automation of the analysis procedure, the matrices of influence coefficients appearing in equations (6.63) for single-edged cap-shaped regions and (6.92) for double-edged frustum-shaped regions would constitute the basic *member flexibility matrices* for the individual regions, while the simultaneous implementation of the external and inter-region boundary conditions would constitute the assembly of the *structure flexibility matrix*, by analogy with the flexibility method of structural analysis.

In view of the considerable computational effort demanded by the rigorous approach that takes into account edge interaction (necessitating expensive computer runs in the case of multi-region shell assemblies), it is suggested that edge effects should always be decoupled wherever possible, the errors introduced by such a simplification having been quantified through the present study. Regardless of whether the shell consists of a single frustum-shaped region, or of several inter-connected regions, only two simultaneous equations need to be considered at a time (permitting hand calculations), if edge effects can be decoupled.

6.4. Practical solutions for the conical shell

For any shell of revolution, the arc length ds of a meridional line element subtending an angle $d\phi$ at the meridional centre of curvature is given by $r_1\, d\phi$, where r_1 is the meridional radius of curvature. Thus,

$$\frac{ds}{d\phi} = r_1 \tag{6.98}$$

Using this relation, we have

$$\frac{d}{d\phi} = \frac{d}{ds}\frac{ds}{d\phi} = r_1 \frac{d}{ds} \tag{6.99}$$

$$\frac{d^2}{d\phi^2} = \frac{d}{d\phi}\left(\frac{d}{d\phi}\right) = r_1 \frac{d}{ds}\left(r_1 \frac{d}{ds}\right) = r_1\left(r_1 \frac{d^2}{ds^2} + \frac{d}{ds}\frac{dr_1}{ds}\right) \tag{6.100}$$

For conical shells, r_1 is constant at infinity, and does not vary with s, so that $dr_1 / ds = 0$. Therefore equation (6.100) reduces to

$$\frac{d^2}{d\phi^2} = r_1^2 \frac{d^2}{ds^2} \tag{6.101}$$

If, for a conical shell, s is the distance (along the straight meridian) from the vertex, then

$$r_2 = s \cot \alpha \tag{6.102}$$

where r_2 is the second principal radius of curvature, and α is the constant value of ϕ for all points on the straight meridian.

From equation (6.102), we can write

$$\frac{dr_2}{ds} = \cot \alpha \quad \text{(constant)} \tag{6.103}$$

When use is made of relations (6.99) and (6.103), we have

$$\frac{dr_2}{d\phi} = r_1 \frac{dr_2}{ds} = r_1 \cot \alpha \tag{6.104}$$

$$\frac{d^2 r_2}{d\phi^2} = \frac{d}{d\phi}\left(\frac{dr_2}{d\phi}\right) = r_1 \frac{d}{ds}(r_1 \cot \alpha) = r_1 \cot \alpha \frac{dr_1}{ds} = 0 \tag{6.105}$$

since $dr_1 / ds = 0$.

Using relations (6.98)–(6.105) in equation (6.19) for a general shell of revolution, noting that $\phi = \alpha$ and r_1 is a constant for a conical shell, we obtain, for a conical shell of constant thickness t (implying that the flexural rigidity D is also constant, so that $dD/d\phi = r_1\left(dD/ds\right) = 0$):

$$r_1 r_2 \left(\sin\alpha\right)\frac{d^2V}{ds^2} + \left[r_2\cos\alpha + r_1\cos\alpha\right]\frac{dV}{ds} - \left[v\sin\alpha + \frac{r_1}{r_2}\frac{\cos^2\alpha}{\sin\alpha}\right]V = -\frac{r_1 r_2}{D}\left(\sin\alpha\right)Q_s$$

First dividing throughout by $r_1 r_2\left(\sin\alpha\right)$, then setting r_1 to infinity (so that terms multiplied with $1/r_1$ vanish), and finally replacing r_2 by $s\cot\alpha$ (equation 6.102), we end up with the equation

$$\frac{d^2V}{ds^2} + \frac{1}{s}\frac{dV}{ds} - \frac{1}{s^2}V = -\frac{Q_s}{D} \tag{6.106}$$

The second equation for the general shell of revolution (i.e. equation (6.22)), when adapted for the conical shell using relations (6.98)–(6.105), with t, r_1 and therefore tr_1 being constant, becomes

$$r_1 r_2^2\frac{d^2Q_s}{ds^2} + \left[r_2^2 + 3r_1 r_2\right]\left(\cot\alpha\right)\frac{dQ_s}{ds} + r_2\left[\left(1-v\right)\cot^2\alpha + \frac{v}{\sin^2\alpha}\right]Q_s = r_1 EtV$$

First dividing throughout by $r_1 r_2^2$, then setting r_1 to infinity (so that terms multiplied with $1/r_1$ vanish), and finally replacing r_2 by $s\cot\alpha$ (equation 6.102), we obtain

$$\frac{d^2Q_s}{ds^2} + \frac{3}{s}\frac{dQ_s}{ds} = \frac{1}{s^2}\left(\tan^2\alpha\right)EtV \tag{6.107}$$

The exact solution to equations (6.106) and (6.107) may be obtained in terms of Kelvin or Bessel functions.[2,3] However, this rigorous approach is not suited to practical engineering calculations of a day-to-day nature, as it is too lengthy, even when tables of Kelvin and/or Bessel functions are available.

Defining a parameter ω as

$$\omega = 2\left[12\left(1-v^2\right)\left(\frac{s^2}{t^2}\right)\tan^2\alpha\right]^{1/4} \tag{6.108}$$

sufficient accuracy is achieved by replacing the Bessel/Kelvin functions with their one-term asymptotic expansions, if $\omega > 15$.[3] When this is done, the following results are obtained. These are applicable (*i*) in the case of a conical shell with one edge (i.e. with a closed apex), or (*ii*) in the vicinity of the larger end of a *long* conical frustum (the term *long* here implying negligibility of edge interaction).

$$Q_s = \frac{e^{\beta}}{s\sqrt{2\pi\omega}}\left[C_1\cos\left(\beta - \frac{\pi}{8}\right) + C_2\sin\left(\beta - \frac{\pi}{8}\right)\right] \tag{6.109}$$

$$N_s = -Q_s\cot\alpha = -\frac{\left(\cot\alpha\right)e^{\beta}}{s\sqrt{2\pi\omega}}\left[C_1\cos\left(\beta - \frac{\pi}{8}\right) + C_2\sin\left(\beta - \frac{\pi}{8}\right)\right] \tag{6.110}$$

$$N_\theta = -\left(\frac{\cot\alpha}{2s}\right)\sqrt{\frac{\omega}{2\pi}}\; e^{\beta}\left[C_1\cos\left(\beta+\frac{\pi}{8}\right)+C_2\sin\left(\beta+\frac{\pi}{8}\right)\right] \tag{6.111}$$

$$M_s = \frac{2e^{\beta}}{\omega\sqrt{2\pi\omega}}\left[C_1\sin\left(\beta+\frac{\pi}{8}\right)-C_2\cos\left(\beta+\frac{\pi}{8}\right)\right] \tag{6.112}$$

$$M_\theta = \nu M_s \tag{6.113}$$

$$V = \sqrt{\frac{12\left(1-\nu^2\right)}{2\pi\omega}}\;\frac{e^{\beta}}{Et^2}\left[C_1\sin\left(\beta-\frac{\pi}{8}\right)-C_2\cos\left(\beta-\frac{\pi}{8}\right)\right] \tag{6.114}$$

$$\delta = \frac{1}{Et}(s\cos\alpha)\left(N_\theta-\nu N_s\right) \tag{6.115}$$

where $\beta = \omega/\sqrt{2}$, and the constants C_1 and C_2 are to be determined from the boundary conditions at the edge of the shell, which for the two cases shown in Figure 6.11, are as follows:

Case (a)

$$\left(N_s\right)_{s=l} = 0 \tag{6.116a}$$

$$\left(M_s\right)_{s=l} = M_e \tag{6.116b}$$

Case (b)

$$\left(N_s\right)_{s=l} = -\,H_e\cos\alpha \tag{6.117a}$$

Figure 6.11 Conical shell subjected to axisymmetric edge actions: (a) bending moment M_e per unit length; (b) horizontal shear force H_e per unit length (axis of revolution of shell is vertical).

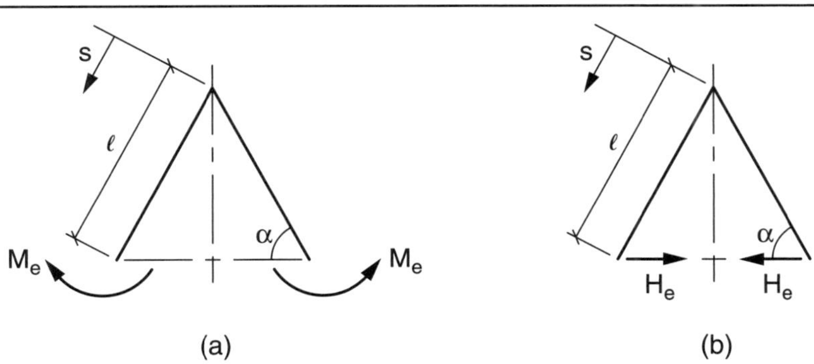

Figure 6.12 Conical shell frustum under the combined action of M_1, H_1, M_2 and H_2.

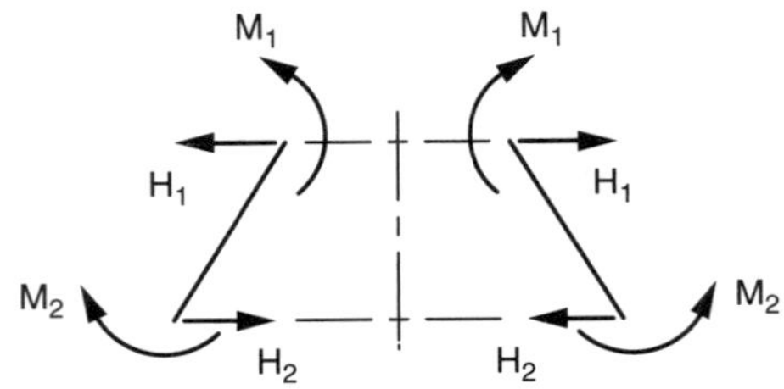

$$(M_s)_{s=l} = 0 \tag{6.117b}$$

The above approach has been applied to obtain good estimates of stresses in the vicinity of the junction of two pressurised cones,[15] where the base of one cone is connected to the base of another (this gives a rhombic pressure vessel if the two cones are identical).

For *short* conical frusta (the term *short* implying that edge interaction cannot be ignored) subjected to axi-symmetric bending moments and shearing forces, $\{M_1, H_1\}$ and $\{M_2, H_2\}$ at edges 1 and 2 respectively (see Figure 6.12), comprehensive formulae for edge deformations and interior actions due to these edge actions are given by Baker *et al.*[16] and Young.[17] Influence coefficients for edge-loaded short conical frusta, derived on the basis of the approximation using one-term asymptotic expansions of the Kelvin functions, have been derived by Baltrukonis.[18] Although such closed-form expressions and tabulated data are available, they may sometimes be inconvenient for quick estimation of stress levels in cones.

Let us investigate a simpler solution for the conical-shell problem, analogous to the Geckeler approximation. Dropping terms in V and dV/ds from the left-hand side of equation (6.106), and the term in dQ_s/ds from the left-hand side of equation (6.107), we obtain the approximate equations

$$\frac{d^2V}{ds^2} \approx -\frac{Q_s}{D} \tag{6.118}$$

$$\frac{d^2Q_s}{ds^2} \approx \frac{1}{s^2}(\tan^2\alpha)\, EtV \tag{6.119}$$

From equation (6.119),

$$V = \frac{\cot^2\alpha}{Et}\left(s^2\frac{d^2Q_s}{ds^2}\right)$$

Successive differentiation of this expression, each time discarding the term in the lower derivative of V in relation to the term in the higher derivative, yields

$$\frac{dV}{ds} = \frac{\cot^2\alpha}{Et}\left(s^2\frac{d^3Q_s}{ds^3} + 2s\frac{d^2Q_s}{ds^2}\right) \approx \frac{\cot^2\alpha}{Et}\left(s^2\frac{d^3Q_s}{ds^3}\right)$$

$$\frac{d^2V}{ds^2} = \frac{\cot^2\alpha}{Et}\left(s^2\frac{d^4Q_s}{ds^4} + 2s\frac{d^3Q_s}{ds^3}\right) \approx \frac{\cot^2\alpha}{Et}\left(s^2\frac{d^4Q_s}{ds^4}\right)$$

Substituting the final result for d^2V / ds^2 into equation (6.118), and rearranging the ensuing expression, we obtain

$$\frac{d^4Q_s}{ds^4} + \frac{Et}{D}(\tan^2\alpha)\left(\frac{1}{s^2}\right)Q_s = 0$$

– that is,

$$\frac{d^4Q_s}{ds^4} + 4\eta^4 Q_s = 0 \tag{6.120}$$

where

$$4\eta^4 = \frac{Et}{D}(\tan^2\alpha)\frac{1}{s^2} = \frac{12(1-\nu^2)}{(s\cot\alpha)^2 t^2} \tag{6.121}$$

or

$$\eta^4 = \frac{3(1-\nu^2)}{s^2(\cot^2\alpha)t^2} \tag{6.122}$$

In equation (6.120), the coefficient of Q_s (i.e. $4\eta^4$) is a function of s, so that an exact solution to the equation is difficult to obtain. However, the distance over which the bending disturbance (emanating from the shell edge) is significant is small in comparison with the dimension s, since we are dealing with non-shallow thin conical shells. As a further approximation, we may therefore assume s is practically constant in the edge zone of the shell, so that equation (6.120) now has constant coefficients.

For the purposes of evaluating η, the value of s corresponding to the edge of the shell (e.g. $s = l$ in Figure 6.11) may be adopted. Equation (6.120), with η now taken as constant, is of the same form as equation (5.25) for the axisymmetric bending of a circular cylinder, or equation (6.25) of the Geckeler approximation for a spherical shell. We may therefore assume a solution of the form

$$Q_s = e^{\eta s}(C_1\cos\eta s + C_2\sin\eta s) + e^{-\eta s}(C_3\cos\eta s + C_4\sin\eta s) \tag{6.123}$$

where

$$\eta = \left[\frac{3(1-\nu^2)}{(l^2\cot^2\alpha)t^2}\right]^{1/4} \tag{6.124}$$

For a conical cap closed over the apex (i.e. at $s = 0$), only the first part of the right-hand side of equation (6.123) is relevant. Letting $x = l - s$ (where l is the value of s at the edge of the cone), we may write the solution in the modified form

$$Q_s = Ce^{-\eta x} \sin(\eta x + \xi) \tag{6.125}$$

where C and ξ are constants to be determined from the boundary conditions at the edge (i.e. at $x = 0$).

The meridional stress resultant N_s follows from

$$\begin{aligned} N_s &= -Q_s \cot\alpha \\ &= -(\cot\alpha)\, Ce^{-\eta x} \sin(\eta x + \xi) \end{aligned} \tag{6.126}$$

The hoop stress resultant N_θ is obtainable from expression (6.21), by ignoring the term in Q_ϕ in relation to the term in $dQ_\phi / d\phi$, and transforming from the variable ϕ to s, then to x, as follows:

$$\begin{aligned} N_\theta &\approx -\frac{r_2}{r_1}\frac{dQ_s}{d\phi} \\ &= -\frac{r_2}{r_1} r_1 \frac{dQ_s}{ds} \\ &= r_2 \frac{dQ_s}{dx} \text{(since } dx = -ds) \\ &\approx -(l\cot\alpha)\, \eta\sqrt{2}\; Ce^{-\eta x} \sin\left(\eta x + \xi - \frac{\pi}{4}\right) \end{aligned} \tag{6.127}$$

Note that, consistent with the approximation that η is constant, r_2 is assumed to be constant and approximately equal to $l \cot\alpha$ in the edge zone.

From equation (6.119), and taking s to be constant and approximately equal to l in the edge zone, we obtain the meridional rotation as

$$\begin{aligned} V &\approx s^2 \frac{\cot^2\alpha}{Et}\frac{d^2Q_s}{ds^2} \\ &\approx l^2 \frac{\cot^2\alpha}{Et}\frac{d^2Q_s}{dx^2} \\ &= -(l^2\cot^2\alpha)\frac{2\eta^2}{Et} Ce^{-\eta x} \cos(\eta x + \xi) \end{aligned} \tag{6.128}$$

The bending moments M_s and M_θ are obtained on the basis of expression (6.18), by ignoring the terms in V in relation to the terms in $dV / d\phi$, and transforming from the variable ϕ to s, then to x, as follows:

$$M_s \approx -D\frac{1}{r_1}\frac{dV}{d\phi}$$

$$= -D \frac{1}{r_1} r_1 \frac{dV}{ds}$$

$$= D \frac{dV}{dx} = \frac{1}{\eta\sqrt{2}} Ce^{-\eta x} \sin\left(\eta x + \xi + \frac{\pi}{4}\right) \tag{6.129}$$

$$M_\theta \approx -D \frac{\nu}{r_1} \frac{dV}{d\phi} = \nu M_s \tag{6.130}$$

The lateral displacement δ is obtained on the basis of expression (2.67), by neglecting νN_ϕ in relation to N_θ (since, by comparing equations (6.126) and (6.127), $N_s \left(\equiv N_\phi\right)$ is proportional to Q_s while N_θ is proportional to dQ_s / ds):

$$\delta \approx \frac{1}{Et}\left(r_2 \sin\alpha\right) N_\theta$$

$$= -\frac{1}{Et}\left(l^2 \cot^2\alpha\right) \eta\sqrt{2}\,(\sin\alpha)\, Ce^{-\eta x} \sin\left(\eta x + \xi - \frac{\pi}{4}\right) \tag{6.131}$$

By referring to Figure 6.11(a), the boundary conditions are

$$\left(N_s\right)_{x=0} = 0 \tag{6.132a}$$

$$\left(M_s\right)_{x=0} = M_e \tag{6.132b}$$

These conditions yield the following values for the constants:

$$C = 2\eta M_e \tag{6.133a}$$

$$\xi = 0 \tag{6.133b}$$

The deformations caused at the edge by M_e become

$$V_e = -\left(l^2 \cot^2\alpha\right) \frac{4\eta^3}{Et} M_e \tag{6.134a}$$

$$\delta_e = \left(l^2 \cot^2\alpha\right) \frac{2\eta^2}{Et} \left(\sin\alpha\right) M_e \tag{6.134b}$$

For the case shown in Figure 6.11(b), the boundary conditions are

$$\left(N_s\right)_{x=0} = -H_e \cos\alpha \tag{6.135a}$$

$$\left(M_s\right)_{x=0} = 0 \tag{6.135b}$$

Figure 6.13 Conical shell with inscribed spherical shell.

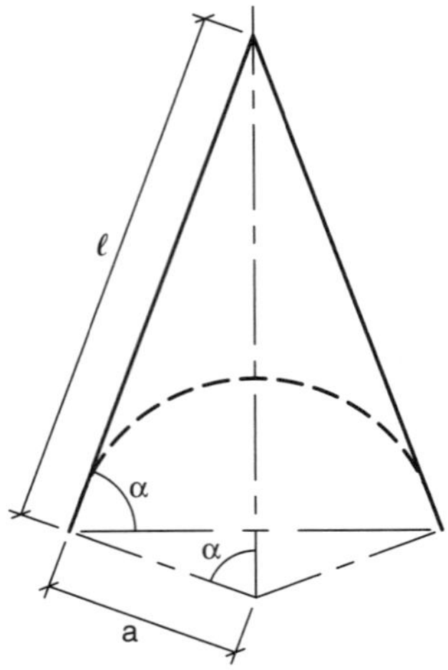

giving

$$C = -\sqrt{2}\; H_e \sin\alpha \tag{6.136a}$$

$$\xi = -\frac{\pi}{4} \tag{6.136b}$$

so that the edge deformations due to H_e become

$$V_e = \left(l^2 \cot^2\alpha\right)\frac{2\eta^2}{Et}\left(\sin\alpha\right)H_e \tag{6.137a}$$

$$\delta_e = -\left(l^2 \cot^2\alpha\right)\frac{2\eta}{Et}\left(\sin^2\alpha\right)H_e \tag{6.137b}$$

Equations (6.126)–(6.137) can be deduced directly from the corresponding Geckeler-approximation expressions for the spherical shell. Consider Figure 6.13, which shows the conical shell with its inscribed spherical shell. Since $a\sin\alpha = l\cos\alpha$, we have

$$a = l\cot\alpha \tag{6.138a}$$

For the spherical shell, assuming the same thickness as the conical shell, we have

$$\lambda^4 = \frac{3\left(1-\nu^2\right)a^2}{t^2}$$

$$= \frac{3\left(1-\nu^2\right)l^2\cot^2\alpha}{t^2} \tag{6.138b}$$

(making use of relation (6.138a)).

For the conical shell, we have

$$\eta^4 = \frac{3\left(1-\nu^2\right)}{\left(l^2 \cot^2 \alpha\right) t^2} \tag{6.138c}$$

Since the governing equations for the spherical shell (equation 6.25) and the conical shell (equation 6.120) are of exactly the same form, λ must correspond to η. From equations (6.138b) and (6.138c), we can write

$$\frac{3\left(1-\nu^2\right)}{t^2} = \frac{\lambda^4}{l^2 \cot^2 \alpha} = \eta^4\, l^2 \cot^2 \alpha$$

from which

$$\lambda = \eta\left(l \cot \alpha\right) \tag{6.138d}$$

Thus, in the Geckeler-approximation expressions for the spherical shell (equations (6.34)–(6.39); (6.51) and (6.52); (6.57) and (6.58)), in order to obtain the corresponding conical-shell equations, we simply replace, in the portion of the right-hand side appearing before C, the parameter a by $l\cot\alpha$, λ by $\eta\, l\cot\alpha$, and $\left(\phi_e - \psi\right)$ or ϕ_e by α; in the exponential and trigonometric terms that appear after the C, $\lambda\psi$ is replaced with ηx.

6.5. Approximate solutions for general shells of revolution

Extending Geckeler's approximation to general shells of revolution that are not too shallow, we may reduce, on the basis of the same arguments as were made for the spherical shell, the general equations for shells of revolution (i.e. equations (6.19) and (6.22)) to the following:

$$\frac{d^2V}{d\phi^2} = -\frac{r_1^2}{D} Q_\phi \tag{6.139a}$$

$$\frac{d^2Q_\phi}{d\phi^2} = \frac{r_1^2}{r_2^2} EtV \tag{6.139b}$$

From equation (6.139b), we may write V in terms of Q_ϕ, E and the geometric variables of the shell. We differentiate this expression for V twice to obtain an expression for $d^2V / d\phi^2$, which we simplify by neglecting all terms except the term in the highest derivative of Q_ϕ. The simplified version of $d^2V / d\phi^2$ is then substituted into equation (6.139a), and on rearranging, we obtain

$$\frac{d^4Q_\phi}{d\phi^4} + \frac{r_1^4 Et}{Dr_2^2} Q_\phi = 0 \tag{6.140}$$

– that is,

$$\frac{d^4Q_\phi}{d\phi^4} + 4\lambda^4 Q_\phi = 0 \tag{6.141}$$

where

$$\lambda^4 = \frac{r_1^4 Et}{4Dr_2^2} = 3\left(1-\nu^2\right)\frac{r_1^4}{r_2^2 t^2} \tag{6.142}$$

If, for a given shell, $r_1^4 / \left(r_2^2 t^2\right)$ happens to be constant along a meridian, then λ is constant, and the solution of equation (6.141) is of the same form as for the spherical shell of constant thickness. This approach has been used to tackle the analysis of egg-shaped liquid-containment vessels with regions of constant r_1 but variable r_2 and t such that $r_2 t$ is constant (giving a constant λ).[19] In general, λ is a function of ϕ, so that an exact solution to equation (6.141) is difficult to obtain. However, owing to the very localised nature of edge effects, λ usually does not vary appreciably over the portion of the shell experiencing significant bending effects, permitting a constant average value for λ to be adopted for the edge zone, and the spherical-shell solution to be applied.

For a general shell of revolution of constant thickness, a reasonable estimate of edge effects is provided by the inscribed (or inscribing) spherical shell having the same thickness, as illustrated for the conical shell in the last section.

If, for any given shell of revolution, the variation of λ is of the form

$$\lambda = \frac{\alpha}{\beta + \phi} \tag{6.143}$$

where α and β are constants, then a practically exact solution to equation (6.141) can be obtained. This solution will be developed in Section 8.2.3 for the purposes of estimating edge effects in spherical domes with gradual thickening in the regions adjacent to the supports.

References

1. Timoshenko S.P. and Goodier J.N. *Theory of Elasticity*. McGraw-Hill, New York, 1982.
2. Timoshenko S.P. and Woinowsky-Krieger S. *Theory of Plates and Shells*. McGraw-Hill, New York, 1959.
3. Flügge W. *Stresses in Shells*. Springer-Verlag, Berlin, 1973.
4. Calladine C.R. and Paskaran N. A re-appraisal of influence coefficients for the edges of thin elastic spherical shells subjected to symmetric loads. *Quarterly Journal of Mechanics and Applied Mathematics*, 1974, **27**, 1–15.
5. Hoff N.J. General formulas for influence coefficients of thin spherical shells. *Journal of the Aerospace Sciences*, 1962, **29**, 174–179 and 225.
6. Hetényi M. Spherical shells subjected to axial symmetrical bending. *Publications of the International Association for Bridge and Structural Engineering*, 1938, **5**, 173–185.
7. Leckie F.A. Asymptotic solutions for the spherical shell subjected to axially symmetric loading. In *Nuclear Reactor Containment Buildings and Pressure Vessels*. Butterworths, London, 1960, 286–297.
8. Nielsen T.B. *Axisymmetric Bending of Shells*. Danish Technical Press, The Institution of Danish Civil Engineers, Copenhagen, 1962.
9. Zingoni A. Bending-disturbance considerations in the axisymmetric response of non-shallow spherical shell structures. PhD Thesis, University of London, 1991.

10. Zingoni A. and Pavlovic M.N. A note on the accuracy of the Geckeler approximation. *Engineering Computations*, 1993, **10** (4), 369–379.
11. Zingoni A. and Pavlovic M.N. On edge-disturbance interaction and decoupling errors in thin-walled nonshallow spherical-shell frusta. *Thin-Walled Structures*, 1992, **13**, 375–386.
12. CBI. *Hortonsphere Pressure Vessels*. Chicago Bridge & Iron Co., Oak Brook (Illinois), 1980.
13. Stern P. and Tsui E.Y.W. On the bending of spherical shells. *Proceedings of the American Society of Civil Engineers: Journal of the Engineering Mechanics Division*, 1966, **92**, 53–66.
14. Kocsis L. Investigation of the edge-influence on spherical shells. *Periodica Polytechnica* (*Mechanical Engineering*), 1975, **19**, 91–99.
15. Zingoni A. Discontinuity effects at cone–cone axisymmetric shell junctions. *Thin-Walled Structures*, 2002, **40** (10), 877–891.
16. Baker E.H., Kovalevsky L. and Rish F.L. *Structural Analysis of Shells*. McGraw-Hill, New York, 1972.
17. Young W.C. *Roark's Formulas for Stress and Strain*. McGraw-Hill, New York, 1989.
18. Baltrukonis J.H. Influence coefficients for edge-loaded short thin conical frustums. *ASME Journal of Applied Mechanics*, 1959, **26** (2), 241–245.
19. Zingoni A. Stresses and deformations in egg-shaped sludge digesters: Discontinuity effects. *Engineering Structures*, 2001, **23** (11), 1373–1382.

Shell Structures in Civil and Mechanical Engineering
ISBN 978-0-7277-6028-9

http://dx.doi.org/10.1680/ssicame.60289.205

Chapter 7
Flexibility analysis of shell-ring systems and multi-shell assemblies

7.1. Pressure vessels
7.1.1 Cylinder-head junction effects

Pressure vessels in the form of thin-walled shells of revolution are often constructed as assemblies of different shell geometries, typically cylindrical, spherical, conical, ellipsoidal and toroidal shells. At the junctions of the different shells, discontinuity stresses will generally occur.

Let us consider a cylindrical pressure vessel of radius a_c and having spherical end closures of radius a_s as shown in Figure 7.1(a). We will take the axis of revolution of the whole assembly as vertical. The membrane solution results in a mismatch of deformations between the cylindrical shell and the spherical caps at the junctions. To correct this, we apply axisymmetric bending moments and horizontal shear forces at the edges of the separated shells; these actions will be denoted by $\{M_o,\ H_o\}$ for the spherical cap, and $\{M'_o,\ H'_o\}$ for the cylindrical shell, as shown in Figure 7.1(b) for one end of the pressure vessel (the other end being identical). Edge interaction may be ignored, since pressure vessels of this type usually have an elongated shape (i.e. the cylindrical portion is a 'long' shell in accordance with the definition of Chapter 5).

The corresponding reactions on an imaginary ring element lying between the cylinder and the cap are also shown in Figure 7.1(b). The additional force H_o^m shown acting on the ring element is simply the combined effect of the membrane meridional forces pulling on the element, resolved in the direction perpendicular to the axis of revolution of the pressure vessel (i.e. the horizontal direction), as illustrated in Figure 7.1(c). The N_x^m stress resultant at the edge of the cylinder, being vertical, has no component in the horizontal direction, so H_o^m is solely due to the N_ϕ^m stress resultant at the edge of the spherical shell. By reference to the free-body diagram of Figure 7.1(c), we can therefore write

$$H_o^m = \left(N_\phi^m\right)_o \cos\phi_o = \frac{pa_s}{2}\cos\phi_o \tag{7.1}$$

where $\left(N_\phi^m\right)_o$ is the value of N_ϕ^m in the spherical shell at its edge (which, from expression (3.12), is simply $pa_s/2$), and ϕ_o is the value of the angular coordinate ϕ of the spherical shell at its edge.

Applying the conditions of equilibrium of moments and horizontal forces to the inter-shell ring element in Figure 7.1(b), we obtain the relationships:

$$M'_o = M_o \tag{7.2a}$$

$$H'_o = H_o - H_o^m \tag{7.2b}$$

In response to the actions that are applied, the shells will experience deformations. The deformations of interest are displacements transverse to the axis of revolution of the assembly, and rotations of the shell meridian. Figure 7.1(d) shows the assumed positive directions for the deformations occurring at the edges of the spherical and the cylindrical shell. In the considerations that follow, we will use primes to denote cylindrical-shell variables, and plain symbols to denote spherical-shell variables, unless otherwise stated or

Figure 7.1 Circular-cylindrical pressure vessel with spherical end closures: (a) general arrangement; (b) free body diagram of inter-shell element and shell edges; (c) action of membrane stress resultants on an inter-shell element; (d) shell-edge deformations.

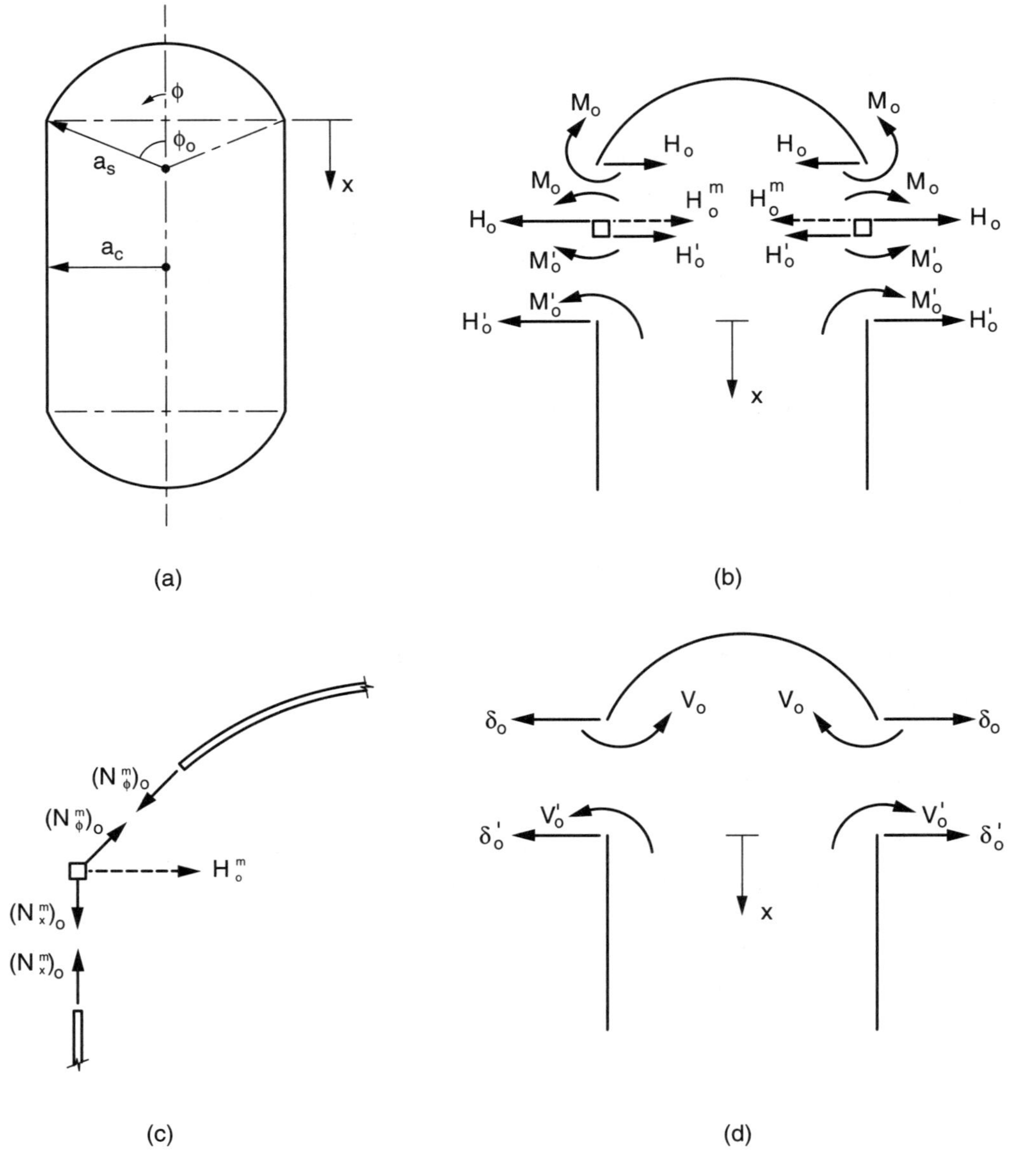

implied. Continuity of deformations requires that the net rotation and the net lateral displacement of the two shells at their junction be the same – that is,

$$V_o^T = V_o'^T \Rightarrow V_o^b + V_o^m = V_o'^b + V_o'^m \tag{7.3a}$$

$$\delta_o^T = \delta_o'^T \Rightarrow \delta_o^b + \delta_o^m = \delta_o'^b + \delta_o'^m \tag{7.3b}$$

As in earlier chapters, superscripts b and m relate to the edge effect and the membrane solution, respectively. Using results (6.63) for the spherical shell (i.e. the Geckeler-approximation version of influence coefficients) and results (5.41) for the cylindrical shell, we may write conditions (7.3) in expanded form, which is given by equation (7.4) below. Note that all edge parameters in the present problem conform to the same positive directions as those assumed in the theoretical derivations of earlier chapters, with the exception of V_o' in Figure 7.1(d), whose direction is opposite to that of the earlier derivations. The parameters β and D on the right-hand side of equation (7.4) are written without primes, as it is quite obvious that they refer to the cylindrical shell.

$$\begin{bmatrix} -\dfrac{4\lambda^3}{Ea_s t} & \dfrac{2\lambda^2}{Et}(\sin\phi_o) \\ \dfrac{2\lambda^2}{Et}(\sin\phi_o) & -\dfrac{2\lambda a_s}{Et}(\sin^2\phi_o) \end{bmatrix} \begin{bmatrix} M_o \\ H_o \end{bmatrix} + \begin{bmatrix} V_o^m \\ \delta_o^m \end{bmatrix} = \begin{bmatrix} \dfrac{1}{\beta D} & \dfrac{1}{2\beta^2 D} \\ \dfrac{1}{2\beta^2 D} & \dfrac{1}{2\beta^3 D} \end{bmatrix} \begin{bmatrix} M_o' \\ H_o' \end{bmatrix} + \begin{bmatrix} -V_o'^m \\ \delta_o'^m \end{bmatrix} \tag{7.4}$$

As shown in Sections 3.1.1 and 3.1.3, for uniform internal pressure p, the membrane rotations of the cylinder and the spherical shell are both zero throughout (i.e. $V_o'^m = V_o^m = 0$). On the other hand, the membrane lateral displacements of the cylinder and the spherical shell under a uniform internal pressure p are given by expressions (3.5a) and (3.14), respectively. At the junction of the two shells, we therefore have

$$\delta_o^m = \frac{pa_s^2}{2Et}(1-\nu)\sin\phi_o \tag{7.5a}$$

for the spherical shell, and

$$\delta_o'^m = \frac{pa_c^2}{2Et'}(2-\nu) \tag{7.5b}$$

for the cylindrical shell, assuming that the material of construction is the same for the two shells (so that Young's modulus E and Poisson's ratio ν are the same).

Let us denote the elements of the flexibility matrix on the left-hand side of equation (7.4) by I_{ij} (for the spherical shell), and those of the flexibility matrix on the right-hand side of the equation by I_{ij}' (for the cylindrical shell), where $\{i = 1,\ 2;\ j = 1,\ 2\}$. Putting $V_o^m = V_o'^m = 0$, equation (7.4) may be written as

$$\begin{bmatrix} (I_{11} - I_{11}') & (I_{12} - I_{12}') \\ (I_{21} - I_{21}') & (I_{22} - I_{22}') \end{bmatrix} \begin{bmatrix} M_o \\ H_o \end{bmatrix} = \begin{bmatrix} -I_{12}' H_o^m \\ \delta_o'^m - \delta_o^m - I_{22}' H_o^m \end{bmatrix} \tag{7.6}$$

where use has been made of relations (7.2) to eliminate M_o' and H_o'.

The solution of equation (7.6) is

$$M_o = \frac{B_1 A_{22} - B_2 A_{12}}{A_{11} A_{22} - A_{12}^2} \tag{7.7a}$$

$$H_o = \frac{B_2 A_{11} - B_1 A_{12}}{A_{11} A_{22} - A_{12}^2} \tag{7.7b}$$

where

$$A_{ij} = I_{ij} - I'_{ij} \quad (i = 1,\ 2;\ j = 1,\ 2) \tag{7.8}$$

noting that $A_{21} = A_{12}$, and

$$B_1 = -I'_{12}\, H_o^m \tag{7.9a}$$

$$B_2 = \delta_o'^{\,m} - \delta_o^m - I'_{22}\, H_o^m \tag{7.9b}$$

with $H_o^m, \{\delta_o^m,\ \delta_o'^{\,m}\}$ and $\{I_{ij},\ I'_{ij}\}$ as defined in equations (7.1), (7.5) and (7.4), respectively. The edge redundants of the cylindrical shell, M'_o and H'_o, follow from relations (7.2).

With the shell-edge redundants $\{M_o,\ H_o\}$ (for the spherical shell) and $\{M'_o,\ H'_o\}$ (for the cylindrical shell) now known, the net stresses on the inner and outer surfaces of the spherical shell follow from expressions (6.70), in which N_ϕ^b, N_θ^b, M_ϕ and M_θ are given by expressions (6.68), and N_ϕ^m and N_θ^m are given by expressions (3.12) and (3.13); those for the cylinder follow from expressions (5.40), in which N_x^b, N_θ^b, M_x and M_θ are given by expressions (5.38) and (5.39), while N_θ^m and N_x^m are given by expressions (3.1) and (3.3), respectively.

Let us consider the special case of a cylindrical pressure vessel with hemispherical closures (i.e. $\phi_o = 90°$, $a_s = a_c = a$), where the thickness of the cylinder and the hemispheres is the same (i.e. $t' = t$). For the hemispherical ends, we have

$$\lambda^4 = 3\left(1 - \nu^2\right)\frac{a^2}{t^2},$$

while for the cylindrical portion, we have

$$\beta^4 = \frac{3\left(1 - \nu^2\right)}{a^2 t^2}.$$

It follows, therefore, that $\beta = \lambda / a$. Also,

$$D = \frac{Et^3}{12\left(1 - \nu^2\right)} = \frac{Eta^2}{4\lambda^4}$$

so that, in the flexibility matrix for the cylinder edge,

$$\frac{1}{\beta D}=\frac{4\lambda^3}{Eat}\;;\quad \frac{1}{2\beta^2 D}=\frac{2\lambda^2}{Et}\;;\quad \frac{1}{2\beta^3 D}=\frac{2\lambda a}{Et}.$$

Evaluating the A_{ij}, we have

$$A_{11}=I_{11}-I'_{11}=-\frac{4\lambda^3}{Eat}-\frac{1}{\beta D}=-\frac{4\lambda^3}{Eat}-\frac{4\lambda^3}{Eat}=-\frac{8\lambda^3}{Eat}$$

$$A_{12}=I_{12}-I'_{12}=\frac{2\lambda^2}{Et}-\frac{1}{2\beta^2 D}=\frac{2\lambda^2}{Et}-\frac{2\lambda^2}{Et}=0$$

$$A_{22}=I_{22}-I'_{22}=-\frac{2\lambda a}{Et}-\frac{1}{2\beta^3 D}=-\frac{2\lambda a}{Et}-\frac{2\lambda a}{Et}=-\frac{4\lambda a}{Et}$$

Evaluating the $B_i\left(i=1,2\right)$, noting from equation (7.1) that H_o^m is now equal to zero, we obtain

$$B_1=0$$

$$B_2=\delta_o'^m-\delta_o^m=\frac{pa^2}{2Et}(2-\nu)-\frac{pa^2}{2Et}(1-\nu)=\frac{pa^2}{2Et}$$

Finally, evaluating the redundants, we obtain

$$M_o=M'_o=0;\quad H_o=H'_o=-\frac{pa}{8\lambda} \qquad (7.10a, b)$$

The cylindrical pressure vessel can, of course, have a closure of some other shape – for example, conical, ellipsoidal, or a shallow spherical cap joined to the cylindrical portion via a circular toroidal transition. The last type of construction is often referred to as a *torispherical* closure. In all cases, the boundary conditions between the cylindrical shell and the end closure remain the same as already stated for the spherical closure. What will be different are the four influence coefficients for the closure, as well as the membrane deformations at the edge of the closure, since these are all functions of shell geometry.

Let us consider a cylindrical pressure vessel with a semi-ellipsoidal head. We know that, owing to symmetry, the elliptical meridian of a pressure vessel in the form of a complete ellipsoid of revolution does not rotate at the poles and at the equator (i.e. $V^m=0$ at $\phi=0,\pm 90°, 180°$). We also know that the straight meridian of a cylinder does not rotate under uniform internal pressure (i.e. $V^m=0$). Therefore, for a cylindrical pressure vessel with a semi-ellipsoidal head, the only discrepancy in the membrane deformations between the cylinder and the semi-ellipsoid is in the lateral displacements δ^m, which differ between the two shells. Such a discrepancy requires only shearing actions H_o (= H'_o, since $H_o^m=0$) to correct it; bending actions M_o $\left(=M'_o\right)$ are not required (i.e. $M_o=0$). In the earlier case of the hemispherical closure (a special case of a semi-ellipsoidal closure), we actually calculated M_o and obtained a value of zero, a result that could have been predicted through the present reasoning.

Let a and b be the semi-axes of the elliptical meridian of the complete ellipsoid of revolution of which the semi-ellipsoidal head forms a part. Furthermore, let us assume that the dimension b is in the direction of the axis of the cylinder, while the dimension a is perpendicular to the axis of the cylinder, b being smaller than a. At the edge of the semi-ellipsoidal shell, the membrane stress resultants under a uniform internal pressure p are obtained from expressions (3.31) and (3.32) with ϕ set equal to 90°. The results are

$$\left(N_\phi^m\right)_{\phi=90^\circ} = \frac{pa}{2} \tag{7.11a}$$

$$\left(N_\theta^m\right)_{\phi=90^\circ} = \frac{pa}{2}\left(\frac{2b^2 - a^2}{b^2}\right) \tag{7.11b}$$

The lateral displacement at the edge of the semi-ellipsoidal head (i.e. at its junction with the cylindrical portion of the pressure vessel) is then given by equation (2.67) with $r_2 = a$ and $\sin\phi = \sin 90° = 1$. The result is as follows:

$$\delta_o^m = \frac{pa^2}{2Et}\left[\frac{b^2(2-\nu) - a^2}{b^2}\right] \tag{7.12}$$

The second of equations (7.6) (which is the condition of compatibility of net lateral displacements between the cylinder and its head), with M_o and H_o^m set equal to zero, reduces to

$$\left(I_{22} - I'_{22}\right)H_o = \delta_o'^m - \delta_o^m \tag{7.13}$$

where the influence coefficient I_{22} now relates to the semi-ellipsoidal shell, and δ_o^m is given by equation (7.12).

Now the edge deformation response of a semi-ellipsoid of revolution to edge bending moments M_o and shearing forces H_o is very nearly like that of a hemispherical shell, since the meridians of these two types of end closures almost coincide as the junction with the cylinder is approached, and tangentially merge into each other at $\phi = 90°$. We may therefore assume, for all practical purposes, that I_{22} for the semi-ellipsoid of revolution is equal to I_{22} for the hemisphere. If we also assume that the thickness t of the semi-ellipsoidal head is equal to that of the cylinder, we can expand expression (7.13) as follows:

$$\left(-\frac{2\lambda a}{Et} - \frac{1}{2\beta^3 D}\right)H_o = \frac{pa^2}{2Et}(2-\nu) - \frac{pa^2}{2Et}\left[\frac{b^2(2-\nu) - a^2}{b^2}\right]$$

– that is,

$$\left(-\frac{2\lambda a}{Et} - \frac{2\lambda a}{Et}\right)H_o = \frac{pa^2}{2Et}\left(\frac{a^2}{b^2}\right)$$

which leads to the result

$$H_o = -\frac{pa}{8\lambda}\left(\frac{a^2}{b^2}\right) \tag{7.14}$$

Noting that H_o is now a^2 / b^2 times larger than the value obtained for the hemispherical closure, we conclude that the price for replacing the hemispherical ends of a cylindrical pressure vessel by relatively flat semi-ellipsoidal ends (which saves on headroom requirements) is the amplification of junction edge effects by a factor of a^2 / b^2.

In the case of a conical closure, the membrane deformations at the edge of the cone are given by relations (3.10) and (3.11), with s set equal to l, i.e.

$$V_o^m = -\frac{3}{2}\frac{pl}{Et}\cot^2\alpha \tag{7.15a}$$

$$\delta_o^m = \frac{pl^2}{2Et}\left(\frac{\cos^2\alpha}{\sin\alpha}\right)(2-\nu) \tag{7.15b}$$

Assuming the conical closure is not 'too flat' (i.e. the angle α of the cone – see Figure 6.11 – is not less than say 30°), and that the shell is relatively thin (i.e. the value of r_2 / t at the base of the cone, equal to $(l\cot\alpha)/t$, is at least 30), we may adopt, for the conical shell, the influence coefficients in equations (6.134) and (6.137) – that is,

$$\begin{bmatrix} V_o^b \\ \delta_o^b \end{bmatrix} = \begin{bmatrix} -(l^2\cot^2\alpha)\dfrac{4\eta^3}{Et} & (l^2\cot^2\alpha)\dfrac{2\eta^2}{Et}\sin\alpha \\ (l^2\cot^2\alpha)\dfrac{2\eta^2}{Et}\sin\alpha & -(l^2\cot^2\alpha)\dfrac{2\eta}{Et}\sin^2\alpha \end{bmatrix}\begin{bmatrix} M_o \\ H_o \end{bmatrix} = \begin{bmatrix} I_{11} & I_{12} \\ I_{21} & I_{22} \end{bmatrix}\begin{bmatrix} M_o \\ H_o \end{bmatrix} \tag{7.16}$$

The solutions for M_o and H_o (at the edge of the closure) are still given by (7.7), with M_o' and H_o' (at the edge of the cylinder) following from relations (7.2). Note, however, that for the conical closure, the horizontal membrane force is evaluated from equation (3.9):

$$H_o^m = \left(N_s^m\right)_o \cos\phi_o = \left(\frac{pl}{2}\cot\alpha\right)\cos\alpha \tag{7.17}$$

Also, the influence coefficients $\{I_{11}, I_{12}, I_{21}, I_{22}\}$ in equation (7.6) now take on the values for the conical shell as given in expression (7.16). Thus the values of the A_{ij} in equation (7.8) are different from those of the spherical closure. The expression for B_2 remains as given by equation (7.9b), but that for B_1 must be modified to

$$B_1 = -I_{12}' H_o^m - V_o^m \tag{7.18}$$

since V_o^m in equation (7.4) is no longer zero in the case of a conical closure. In calculating B_1 and B_2, the parameters of the cylindrical shell remain, of course, the same as before, but $\{V_o^m, \delta_o^m\}$ and H_o^m for the conical head are now as given by equations (7.15) and (7.17).

The bending-related internal actions for the cone then follow from relations (6.126, 6.127, 6.129, 6.130), with C and ξ being given by equations (6.133) for M_o and (6.136) for H_o. The stresses associated with the bending-related actions N_s^b, N_θ^b, M_s and M_θ are superimposed with those associated with the membrane actions N_s^m and N_θ^m (equations (3.6) and (3.9)) in the usual manner. The stresses in the cylinder are determined as before, using the M_o' and H_o' values that now apply.

More accurate assessments of influence coefficients for various geometries of pressure-vessel shells may be seen in the publications of Galletly[1,2] and Baker and Cline.[3] For shell-of-revolution frusta that have a symmetry plane perpendicular to the axis of revolution of the shell, and as shown in a recent publication,[4] it is possible to derive exact influence coefficients via a computationally efficient procedure that takes into account this type of symmetry.

7.1.2 Ring-stiffened cylindrical shells

Figure 7.2(a) shows a cylindrical pressure vessel that is stiffened along its length by a series of rings. For simplicity, and with only a little loss of generality, let us assume the distance between ring stiffeners is not too small, to allow us to use the solution for a long cylinder (equation 5.37).

Figure 7.2(b) shows a free-body diagram of a ring and adjoining shell portions. The positive directions for shell-edge actions $\{M_o,\ H_o\}$ for the shell portion to the right of the ring (which is a mirror image of the portion to the left) exactly match those of Figure 5.5, so that the results of Section 5.4 are directly applicable here. Using expressions (5.42), with V_o^m and δ_o^m given by equations (3.5), we have, for this shell edge,

$$V_o^T = V_o^b + V_o^m = -\frac{1}{2\beta D}\left(2M_o + \frac{1}{\beta}H_o\right) + 0 \tag{7.19a}$$

Figure 7.2 Ring-stiffened cylindrical pressure vessel: (a) general arrangement; (b) actions at shell edges and corresponding reactions on the ring.

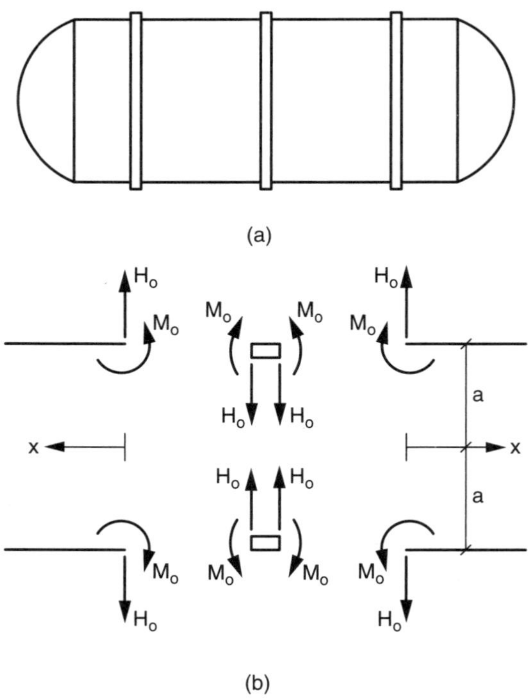

$$\delta_o^T = \delta_o^b + \delta_o^m = \frac{1}{2\beta^2 D}\left(M_o + \frac{1}{\beta}H_o\right) + \frac{pa^2}{2Et}(2-\nu) \tag{7.19b}$$

Owing to the symmetry of the arrangement depicted in Figure 7.2, the ring suffers no net rotation. For compatibility, it follows that the shell edge also suffers no net rotation. Thus for the shell, $V_o^T = 0$. By reference to equation (7.19a), it means that

$$H_o = -2\beta M_o \tag{7.20}$$

Using this result in (7.19b), we obtain

$$\delta_o^T = -\frac{1}{2\beta^2 D}M_o + \frac{pa^2}{2Et}(2-\nu) \tag{7.21}$$

While the ring does not rotate under the system of actions in Figure 7.2(b), it will move inwards under the compressive effect of the net force $2H_o$ acting on it. The self-equilibrating moments M_o on the ring do not, of course, induce any lateral movement. The membrane N_x actions at the edges of the shell portions adjoining the ring are also self-equilibrating and, in any case, do not have lateral components perpendicular to the axis of the shell. Therefore $H_o^m = 0$. The net lateral displacement of the ring is therefore due solely to the H_o forces, and is given by

$$\delta_r^T = (\text{hoop strain due to } 2H_o) \times (\text{ring radius})$$

Now the hoop tension in a ring, due to a radial external force $2H_o$ (per unit length of ring circumference, pointing outwards) in the plane of the ring, is obtained from elementary statics as $2H_o a$, where a is the ring radius (equal, of course, to the radius of the cylinder). The corresponding hoop stress in the ring is therefore $2H_o a / A_r$, where A_r is the cross-sectional area of the ring. Applying Hooke's law, the hoop strain in the ring follows as $2H_o a / (E_r A_r)$, where E_r is the Young modulus of the ring. We can now write the expression for δ_r^T as

$$\delta_r^T = -\frac{2H_o a^2}{E_r A_r} = \frac{4\beta a^2}{E_r A_r}M_o \tag{7.22}$$

in which use has been made of relation (7.20) to eliminate H_o. The negative sign in the intermediate term denotes that the ring moves *inwards* under the (inward-pointing) force $2H_o$, whereas the positive direction of δ is *away* from the axis of revolution of the shell.

Imposing the condition of compatibility of lateral displacement between the ring and the adjoining shell, we have

$$\delta_o^T = \delta_r^T \tag{7.23}$$

which, when use is made of the results of equations (7.21) and (7.22), becomes

$$-\frac{1}{2\beta^2 D}M_o + \frac{pa^2}{2Et}(2-\nu) = \frac{4\beta a^2}{E_r A_r}M_o$$

This gives the result

$$M_o = \frac{pa^2(2-\nu)}{Et}\left(\frac{\beta^2 DE_r A_r}{8a^2\beta^3 D + E_r A_r}\right) \tag{7.24a}$$

From expression (7.20), we obtain the other unknown as

$$H_o = -\frac{2pa^2(2-\nu)}{Et}\left(\frac{\beta^3 DE_r A_r}{8a^2\beta^3 D + E_r A_r}\right) \tag{7.24b}$$

If the ring is of the same material as the shell (i.e. $E_r = E$), and using the relation

$$D = \frac{Et}{4a^2\beta^4} \tag{7.25}$$

(which the reader may easily verify), the above results simplify to

$$M_o = \frac{p(2-\nu)A_r}{4\beta(2t+\beta A_r)} \tag{7.26a}$$

$$H_o = -\frac{p(2-\nu)A_r}{2(2t+\beta A_r)} \tag{7.26b}$$

7.2. Intze tanks

Let us revisit the liquid-filled Intze tank of Figure 3.11 in Section 3.2.4. We will assume that ring beams are provided at the spherical-top/cylinder, cylinder/cone and cone/spherical-bottom junctions; these ring beams will be distinguished by the subscripts $r1$, $r2$ and $r3$, respectively. Each ring beam has two interfaces with the containment shell. Shell-edge and ring-beam variables pertaining to the upper and lower interfaces of the *top ring beam* will be denoted by subscripts A and B, respectively; those pertaining to the upper and lower interfaces of the *middle ring beam* will be denoted by subscripts C and D, respectively; finally, the variables associated with the left and right interfaces of the *bottom ring beam* will be denoted by subscripts E and F, respectively.

Figure 7.3(a) shows the corrective bending-moment and shear-force actions at the edges of the top spherical shell, the cylindrical shell, the conical shell and the bottom spherical shell, as well as the corresponding reactions on the ring beams. The forces $H_{ri}^{m}\ (i = 1,\ 2,\ 3)$ shown acting upon the ring beams are the resultants of the membrane actions on the ring beams, in the direction perpendicular to the vertical axis of the Intze tank. These forces may readily be written down, by reference to Figure 7.3(b), which shows free-body diagrams of the ring beams under the membrane actions of adjacent shell portions. The assumed positive directions for rotations and lateral displacements of shell edges and ring-beam interfaces are shown in Figure 7.3(c).

Figure 7.3 Intze tank: (a) free-body diagram of ring beams and shell edges; (b) net horizontal thrust on ring beams stemming from shell membrane actions; (c) assumed positive directions for rotations and lateral displacements of shell edges and ring interfaces.

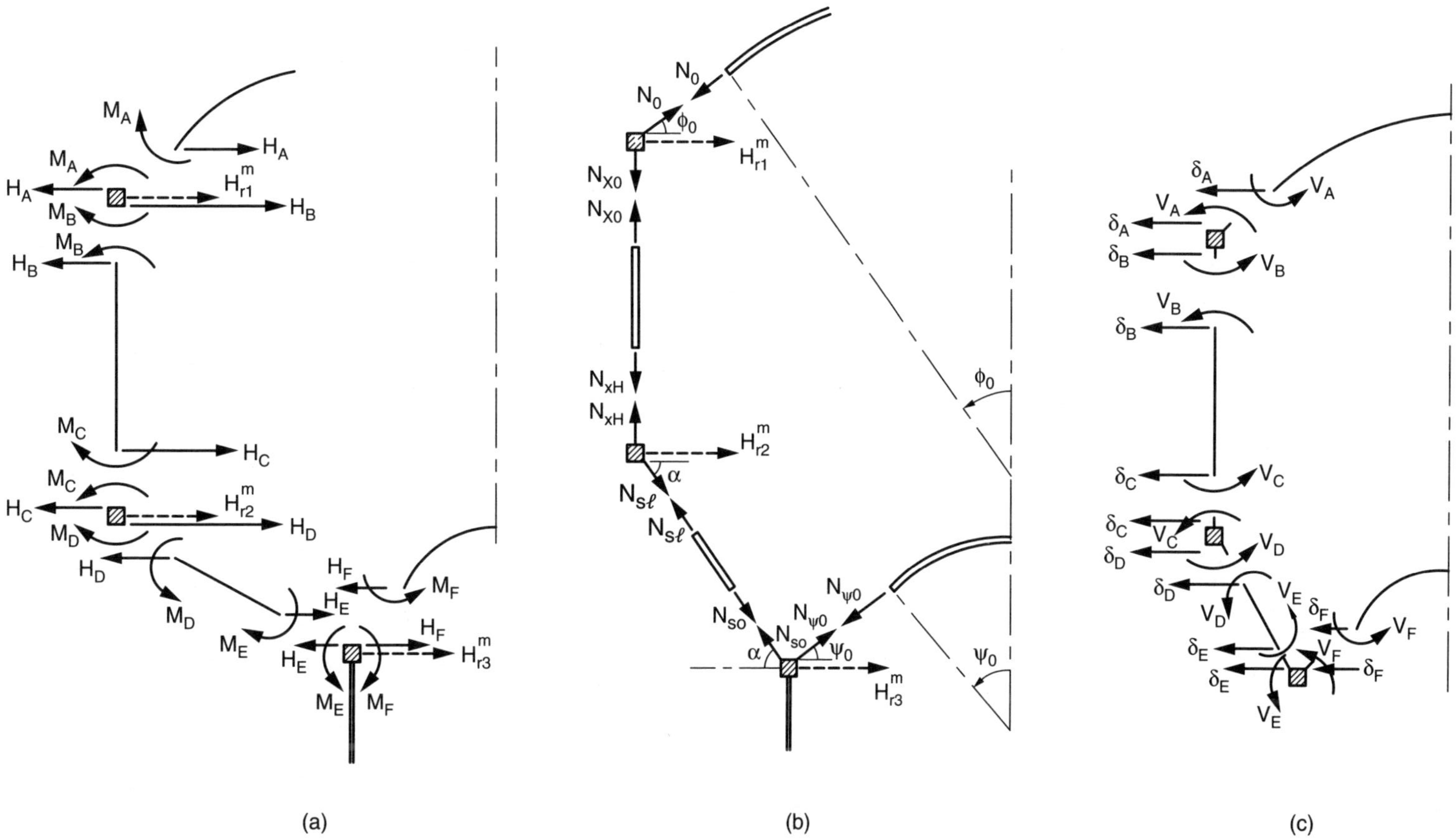

For the ring beam at the junction of the spherical top and the cylindrical sides of the tank, we have

$$H_{r1}^{m} = N_o \cos \phi_o \tag{7.27}$$

where ϕ_o is the value of the ϕ coordinate of the spherical shell at its edge (refer to Figure 3.11), and N_o is the value of N_ϕ in the spherical shell at $\phi = \phi_o$, as given by equation (3.87).

For the ring beam at the junction of the cylindrical shell and the conical shell, we have

$$H_{r2}^{m} = N_{sl} \cos \alpha \tag{7.28}$$

where α is the angle of the cone (as shown in Figure 3.11), and N_{sl} is the value of N_s in the conical shell at its upper edge (i.e. at $s = l$). From equation (3.94), we obtain

$$N_{sl} = N_o \frac{\sin \phi_o}{\sin \alpha} \tag{7.29}$$

so that

$$H_{r2}^{m} = N_o \sin \phi_o \cot \alpha \tag{7.30}$$

For the ring beam at the junction of the conical portion and the spherical bottom, we have

$$H_{r3}^{m} = N_{\psi o} \cos \psi_o - N_{so} \cos \alpha \tag{7.31}$$

where, by reference to Figure 3.11, N_{so} is the value of N_s in the conical shell at its lower edge (i.e. at $s = s_o$), while $N_{\psi o}$ is the value of N_ψ in the spherical bottom at its edge (i.e. at $\psi = \psi_o$). From expression (3.96a), the conical-shell parameter N_{so} is obtained as

$$N_{so} = \frac{1}{s_o}\left[\gamma(\cot \alpha)\left\{\frac{\sin \alpha}{3}\left(l^3 - s_o^3\right) - \frac{K_2}{2}\left(l^2 - s_o^2\right)\right\} + l\, N_o\left(\frac{\sin \phi_o}{\sin \alpha}\right)\right] \tag{7.32}$$

where K_2 is given by expression (3.83) and N_o by expression (3.87). The spherical-shell parameter $N_{\psi o}$ is given by expression (3.101a) with ψ set equal to ψ_o:

$$N_{\psi o} = \frac{\gamma c}{6\left(1 + \cos \psi_o\right)}\left\{3K_3\left(1 + \cos \psi_o\right) - 2c\left(1 + \cos \psi_o + \cos^2 \psi_o\right)\right\} \tag{7.33}$$

Here, c is the radius of the spherical bottom, while K_3 is as defined by equation (3.85).

To obtain the deformation response of a ring beam under the actions in Figure 7.3(a), let us consider the generalised arrangement in Figure 7.4(a), showing a rectangular ring beam of depth d and breadth b, subjected

Figure 7.4 Derivation of ring-beam deformations.

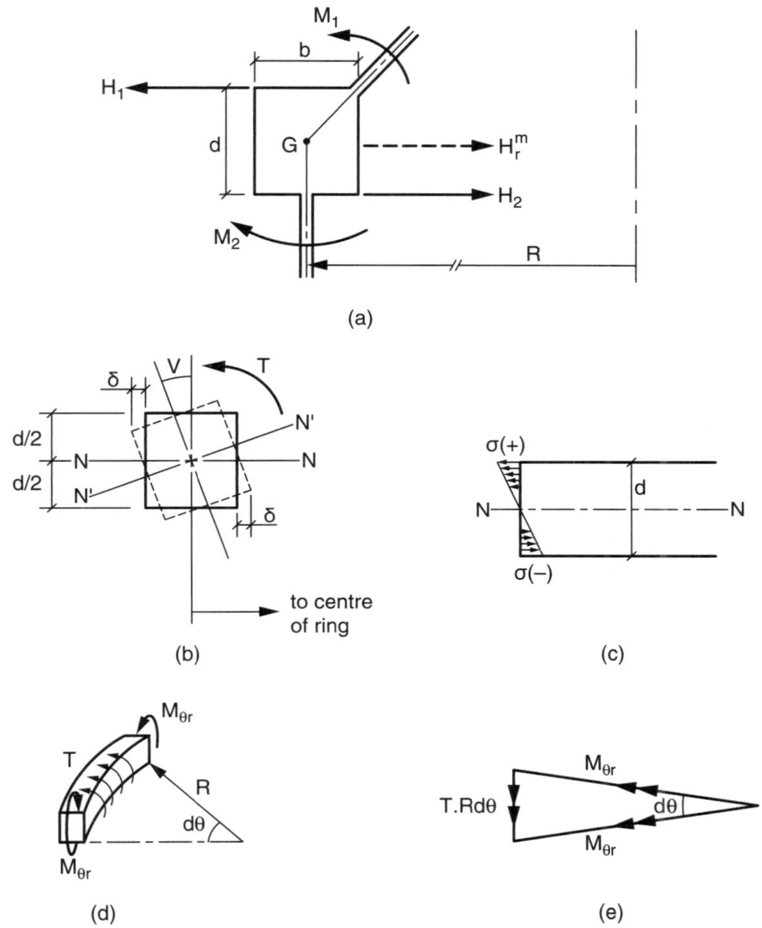

to (*a*) moments and horizontal forces $\{M_1,\ H_1,\ M_2,\ H_2\}$ (which are reactions to shell-edge corrective actions), and (*b*) the horizontal force H_r^m (stemming from membrane action in the adjoining shell portions). The ring is of radius *R*. We assume that the midsurfaces of the shell portions connected to the ring beam, when extended, pass through the centroid *G* of the ring-beam cross-section. Thus, H_r^m has no eccentricity. On the other hand, both H_1 and H_2 have eccentricities about *G* equal to $d/2$.

From similar considerations as those leading to equation (7.22), the lateral-displacement components of the ring beam, namely δ_r^m (due to the membrane resultant H_r^m) and δ_r^H (due to the resultant of the edge forces H_1 and H_2), are given by

$$\delta_r^m = -\frac{H_r^m R^2}{E_r\, bd} \tag{7.34}$$

$$\delta_r^H = \frac{(H_1 - H_2)R^2}{E_r\, bd} \tag{7.35}$$

where, as before, E_r is the Young modulus of the ring beam.

The net torque T on the ring-beam cross-section is due to the difference between M_1 and M_2, and the eccentricities of the forces H_1 and H_2 – that is,

$$T = M_1 - M_2 + H_1\,\frac{d}{2} + H_2\,\frac{d}{2} \tag{7.36}$$

As a result of a small rigid rotation V of the ring-beam cross-section caused by an axisymmetric torque T (per unit length of the ring-beam circumference), points above the horizontal neutral plane *NN* (analogous to the neutral axis of a straight beam undergoing transverse bending) move outwards, while those below *NN* move inwards (refer to Figure 7.4(b)). The magnitude of the lateral displacement of such points is proportional to their distance above or below the plane *NN*. Thus, hoop tension is induced above *NN*, while hoop compression occurs below *NN*.

The associated direct stresses, which act perpendicularly to the vertical cross-section of the ring beam (i.e. in the hoop direction), have a triangular distribution across the depth d of the ring-beam cross-section, as depicted in the side view of Figure 7.4(c). This is the stress distribution of the internal bending moment $M_{\theta r}$ that resists the externally applied torque T, as shown in Figure 7.4(d), depicting a portion of the ring beam of length $ds = R\ d\theta$. The total external torque $TR\ d\theta$ on the ring element has to be equilibrated by the bending moments $M_{\theta r}$ at its ends, which are not in parallel planes owing to the curvature of the ring element. By reference to the vector sketch of the self-equilibrating moments on the ring element (Figure 7.4(e)), we establish the relation

$$T\ R\ d\theta = M_{\theta r}\ d\theta$$

– that is,

$$M_{\theta r} = TR \tag{7.37}$$

For the ring-beam cross-section under the direct stresses associated with $M_{\theta r}$, we have, from simple beam theory,

$$\sigma_{(\max)r} = \frac{M_{\theta r}}{I_r}\, y_{(\max)r} = \pm\ \frac{M_{\theta r}}{(bd^3/12)}\left(\frac{d}{2}\right) = \pm\ \frac{6M_{\theta r}}{bd^2} = \pm\ \frac{6TR}{bd^2} \tag{7.38}$$

in which use has been made of relation (7.37) to eliminate $M_{\theta r}$. The top-surface and bottom-surface strains then follow by application of Hooke's law:

$$\varepsilon_{(\max)r} = \frac{\sigma_{(\max)r}}{E_r} = \pm\ \frac{6TR}{E_r\ bd^2} \tag{7.39}$$

To convert these strains into displacements, we multiply by the ring radius R, so that the top-surface and bottom-surface lateral displacements due to the torque T are

$$\delta^{Tq}_{(\max)r} = \varepsilon_{(\max)r}\ R = \pm\frac{6TR^2}{E_r\ bd^2} \tag{7.40}$$

The rotation of the ring-beam cross-section under the torque T will be denoted by V^{Tq}_r. By reference to Figure 7.4(b), we see that this rotation is given by

$$V^{Tq}_r = \frac{\delta^{Tq}_{(\max)r}}{(\pm d/2)} = \frac{12TR^2}{E_r\ bd^3} \tag{7.41}$$

We are now ready to apply the general results (7.34)–(7.41) to the three ring beams of the Intze tank. Let us denote the dimensions of ring beams 1, 2 and 3 by $\{b_1,\ d_1\}$, $\{b_2,\ d_2\}$ and $\{b_3,\ d_3\}$ respectively. The net lateral displacement δ^T_{rjI} of ring beam $j\,(j = 1,\ 2,\ 3)$ at interface I (I being either A, B, C, D, E or F) is obtained by superimposing δ^m_r, δ^H_r and $\delta^{Tq}_{(\max)r}$, which are given by the general expressions (7.34), (7.35) and (7.40). The net rotation V^T_{rjI} of ring beam j at interface I is simply given by expression (7.41). The generalised parameters $\{M_1,\ H_1,\ M_2,\ H_2,\ H^m_r,\ R,\ b,\ d,\ E_r\}$ appearing in the expressions for torque T (equation 7.36) and deformation parameters δ^m_r, δ^H_r, $\delta^{Tq}_{(\max)r}$ and V^{Tq}_r now assume the specific values corresponding to the ring beam in question. With reference to Figures 7.3(a) and (c), the δ^T_{rjI} and V^T_{rjI} for each of the six ring-beam interfaces (i.e. two interfaces per ring beam) are as follows:

$$\delta^T_{r1I} = -\frac{H^m_{r1}\left(a^2\sin^2\phi_o\right)}{E_{r1}\ b_1d_1} + \frac{\left(H_A - H_B\right)\left(a^2\sin^2\phi_o\right)}{E_{r1}\ b_1d_1}$$

$$\pm\frac{6}{E_{r1}\ b_1d_1^2}\left(M_A - M_B + H_A\ \frac{d_1}{2} + H_B\ \frac{d_1}{2}\right)\left(a^2\sin^2\phi_o\right) \tag{7.42a}$$

(for both $I = A$ and $I = B$, with the upper sign of the $\pm$ symbol applicable to $I = A$ and the lower sign applicable to $I = B$)

$$V^T_{r1A} = V^T_{r1B} = \frac{12}{E_{r1}\ b_1d_1^3}\left(M_A - M_B + H_A\ \frac{d_1}{2} + H_B\ \frac{d_1}{2}\right)\left(a^2\sin^2\phi_o\right) \tag{7.42b}$$

$$\delta^T_{r2I} = -\frac{H^m_{r2}\left(a^2\sin^2\phi_o\right)}{E_{r2}\ b_2d_2} + \frac{\left(H_C - H_D\right)\left(a^2\sin^2\phi_o\right)}{E_{r2}\ b_2d_2}$$

$$\pm\frac{6}{E_{r2}\ b_2d_2^2}\left(M_C - M_D + H_C\ \frac{d_2}{2} + H_D\ \frac{d_2}{2}\right)\left(a^2\sin^2\phi_o\right) \tag{7.43a}$$

(for both $I = C$ and $I = D$, with the upper sign of the $\pm$ symbol applicable to $I = C$ and the lower sign applicable to $I = D$)

$$V^T_{r2C} = V^T_{r2D} = \frac{12}{E_{r2}\ b_2 d_2^3}\left(M_C - M_D + H_C\ \frac{d_2}{2} + H_D\ \frac{d_2}{2}\right)\left(a^2 \sin^2 \phi_o\right) \tag{7.43b}$$

$$\delta^T_{r3E} = \delta^T_{r3F} = -\frac{H^m_{r3}\left(c^2 \sin^2 \psi_o\right)}{E_{r3}\ b_3 d_3} + \frac{\left(H_E - H_F\right)\left(c^2 \sin^2 \psi_o\right)}{E_{r3}\ b_3 d_3}$$

$$+ \frac{6}{E_{r3}\ b_3 d_3^2}\left(M_E - M_F + H_E\ \frac{d_3}{2} - H_F\ \frac{d_3}{2}\right)\left(c^2 \sin^2 \psi_o\right) \tag{7.44a}$$

$$V^T_{r3E} = V^T_{r3F} = \frac{12}{E_{r3}\ b_3 d_3^3}\left(M_E - M_F + H_E\ \frac{d_3}{2} - H_F\ \frac{d_3}{2}\right)\left(c^2 \sin^2 \psi_o\right) \tag{7.44b}$$

The net rotations and net lateral displacements of shell edges will be denoted by V^T_{siI} and δ^T_{siI}, respectively, where $i = 1$ for the spherical top, $i = 2$ for the cylindrical portion, $i = 3$ for the conical portion, and $i = 4$ for the spherical bottom. As before, I is either A, B, C, D, E or F, depending on the ring-shell interface in question. For the six shell edges of the present problem, V^T_{siI} and δ^T_{siI} are obtained in the usual way, by superimposing the membrane-solution component with the edge-effect component. Geometric parameters t, λ, β, D and η will be assigned the subscript $i\left(i = 1,\ 2,\ 3,\ 4\right)$, to denote the various shell portions of the Intze tank.

In the case of the cylindrical and the conical portions, which have two edges each, edge interaction will be neglected since, in practice, these portions are usually not 'too short'. We may also assume that both the spherical top and the spherical bottom are not 'too shallow' (i.e. both ϕ_o and ψ_o are at least 30°, a condition that is satisfied by many practical constructions of Intze tanks), so that the Geckeler approximation provides a satisfactory assessment of edge effects in these shells. The inclination of the conical sides is usually in the range $30° \le \alpha \le 60°$, in which the Geckeler approximation for conical shells yields reasonably satisfactory results.

In the expressions for V^T_{siI} and δ^T_{siI} given below, the parameters $\left\{V^m_{s1A},\ \delta^m_{s1A}\right\}$, $\left\{V^m_{s2B},\ \delta^m_{s2B}\right\}$, $\left\{V^m_{s2C},\ \delta^m_{s2C}\right\}$, $\left\{V^m_{s3D},\ \delta^m_{s3D}\right\}$, $\left\{V^m_{s3E},\ \delta^m_{s3E}\right\}$ and $\left\{V^m_{s4F},\ \delta^m_{s4F}\right\}$ are given by expressions (3.88), (3.91), (3.92), (3.98), (3.97) with $s = s_o$, and (3.102) with $\psi = \psi_o$, respectively. The expressions for $\left\{V^b_{s1A},\ \delta^b_{s1A},\ V^b_{s4F},\ \delta^b_{s4F}\right\}$ are based on equations (6.62). Those for $\left\{V^b_{s2B},\ \delta^b_{s2B},\ V^b_{s2C},\ \delta^b_{s2C}\right\}$ are based on equations (5.41), with due regard to the positive directions for edge actions and deformations adopted for the present problem. Finally, expressions for $\left\{V^b_{s3D},\ \delta^b_{s3D}\right\}$ and $\left\{V^b_{s3E},\ \delta^b_{s3E}\right\}$ are based on equations (6.134) and (6.137), with the parameter l in those equations now replaced with the values l and s_o applicable to interfaces D and E respectively.

$$V^T_{s1A} = V^b_{s1A} + V^m_{s1A} = -\frac{4\lambda_1^3}{Eat_1} M_A + \frac{2\lambda_1^2}{Et_1}\left(\sin \phi_o\right) H_A + V^m_{s1A} \tag{7.45a}$$

$$\delta^T_{s1A} = \delta^b_{s1A} + \delta^m_{s1A} = \frac{2\lambda_1^2}{Et_1}\left(\sin \phi_o\right) M_A - \frac{2\lambda_1 a}{Et_1}\left(\sin^2 \phi_o\right) H_A + \delta^m_{s1A} \tag{7.45b}$$

$$V_{s2B}^{T} = V_{s2B}^{b} + V_{s2B}^{m} = \frac{1}{\beta_2 D_2} M_B + \frac{1}{2\beta_2^2 D_2} H_B + V_{s2B}^{m} \tag{7.45c}$$

$$\delta_{s2B}^{T} = \delta_{s2B}^{b} + \delta_{s2B}^{m} = \frac{1}{2\beta_2^2 D_2} M_B + \frac{1}{2\beta_2^3 D_2} H_B + \delta_{s2B}^{m} \tag{7.45d}$$

$$V_{s2C}^{T} = V_{s2C}^{b} + V_{s2C}^{m} = -\frac{1}{\beta_2 D_2} M_C + \frac{1}{2\beta_2^2 D_2} H_C + V_{s2C}^{m} \tag{7.46a}$$

$$\delta_{s2C}^{T} = \delta_{s2C}^{b} + \delta_{s2C}^{m} = \frac{1}{2\beta_2^2 D_2} M_C - \frac{1}{2\beta_2^3 D_2} H_C + \delta_{s2C}^{m} \tag{7.46b}$$

$$V_{s3D}^{T} = V_{s3D}^{b} + V_{s3D}^{m} = \left(l^2 \cot^2 \alpha\right) \frac{4\eta_3^3}{Et_3} M_D + \left(l^2 \cot^2 \alpha\right) \frac{2\eta_3^2}{Et_3} (\sin \alpha) H_D + V_{s3D}^{m} \tag{7.46c}$$

$$\delta_{s3D}^{T} = \delta_{s3D}^{b} + \delta_{s3D}^{m} = \left(l^2 \cot^2 \alpha\right) \frac{2\eta_3^2}{Et_3} (\sin \alpha) M_D + \left(l^2 \cot^2 \alpha\right) \frac{2\eta_3}{Et_3} (\sin^2 \alpha) H_D + \delta_{s3D}^{m} \tag{7.46d}$$

$$V_{s3E}^{T} = V_{s3E}^{b} + V_{s3E}^{m} = -\left(s_o^2 \cot^2 \alpha\right) \frac{4\eta_3^3}{Et_3} M_E + \left(s_o^2 \cot^2 \alpha\right) \frac{2\eta_3^2}{Et_3} (\sin \alpha) H_E + V_{s3E}^{m} \tag{7.47a}$$

$$\delta_{s3E}^{T} = \delta_{s3E}^{b} + \delta_{s3E}^{m} = \left(s_o^2 \cot^2 \alpha\right) \frac{2\eta_3^2}{Et_3} (\sin \alpha) M_E - \left(s_o^2 \cot^2 \alpha\right) \frac{2\eta_3}{Et_3} (\sin^2 \alpha) H_E + \delta_{s3E}^{m} \tag{7.47b}$$

$$V_{s4F}^{T} = V_{s4F}^{b} + V_{s4F}^{m} = \frac{4\lambda_4^3}{Ect_4} M_F - \frac{2\lambda_4^2}{Et_4} (\sin \psi_o) H_F + V_{s4F}^{m} \tag{7.47c}$$

$$\delta_{s4F}^{T} = \delta_{s4F}^{b} + \delta_{s4F}^{m} = -\frac{2\lambda_4^2}{Et_4} (\sin \psi_o) M_F + \frac{2\lambda_4 c}{Et_4} (\sin^2 \psi_o) H_F + \delta_{s4F}^{m} \tag{7.47d}$$

Imposing the conditions of compatibility of ring-beam and shell-edge deformations, we obtain three independent sets of simultaneous equations, as follows:

$$V_{r1A}^{T} = V_{s1A}^{T} \tag{7.48a}$$

$$\delta_{r1A}^{T} = \delta_{s1A}^{T} \tag{7.48b}$$

$$V_{r1B}^{T} = V_{s2B}^{T} \tag{7.48c}$$

$$\delta_{r1B}^{T} = \delta_{s2B}^{T} \tag{7.48d}$$

(with the V^T's and δ^T's as defined in expressions (7.42) and (7.45), giving four equations in the four unknowns $\{M_A, H_A, M_B, H_B\}$);

$$V_{r2C}^T = V_{s2C}^T \tag{7.49a}$$

$$\delta_{r2C}^T = \delta_{s2C}^T \tag{7.49b}$$

$$V_{r2D}^T = V_{s3D}^T \tag{7.49c}$$

$$\delta_{r2D}^T = \delta_{s3D}^T \tag{7.49d}$$

(with the V^T's and δ^T's as defined in expressions (7.43) and (7.46), giving four equations in the four unknowns $\{M_C, H_C, M_D, H_D\}$);

$$V_{r3E}^T = V_{s3E}^T \tag{7.50a}$$

$$\delta_{r3E}^T = \delta_{s3E}^T \tag{7.50b}$$

$$V_{r3F}^T = V_{s4F}^T \tag{7.50c}$$

$$\delta_{r3F}^T = \delta_{s4F}^T \tag{7.50d}$$

(with the V^T's and δ^T's as defined in expressions (7.44) and (7.47), giving four equations in the four unknowns $\{M_E, H_E, M_F, H_F\}$).

If no ring beams are provided, we first impose the conditions of equilibrium of an element at the junction of two shells. By reference to the generalised arrangement in Figure 7.4(a), the net horizontal force $(H_1 - H_2 - H_r^m)$ and net torque $(M_1 - M_2)$ in the limiting case of vanishing d must both be equal to zero. Applying these conditions of equilibrium (of a ring element of zero hoop and rotational rigidities) to the three shell junctions of the Intze tank, we obtain the relations

$$M_B = M_A \tag{7.51a}$$

$$H_B = H_A - H_{r1}^m \tag{7.51b}$$

$$M_D = M_C \tag{7.52a}$$

$$H_D = H_C - H_{r2}^m \tag{7.52b}$$

$$M_F = M_E \tag{7.53a}$$

$$H_F = H_E - H_{r3}^m \tag{7.53b}$$

We then impose the conditions of compatibility of deformations between adjacent shell edges, to obtain three independent sets of simultaneous equations:

$$V_{s1A}^{T} = V_{s2B}^{T} \tag{7.54a}$$

$$\delta_{s1A}^{T} = \delta_{s2B}^{T} \tag{7.54b}$$

(in the two unknowns $\{M_A,\ H_A\}$, after using relations (7.51) to eliminate M_B and H_B);

$$V_{s2C}^{T} = V_{s3D}^{T} \tag{7.55a}$$

$$\delta_{s2C}^{T} = \delta_{s3D}^{T} \tag{7.55b}$$

(in the two unknowns $\{M_C,\ H_C\}$, after using relations (7.52) to eliminate M_D and H_D);

$$V_{s3E}^{T} = V_{s4F}^{T} \tag{7.56a}$$

$$\delta_{s3E}^{T} = \delta_{s4F}^{T} \tag{7.56b}$$

(in the two unknowns $\{M_E,\ H_E\}$, after using relations (7.53) to eliminate M_F and H_F).

Once the shell-edge redundants (unknowns) have been solved for, the interior actions associated with edge effects are given by expressions (6.68) for the top and bottom spherical closures, with the parameters $\{\phi_e,\ a,\ \lambda,\ M_e,\ H_e\}$ in relations (6.68) now assuming the values $\{\phi_o,\ a,\ \lambda_1,\ M_A,\ H_A\}$ for the top closure and $\{\psi_o,\ c,\ \lambda_4,\ -M_F,\ -H_F\}$ for the bottom closure.

For the vicinity of the cylinder around its upper edge, expressions (5.38) and (5.39) directly give the interior actions of the edge effect, if the origin of the x axis is taken as the upper edge (so that the x axis points downwards). The general parameters $\{t,\ a,\ D,\ \beta,\ M_e,\ H_e\}$ in expressions (5.38) and (5.39) simply take on the specific values $\{t_2,\ a\sin\phi_o,\ D_2,\ \beta_2,\ M_B,\ H_B\}$. For the vicinity of the cylinder around its lower edge, the same expressions give the interior actions of the edge effect, if the origin of the x axis is now taken as the lower edge (so that the x axis points upwards), and the general parameters $\{t,\ a,\ D,\ \beta,\ M_e,\ H_e\}$ are assigned the values $\{t_2,\ a\sin\phi_o,\ D_2,\ \beta_2,\ M_C,\ -H_C\}$.

In the case of the conical shell, the interior actions of the edge effect are given by expressions (6.126), (6.127), (6.129) and (6.130). The general constants $\{C,\ \xi\}$ in these expressions must, of course, first be replaced with $\{2\eta M_e,\ 0\}$ corresponding to the application of an arbitrary edge moment M_e directed as shown in Figure 6.11(a), and $\{-\sqrt{2}\ H_e\sin\alpha,\ -\pi/4\}$ corresponding to the application of an arbitrary edge force H_e directed as shown in Figure 6.11(b), and the expressions for N_s^b, N_θ^b, M_s and M_θ corresponding to the simultaneous application of M_e and H_e obtained by superposition. In these ensuing expressions for net internal actions of the edge effect, the parameters $\{\alpha,\ l,\ \eta,\ M_e,\ H_e\}$ must be assigned the values $\{\alpha,\ l,\ \eta_3,\ M_D,\ -H_D\}$ for the vicinity of the conical shell around its upper edge (with the origin of the x distance coordinate at this upper edge, so that x is measured downwards), and $\{(\pi-\alpha),\ s_o,\ \eta_3,\ M_E,\ H_E\}$ for the vicinity of the conical shell around its lower edge (with the origin of the x distance coordinate being taken at the lower edge, so that x is measured upwards).

To complete the stress analysis of the shell portions of the Intze tank, we must superimpose the stresses of the membrane solution with those of the edge effect, as indicated by expressions (6.70) (applicable to the upper and lower spherical closures), and (5.40) (applicable to the cylindrical portion), using the value of t applicable to the shell portion in question, as well as the expressions

$$\sigma_s^T = \frac{N_s^m}{t_3} + \frac{N_s^b}{t_3} \pm \frac{6M_s}{t_3^2} \tag{7.57a}$$

$$\sigma_\theta^T = \frac{N_\theta^m}{t_3} + \frac{N_\theta^b}{t_3} \pm \frac{6M_\theta}{t_3^2} \tag{7.57b}$$

applicable to the conical shell. The last two terms of the right-hand sides of these expressions are, of course, associated with the internal actions of the edge effect, evaluated as already indicated earlier. The first terms of the right-hand sides of expressions (5.40), (6.70) and (7.57) correspond, of course, to the membrane solution, and are evaluated from expressions (3.86) (for the spherical top), (3.89) (for the cylindrical sides), (3.96) (for the conical shell) and (3.101) (for the spherical bottom), using the value of t applicable to the shell portion in question.

Finally, the net stresses on the upper and lower soffits of the ring beams are obtained by superimposing the extensional effects of the net horizontal force with the flexural effects of the net couple. By reference to the generalised arrangement shown in Figure 7.4(a), we have

$$\sigma_r^T = \frac{(H_1 - H_2 - H_r^m)R}{bd} \pm \frac{6}{bd^2}\left(M_1 - M_2 + H_1\,\frac{d}{2} \pm H_2\,\frac{d}{2}\right)R \tag{7.58}$$

In this expression, the first term of the right-hand side is written down from the considerations leading to expression (7.22), while the second term is written down from expression (7.38). The general parameters $\{H_1,\ H_2,\ H_r^m,\ M_1,\ M_2\}$ appearing in equation (7.58) assume the values $\{H_A,\ H_B,\ H_{r1}^m,\ M_A,\ M_B\}$, $\{H_C,\ H_D,\ H_{r2}^m,\ M_C,\ M_D\}$ and $\{H_E,\ H_F,\ H_{r3}^m,\ M_E,\ M_F\}$ for the top, middle and bottom ring beams, respectively. Similarly, the geometric parameters R, b and d assume the values of the ring beam in question. The upper and lower signs of the ± symbol *between* the two terms of the right-hand side of equation (7.58) refer to the upper and lower surfaces of the ring beam, respectively; the upper sign of the ± symbol *within* the second term of the right-hand side refers to the top and middle ring beams, while the lower sign refers to the bottom ring beam.

7.3. Domes with edge ring beams

Let us consider a spherical dome of constant thickness t and radius a, subjected to a uniform vertical loading q per unit area of the shell midsurface, in the form of self-weight and any superimposed dead load. The dome is assumed to be axisymmetrically supported on horizontal rollers or closely spaced vertical columns, via a ring beam of rectangular section, as shown in Figure 7.5(a). The redundant actions between the shell and the ring beam are as shown in Figure 7.5(b), while the deformations of the shell edge and the ring beam are as shown in Figure 7.5(c).

With the assessment of edge effects based on the Geckeler approximation, and following the considerations of previous sections, the net deformations at the edge of the shell are given by

$$V_s^T = -\frac{4\lambda^3}{Eat}M_s + \frac{2\lambda^2}{Et}(\sin\phi_s)H_s + V_s^m \tag{7.59a}$$

Figure 7.5 Spherical dome supported on a ring beam: (a) general arrangement; (b) actions at the shell edge and reactions on the ring; (c) deformations of the shell edge and the ring.

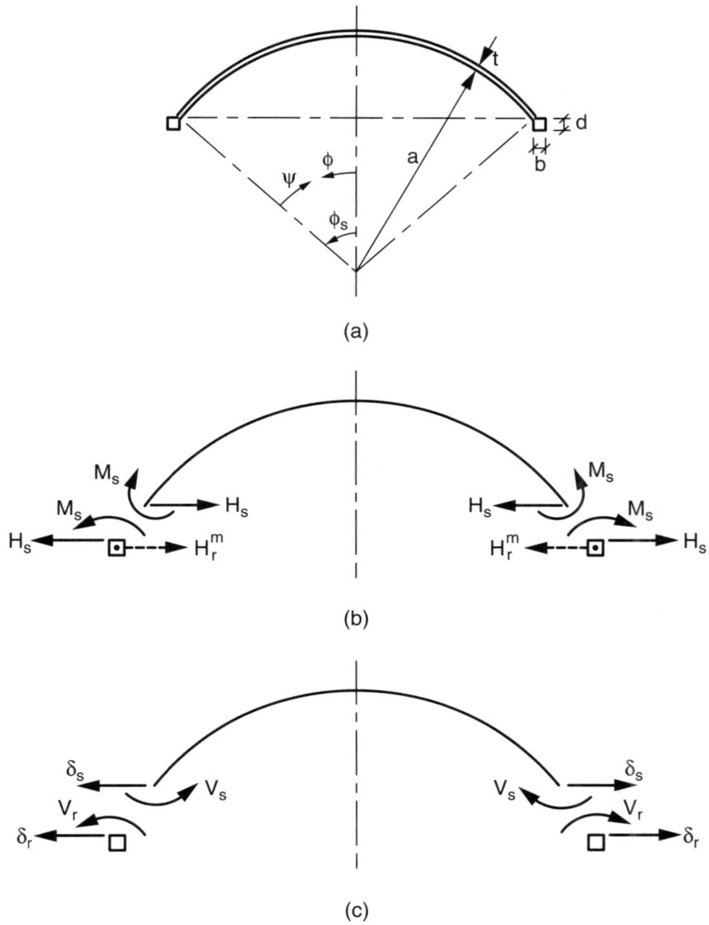

$$\delta_s^T = \frac{2\lambda^2}{Et}(\sin\phi_s)M_s - \frac{2\lambda a}{Et}(\sin^2\phi_s)H_s + \delta_s^m \tag{7.59b}$$

where the membrane deformations V_s^m and δ_s^m are given by results (3.192), with ϕ set equal to ϕ_s.

The net deformations of the ring beam are readily written down by adapting expressions (7.34), (7.35), (7.40) and (7.41) to the present problem:

$$V_r^T = \frac{12}{E_r bd^3}\left(M_s + H_s\frac{d}{2}\right)(a^2\sin^2\phi_s) \tag{7.60a}$$

$$\delta_r^T = -\frac{H_r^m\left(a^2\sin^2\phi_s\right)}{E_r\,bd} + \frac{H_s\left(a^2\sin^2\phi_s\right)}{E_r\,bd} + \frac{6}{E_r\,bd^2}\left(M_s + H_s\,\frac{d}{2}\right)\left(a^2\sin^2\phi_s\right) \tag{7.60b}$$

where, by adapting expression (7.27) and making use of result (3.191a),

$$H_r^m = \left(N_\phi^m\right)_{\phi=\phi_s}\cos\phi_s = -\,qa\left(\frac{\cos\phi_s}{1+\cos\phi_s}\right) \tag{7.61}$$

Imposing the conditions of compatibility of deformations between the shell and the ring beam, we have

$$V_s^T = V_r^T \tag{7.62a}$$

$$\delta_s^T = \delta_r^T \tag{7.62b}$$

These conditions yield two simultaneous equations in the two unknowns M_s and H_s. Once M_s and H_s are known, internal actions due to the edge effect follow from expressions (6.68), and net stresses on the inner and outer surface of the dome in turn follow from expressions (6.70). The stresses on the upper and lower surfaces of the ring beam are obtained by combining the uniform extensional stresses due to the thrust $\left(H_s - H_r^m\right)$, with the depthwise varying flexural stresses due to the torque $\left(M_s + H_s(d/2)\right)$:

$$\sigma_r^T = \frac{\left(H_s - H_r^m\right)\left(a\sin\phi_s\right)}{bd} \pm \frac{6\left(M_s + H_s\,\frac{d}{2}\right)\left(a\sin\phi_s\right)}{bd^2} \tag{7.63}$$

where the upper and lower signs of the ± symbol refer to the upper and lower surfaces, respectively, of the ring beam.

7.4. Multi-segmented spherical vessels

7.4.1 Shell geometry

Here we illustrate how the tools developed in Chapters 2, 3 and 6 can be used to explore the feasibility of new shell forms. Figure 7.6 shows a liquid-storage vessel comprising an assembly of spherical shell segments of different radii, whose centres of curvatures all lie on the axis of revolution of the vessel. The arrangement is therefore axisymmetric, where the uppermost and lowermost shell segments are spherical caps, and the rest of the shell segments are spherical frusta. This type of vessel has recently been proposed for high-capacity liquid containment.[5]

As one moves from top to bottom, we may denote the various shell regions or segments by $\{S1, S2, S3, \ldots\}$ and the junctions between these shell segments by {J1, J2, J3, …}. The radii of shells {S1, S2, S3} are denoted by $\{a_1, a_2, a_3\}$ and so forth. The angular coordinate ϕ (which is the angle between the normal to the shell midsurface and the axis of revolution of the entire shell assembly) defines the position of any point on a given shell segment. For the shell cap S1 (uppermost portion of the assembly), we will denote the angular coordinate of the edge of the cap by ϕ_{10}. For all other segments Si ($i = 2, 3, 4$ etc.) below this, we will define the upper and lower edges of segment Si by the coordinates ϕ_{i1} and ϕ_{i2} respectively.

Figure 7.6 Multi-segmented spherical vessel: (a) shell segments, junctions and corresponding geometric parameters; (b) external appearance of the vessel.

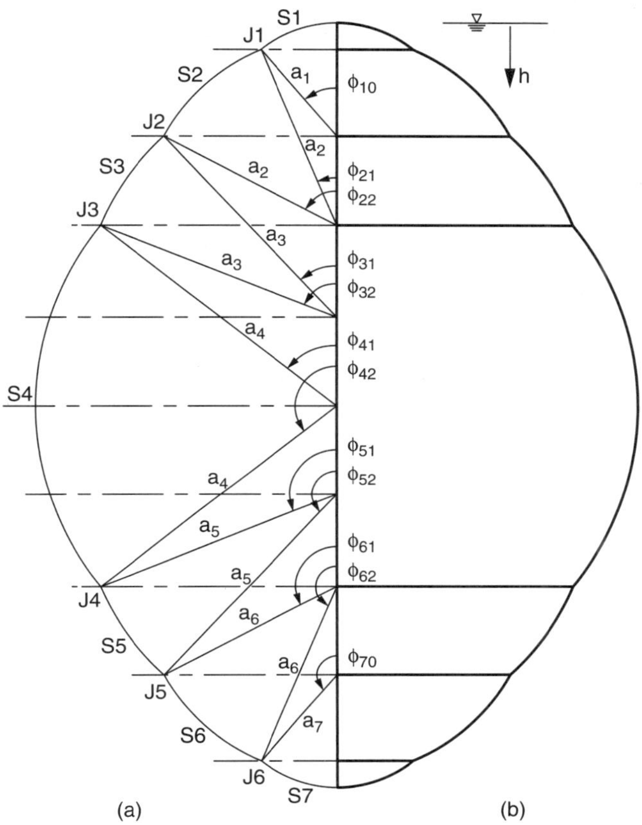

Let us consider the central segment (S4 in our illustration). Progressive addition of segments S3, S2 and Sl, with slope enhancements of $(\phi_{32} - \phi_{41})$, $(\phi_{22} - \phi_{31})$ and $(\phi_{10} - \phi_{21})$ at junctions J3, J2 and Jl respectively, adds height and additional storage capacity to the basic spherical vessel of radius a_4. Similar enhancement of capacity is achieved by the addition of segments S5, S6 and S7 in the lower part of the vessel. The overall result is a spherical assembly of relatively large storage capacity. It may be noted that, if this was a pressure vessel, to achieve the same storage capacity while keeping the radius of the vessel constant would require a sphere of radius bigger than a_4, which would attract higher shell stresses since these are proportional to the radius. However, in the present case of liquid containment, the stress-reducing benefit of radius limitation is likely to be offset by the higher hydrostatic pressures associated with the taller segmented vessel.

In accordance with the considerations of Chapter 1, the slope and curvature discontinuities at the shell junctions attract bending disturbances, but the inward-pointing intersections in the profile could have the effect of stiffening the shell response there, possibly resulting in a beneficial lowering of hoop stresses. From an aesthetics point of view, the segmented geometry of the storage vessel has a rather pleasing appearance, which might favour the adoption of this type of vessel in locations where appearance is a major consideration.

We may use the membrane theory to calculate the state of stress of the vessel under internal hydrostatic pressure. However, the results become inadequate in the vicinity of the geometric discontinuities J1, J2, J3 and so on, and the more comprehensive bending theory of shells must be invoked. As before, we will regard the membrane solution as the particular solution of the differential equations describing the behaviour of the shell, and take the homogeneous solution in the form of a system of axisymmetric bending moments and shearing forces applied at the shell edges. The net effects in the shell are then simply the sum of the membrane effects and the edge effects.

7.4.2 Loading formulation

Consider the case of the vessel completely filled with liquid of weight γ per unit volume. By reference to Figure 7.6, the hydrostatic pressure p_r on the inside surface of the shell at a depth h from the top is given by

$$p_r = \gamma h \tag{7.64}$$

Let us denote by $h_i(\phi)$ the depth of liquid at the coordinate ϕ of shell S*i*. Notice that each shell segment S*i* has its own range of ϕ values, which is different to that of other shell segments. For the very top shell (S1), which has a radius a_1, the depth of liquid at any given point is given by

$$h_1(\phi) = a_1(1-\cos\phi) \quad [0 \le \phi \le \phi_{10}] \tag{7.65a}$$

At $\phi = \phi_{10}$, we have

$$h_{10} = a_1(1-\cos\phi_{10}) \tag{7.65b}$$

For shell S2, which has a radius of a_2, the depth of liquid at the coordinate ϕ of this shell is given by

$$h_2(\phi) = h_{10} + a_2(\cos\phi_{21} - \cos\phi) \quad [\phi_{21} \le \phi \le \phi_{22}] \tag{7.66a}$$

At $\phi = \phi_{22}$, we have

$$h_{22} = h_{10} + a_2(\cos\phi_{21} - \cos\phi_{22}) \tag{7.66b}$$

For shell S3, which has a radius of a_3, the depth of liquid at the coordinate ϕ of this shell is given by

$$h_3(\phi) = h_{22} + a_3(\cos\phi_{31} - \cos\phi) \quad [\phi_{31} \le \phi \le \phi_{32}] \tag{7.67a}$$

At $\phi = \phi_{32}$, we have

$$h_{32} = h_{22} + a_3(\cos\phi_{31} - \cos\phi_{32}) \tag{7.67b}$$

In general, for shell S*i* ($i > 2$), which has a radius of a_i, the depth of liquid at the coordinate ϕ of that shell is given by

$$h_i(\phi) = h_{(i-1)2} + a_i(\cos\phi_{i1} - \cos\phi) \quad [\phi_{i1} \le \phi \le \phi_{i2}] \tag{7.68a}$$

At $\phi = \phi_{i2}$, we have

$$h_{i2} = h_{(i-1)2} + a_i\left(\cos\phi_{i1} - \cos\phi_{i2}\right) \tag{7.68b}$$

A more general way of expressing the results for all shell regions Si, where this time $i = 1, 2, \ldots, I$ (I being the total number of shell segments, e.g. $I = 7$ for the vessel in Figure 7.6), is achieved by writing $h_i(\phi)$ as follows:

$$h_i(\phi) = k_i - a_i\cos\phi \tag{7.69}$$

where the parameter k_i is a constant defined differently for the various regions as follows:

for $i = 1$: $\quad k_i = k_1 = a_1 \qquad$ (7.70a)

for $i = 2$: $\quad k_i = k_2 = h_{10} + a_2\cos\phi_{21} \qquad$ (7.70b)

for $2 < i < I$: $\quad k_i = h_{(i-1)2} + a_i\cos\phi_{i1} \qquad$ (7.70c)

for $i = I$: $\quad k_i = k_I = h_{(I-1)2} + a_I\cos\phi_{I0} \qquad$ (7.70d)

Substituting the above generalised expression for $h_i(\phi)$ in equation (7.64), the hydrostatic pressure p_r at any given point of any given shell Si ($i = 1, 2, \ldots, I$) becomes

$$p_r(\phi) = \gamma h_i(\phi) = \gamma k_i - \gamma a_i\cos\phi \tag{7.71}$$

7.4.3 Membrane stress resultants

We can obtain the meridional and hoop membrane stress resultants in the various segments of the vessel by applying the general solutions of Chapter 2. For a spherical shell, $r_1 = r_2 = a$ (radius of the spherical shell). If hydrostatic pressure is the only loading applied, then $p_\phi = 0$. We may write equation (2.15) as follows

$$N_\phi = \frac{1}{a_i\sin^2\phi}\left[a_i^2\int p_r\cos\phi\sin\phi\,d\phi + B_i\right] \tag{7.72}$$

for every spherical shell region Si of the vessel in Figure 7.6. Substituting the expression for p_r (equation 7.71) into this equation and evaluating the integral, we obtain

$$N_\phi = \frac{1}{6}\frac{\gamma a_i^2}{\sin^2\phi}\left(2\cos^3\phi - 3\frac{k_i}{a_i}\cos^2\phi + C_i\right) \tag{7.73}$$

where the constant of integration C_i has to be evaluated from a suitable boundary condition.

The constant of integration C_i for the end shells (shell S1 and shell SI) may be evaluated from the finiteness condition for N_ϕ in these regions. For shell S1 (top of the vessel),

$$N_\phi^{(1)} = \frac{1}{6}\frac{\gamma a_1^2}{\sin^2\phi}\left(2\cos^3\phi - 3\cos^2\phi + C_1\right) \tag{7.74}$$

(the notation $N_\phi^{(1)}$ denoting N_ϕ for shell S1). The finiteness condition for N_ϕ at $\phi = 0$ (apex of the vessel) gives $C_1 = 1$, so that

$$N_\phi^{(1)} = \frac{\gamma a_1^2}{6}\left(\frac{1-\cos\phi}{1+\cos\phi}\right)(1+2\cos\phi) \tag{7.75a}$$

The hoop stress resultant follows from equation (2.12):

$$N_\theta^{(1)} = \frac{\gamma a_1^2}{6}\left(\frac{1-\cos\phi}{1+\cos\phi}\right)(5+4\cos\phi) \tag{7.75b}$$

These are well-known results for spherical tanks of constant radius (see Chapter 3).

Let us assume that the vessel is supported at a level somewhere above the edge of the bottom shell cap SI (the exact position of the supports does not matter). For the shell cap SI, with the parameter k_I as defined by expression (7.70d), equation (7.73) may be written as

$$N_\phi^{(I)} = \frac{1}{6}\frac{\gamma a_I^2}{\sin^2\phi}\left(2\cos^3\phi - 3\frac{k_I}{a_I}\cos^2\phi + C_I\right) \tag{7.76}$$

The finiteness condition for N_ϕ at $\phi = \pi$ (the lowest point of the vessel) gives

$$C_I = 2 + 3\frac{k_I}{a_I} \tag{7.77}$$

Substituting this value for C_I into equation (7.75a), and making use of equation (2.12), we obtain the meridional and hoop stress resultants in shell SI as:

$$N_\phi^{(I)} = \frac{1}{6}\frac{\gamma a_I^2}{\sin^2\phi}\left(2 + 3\frac{k_I}{a_I}\sin^2\phi + 2\cos^3\phi\right) \tag{7.78a}$$

$$N_\theta^{(I)} = \frac{1}{6}\gamma a_I^2\left\{3\frac{k_I}{a_I} + \left(\frac{2}{1-\cos\phi}\right)(2\cos^2\phi - 2\cos\phi - 1)\right\} \tag{7.78b}$$

Consider the interface of shell S1 and shell S2 (i.e. junction J1 in Figure 7.6). For shell S1, the value of N_ϕ at junction J1 (where ϕ for this shell is equal to ϕ_{10}) is given by:

$$\left(N_\phi^{(1)}\right)_{\phi=\phi_{10}} = N_{\phi 1}^{(1)} = \frac{1}{6}\frac{\gamma a_1^2}{\sin^2\phi_{10}}\left(1 - 3\cos^2\phi_{10} + 2\cos^3\phi_{10}\right) \tag{7.79}$$

For shell S2, the value of N_ϕ at junction J1 (where ϕ for this shell is equal to ϕ_{21}) is given by:

$$\left(N_\phi^{(2)}\right)_{\phi=\phi_{21}} = N_{\phi 1}^{(2)} = \frac{1}{6}\frac{\gamma a_2^2}{\sin^2\phi_{21}}\left(2\cos^3\phi_{21} - 3\frac{k_2}{a_2}\cos^2\phi_{21} + C_2\right) \tag{7.80}$$

Figure 7.7 Forces acting upon a spherical cap above a horizontal section Y – Y.

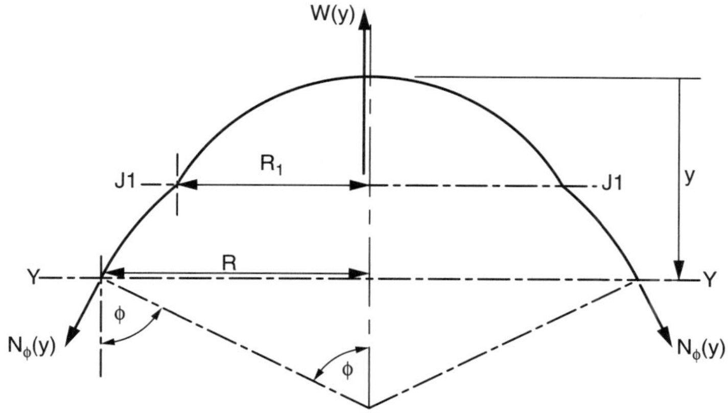

Figure 7.7 shows the forces acting upon the portion of the vessel above a horizontal cross-section $Y-Y$ located at an arbitrary depth y below the apex of the vessel and within the domain of shell S2. We will refer to such a portion of the vessel as a *vessel cap*. Let the radius of the circular edge of the vessel cap be denoted by R. The force $W(y)$ represents the vertical resultant of the hydrostatic pressure acting over the entire surface of the vessel cap.

Treating the vessel cap as a free-body diagram, we see that $W(y)$ is balanced by the vertical component of the stress resultants $N_\phi(y)$ acting along the circular edge of the vessel cap. We may therefore write

$$W(y) = N_\phi(y)\sin\phi\,(2\pi R) \tag{7.81}$$

Let the vertical resultant on the portion of the vessel above junction J1 (i.e. on shell S1) be denoted by W_o. Let the value of R at junction J1 be denoted by R_1. At junction J1, just within shell S1, we may write

$$W_o = N_{\phi 1}^{(1)}\sin\phi_{10}\,(2\pi R_1) \tag{7.82a}$$

and just within shell S2, we may write

$$W_o = N_{\phi 1}^{(2)}\sin\phi_{21}\,(2\pi R_1) \tag{7.82b}$$

From these two expressions for W_o, we deduce that

$$N_{\phi 1}^{(2)} = \left(\frac{\sin\phi_{10}}{\sin\phi_{21}}\right)N_{\phi 1}^{(1)} \tag{7.83a}$$

which is the condition of vertical equilibrium at junction J1. Substituting the expressions for $N_{\phi 1}^{(1)}$ and $N_{\phi 1}^{(2)}$ (equations (7.79) and (7.80)) into equation (7.83a), we obtain

$$\frac{1}{6}\frac{\gamma a_2^2}{\sin^2\phi_{21}}\left(2\cos^3\phi_{21} - 3\frac{k_2}{a_2}\cos^2\phi_{21} + C_2\right) = \left(\frac{\sin\phi_{10}}{\sin\phi_{21}}\right)\frac{1}{6}\frac{\gamma a_1^2}{\sin^2\phi_{10}}\left(1 - 3\cos^2\phi_{10} + 2\cos^3\phi_{10}\right) \tag{7.83b}$$

giving the solution

$$C_2 = \frac{a_1^2}{a_2^2}\left(\frac{\sin\phi_{21}}{\sin\phi_{10}}\right)\left(1 - 3\cos^2\phi_{10} + 2\cos^3\phi_{10}\right) - 2\cos^3\phi_{21} + 3\frac{k_2}{a_2}\cos^2\phi_{21} \tag{7.84}$$

Using this result in equation (7.73) (to eliminate C_i for $i = 2$), we can write N_ϕ for shell S2 in explicit form, and in turn using the ensuing expression for N_ϕ in equation (2.12), we also obtain the hoop stress resultant N_θ in explicit form. The results are:

$$N_\phi^{(2)} = \frac{1}{6}\frac{\gamma a_2^2}{\sin^2\phi}\left\{\left(2\cos^3\phi - 3\frac{k_2}{a_2}\cos^2\phi\right) - \left(2\cos^3\phi_{21} - 3\frac{k_2}{a_2}\cos^2\phi_{21}\right)\right.$$
$$\left. + \frac{a_1^2}{a_2^2}\left(\frac{\sin\phi_{21}}{\sin\phi_{10}}\right)\left(1 - 3\cos^2\phi_{10} + 2\cos^3\phi_{10}\right)\right\} \tag{7.85a}$$

$$N_\theta^{(2)} = a_2 p_r - N_\phi^{(2)}$$

$$= \frac{\gamma a_2^2}{6}\left[6\left(\frac{k_2}{a_2} - \cos\phi\right) - \frac{1}{\sin^2\phi}\left\{\left(2\cos^3\phi - 3\frac{k_2}{a_2}\cos^2\phi\right) - \left(2\cos^3\phi_{21} - 3\frac{k_2}{a_2}\cos^2\phi_{21}\right)\right.\right.$$
$$\left.\left. + \frac{a_1^2}{a_2^2}\left(\frac{\sin\phi_{21}}{\sin\phi_{10}}\right)\left(1 - 3\cos^2\phi_{10} + 2\cos^3\phi_{10}\right)\right\}\right] \tag{7.85b}$$

where, from equations (7.70b) and (7.65b), k_2 has the explicit form

$$k_2 = h_{10} + a_2\cos\phi_{21} = a_1\left(1 - \cos\phi_{10}\right) + a_2\cos\phi_{21} \tag{7.86}$$

For any subsequent shell Si ($i > 2$), the C_i in equation (7.73) is obtained from

$$N_{\phi(i-1)}^{(i)} = \left(\frac{\sin\phi_{(i-1)2}}{\sin\phi_{i1}}\right)N_{\phi(i-1)}^{(i-1)} \tag{7.87}$$

which is the condition of vertical equilibrium at junction J($i-1$), noting that the $(i-1)$ in the subscript for N_ϕ refers to junction J($i-1$) between shell S($i-1$) and shell Si. Expanding the N_ϕ terms in equation (7.87) on the basis of equation (7.73) and rearranging, we obtain

$$C_i = \frac{a_{i-1}^2}{a_i^2}\left(\frac{\sin\phi_{i1}}{\sin\phi_{(i-1)2}}\right)\left(2\cos^3\phi_{(i-1)2} - 3\frac{k_{i-1}}{a_{i-1}}\cos^2\phi_{(i-1)2} + C_{i-1}\right)$$
$$- \left(2\cos^3\phi_{i1} - 3\frac{k_i}{a_i}\cos^2\phi_{i1}\right) \tag{7.88}$$

where C_{i-1} is known from the preceding step. With C_i now known, the stress resultants $N_\phi^{(i)}$ and $N_\theta^{(i)}$ for shell Si (where $i = 3, 4, ..., (I-1)$) then follow:

$$N_\phi^{(i)} = \frac{1}{6}\frac{\gamma a_i^2}{\sin^2\phi}\left\{\left(2\cos^3\phi - 3\frac{k_i}{a_i}\cos^2\phi\right) - \left(2\cos^3\phi_{i1} - 3\frac{k_i}{a_i}\cos^2\phi_{i1}\right)\right.$$
$$\left. + \frac{a_{i-1}^2}{a_i^2}\left(\frac{\sin\phi_{i1}}{\sin\phi_{(i-1)2}}\right)\left(2\cos^3\phi_{(i-1)2} - 3\frac{k_{i-1}}{a_{i-1}}\cos^2\phi_{(i-1)2} + C_{i-1}\right)\right\} \tag{7.89a}$$

$$N_\theta^{(i)} = \frac{\gamma a_i^2}{6}\left[6\left(\frac{k_i}{a_i} - \cos\phi\right) - \frac{1}{\sin^2\phi}\left\{\left(2\cos^3\phi - 3\frac{k_i}{a_i}\cos^2\phi\right) - \left(2\cos^3\phi_{i1} - 3\frac{k_i}{a_i}\cos^2\phi_{i1}\right)\right.\right.$$
$$\left.\left. + \frac{a_{i-1}^2}{a_i^2}\left(\frac{\sin\phi_{i1}}{\sin\phi_{(i-1)2}}\right)\left(2\cos^3\phi_{(i-1)2} - 3\frac{k_{i-1}}{a_{i-1}}\cos^2\phi_{(i-1)2} + C_{i-1}\right)\right\}\right] \tag{7.89b}$$

Note that the above expressions are only valid for regions of the shell above the level of the supports. For regions below the level of the supports, the constants have to be evaluated successively from the bottom upwards (since C_I is known) until we get to the supports.

The calculation of edge effects around the junction locations (which we will deal with later) requires the horizontal resultant of the membrane meridional stress resultants at a junction. Consider junction Ji between shell Si and shell S$(i+1)$. The horizontal resultant H_i^m at this junction (the superscript m denoting association with the membrane solution) is simply given by the difference between the horizontal components of the membrane meridional stress resultants at the edges of the two shells:

$$H_i^m = N_{\phi i}^{(i)}\cos\phi_{i2} - N_{\phi i}^{(i+1)}\cos\phi_{(i+1)1}$$
$$= \frac{1}{6}\frac{\gamma a_i^2}{\sin^2\phi_{i2}}\left(2\cos^3\phi_{i2} - 3\frac{k_i}{a_i}\cos^2\phi_{i2} + C_i\right)\cos\phi_{i2}$$
$$- \frac{1}{6}\frac{\gamma a_{i+1}^2}{\sin^2\phi_{(i+1)1}}\left(2\cos^3\phi_{(i+1)1} - 3\frac{k_{i+1}}{a_{i+1}}\cos^2\phi_{(i+1)1} + C_{i+1}\right)\cos\phi_{(i+1)1} \tag{7.90}$$

7.4.4 Membrane deformations

The deformations δ (horizontal displacement of any point on the vessel) and V (rotation of the meridian) may be evaluated from equations (2.67) and (2.72). For the present problem, and considering shell Si, these equations become

$$\delta^m = \frac{\gamma a_i^3}{6Et}(\sin\phi)\left\{6\left(\frac{k_i}{a_i} - \cos\phi\right) - \frac{1+\nu}{\sin^2\phi}\left(2\cos^3\phi - 3\frac{k_i}{a_i}\cos^2\phi + C_i\right)\right\} \tag{7.91a}$$

$$V^m = -\frac{\gamma a_i^2}{Et}\sin\phi \tag{7.91b}$$

Figure 7.8 Arbitrary junction O and geometric parameters for the upper and lower shells.

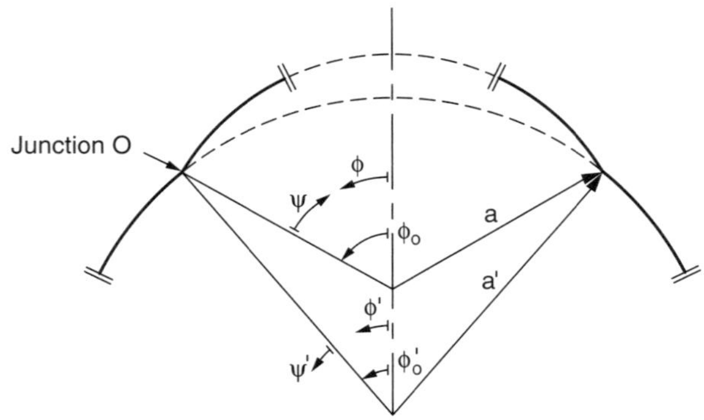

Let us denote an arbitrary shell junction (J1, J2, J3, etc.) by O. Such a junction is depicted in Figure 7.8. Furthermore, let us denote the parameters pertaining to the upper shell by plain symbols (without a prime), and parameters pertaining to the lower shell by primed symbols. Thus, at a junction where shell Si is connected to shell S($i+1$), the angle ϕ_{i2} (corresponding to the lower edge of shell Si) becomes simply ϕ_o, while the angle $\phi_{(i+1)1}$ (corresponding to the upper edge of shell S($i+1$)) becomes simply ϕ'_o. In this notation, the radii of the upper and lower shells become simply a and a', respectively, and so forth. This notation greatly simplifies the writing down of results for the junction effects, since the same set of expressions becomes applicable for all junctions of the vessel.

Applying this notation to equation (7.90), we may write down the expression for the horizontal resultant of the membrane meridional stress resultants at junction O as follows:

$$H_o^m = \frac{1}{6}\frac{\gamma a^2}{\sin^2\phi_o}\left(2\cos^3\phi_o - 3\frac{k}{a}\cos^2\phi_o + C\right)\cos\phi_o - \frac{1}{6}\frac{\gamma a'^2}{\sin^2\phi'_o}\left(2\cos^3\phi'_o - 3\frac{k'}{a'}\cos^2\phi'_o + C'\right)\cos\phi'_o \tag{7.92}$$

where the parameters k and C (denoted earlier as k_i and C_i for shell Si) relate to the upper shell, and k' and C' relate to the lower shell.

The membrane edge deformations for the upper and lower shells at junction O become:

$$\delta_o^m = \frac{\gamma a^3}{6Et}(\sin\phi_o)\left\{6\left(\frac{k}{a} - \cos\phi_o\right) - \frac{1+\nu}{\sin^2\phi_o}\left(2\cos^3\phi_o - 3\frac{k}{a}\cos^2\phi_o + C\right)\right\} \tag{7.93a}$$

$$\delta_o'^m = \frac{\gamma a'^3}{6Et'}(\sin\phi'_o)\left\{6\left(\frac{k'}{a'} - \cos\phi'_o\right) - \frac{1+\nu}{\sin^2\phi'_o}\left(2\cos^3\phi'_o - 3\frac{k'}{a'}\cos^2\phi'_o + C'\right)\right\} \tag{7.93b}$$

$$V_o^m = -\frac{\gamma a^2}{Et}\sin\phi_o \tag{7.94a}$$

$$V_o'^m = -\frac{\gamma a'^2}{Et'} \sin \phi_o' \qquad (7.94b)$$

where the thickness parameters t and t' refer to the upper and lower shells respectively. The positive directions of these are shown in Figure 7.9(a).

7.4.5 Bending solution

Let us assume that the edges of the upper and lower shells at junction O are subjected to pairs of axisymmetric bending moments and horizontal shear forces, as shown in Figure 7.9(b). Each shell will experience interior bending moments, direct stress resultants and deformations in the vicinity of the edge, in response to the application of these axisymmetric edge actions. To quantify the edge effects, we may adopt the asymptotic solution of Hetényi for the axisymmetric bending of spherical shells, as presented in Chapter 6. For this solution approach, general expressions for interior actions and deformations due to axisymmetric bending moments and shear forces applied at the edge of the shell were developed in Chapter 6. We may readily apply the results to the problem depicted in Figure 7.9(b).

Figure 7.9 Edge deformations and edge actions at junction O: (a) horizontal displacements and meridional rotations; (b) horizontal shear forces and bending moments.

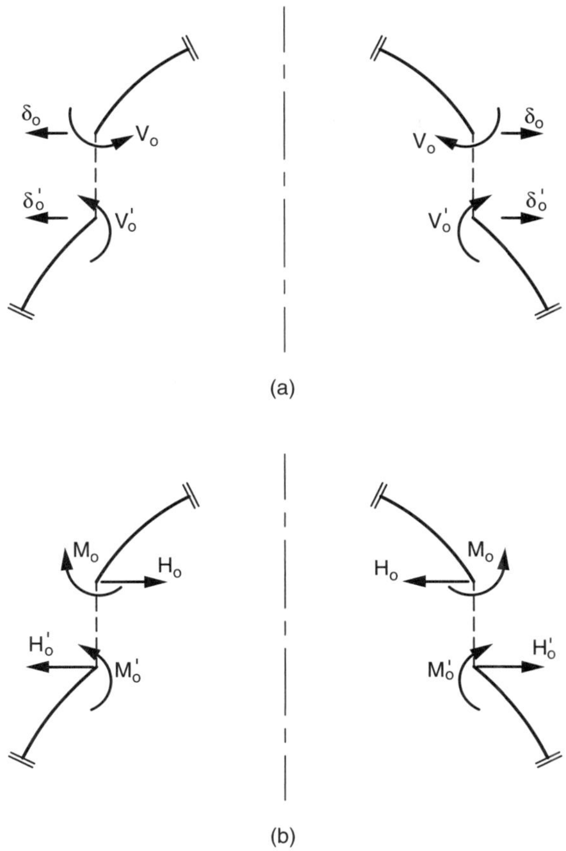

By reference to Figure 7.8, if the meridional angle measured from the shell edge is denoted by ψ for the upper shell and by ψ' for the lower shell, it is clear that $\psi = \phi_o - \phi$ for the upper shell, and $\psi' = \phi' - \phi_o'$ for the lower shell. Let λ and λ' denote the shell slenderness parameters; for the upper shell, $\lambda = \left[3(1-\nu^2)(a^2/t^2)\right]^{1/4}$, and for the lower shell, $\lambda' = \left[3(1-\nu^2)(a'^2/t'^2)\right]^{1/4}$.

The results for actions in the interior of the shell and deformations at the shell edge are as follows (the superscript b denotes that these quantities are associated with the bending disturbance):

Upper shell

$$N_\phi^b = -\cot(\phi_o - \psi)\frac{e^{-\lambda\psi}}{\sqrt{\sin(\phi_o - \psi)}}\left\{\frac{2\lambda}{aK_1}(\sin\phi_o)^{1/2} M_o \sin\lambda\psi - \frac{(1+K_1^2)^{1/2}}{K_1}(\sin\phi_o)^{3/2} H_o \sin(\lambda\psi+\beta)\right\} \tag{7.95a}$$

$$N_\theta^b = \frac{\lambda}{2}\frac{e^{-\lambda\psi}}{\sqrt{\sin(\phi_o - \psi)}}\left[\frac{2\lambda}{aK_1}(\sin\phi_o)^{1/2} M_o\left\{2\cos\lambda\psi - (k_1+k_2)\sin\lambda\psi\right\} - \frac{(1+K_1^2)^{1/2}}{K_1}(\sin\phi_o)^{3/2} H_o\left\{2\cos(\lambda\psi+\beta) - (k_1+k_2)\sin(\lambda\psi+\beta)\right\}\right] \tag{7.95b}$$

$$M_\phi = \frac{a}{2\lambda}\frac{e^{-\lambda\psi}}{\sqrt{\sin(\phi_o - \psi)}}\left[\frac{2\lambda}{aK_1}(\sin\phi_o)^{1/2} M_o\left\{k_1\cos\lambda\psi + \sin\lambda\psi\right\} - \frac{(1+K_1^2)^{1/2}}{K_1}(\sin\phi_o)^{3/2} H_o\left\{k_1\cos(\lambda\psi+\beta) + \sin(\lambda\psi+\beta)\right\}\right] \tag{7.96a}$$

$$M_\theta = \frac{a}{4\lambda\nu}\frac{e^{-\lambda\psi}}{\sqrt{\sin(\phi_o - \psi)}}\left[\frac{2\lambda}{aK_1}(\sin\phi_o)^{1/2} M_o \times \left[\left\{(1+\nu^2)(k_1+k_2) - 2k_2\right\}\cos\lambda\psi + 2\nu^2\sin\lambda\psi\right] - \frac{(1+K_1^2)^{1/2}}{K_1}(\sin\phi_o)^{3/2} H_o \times \left[\left\{(1+\nu^2)(k_1+k_2) - 2k_2\right\}\cos(\lambda\psi+\beta) + 2\nu^2\sin(\lambda\psi+\beta)\right]\right] \tag{7.96b}$$

$$\delta_o^b = \frac{2\lambda^2}{EtK_1}(\sin\phi_o)\, M_o - \frac{\lambda a}{Et}\left(\frac{1+K_1K_2}{K_1}\right)(\sin^2\phi_o)\, H_o \tag{7.97a}$$

$$V_o^b = -\frac{4\lambda^3}{EatK_1} M_o + \frac{2\lambda^2}{EtK_1}(\sin\phi_o)\, H_o \tag{7.97b}$$

where

$$k_1 = 1 - \left(\frac{1-2\nu}{2\lambda}\right)\cot(\phi_o - \psi) \tag{7.98a}$$

$$k_2 = 1 - \left(\frac{1+2\nu}{2\lambda}\right)\cot(\phi_o - \psi) \tag{7.98b}$$

$$K_1 = 1 - \left(\frac{1-2\nu}{2\lambda}\right)\cot\phi_o \tag{7.99a}$$

$$K_2 = 1 - \left(\frac{1+2\nu}{2\lambda}\right)\cot\phi_o \tag{7.99b}$$

$$\beta = -\tan^{-1}(K_1) \tag{7.100}$$

Lower shell

$$N_\phi'^b = -\cot(\phi_o' + \psi')\frac{e^{-\lambda'\psi'}}{\sqrt{\sin(\phi_o'+\psi')}}\left\{\frac{2\lambda'}{a'K_1'}(\sin\phi_o')^{1/2} M_o' \sin\lambda'\psi' + \frac{(1+K_1'^2)^{1/2}}{K_1'}(\sin\phi_o')^{3/2} H_o' \sin(\lambda'\psi'+\beta')\right\} \tag{7.101a}$$

$$N_\theta'^b = \frac{\lambda'}{2}\frac{e^{-\lambda'\psi'}}{\sqrt{\sin(\phi_o'+\psi')}}\left[\frac{2\lambda'}{a'K_1'}(\sin\phi_o')^{1/2} M_o'\left\{2\cos\lambda'\psi' - (k_1'+k_2')\sin\lambda'\psi'\right\} + \frac{(1+K_1'^2)^{1/2}}{K_1'}(\sin\phi_o')^{3/2} H_o'\left\{2\cos(\lambda'\psi'+\beta') - (k_1'+k_2')\sin(\lambda'\psi'+\beta')\right\}\right] \tag{7.101b}$$

$$M'_\phi = \frac{a'}{2\lambda'} \frac{e^{-\lambda'\psi'}}{\sqrt{\sin(\phi'_o+\psi')}} \left[\frac{2\lambda'}{a'K'_1} (\sin\phi'_o)^{1/2} M'_o \left\{ k'_1 \cos\lambda'\psi' + \sin\lambda'\psi' \right\} \right.$$

$$\left. + \frac{(1+K'^2_1)^{1/2}}{K'_1} (\sin\phi'_o)^{3/2} H'_o \left\{ k'_1 \cos(\lambda'\psi'+\beta') + \sin(\lambda'\psi'+\beta') \right\} \right] \tag{7.102a}$$

$$M'_\theta = \frac{a'}{4\lambda'\nu} \frac{e^{-\lambda'\psi'}}{\sqrt{\sin(\phi'_o+\psi')}} \left[\frac{2\lambda'}{a'K'_1} (\sin\phi'_o)^{1/2} M'_o \times \right.$$

$$\left[\left\{ (1+\nu^2)(k'_1+k'_2) - 2k'_2 \right\} \cos\lambda'\psi' + 2\nu^2 \sin\lambda'\psi' \right]$$

$$+ \frac{(1+K'^2_1)^{1/2}}{K'_1} (\sin\phi'_o)^{3/2} H'_o \times$$

$$\left. \left[\left\{ (1+\nu^2)(k'_1+k'_2) - 2k'_2 \right\} \cos(\lambda'\psi'+\beta') + 2\nu^2 \sin(\lambda'\psi'+\beta') \right] \right] \tag{7.102b}$$

$$\delta'^b_o = \frac{2\lambda'^2}{Et'K'_1} (\sin\phi'_o)\, M'_o + \frac{\lambda' a'}{Et'} \left(\frac{1+K'_1K'_2}{K'_1} \right) (\sin^2\phi'_o)\, H'_o \tag{7.103a}$$

$$V'^b_o = \frac{4\lambda'^3}{Ea't'K'_1} M'_o + \frac{2\lambda'^2}{Et'K'_1} (\sin\phi'_o)\, H'_o \tag{7.103b}$$

where

$$k'_1 = 1 + \left(\frac{1-2\nu}{2\lambda'} \right) \cot(\phi'_o + \psi') \tag{7.104a}$$

$$k'_2 = 1 + \left(\frac{1+2\nu}{2\lambda'} \right) \cot(\phi'_o + \psi') \tag{7.104b}$$

$$K'_1 = 1 + \left(\frac{1-2\nu}{2\lambda'} \right) \cot\phi'_o \tag{7.105a}$$

$$K'_2 = 1 + \left(\frac{1+2\nu}{2\lambda'} \right) \cot\phi'_o \tag{7.105b}$$

$$\beta' = -\tan^{-1}(K'_1) \tag{7.106}$$

7.4.6 Evaluation of edge actions and net stresses

To evaluate the edge actions M_o, H_o, M'_o, H'_o, we apply the boundary conditions of continuity of net deformations across the junction of the two shells

$$\delta^T_o = \delta'^T_o \tag{7.107a}$$

$$V^T_o = V'^T_o \tag{7.107b}$$

and the boundary conditions expressing the equilibrium of an element at the interface of the two shells, namely that the net torque and the net horizontal thrust upon such an element must be zero:

$$M_o - M_o' = 0 \tag{7.108a}$$

$$H_o^m + H_o' - H_o = 0 \tag{7.108b}$$

Net deformations δ_o^T, $\delta_o'^T$, V_o^T, $V_o'^T$ in the shell are given by combining the components due to the bending and shear actions – expressions (7.97) and (7.103) – with their membrane-solution counterparts. The latter have already been obtained in explicit form (see equations (7.93) and (7.94)), and may therefore be taken as known quantities. Thus

$$\delta_o^T = \delta_o^m + \frac{2\lambda^2}{EtK_1}(\sin\phi_o)\,M_o - \frac{\lambda a}{Et}\left(\frac{1+K_1K_2}{K_1}\right)(\sin^2\phi_o)\,H_o \tag{7.109a}$$

$$\delta_o'^T = \delta_o'^m + \frac{2\lambda'^2}{Et'K_1'}(\sin\phi_o')\,M_o' + \frac{\lambda' a'}{Et'}\left(\frac{1+K_1'K_2'}{K_1'}\right)(\sin^2\phi_o')\,H_o' \tag{7.109b}$$

$$V_o^T = V_o^m - \frac{4\lambda^3}{EatK_1}M_o + \frac{2\lambda^2}{EtK_1}(\sin\phi_o)\,H_o \tag{7.110a}$$

$$V_o'^T = V_o'^m + \frac{4\lambda'^3}{Ea't'K_1'}M_o' + \frac{2\lambda'^2}{Et'K_1'}(\sin\phi_o')\,H_o' \tag{7.110b}$$

From equations (7.108), we obtain the relations

$$M_o' = M_o \tag{7.111a}$$

$$H_o' = H_o - H_o^m \tag{7.111b}$$

where H_o^m is a known quantity (equation (7.92)). Using relations (7.109) and (7.110) to write out equations (7.107) in expanded form, and eliminating M_o' and H_o' from the ensuing pair of equations (using relations (7.111)), we obtain a pair of simultaneous equations in two unknowns M_o and H_o, which upon solving yield

$$M_o = \frac{f_1g_1 - f_2g_2}{f_1^2 - f_2f_3} \tag{7.112a}$$

$$H_o = \frac{f_1g_2 - f_3g_1}{f_1^2 - f_2f_3} \tag{7.112b}$$

where

$$f_1 = \frac{2}{E}\left\{\frac{\lambda^2}{tK_1}(\sin\phi_o) - \frac{\lambda'^2}{t'K_1'}(\sin\phi_o')\right\} \tag{7.113a}$$

$$f_2 = -\frac{1}{E}\left\{\frac{\lambda a}{t}\left(\frac{1+K_1K_2}{K_1}\right)(\sin^2\phi_o) + \frac{\lambda' a'}{t'}\left(\frac{1+K_1'K_2'}{K_1'}\right)(\sin^2\phi_o')\right\} \tag{7.113b}$$

$$f_3 = -\frac{4}{E}\left(\frac{\lambda^3}{atK_1} + \frac{\lambda'^3}{a't'K_1'}\right) \tag{7.113c}$$

$$g_1 = (\delta_o')^m - (\delta_o)^m - \frac{\lambda' a'}{Et'}\left(\frac{1+K_1'K_2'}{K_1'}\right)(\sin^2\phi_o')\,H_o^m \tag{7.113d}$$

$$g_2 = (V_o')^m - (V_o)^m - \frac{2\lambda'^2}{Et'K_1'}(\sin\phi_o')\,H_o^m \tag{7.113e}$$

The other two edge actions M_o' and H_o' then follow from relations (7.111).

Substitution of the now known results for shell-edge actions $\{M_o, H_o, M_o', H_o'\}$ into expressions (7.95), (7.96), (7.101) and (7.102) allows the bending-related stress resultants $\{N_\phi^b, N_\theta^b, N_\phi'^b, N_\theta'^b\}$ and bending moments $\{M_\phi, M_\theta, M_\phi', M_\theta'\}$ in the two shells to be evaluated. The net stresses on the inner and outer surfaces of the two shells then follow by superimposing the membrane-solution stresses with those stemming from the edge effects:

$$\sigma_\phi^T = \frac{N_\phi^m}{t} + \frac{N_\phi^b}{t} \pm \frac{6M_\phi}{t^2} \tag{7.114a}$$

$$\sigma_\theta^T = \frac{N_\theta^m}{t} + \frac{N_\theta^b}{t} \pm \frac{6M_\theta}{t^2} \tag{7.114b}$$

$$\sigma_\phi'^T = \frac{N_\phi'^m}{t'} + \frac{N_\phi'^b}{t'} \pm \frac{6M_\phi'}{t'^2} \tag{7.115a}$$

$$\sigma_\theta'^T = \frac{N_\theta'^m}{t'} + \frac{N_\theta'^b}{t'} \pm \frac{6M_\theta'}{t'^2} \tag{7.115b}$$

7.4.7 Numerical example

Figure 7.10 shows a vessel consisting of three spherical shells of radii $a_1 = 10$m (shell A), $a_2 = 13.473$m (shell B) and $a_3 = 10$m (shell C). In the notation of the present formulation, the shells A, B, C will be denoted by S1, S2, S3 respectively. These shells meet at junctions J1 and J2 as shown. The uppermost shell (shell S1) has an angle of opening $\phi_{10} = 60°$; the bottom shell (shell S3) has an angle of opening $\phi_{30} = 120°$; the middle shell (shell S2) has angles of opening $\phi_{21} = 40°$ at junction J1 and $\phi_{22} = 140°$ at junction J2. The shells are assumed to be fabricated from steel plate of constant thickness $t = 0.05$m throughout, giving a minimum radius-to-thickness ratio of 200 for the vessel. Young's modulus will be taken as $E = 200 \times 10^9$ N/m^2, and Poisson's ratio as $\nu = 0.3$. This example is taken from Ref. 5.

The vessel is assumed to be axisymmetrically supported in the lower region of shell S3, the distance of the supports (from junction J2) being such that edge effects at the supports do not significantly interact with

Figure 7.10 Geometric parameters of the numerical example.

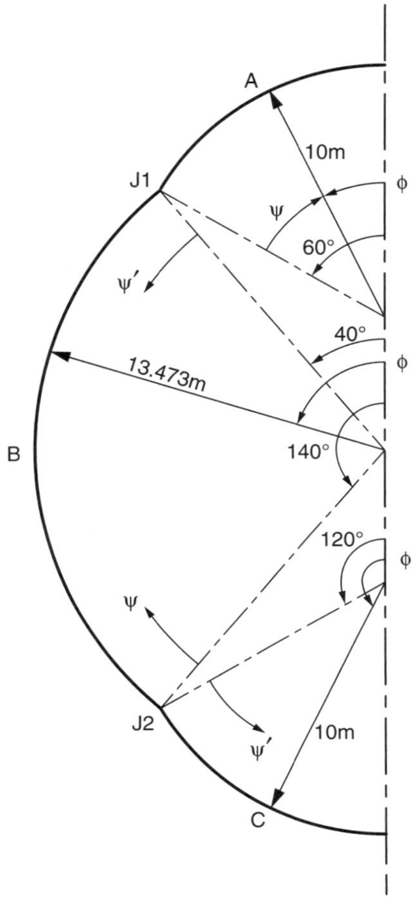

edge effects at junction J2 (which is valid if the location of the supports is below the level $\phi = 150°$). It is assumed that the vessel is completely filled with water of unit weight $\gamma = 10 \times 10^3\,\text{N/m}^3$. The recommended sequence of calculations is as follows:

First, we calculate the depth parameters h_{10} and h_{22} using equations (7.65b) and (7.66b) respectively, then the parameters k_1, k_2 and k_3 for the three shells using equations (7.70a), (7.70b) and (7.70d) respectively. The constant of integration of the membrane solution for the top cap (shell S1) is $C_1 = 1$. The constants of integration C_2 and C_3 (for shells S2 and S3 respectively) are evaluated from equations (7.84) and (7.88) using the relevant geometric parameters. The variations of membrane stress resultants in shells S1, S2 and S3 then follow from equations (7.75), (7.85) and (7.89). We then use equation (7.92) to evaluate the horizontal resultant of the membrane meridional stress resultants at the junctions, and equations (7.93) and (7.94) to evaluate the membrane deformations at the junctions. This completes the calculation of all relevant membrane quantities.

The quantities associated with the bending effects are evaluated next. Knowing the slenderness parameter λ for each of the three shells, we evaluate the bending-solution parameters represented by equations (7.98) to

(7.100) for the upper side of a given junction, and equations (7.104) to (7.106) for the lower side of a given junction. The parameters $\{f_1, f_2, f_3, g_1, g_2\}$ are evaluated from equations (7.113); the actions $\{M_o, H_o\}$ at the upper edge of a given junction follow from equations (7.112), while the actions $\{M'_o, H'_o\}$ at the lower edge of a given junction follow from equations (7.111). Using these values of $\{M_o, H_o, M'_o, H'_o\}$, the bending-related stress resultants $\{N^b_\phi, N^b_\theta, N'^b_\phi, N'^b_\theta\}$ and bending moments $\{M_\phi, M_\theta, M'_\phi, M'_\theta\}$ in the junction zones are evaluated on the basis of equations (7.95), (7.96), (7.101) and (7.102). Combining the membrane stresses with the bending-related stresses in accordance with equations (7.114) and (7.115) then gives the final stresses in the shell.

Figures 7.11(a) and 7.11(b) show meridional and hoop stresses (in MPa), respectively, plotted versus the coordinate s, this being the *cumulative* distance travelled from the apex of the vessel along the curved meridian up to the point in question. For points lying on shell A, this distance is simply the arc length over shell A up to the point in question. For points lying on shell B, it will be the full arc length of shell A, plus the additional arc length from junction J1 over shell B up to the point in question. For points lying on shell C, the distance s will be the sum of the full arc lengths of shells A and B, plus the additional arc length from junction J2 over shell C up to the point in question. The range of s covered by the plots is $0 \le s \le 37.5$m.

The gradually varying curves are the membrane-stress variations $\{\sigma^m_\phi, \sigma^m_\theta\}$, while the sharply oscillating curves in the vicinity of the junction locations are the total-stress variations for the inner and outer shell surfaces of the shell – that is, $\{\sigma^T_\phi\ (i);\ \sigma^T_\phi\ (o)\}$ for the meridional stress variations in Figure 7.11(a), and $\{\sigma^T_\theta\ (i);\ \sigma^T_\theta\ (o)\}$ for the hoop stress variations in Figure 7.11(b).

To see more clearly what is happening in the vicinity of the junctions, the meridional stress variations around junction J1 have been magnified in Figure 7.12(a), while those around junction J2 have been magnified in Figure 7.12(b). Similarly, Figure 7.13(a) shows magnified hoop stress variations around junction J1, while Figure 7.13(b) shows magnified hoop stress variations around junction J2.

Here is a summary of some observations:

(*a*) As we would expect, the membrane-stress variations show discontinuities (or jumps) at both junctions J1 and J2. On combining the membrane stresses with the stresses due to the edge effect, the total-stress variations show continuity in values (but not in slope) across the junctions.

(*b*) The discontinuity stresses at the lower junction are several times larger than those at the upper junction. This is because the larger hydrostatic pressures in the lower part of the vessel cause bigger membrane-deformation incompatibilities, in turn inducing larger bending effects.

(*c*) The discontinuity stresses are localised to the shell junctions, dying out within a distance of only 2.5m on either side of the junctions. However, within these narrow zones, the discontinuity stresses sharply rise to magnitudes that are several times larger than the membrane stresses.

(*d*) At both the upper and the lower junctions, the discontinuity effect results in net meridional stresses in both tension and compression that are relatively large in comparison with the surrounding membrane stresses.

(*e*) On the other hand, while the discontinuity effect increases the relatively modest membrane hoop tension at the upper junction, it has the beneficial effect of sharply lowering the much larger membrane hoop tension at the lower junction, with the result that outer parts of the shell end up being slightly in compression.

(*f*) The inward-pointing kink in the vessel profile at junction J2 has the same effect as circumferentially prestressing the smooth shell at this level (in order to counter hoop tension due to outward hydrostatic pressure). However, the benefit is only very localised.

Figure 7.11 Variations of stresses with arc length *s* over the full profile of the vessel (top to bottom): (a) meridional stresses; (b) hoop stresses. In the plots, the symbol *m* denotes membrane stresses, *i* denotes total stresses on the inner surface of the shell, and *o* denotes total stresses on the outer surface of the shell.

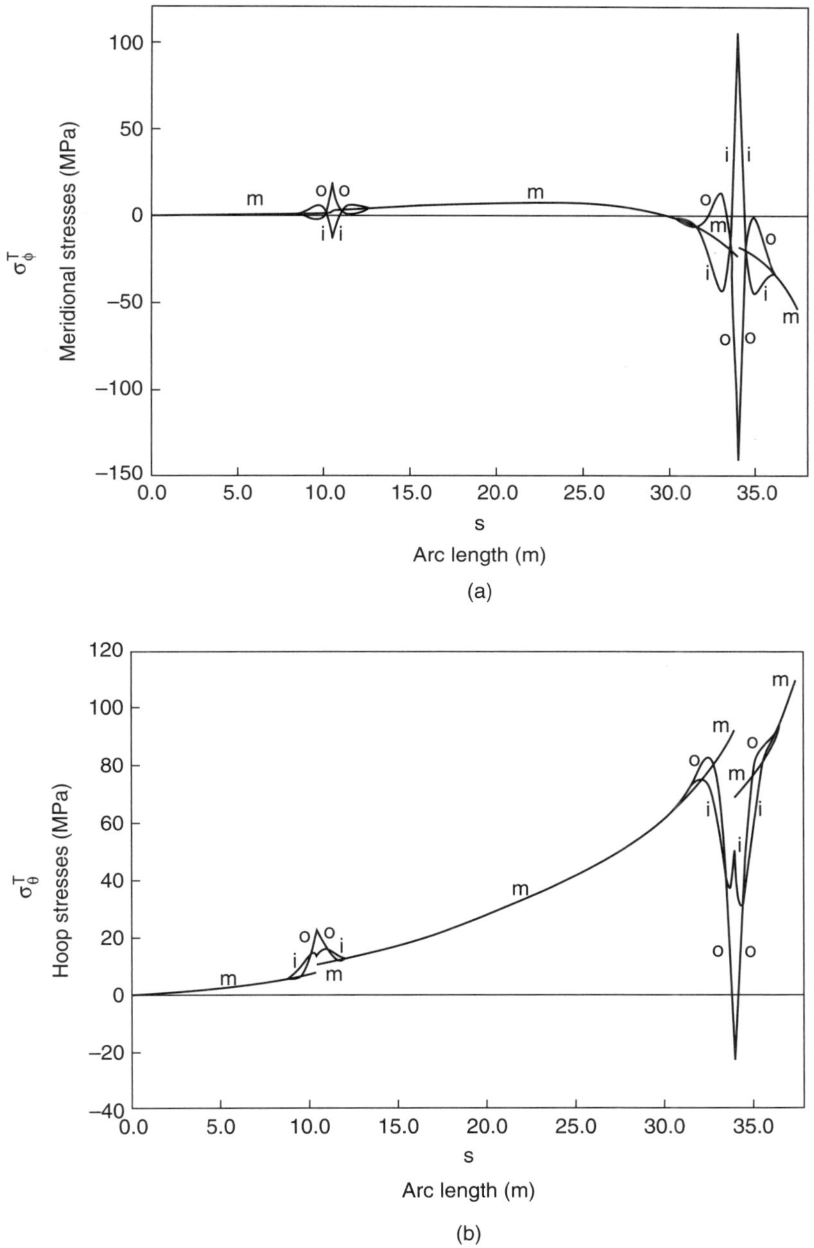

Figure 7.12 Meridional stress variations in the vicinity of the shell junctions: (a) junction J1; (b) junction J2. In the plots, the symbol *m* denotes membrane stresses, *i* denotes total stresses on the inner surface of the shell, and *o* denotes total stresses on the outer surface of the shell.

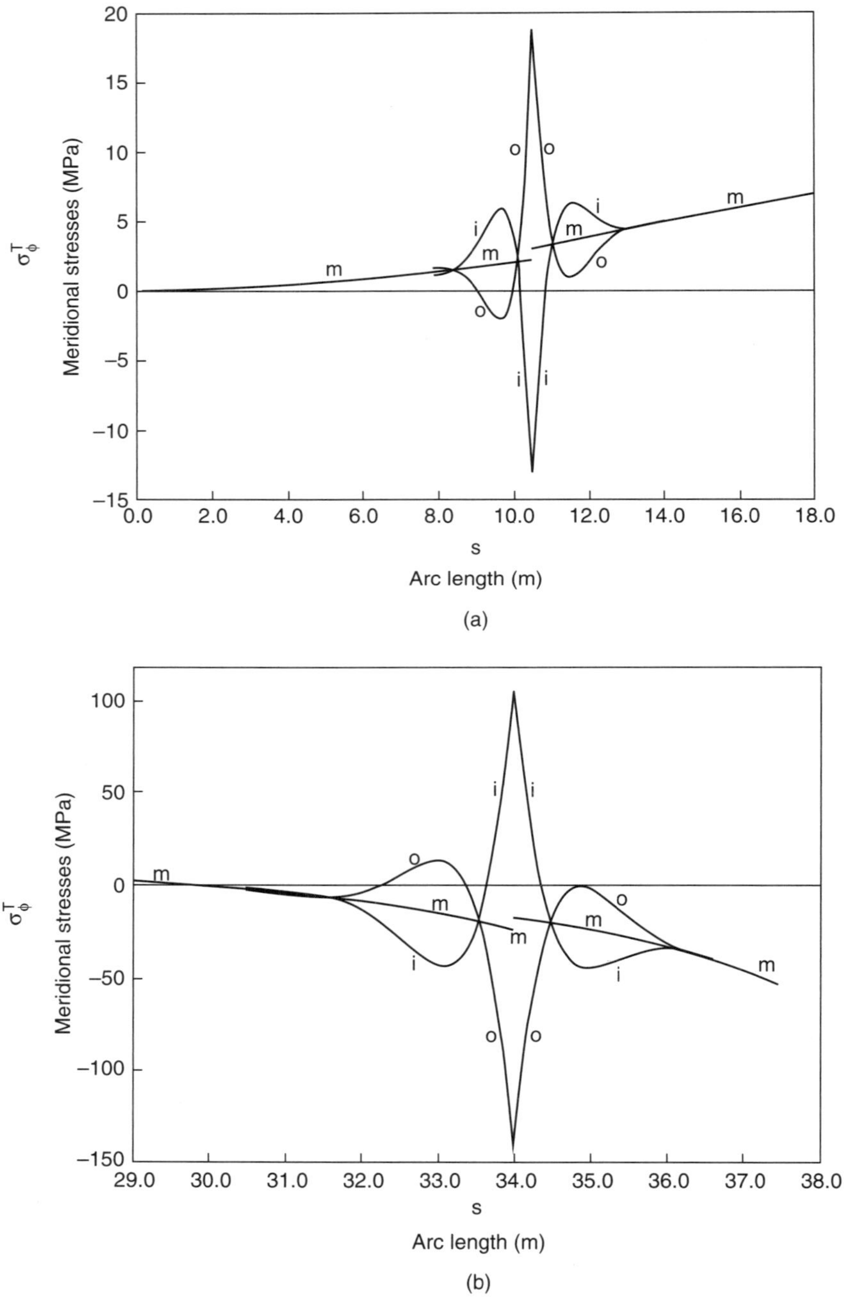

Figure 7.13 Hoop stress variations in the vicinity of the shell junctions: (a) junction J1; (b) junction J2. In the plots, the symbol *m* denotes membrane stresses, *i* denotes total stresses on the inner surface of the shell, and *o* denotes total stresses on the outer surface of the shell.

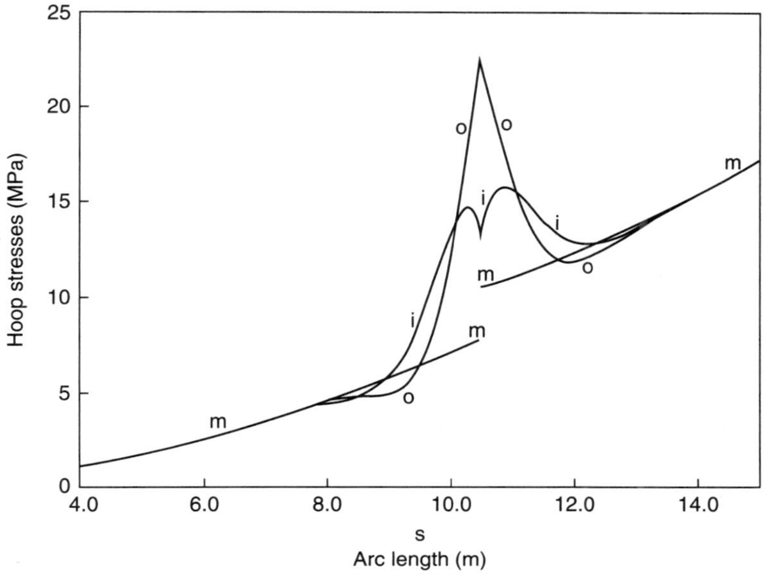

(a)

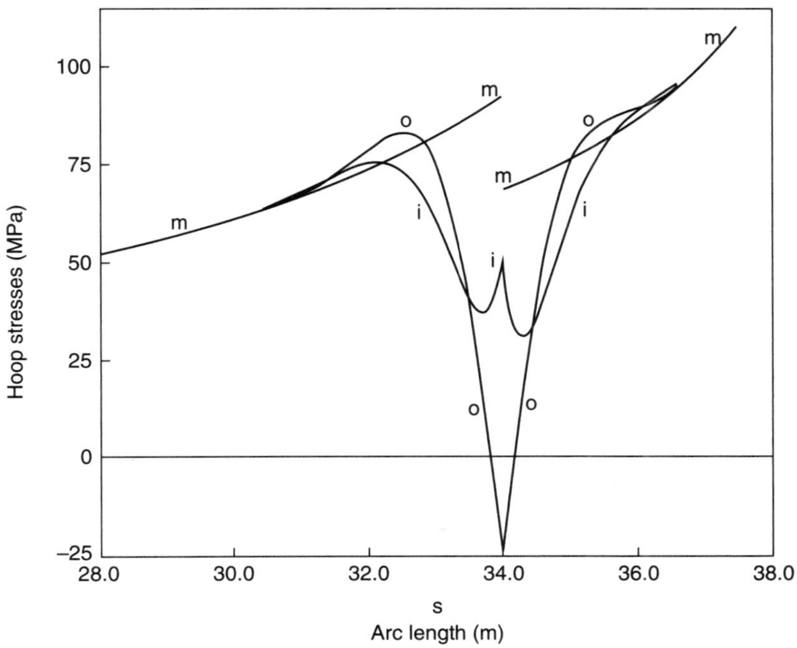

(b)

The present formulation for multi-shell assemblies has assumed that the transfer of membrane forces between adjacent shell segments is such that only vertical equilibrium of stress resultants needs to be preserved. It has been left to the bending theory to ensure that all other conditions are satisfied. The formulation may easily be programmed for ease of computation.

References

1. Galletly G.D. Edge influence coefficients for toroidal shells of positive Gaussian curvature. *Proceedings of the American Society of Mechanical Engineers: Journal of Engineering for Industry*, 1960 (Feb), 82, 60–68.
2. Galletly G.D. Influence coefficients and pressure vessel analysis. *Proceedings of the American Society of Mechanical Engineers: Journal of Engineering for Industry*, 1960 (Aug), 82, 259–267.
3. Baker B.R. and Cline G.B. Jr. Influence coefficients for thin smooth shells of revolution subjected to symmetric loads. *Proceedings of the American Society of Mechanical Engineers: Journal of Applied Mechanics*, 1962 (June), 29 (2), 335–339.
4. Zingoni A. Simplification of the derivation of influence coefficients for symmetric frusta of shells of revolution. *Thin-Walled Structures*, 2009, **47**, 912–918.
5. Zingoni A., Mokhothu B. and Enoma N. A theoretical formulation for the stress analysis of multi-segmented spherical shells for high-volume liquid containment. *Engineering Structures*, 2015, **87**, 21–31.

Shell Structures in Civil and Mechanical Engineering
ISBN 978-0-7277-6028-9

http://dx.doi.org/10.1680/ssicame.60289.247

Chapter 8
Parametric studies of liquid-containment shells of revolution and roof domes

8.1. Elevated liquid-filled spherical vessels

8.1.1 Introduction

Worldwide, a large number of spherical vessels have been built to date for various applications, particularly for the containment of industrial fluids such as volatile grades of oil at loading terminals and refineries, and chemical products like propane, butane and ammonia, which are normally stored under pressure at temperatures above their boiling points.[1] A typical form of storage is the hortonsphere, a large spherical fluid-retaining vessel supported above the ground by a series of equi-spaced vertical columns.[2] Apart from good aesthetics, the spherical shape enjoys the highest ratio of internal volume to surface area of all vessel shapes, which in turn minimises usage of the material of construction of the shell, and the amount of heat transferred from the surroundings into the contained liquid.

As early as 1949, Zick and Carlson[3] studied experimentally the stresses around the supports of a hortonsphere, and confirmed the benefit of locally thickening the shell in the region of its attachment to the columns, which was already common design practice. The stresses induced in a spherical pressure vessel due to the constraining effect of a cylindrical skirt, this arrangement comprising the containment and support structure for the graphite core of a nuclear reactor, were investigated in 1958 by Hicks,[4] who presented closed-form solutions for the relevant junction redundants, and in 1960 by Bailey and Hicks,[5] who produced plots of the ensuing bending stresses. About the same time, Houghton[6] treated essentially the same problem in a paper on various discontinuity problems in shell structures, and established the good agreement between theoretically determined strains and experimentally measured ones. In a survey published in 1982,[1] Tooth outlined some practical design aspects of spherical and related vessels, including the then current practice as reflected in British and American codes.

Parametric studies are an indispensable design tool for complex structures such as shells. In the sections that follow, we will outline some parametric studies that have been undertaken on elevated liquid-filled spherical vessels;[7] these studies have been published in full elsewhere.[8–11] Here, we will summarise the main findings, conclusions and recommendations.

8.1.2 Effect of ring-beam stiffnesses on shell stresses

8.1.2.1 Introduction

We seek to investigate the effect upon shell stresses of the stiffnesses of the ring beam provided at the support level, so as to establish the relative importance of *axial* and *axi-rotational* rigidities of the support ring. *Axial rigidity* refers to the purely extensional resistance of a ring beam to circumferential stretching or contracting due to a hoop force within the ring beam induced by an axisymmetric radial force acting on the ring beam in the plane of the ring beam. On the other hand, *axi-rotational rigidity* refers to the essentially flexural resistance of a ring beam to the axisymmetric rotation of the cross-section about its centroid, caused by an axisymmetric torque acting on the ring beam. Four possible combinations of limiting values of the stiffnesses of the ring beam will be considered. In addition, we will also compute the displacements in the shell, to give us a complete picture of the structural response of the shell.

In Chapter 7, we saw how an axisymmetric lateral force H_r acting upon a ring beam of finite axial rigidity causes a lateral movement δ_r of the ring cross-section, and how an axisymmetric torque T_r acting upon a ring beam of finite axi-rotational rigidity causes a rotation V_r of the ring cross-section. In this section, we are interested in predicting shell response when ring-beam axial and axi-rotational rigidities approach the limiting values of zero and infinity. The four combinations of limiting ring-beam stiffnesses are {**z**ero **a**xial, **z**ero axi-ro**t**ational}, {**i**nfinite **a**xial, **i**nfinite axi-ro**t**ational}, {**z**ero **a**xial, **i**nfinite axi-ro**t**ational} and {**i**nfinite **a**xial, **z**ero axi-ro**t**ational}, henceforth abbreviated as ZAZT, IAIT, ZAIT and IAZT, respectively.

In the present investigation, we will assume the vessel to be completely filled to capacity with liquid of weight γ per unit volume, so that the loading on the shell is essentially hydrostatic. Furthermore, we will assume the vessel is supported on axially stiff (i.e. relatively incompressible) vertical columns, and that the reactions of these on the shell are uniformly distributed around the support circumference. We may justify the last assumption on the basis that the vertical columns are usually closely spaced (typically 8 to 10 around the circumference); if a ring beam or skirt is provided at the interface of the vertical columns and the shell, then the assumption of uniformly distributed reactions becomes even more valid.

Unlike gas-filled pressure vessels, which may be supported directly on discrete columns without the provision of a ring beam,[12] liquid-filled vessels may require a ring beam owing to the relatively large weight of the contents; the absence of such a ring beam would result in severe stress concentrations around the points of attachment of the columns to the shell, unless, as will be seen in due course, the columns meet the shell midsurface tangentially. Clearly, the size of the ring beam necessary to ensure effective dispersal of the vertical reactions gets smaller with increase in the number of support columns provided.

The present study of infinitely low and infinitely high ring-beam stiffnesses is intended to establish limits of shell response, within which the behaviour of shells attached to ring beams of finite stiffnesses would be bound. This information provides a guide to the shell designer in as much as simply supported and encastré supports constitute useful and safe limiting conditions for actual, and often not easily quantifiable, fixity details for beams and slabs. Alternatively, if we can show that particular combinations of limiting ring-beam stiffnesses result in relatively low shell stresses in the support region, this would allow the designer to provide actual support details that approach the ideal conditions as closely as possible.

As in the preceding chapters, the membrane hypothesis is first invoked, and the stress resultants $\left\{N_\phi^m, N_\theta^m\right\}$ calculated. The results for these were given in Section 3.2.3. The edge effects will be evaluated on the basis of the Geckeler approximation.

8.1.2.2 Boundary conditions at shell edges

Figure 8.1 shows the structural layout of the present problem, while Figure 8.2 indicates the positive sign convention (based on the left half of the vessel) for (*a*) actions on the shell edges and reactions on the ring beam, with forces and moments shown separately for clarity, and (*b*) deformations of the shell edges. The actions $\left\{M_t, H_t, M_b, H_b\right\}$ are the redundants required at the shell edges to satisfy boundary conditions there. Subscripts *t* (for *top*) and *b* (for *bottom*) are used to denote the shell edges above and below the ring beam, respectively.

Figure 8.1 Spherical vessel supported on vertical columns.

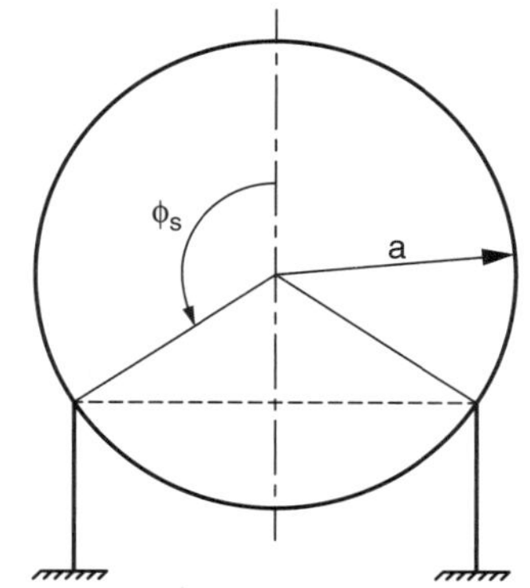

Figure 8.2 Positive sign conventions for: (a) actions on the edges of the shell domains and reactions on the ring beam; (b) deformations of the edges of the shell domains.

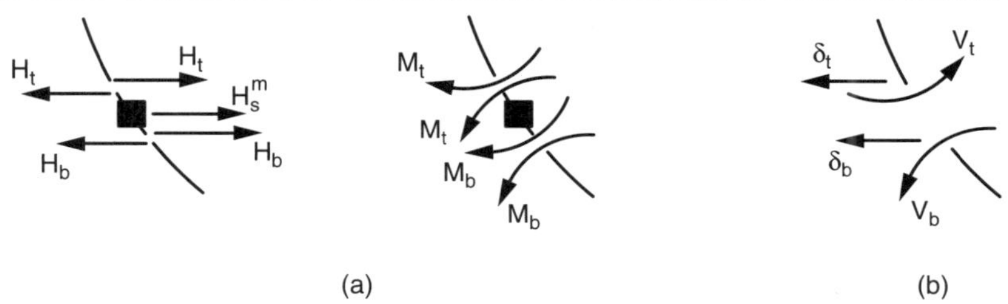

In Figure 8.2(a), the action H_s^m shown acting on the ring beam represents a horizontal force per unit length stemming from the difference in the values of the membrane meridional stress resultants in the shell just above and just below the ring beam. In general, and following the usual notation, superscripts m, b and T are used to denote quantities associated with the membrane solution, the bending solution (i.e. edge effect) and the total solution, respectively.

Subscripts u and l are used to denote *interior* variables of the upper and lower domains, respectively. When the edge of the shell is the location of interest, shell variables assume the subscripts t and b for the upper and lower domains, respectively, as already explained. Following the usual notation, subscript s is used to denote the support location, while subscript r specifically refers to the ring beam.

For the purposes of this study, we will assume that the net horizontal force $H_r^T \left(= H_t - H_b - H_s^m\right)$ acting *outwards* upon the ring beam (refer to Figure 8.2(a)) essentially passes through the centroid of the cross-section of the ring beam. Thus, this resultant has negligible eccentricity about the centroid of the ring cross-section, and therefore causes an outward expansion of the ring beam without rotation of the cross-section. Strictly speaking, and as shown in Chapter 7, the actions H_t and H_b will cause the ring cross-section to rotate, but here we consider these rotations to be small in comparison with those due to M_t and M_b. The resultant torque T_r^T on the ring beam is simply equal to the difference between the two moments acting on it – that is, $T_r^T = M_t - M_b$. The deformations of the ring beam under the resultant actions H_r^T and T_r^T may readily be written down from the results of Chapter 7.

For continuity of the shell midsurface, the net horizontal displacements of the edges of the upper and lower domains of the vessel must be equal (i.e. $\delta_t^T = \delta_b^T$), as must the net meridional rotations of these edges (i.e. $V_t^T = V_b^T$), regardless of the actual values of ring-beam stiffnesses. While the second condition always holds within the framework of small-deformation theory, the first is only valid if the overall depth of the ring beam or the rotation of the ring beam is small. In either case, the rotation of the ring beam has negligible contribution to the lateral displacement of the shell edges. Let us assume that the physical dimensions of the ring beam are small. This automatically ensures that the condition $\delta_t^T = \delta_b^T$ and the earlier one of zero eccentricity of the net force H_r^T are simultaneously satisfied.

If the axial stiffness of the ring beam happens to be infinite, then $\delta_t^T = \delta_b^T = 0$; if the rotational stiffness is infinite, then $V_t^T = V_b^T = 0$. As regards equilibrium implications of limiting support conditions, a ring beam of zero axial rigidity cannot resist any lateral force (i.e. $H_r^T = 0$), while a ring beam of zero axi-rotational rigidity cannot resist any torque (i.e. $T_r^T = 0$). Summarising, the sets of boundary conditions associated with the ZAZT, IAIT, IAZT and ZAIT combinations of limiting ring-beam stiffnesses are, in that order, as follows:

$$\left\{\delta_t^T = \delta_b^T;\ V_t^T = V_b^T;\ T_r^T = M_t - M_b = 0;\ H_r^T = H_t - H_b - H_s^m = 0\right\} \tag{8.1}$$

$$\left\{\delta_t^T = 0;\ \delta_b^T = 0;\ V_t^T = 0;\ V_b^T = 0\right\} \tag{8.2}$$

$$\left\{\delta_t^T = 0;\ \delta_b^T = 0;\ V_t^T = V_b^T;\ T_r^T = M_t - M_b = 0\right\} \tag{8.3}$$

$$\left\{\delta_t^T = \delta_b^T;\ V_t^T = 0;\ V_b^T = 0;\ H_r^T = H_t - H_b - H_s^m = 0\right\} \tag{8.4}$$

In these equations, the net horizontal displacements $\left\{\delta_t^T, \delta_b^T\right\}$ and net meridional rotations $\left\{V_t^T, V_b^T\right\}$ of the shell edges are, of course, the sum of the contributions of the edge effect (as given by expressions (6.62), with due regard to the positive sign conventions for shell-edge redundants and shell-edge deformations as defined in Figure 8.2) and the membrane solution:

$$\delta_t^T = \frac{2\lambda^2}{Et}(\sin\phi_s)M_t - \frac{2\lambda a}{Et}(\sin^2\phi_s)H_t + \delta_t^m \tag{8.5}$$

$$\delta_b^T = \frac{2\lambda^2}{Et}(\sin\phi_s)M_b + \frac{2\lambda a}{Et}(\sin^2\phi_s)H_b + \delta_b^m \tag{8.6}$$

$$V_t^T = -\frac{4\lambda^3}{Eat}M_t + \frac{2\lambda^2}{Et}(\sin\phi_s)H_t + V_t^m \tag{8.7}$$

$$V_b^T = \frac{4\lambda^3}{Eat} M_b + \frac{2\lambda^2}{Et} (\sin \phi_s) H_b + V_b^m \tag{8.8}$$

The membrane components of the deformations are given by expressions (3.67) and (3.68) for the upper part of the vessel, and (3.72) and (3.73) for the lower part of the vessel, with ϕ set equal to ϕ_s. As stated earlier, the quantity H_s^m is simply the difference between the membrane meridional stress resultants occurring just above the support level (i.e. at the edge of the upper domain) and just below the support level (i.e. at the edge of the lower domain), resolved horizontally:

$$H_s^m = \left\{ \left(N_\phi^m\right)_t - \left(N_\phi^m\right)_b \right\} \cos \phi_s \tag{8.9}$$

The parameters $\left(N_\phi^m\right)_t$ and $\left(N_\phi^m\right)_b$ are given by expressions (3.65) and (3.70), respectively, with ϕ set equal to ϕ_s.

Substituting expressions (8.5)–(8.9) into equation sets (8.1)–(8.4), we obtain for each set four simultaneous equations in the unknowns $\{M_t, H_t, M_b, H_b\}$, which may readily be solved. Once M_t, H_t, M_b and H_b for each of the four cases (ZAZT, IAIT, IAZT and ZAIT) have been evaluated, final shell stresses may then be computed for a vessel of any given combination of parameters $\{a, t, \phi_s, a/t, \lambda, E, \nu\}$, following the procedure illustrated in earlier chapters.

8.1.2.3 Displacements

8.1.2.3.1 DISPLACEMENTS OF THE MEMBRANE SOLUTION

For any arbitrary smoothly varying axisymmetric surface loading, the basic membrane-solution displacements in a thin shell of revolution are given by expressions (2.64) and (2.65). Upon specialisation of these expressions to spherical shells of constant thickness, we obtain

$$v^m = \frac{a}{Et}(1+\nu)(\sin\phi)\left\{ \int \left(\frac{N_\phi^m - N_\theta^m}{\sin\phi} \right) d\phi + k^m \right\} \tag{8.10a}$$

$$w^m = -v^m \cot\phi + \frac{a}{Et}\left(N_\theta^m - \nu N_\phi^m\right) \tag{8.10b}$$

where v is the displacement in the direction of the tangent to the meridian, w is the displacement in the direction of the normal to the shell midsurface, and k is a constant of integration. The positive sign convention for these displacements is shown in Figure 8.3 for each part of the vessel. The horizontal and vertical resultants of v and w are shown in the figure as δ and y respectively.

To evaluate the constant of integration k^m, we impose the boundary condition

$$\left(v^m\right)_{\phi=\phi_s} = v_s^m = 0 \tag{8.11}$$

Figure 8.3 Positive sign convention for displacements in the upper and lower shell domains.

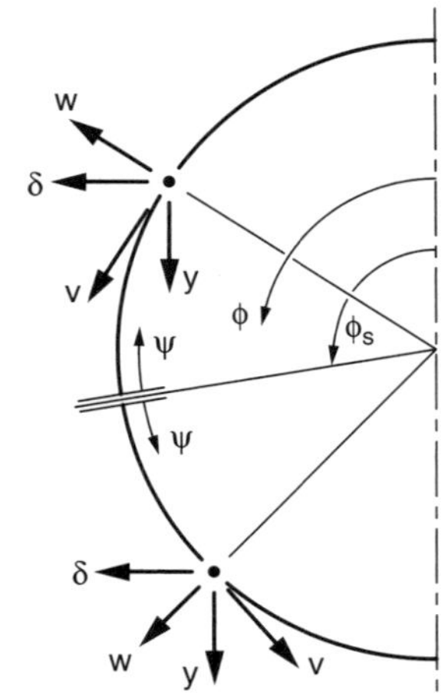

which is consistent with the requirements of the membrane hypothesis (see Section 1.5.3). For the present problem of a spherical vessel completely filled with a liquid of unit weight γ, evaluation of k^m on the basis of condition (8.11), with N_ϕ^m and N_θ^m given by expressions (3.65) and (3.66) for the upper domain of the vessel and expressions (3.70) and (3.71) for the lower domain, leads to the following results:

$$v_u^m = \frac{1}{3}\frac{\gamma a^3}{Et}(1+\nu)(\sin\phi)\left\{\ln\left(\frac{1+\cos\phi}{1+\cos\phi_s}\right)-\left(\frac{1}{1+\cos\phi}\right)+\left(\frac{1}{1+\cos\phi_s}\right)\right\} \tag{8.12a}$$

$$w_u^m = -\frac{1}{3}\frac{\gamma a^3}{Et}(1+\nu)(\cos\phi)\left\{\ln\left(\frac{1+\cos\phi}{1+\cos\phi_s}\right)-\left(\frac{1}{1+\cos\phi}\right)+\left(\frac{1}{1+\cos\phi_s}\right)\right\}$$

$$+\frac{1}{6}\frac{\gamma a^3}{Et}\left(\frac{1-\cos\phi}{1+\cos\phi}\right)\{(5+4\cos\phi)-\nu(1+2\cos\phi)\} \tag{8.12b}$$

$$v_l^m = \frac{1}{3}\frac{\gamma a^3}{Et}(1+\nu)(\sin\phi)\left\{\ln\left(\frac{1-\cos\phi}{1-\cos\phi_s}\right)-\left(\frac{1}{1-\cos\phi}\right)+\left(\frac{1}{1-\cos\phi_s}\right)\right\} \tag{8.13a}$$

$$w_l^m = -\frac{1}{3}\frac{\gamma a^3}{Et}(1+\nu)(\cos\phi)\left\{\ln\left(\frac{1-\cos\phi}{1-\cos\phi_s}\right)-\left(\frac{1}{1-\cos\phi}\right)+\left(\frac{1}{1-\cos\phi_s}\right)\right\}$$

$$+\frac{1}{6}\frac{\gamma a^3}{Et}\left(\frac{1}{1-\cos\phi}\right)\{(1-7\cos\phi+4\cos^2\phi)-\nu(5-5\cos\phi+2\cos^2\phi)\} \tag{8.13b}$$

where, as already pointed out, subscripts u and l denote upper and lower domains of the vessel, respectively.

8.1.2.3.2 DISPLACEMENTS OF THE EDGE EFFECT

From expressions (6.20) and (6.21), with $r_1 = r_2 = a$, we have

$$N_\phi^b - N_\theta^b = -Q_\phi \cot\phi + \frac{dQ_\phi}{d\phi}$$

so that

$$\frac{N_\phi^b - N_\theta^b}{\sin\phi} = -Q_\phi \frac{\cos\phi}{\sin^2\phi} + \frac{1}{\sin\phi}\frac{dQ_\phi}{d\phi} = \frac{d}{d\phi}\left(\frac{Q_\phi}{\sin\phi}\right)$$

Therefore,

$$\int\left(\frac{N_\phi^b - N_\theta^b}{\sin\phi}\right)d\phi = \int d\left(\frac{Q_\phi}{\sin\phi}\right) = \frac{Q_\phi}{\sin\phi}$$

Substituting this result into equation (8.10a), with bending-related in place of membrane-solution variables, we may write

$$v^b = \frac{a}{Et}(1+\nu)Q_\phi + k^b \sin\phi \tag{8.14a}$$

The displacement w^b normal to the shell midsurface then follows from relation (8.10b), when we replace membrane-solution quantities with their bending-disturbance counterparts:

$$w^b = -v^b \cot\phi + \frac{a}{Et}\left(N_\theta^b - \nu N_\phi^b\right) \approx -v^b \cot\phi + \frac{a}{Et}N_\theta^b \tag{8.14b}$$

Note that, in arriving at the final form of expression (8.14b), νN_ϕ^b is neglected in comparison with N_θ^b, consistent with the accuracy of the Geckeler approximation.

From relation (6.20), we may write

$$Q_\phi = -N_\phi^b \tan\phi$$

Using expression (6.68a) to eliminate N_ϕ^b, we obtain

$$Q_\phi = e^{-\lambda\psi}\left\{\frac{2\lambda}{a}M_e \sin\lambda\psi - H_e\left(\sin\phi_e\right)\left(\sin\lambda\psi - \cos\lambda\psi\right)\right\} \tag{8.15}$$

Substituting this expression for Q_ϕ into the right-hand side of equation (8.14a), and also using relation (6.68b) to eliminate N_θ^b from the right-hand side of equation (8.14b), we may write

$$v^b = \frac{1}{Et}(1+\nu)e^{-\lambda\psi}\left\{2\lambda M_e\left(\sin\lambda\psi\right) - aH_e\sin\phi_e\left(\sin\lambda\psi - \cos\lambda\psi\right)\right\} + k^b\sin\left(\phi_e - \psi\right) \tag{8.16a}$$

$$w^b = -\left[v^b\cot\left(\phi_e - \psi\right) + \frac{2\lambda}{Et}e^{-\lambda\psi}\left\{\lambda M_e\left(\sin\lambda\psi - \cos\lambda\psi\right) + aH_e\sin\phi_e\left(\cos\lambda\psi\right)\right\}\right] \tag{8.16b}$$

The use of the subscript e (for *edge*) in expressions (8.15) and (8.16) implies that these are general results based on a shell cap with the apex at $\phi = 0$, with the positive directions for M_e and H_e as shown in Figure 6.6. In the present problem, the supporting columns are vertical and assumed to be incompressible. The constant of integration k^b must therefore be evaluated from the condition that the net vertical displacement (i.e. membrane + edge effect) of each shell domain must be zero at the support – that is,

$$y_s^T \left(= y_s^m + y_s^b\right) = 0 \tag{8.17}$$

The vertical displacement y is given by the resultant of the basic shell displacements in the downward vertical direction (refer to Figure 8.3) – that is,

$$y = v\sin\phi - w\cos\phi \tag{8.18}$$

We may write the membrane and bending components of y in terms of v and w as given by equation (8.18). When we substitute these into condition (8.17) with v_s^m set equal to zero (in accordance with the membrane condition of equation (8.11)), we obtain

$$-w_s^m\cos\phi_s + v_s^b\sin\phi_s - w_s^b\cos\phi_s = 0 \tag{8.19}$$

From equation (8.10b), we may write w_s^m as

$$w_s^m = \frac{a}{Et}\left(N_\theta^m - \nu N_\phi^m\right)_s \tag{8.20}$$

(since $v_s^m = 0$). In equation (8.19), we then substitute for v_s^b and w_s^b using the right-hand sides of expressions (8.16a) and (8.16b) with ψ set equal to zero, and for w_s^m using the right-hand side of expression (8.20). This leads to the following solution for the constant of integration k_u^b associated with the upper domain of the vessel:

$$k_u^b = \frac{1}{Et}\left[\left\{a\left(N_\theta^m - \nu N_\phi^m\right)_t + 2\lambda\left(\lambda M_t - aH_t\sin\phi_s\right)\right\}\cos\phi_s - (1+\nu)aH_t\right] \tag{8.21}$$

Note that for the upper domain, the edge redundants $\{M_t, H_t\}$ have the same positive directions as the general redundants $\{M_e, H_e\}$ that were depicted in Figure 6.6.

Thus, for the upper domain ($0 \le \phi \le \phi_s$; $\psi = \phi_s - \phi$), expressions (8.16) become

$$v_u^b = \frac{1}{Et}(1+\nu)e^{-\lambda\psi}\left\{2\lambda M_t\left(\sin\lambda\psi\right) - aH_t\sin\phi_s\left(\sin\lambda\psi - \cos\lambda\psi\right)\right\}$$

$$+\frac{1}{Et}\left[\left\{a\left(N_\theta^m - \nu N_\phi^m\right)_t + 2\lambda\left(\lambda M_t - aH_t\sin\phi_s\right)\right\}\cos\phi_s - (1+\nu)aH_t\right]\sin\left(\phi_s - \psi\right) \tag{8.22a}$$

$$w_u^b = -\left[v_u^b\cot\left(\phi_s - \psi\right) + \frac{2\lambda}{Et}e^{-\lambda\psi}\left\{\lambda M_t\left(\sin\lambda\psi - \cos\lambda\psi\right) + aH_t\sin\phi_s\left(\cos\lambda\psi\right)\right\}\right] \tag{8.22b}$$

The expressions for k_l^b, v_l^b and w_l^b corresponding to the lower domain of the vessel ($\phi_s \le \phi \le \pi$; $\psi = \phi - \phi_s$) may be deduced directly from expressions (8.21), (8.22a) and (8.22b), respectively. Let us take the horizontal equatorial plane as a reflection plane, and compare the positive senses of the mirror images of the upper-domain variables with the positive sign convention actually adopted for the lower domain. We note that the mirror images of the parameters of the upper domain (which we will denote with primes) are brought into coincidence with the parameters of the lower domain via the following interchanges:

$$M_t' \leftrightarrow M_b \tag{8.23a}$$

$$H_t' \leftrightarrow -H_b \tag{8.23b}$$

$$v_u' \leftrightarrow -v_l \tag{8.23c}$$

$$w_u' \leftrightarrow w_l \tag{8.23d}$$

$$\delta_u' \leftrightarrow \delta_l \tag{8.23e}$$

$$(N_\theta^m - \nu N_\phi^m)_u' \leftrightarrow (N_\theta^m - \nu N_\phi^m)_l \tag{8.23f}$$

$$y_u' \leftrightarrow -y_l \tag{8.23g}$$

$$k_u' \leftrightarrow -k_l \tag{8.23h}$$

$$\psi_u' \leftrightarrow \psi_l \tag{8.23i}$$

$$\phi'_u \leftrightarrow \pi - \phi_l \tag{8.23j}$$

$$V'_u \leftrightarrow -V_l \tag{8.23k}$$

Expressions (8.23f) and (8.23h) may be more readily understood by noting that $(N_\theta - \nu N_\phi)$ is proportional to the hoop strain, and hence to the lateral displacement δ, while k represents a vertical rigid-body downward movement, and hence has the same positive convention as the displacement y. The required transformed expressions are

$$k_l^b = \frac{1}{Et}\left[\left\{a\left(N_\theta^m - \nu N_\phi^m\right)_b + 2\lambda\left(\lambda M_b + aH_b \sin\phi_s\right)\right\}\cos\phi_s - (1+\nu)aH_b\right] \tag{8.24}$$

$$v_l^b = -\frac{1}{Et}(1+\nu)e^{-\lambda\psi}\left\{2\lambda M_b\left(\sin\lambda\psi\right) + aH_b \sin\phi_s\left(\sin\lambda\psi - \cos\lambda\psi\right)\right\}$$

$$+\frac{1}{Et}\left[\left\{a\left(N_\theta^m - \nu N_\phi^m\right)_b + 2\lambda\left(\lambda M_b + aH_b \sin\phi_s\right)\right\}\cos\phi_s - (1+\nu)aH_b\right]\sin\left(\phi_s + \psi\right) \tag{8.25a}$$

$$w_l^b = -\left[v_l^b \cot\left(\phi_s + \psi\right) + \frac{2\lambda}{Et}e^{-\lambda\psi}\left\{\lambda M_b\left(\sin\lambda\psi - \cos\lambda\psi\right) - aH_b \sin\phi_s\left(\cos\lambda\psi\right)\right\}\right] \tag{8.25b}$$

The total displacements v^T and w^T are simply obtained by superimposing membrane components v^m and w^m for the domain in question (as given by equations (8.12) and (8.13)) with the bending-disturbance components v^b and w^b (as given by equations (8.22) for the upper domain and equations (8.25) for the lower domain), as follows:

$$v^T = v^m + v^b \tag{8.26a}$$

$$w^T = w^m + w^b \tag{8.26b}$$

The resultants of the above displacements in the outward horizontal and downward vertical directions, if required, are obtained from the relationships

$$\delta^T = v^T \cos\phi + w^T \sin\phi \tag{8.27a}$$

$$y^T = v^T \sin\phi - w^T \cos\phi \tag{8.27b}$$

8.1.2.4 Findings of a parametric study

Based on the ZAZT, IAIT, IAZT and ZAIT limiting cases, a parametric study[8] has been undertaken to better understand the effect of ring-beam stiffnesses on the structural response of the spherical vessel. The results provide a 'feel' for the relative magnitudes of peak stresses, and useful insights on the displacement behaviour of the vessel throughout its two domains. The main conclusions and design recommendations may be summarised as follows.

For practical reasons, let us take the possible range for vessel supports as $90° \le \phi_s \le 120°$, and observe shell behaviour within this range. We note that the positioning of the supports below the equatorial level ($\phi_s = 90°$) leads to relatively large shell stresses in the vicinity of the supports. As the support location approaches 120°, the shell stresses in the vicinity of the supports rapidly become critical, unless the ring beam is endowed with sufficient axial stiffness to prevent lateral movement of the shell at the supports. In particular, the combination of low ring-beam hoop stiffness and low support location generally results in unacceptably high shell stresses in both tension and compression for slender steel tanks; for concrete vessels, while the relatively high compressive stresses may still be within permissible limits, the accompanying tensile stresses may give rise to severe cracking in the vicinity of the supports, which, of course, is not desirable for liquid-retaining structures.

For such low positions of the supports (i.e. $\phi_s \to 120°$), axi-rotational rigidity of the ring beam does not significantly affect the peak stresses in the shell. Hence, while the accurate determination of the axi-rotational stiffness of practical shell-to-support connections may be difficult, this finding shows that any uncertainty in this design parameter will, for all practical purposes, be of no significant consequence upon peak shell stresses.

In summary, it may be stated that lack of adequate lateral restraint upon the shell at the supports (as in the case when a ring beam with negligible hoop stiffness is provided) results in relatively large shell stresses in the vicinity of the supports as the location of the ring beam is lowered towards the practical limit of $\phi_s = 120°$. The effects are most severe for vessels with high a/t ratios. This is so regardless of the amount of restraint provided against rotation of the shell edges.

On the other hand, if the supports are located at the equator (i.e. $\phi_s = 90°$), the magnitude of shell stresses in the vicinity of the supports becomes significantly reduced. In fact, the peak values of net shell stresses in the vicinity of the supports reduce to the same order of magnitude as the maximum membrane stresses noted elsewhere in the vessel, so that neither the axial nor the axi-rotational stiffnesses of the ring beam are likely to influence the design of the shell for vessels supported at the equator. For this location of supports, lack of lateral restraint upon the shell edges (i.e. low axial stiffness of the ring beam) now becomes somewhat beneficial, while the influence of ring axi-rotational stiffness on local shell stresses becomes comparable with that of ring axial stiffness. However, since the general level of support-zone stresses is low, the stiffness properties of the ring do not play a significant role in the choice of shell thickness.

8.1.3 Effect of column inclination on shell stresses

8.1.3.1 Introduction

In the last section, we concluded that, for liquid-filled spherical tanks conventionally supported on vertical columns, the optimum location of supports is the equator. For this location, bending disturbances are minimised. Lowering of the vertical-support system (ϕ_s approaching 120°) results in a rapid increase in bending actions in the support zone of the shell, particularly for relatively low values of ring-beam axial stiffness. For relatively high values of ring-beam axial stiffness, bending actions in the shell do not increase as rapidly. However, in order to achieve such a high degree of lateral restraint, a fairly robust (i.e. expensive) ring beam may be necessary.

For spherical vessels in close proximity with each other (or other structures), considerations of accessibility may dictate a lower positioning of the ring beam, as the corresponding clustering of columns would provide more space between adjacent vessels. This space may be required for vehicle access during filling operations. In such instances, a change in the inclination of the support columns from vertical to tangential may possibly achieve a significant reduction in the magnitude of the bending-disturbance shell stresses associated with low support locations (such as $\phi_s = 120°$) without the need for providing a costly ring beam with high hoop stiffness. It is the purpose of the present study to investigate how effective such a solution would be.

It may be argued that, from a practical standpoint, raked columns represent a less straightforward arrangement than their vertical counterparts, but this disadvantage might well be outweighed in certain cases by the combined beneficial effects of improved accessibility, low bending stresses in the vessel (rendering any local shell thickening in the support zone unnecessary), and reduction in both ring-beam circumference and ring stiffness requirements.

As before, we assume the vessel to be completely filled with liquid of weight γ per unit volume. It is axisymmetrically supported, but this time on columns that meet the shell tangentially at locations below the equatorial level. The columns are again taken to be axially and flexurally rigid; the top end of a column may only displace along a direction perpendicular to the longitudinal axis of the column.

For the present problem, the hoop stiffness of the ring beam must not be infinite; the combination of a ring beam of infinite axial rigidity and inclined columns of infinite axial rigidity would imply a rigid spatial configuration whose effect upon the shell would be identical to that of its vertical-column counterpart for which the structural response of the shell is already known (from the previous study). In fact, we will assume the hoop stiffness of the ring beam is negligible. This assumption is consistent with what we are hoping to achieve by inclining the supports, namely to do away altogether with the requirement of a ring beam.

We start by developing a suitable analytical approach for the problem of the spherical vessel axisymmetrically supported on tangentially inclined columns. We depart from the usual approach of assuming the force redundant at the edge of the shell in the form of a horizontal shear H acting in a direction perpendicular to the axis of revolution of the shell, and instead, adopt the force redundant at the edge of the shell in the form of a normal shear S acting in a direction normal to the shell midsurface. Similarly, we discard the horizontal displacement δ in the direction perpendicular to the axis of revolution of the shell, and instead, adopt the normal displacement W in the direction normal to the shell midsurface. This will be the translational deformation of interest. The moment (M) and rotation (V) variables are as in the usual approach.

Figure 8.4 compares the conventional system of shell-edge actions and shell-edge deformations for vertical columns (a), versus the new system that we are adopting for the present problem of tangentially inclined columns (b). Positive directions for shell-edge actions are indicated by solid arrows, while those for shell-edge deformations are indicated by dashed arrows.

In prescribing the directions of the shear force redundants for these systems, it may be remarked that, if we were to apply a shear force upon the shell edge that is not perpendicular to the longitudinal axis of the columns, such a force would have a component that is absorbed by the axial stiffness of the columns, and would therefore not have any effect upon the shell.

The normal shearing forces S (Figure 8.4b) applied at the edge of either the upper or lower domain of the vessel appear to violate the overall equilibrium of the domain in question, since, unlike their horizontal counterparts H (Figure 8.4a), which are self-equilibrating, the S forces have a net resultant in the vertical direction. However, such a violation of basic principles does not, in fact, occur, since the vertical resultant

Figure 8.4 Systems of shell-edge redundants and shell-edge deformations for the spherical vessel supported on: (a) vertical columns; (b) tangentially inclined columns.

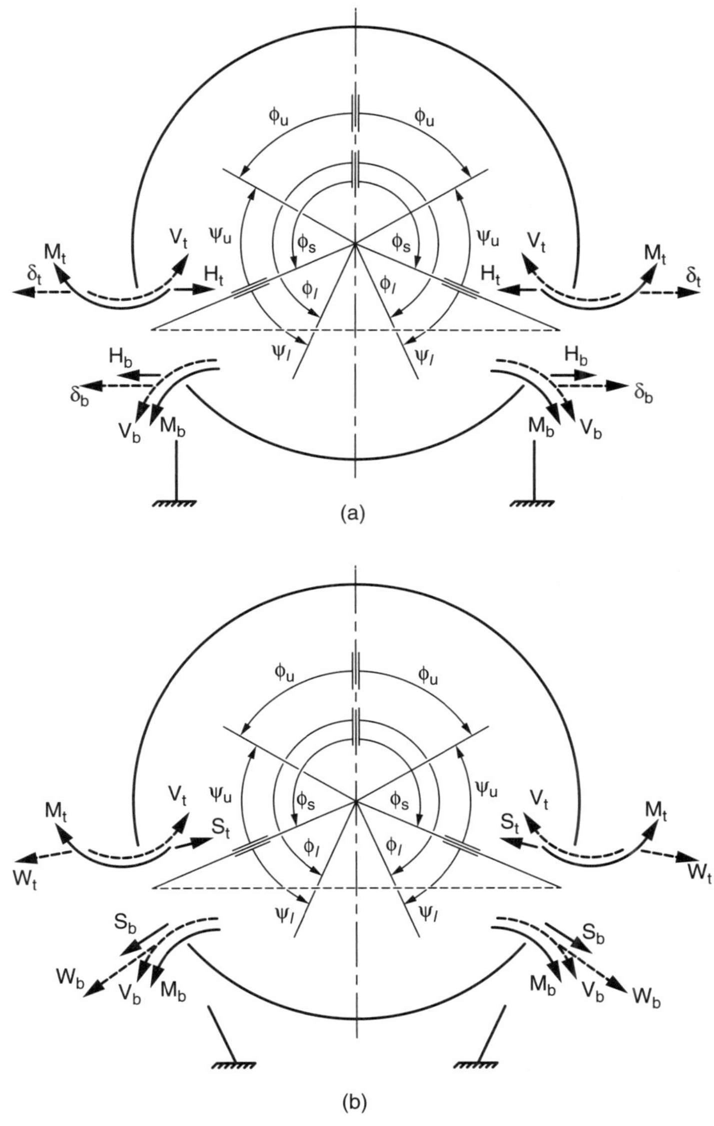

of the S actions is equilibrated by the vertical resultant of the axial forces occurring in the inclined columns simultaneously with the application of the S forces.

This situation is best understood if we begin with a horizontal force H' applied along the shell edge, which is self-equilibrating. The components of H' in the direction normal to the shell midsurface are then the shears S, while the components of H' in the direction tangential to the shell midsurface are wholly resisted by the axial stiffness of the columns. In the present scheme, no mention will be made anymore of the latter components, since they have no effect upon the shell; it is enough to appreciate their equilibrating role. Treatment of the problem will therefore proceed with S directly applied upon the shell edges.

With the approach to the present problem now clear, the basis for evaluating the edge effects is the fourth-order differential equation of Geckeler (i.e. equation (6.25) of Chapter 6).

8.1.3.2 Membrane displacements and deformations

Basic membrane displacements v^m and w^m corresponding to the full spherical vessel remain, of course, the same as those given in Section 8.1.2 for the case of vertical supports.

The deformation W^m is identical to the basic displacement w^m (and independent of v^m):

$$W^m = w^m \tag{8.28}$$

Using this result in equation (8.10b) for w^m, the value of W^m at the support (where $v^m = v_s^m = 0$) is obtained as

$$W_s^m = \frac{a}{Et}\left(N_\theta^m - \nu N_\phi^m\right)_s \tag{8.29}$$

Evaluating equation (8.29) for the upper and lower domains of the vessel in turn (using results (3.65), (3.66), (3.70) and (3.71) with ϕ set equal to ϕ_s), and using expressions (3.68) and (3.73) to evaluate V at the edges of the upper and lower domains respectively, we obtain the following explicit results for the edge deformations of present relevance:

$$W_t^m = \frac{1}{6}\frac{\gamma a^3}{Et}\left(\frac{1-\cos\phi_s}{1+\cos\phi_s}\right)\left\{\left(5+4\cos\phi_s\right) - \nu\left(1+2\cos\phi_s\right)\right\} \tag{8.30a}$$

$$W_b^m = \frac{1}{6}\frac{\gamma a^3}{Et}\left(\frac{1}{1-\cos\phi_s}\right)\left\{\left(1-7\cos\phi_s+4\cos^2\phi_s\right) - \nu\left(5-5\cos\phi_s+2\cos^2\phi_s\right)\right\} \tag{8.30b}$$

$$V_t^m = V_b^m = -\frac{\gamma a^2}{Et}\sin\phi_s \tag{8.31}$$

As before, subscripts t and b refer to the edges of the upper and lower domains of the vessel, respectively.

8.1.3.3 Edge deformations due to arbitrary edge actions

We already have expressions for stress resultants $\{N_\phi^b, N_\theta^b\}$, bending moments $\{M_\phi, M_\theta\}$ and deformations $\{\delta^b, V^b\}$ in terms of the transverse shear Q_ϕ. These were given in Chapter 6 as equations (6.34)–(6.39). However, and as already explained, the translational deformation of interest is W^b rather than δ^b. For the purposes of evaluating the shell-edge redundants, only the value of W^b at the edge (i.e. W_e^b) needs to be known. This may easily be written down from expression (8.29) by replacing superscript m with b (since variables now relate to the bending-disturbance solution), and ignoring the second term (since $\nu N_\phi^b << N_\theta^b$):

$$W_e^b \approx \frac{a}{Et}\left(N_\theta^b\right)_e = \frac{a}{Et}\left(-\frac{dQ_\phi}{d\phi}\right)_e = -\sqrt{2}\,\frac{\lambda a}{Et}\,C\sin\left(\beta - \frac{\pi}{4}\right) \tag{8.32}$$

In the above, $\left(N_\theta^b\right)_e$ has been evaluated from expression (6.35) with ψ set equal to zero. The parameters C and β are, of course, the constants of integration featuring in the general solution for Q_ϕ (equation 6.28).

The rotational deformation of interest is simply given by equation (6.36) with ψ set equal to zero:

$$V_e^b = \frac{1}{Et}\left(\frac{d^2Q_\phi}{d\phi^2}\right)_e = -\frac{2\lambda^2}{Et}\,C\cos\beta \tag{8.33}$$

We also require values of Q_ϕ and M_ϕ at the edge of the shell in order to evaluate the constants C and β. The relevant expressions are deduced from equations (6.28) and (6.37), with ψ set equal to zero:

$$\left(Q_\phi\right)_e = C\sin\beta \tag{8.34}$$

$$\left(M_\phi\right)_e = -\frac{a}{4\lambda^4}\left(\frac{d^3Q_\phi}{d\phi^3}\right)_e = \frac{1}{\sqrt{2}}\,\frac{a}{\lambda}\,C\sin\left(\beta + \frac{\pi}{4}\right) \tag{8.35}$$

Following a procedure similar to the standard approach of Chapter 6, the constants C and β are obtained by considering the static boundary conditions at the edge of a domain when that edge is subjected to the axi-symmetric actions M_e and S_e considered separately, and assuming that no surface loading acts upon the shell. Considering first the application of M_e, substitution of the boundary conditions

$$\left(M_\phi\right)_e = M_e\,;\quad \left(Q_\phi\right)_e = 0 \tag{8.36}$$

into expressions (8.34) and (8.35) gives

$$C = \frac{2\lambda M_e}{a}\,;\quad \beta = 0 \tag{8.37}$$

For the application of S_e, the relevant boundary conditions are

$$\left(M_\phi\right)_e = 0\,;\quad \left(Q_\phi\right)_e = S_e \tag{8.38}$$

Using the above values of $\left(M_\phi\right)_e$ and $\left(Q_\phi\right)_e$ in expressions (8.34) and (8.35), we obtain

$$C = -\sqrt{2}\,S_e\,;\quad \beta = -\frac{\pi}{4} \tag{8.39}$$

Substituting the values of C and β (as given by solutions (8.37) for M_e and solutions (8.39) for S_e) into expressions (8.32) for W_e^b and (8.33) for V_e^b, and superimposing the deformation contributions of M_e and S_e, we obtain the following expressions for the edge deformations due to the arbitrary edge redundants M_e and S_e, where a distinction has now been made between variables of the upper domain and those of the lower domain (refer to Figure 8.4):

$$W_t^b = \frac{2\lambda^2}{Et}M_t - \frac{2\lambda a}{Et}S_t \tag{8.40}$$

$$W_b^b = \frac{2\lambda^2}{Et}M_b + \frac{2\lambda a}{Et}S_b \tag{8.41}$$

$$V_t^b = -\frac{4\lambda^3}{Eat}M_t + \frac{2\lambda^2}{Et}S_t \tag{8.42}$$

$$V_b^b = \frac{4\lambda^3}{Eat}M_b + \frac{2\lambda^2}{Et}S_b \tag{8.43}$$

It may be noted that the influence coefficients in the above expressions (i.e. the coefficients of the terms on the right) are all independent of the parameter ϕ_s owing to the symmetry associated with the presently adopted system of variables (Figure 8.4b). The conventional system depicted in Figure 8.4(a) clearly lacks this type of symmetry, hence some of its influence coefficients depend upon ϕ_s. The reader may like to compare the coefficients of the right-hand terms of equations (8.40)–(8.43) with the corresponding coefficients in equations (8.5)–(8.8).

8.1.3.4 Variation of total stresses with distance from the edge

Membrane stress resultants in the vessel are still given by expressions (3.65), (3.66), (3.70) and (3.71). Bending-related internal actions N_ϕ^b, N_θ^b, M_ϕ and M_θ may easily be written down in terms of the arbitrary edge actions M_e and S_e since the constants C and β (which feature in the expressions for N_ϕ^b, N_θ^b, M_ϕ and M_θ as given by equations (6.34), (6.35), (6.37) and (6.38)) are now known in terms of M_e and S_e (solutions (8.37) and (8.39)).

Combining membrane stresses with bending-disturbance stresses in accordance with relations (6.70), variations of σ_ϕ^T and σ_θ^T with ψ for the upper and lower domains of the vessel are obtained in terms of arbitrary M_t, S_t, M_b and S_b as follows (the upper and lower signs of the ± symbol refer to the inner and outer surfaces of the shell respectively):

$$\left(\sigma_\phi^T\right)_u = \frac{1}{t}\left[\ \frac{\gamma a^2}{6}\left(\frac{1-\cos(\phi_s-\psi)}{1+\cos(\phi_s-\psi)}\right)\{1+2\cos(\phi_s-\psi)\}\right.$$

$$\left. +\cot(\phi_s-\psi)\,e^{-\lambda\psi}\left\{(\sin\lambda\psi-\cos\lambda\psi)S_t - 2(\sin\lambda\psi)\frac{\lambda M_t}{a}\right\}\ \right]$$

$$\pm\frac{6e^{-\lambda\psi}}{t^2}\left\{(\sin\lambda\psi+\cos\lambda\psi)M_t-(\sin\lambda\psi)\frac{aS_t}{\lambda}\right\} \tag{8.44a}$$

$$\left(\sigma_\theta^T\right)_u=\frac{1}{t}\left[\ \frac{\gamma a^2}{6}\left(\frac{1-\cos(\phi_s-\psi)}{1+\cos(\phi_s-\psi)}\right)\{5+4\cos(\phi_s-\psi)\}\right.$$

$$\left.-2\lambda\, e^{-\lambda\psi}\left\{(\cos\lambda\psi)S_t+(\sin\lambda\psi-\cos\lambda\psi)\frac{\lambda M_t}{a}\right\}\ \right]$$

$$\pm\frac{6\,\nu e^{-\lambda\psi}}{t^2}\left\{(\sin\lambda\psi+\cos\lambda\psi)M_t-(\sin\lambda\psi)\frac{aS_t}{\lambda}\right\} \tag{8.44b}$$

$$\left(\sigma_\phi^T\right)_l=\frac{1}{t}\left[\ \frac{\gamma a^2}{6}\left(\frac{5-5\cos(\phi_s+\psi)+2\cos^2(\phi_s+\psi)}{1-\cos(\phi_s+\psi)}\right)\right.$$

$$\left.+\cot(\phi_s+\psi)\, e^{-\lambda\psi}\left\{(\sin\lambda\psi-\cos\lambda\psi)S_b-2(\sin\lambda\psi)\frac{\lambda M_b}{a}\right\}\ \right]$$

$$\pm\frac{6e^{-\lambda\psi}}{t^2}\left\{(\sin\lambda\psi+\cos\lambda\psi)M_b+(\sin\lambda\psi)\frac{aS_b}{\lambda}\right\} \tag{8.45a}$$

$$\left(\sigma_\theta^T\right)_l=\frac{1}{t}\left[\ \frac{\gamma a^2}{6}\left(\frac{1-7\cos(\phi_s+\psi)+4\cos^2(\phi_s+\psi)}{1-\cos(\phi_s+\psi)}\right)\right.$$

$$\left.+2\lambda\, e^{-\lambda\psi}\left\{(\cos\lambda\psi)S_b-(\sin\lambda\psi-\cos\lambda\psi)\frac{\lambda M_b}{a}\right\}\ \right]$$

$$\pm\frac{6\,\nu e^{-\lambda\psi}}{t^2}\left\{(\sin\lambda\psi+\cos\lambda\psi)M_b+(\sin\lambda\psi)\frac{aS_b}{\lambda}\right\} \tag{8.45b}$$

8.1.3.5 Displacements

To obtain expressions for v^b and w^b in terms of the presently adopted pair of variables $\{M_e, S_e\}$, we may simply go to equations (8.16) and replace the term $H_e \sin\phi_e$ (the transverse shear Q_ϕ at the edge due to the application of H_e) by S_e (the transverse shear Q_ϕ at the edge due to the application of S_e). This substitution yields:

$$v^b=\frac{1}{Et}(1+\nu)\, e^{-\lambda\psi}\left\{2\lambda M_e(\sin\lambda\psi)-aS_e(\sin\lambda\psi-\cos\lambda\psi)\right\}+k^b\sin(\phi_e-\psi) \tag{8.46a}$$

$$w^b = -\left[v^b \cot(\phi_e - \psi) + \frac{2\lambda}{Et} e^{-\lambda\psi} \left\{\lambda M_e (\sin\lambda\psi - \cos\lambda\psi) + aS_e \cos\lambda\psi\right\}\right] \tag{8.46b}$$

As with their equivalents in Section 8.1.2, expressions (8.46) are based on a spherical cap with the crown at $\phi = 0$. For the present problem of tangentially inclined columns, tangential movement of the shell at the support is completely inhibited by the axially rigid columns, so that

$$v_s^T \left(= v_s^m + v_s^b\right) = 0 \tag{8.47}$$

Since $v_s^m = 0$, it follows that $v_s^b = 0$. Applying this result to expression (8.46a) with ψ set equal to zero, we readily obtain the constant of integration for the upper domain of the vessel as follows:

$$k_u^b = -\frac{1}{Et}\left(\frac{1+\nu}{\sin\phi_s}\right) aS_t \tag{8.48}$$

Thus, for the upper domain ($0 \le \phi \le \phi_s$; $\psi = \phi_s - \phi$), expressions (8.46) become

$$v_u^b = \frac{1}{Et}(1+\nu)\, e^{-\lambda\psi} \left\{2\lambda M_t (\sin\lambda\psi) - aS_t (\sin\lambda\psi - \cos\lambda\psi)\right\} - \frac{1}{Et}\left(\frac{1+\nu}{\sin\phi_s}\right) aS_t \sin(\phi_s - \psi) \tag{8.49a}$$

$$w_u^b = -\left[v_u^b \cot(\phi_s - \psi) + \frac{2\lambda}{Et} e^{-\lambda\psi} \left\{\lambda M_t (\sin\lambda\psi - \cos\lambda\psi) + aS_t \cos\lambda\psi\right\}\right] \tag{8.49b}$$

The expressions for k_l^b, v_l^b and w_l^b corresponding to the lower domain of the vessel ($\phi_s \le \phi \le \pi$; $\psi = \phi - \phi_s$) are obtained directly from expressions (8.48), (8.49a) and (8.49b), by invoking the technique that was used in Section 8.1.2.3.2 for the case of vertical supports. The relevant transformations are still given by equations (8.23), and the additional relations

$$S_t' = -S_b \tag{8.50a}$$

$$W_u' = W_l \tag{8.50b}$$

in which, as before, primes denote mirror images of the upper-domain variables in the opposite (i.e. lower) domain. The resulting expressions are as follows:

$$k_l^b = -\frac{1}{Et}\left(\frac{1+\nu}{\sin\phi_s}\right) aS_b \tag{8.51}$$

$$v_l^b = -\frac{1}{Et}(1+\nu)\,e^{-\lambda\psi}\left\{2\lambda M_b\left(\sin\lambda\psi\right)+aS_b\left(\sin\lambda\psi-\cos\lambda\psi\right)\right\}-\frac{1}{Et}\left(\frac{1+\nu}{\sin\phi_s}\right)aS_b\sin\left(\phi_s+\psi\right) \tag{8.52a}$$

$$w_l^b = -\left[v_l^b\cot\left(\phi_s+\psi\right)+\frac{2\lambda}{Et}e^{-\lambda\psi}\left\{\lambda M_b\left(\sin\lambda\psi-\cos\lambda\psi\right)-aS_b\cos\lambda\psi\right\}\right] \tag{8.52b}$$

Superposition of membrane displacements (equations (8.12) for the upper domain and (8.13) for the lower domain) with their bending-disturbance counterparts (equations (8.49) for the upper domain and (8.52) for the lower domain) then gives the net displacements. If required, these may be resolved into horizontal and vertical components in accordance with equations (8.27).

8.1.3.6 Application of developed approach

8.1.3.6.1 INTRODUCTION

Let us now focus attention on the ZAZT case of Section 8.1.2, for which shell stresses in the vicinity of the supports were noted to be large when vertical columns are deployed in relatively low positions ($\phi_s \rightarrow 120°$). Since it is mainly the hoop rigidity of the ring beam that governs the magnitude of peak shell stresses for such positions of the supports (torsional rigidity of the ring beam has insignificant influence), we need not consider the ZAIT case.

8.1.3.6.2 BOUNDARY CONDITIONS AND SHELL-EDGE REDUNDANTS FOR THE ZAZT CASE

For ring-beam properties of zero axial and zero axi-rotational rigidities, the boundary conditions governing the magnitude of shell-edge redundants are as follows:

$$W_t^T\;\left(W_t^m+W_t^b\right)=W_b^T\;\left(W_b^m+W_b^b\right) \tag{8.53a}$$

$$V_t^T\;\left(V_t^m+V_t^b\right)=V_b^T\;\left(V_b^m+V_b^b\right) \tag{8.53b}$$

$$M_b-M_t=0 \tag{8.54a}$$

$$S_b-S_t=0 \tag{8.54b}$$

Conditions (8.53) enforce continuity of radial displacement and meridional rotation between the upper and lower domains of the vessel at their junction, while conditions (8.54) express that the ring beam cannot resist any net torque or net thrust, since it is completely flexible. Note the absence of a membrane contribution in equation (8.54b), analogous to H_s^m occurring in the corresponding equation for the vertical-column system (refer to equations (8.1) and (8.9)). This is because the present formulation considers forces and displacements in the direction normal to the shell midsurface, a direction in which membrane stress resultants do not have a component.

Expanding equation (8.53a) (using expressions (8.30), (8.40) and (8.41)) and equation (8.53b) (using expressions (8.31), (8.42) and (8.43)), in the light of relations (8.54), we obtain the solutions

$$M_t = M_b = 0 \tag{8.55}$$

$$S_t = S_b = \frac{1}{6\lambda}\left(\frac{1+\nu}{\sin^2\phi_s}\right)\gamma a^2 \tag{8.56}$$

As expected, expressions (8.55) and (8.56) coincide with the corresponding ones for vertical columns when $\phi_s = 90°$.

8.1.3.6.3 NET STRESSES AND DISPLACEMENTS

Substitution of the obtained values for M_t, S_t, M_b and S_b into expressions (8.44) and (8.45) enables total shell stresses $\left(\sigma_\phi^T\right)_u$, $\left(\sigma_\theta^T\right)_u$, $\left(\sigma_\phi^T\right)_l$ and $\left(\sigma_\theta^T\right)_l$ to be explicitly obtained for the ZAZT support conditions. Similarly, substitution of the edge redundants into expressions (8.49) and (8.52) enables bending-disturbance displacements v_u^b, w_u^b, v_l^b and w_l^b to be evaluated. Adding these to membrane displacements then yields total displacements.

8.1.3.7 Results of a parametric study

The findings of a parametric study[9] aimed at evaluating the benefits of inclining columns may be summarised as follows:

- The displacements associated with tangentially inclined supports for ϕ_s approaching 120°, when the ring beam has no hoop rigidity at all, practically coincide with those of vertical columns and a rigid beam located at the same level, so that the former arrangement reproduces the low deformations of the latter while dispensing with ring-beam stiffness requirements altogether.
- More importantly, the often prohibitively high shell stresses associated with low locations of vertical supports combined with low hoop stiffness of the support ring, are drastically reduced by simply changing the inclination of the supports from vertical to tangential. The resulting stresses are even considerably lower than those associated with vertical supports in combination with a ring beam of infinite hoop rigidity.
- Peak shell stresses for low positions of supports (i.e. ϕ_s approaching 120°), when the supports are tangentially inclined to the shell midsurface and the ring beam is assumed to be axially flexible, have essentially the same magnitude as those for vertical supports located at the equator, the latter case being rather insensitive to ring-beam stiffness properties.
- When the arrangement of supports is tangential, peak stresses in the support zone of the vessel are practically insensitive to support location ϕ_s (at least in the practical range $90° \le \phi_s \le 120°$), provided that the hoop stiffness of the support ring is negligible. This behaviour differs from that for vertical supports, where peak shell stresses in the support zone are relatively small when the supports are located at the equator, and rise sharply with gradual lowering of the level of the supports towards $\phi_s = 120°$.

It may be remarked that, although inclined supports optimise the design of the ring-beam/shell assembly for sub-equatorial locations of the supports, large bending stresses may occur in the raked supports themselves. Nevertheless, these stresses can easily be countered by enhancing the depth of the column sections in the direction towards the axis of revolution of the shell, or by tying together the individual inclined columns by means of horizontal circular hoop rings at a number of levels along the height of the columns, thus effectively affording intermediate *elastic supports* along the column lengths.

8.1.4 Thickness-discontinuity effects

8.1.4.1 Introduction

In Section 8.1.2, it was found that, for the conventional arrangement of supports for elevated liquid-filled spherical vessels, where vertical columns are provided, certain combinations of ring-beam stiffnesses and ring-beam position can lead to excessively high shell stresses in the vicinity of the supports. This has been noted for low positions of the supports in combination with low extensional (i.e. hoop) stiffness of the ring beam.

Excessively high local shell stresses can be avoided without resort to local shell thickening by simply positioning the supports at the equatorial level. If a low support position is unavoidable, then we must ensure that the support ring is sufficiently inextensible; alternatively, we must incline the supports to meet the shell tangentially. However, the use of inclined columns may lead to greater overall costs, so the option of local shell thickening in the vicinity of the supports must be kept open, if the support ring has to be positioned below the equatorial level and effective hoop rigidity of the ring beam cannot be easily attained.

In this section, and as a final measure for countering excessive shell stresses, we investigate the benefit of local shell thickening in the vicinity of supports, at the same time quantifying the associated discontinuity stresses. We consider a spherical steel vessel filled to capacity with a liquid of weight γ per unit volume, and also allow for the existence of a uniform surchage pressure p over and above the hydrostatic pressure of the contained liquid. We go back to the more common practice of supporting spherical vessels via vertical columns.

In metal-plate construction, local shell thickening is usually effected abruptly at circles of latitude above and below the support level. We will refer to these circles (where the thickness suddenly changes) as Junction *O*

Figure 8.5 Notation for shell regions and inter-region junctions.

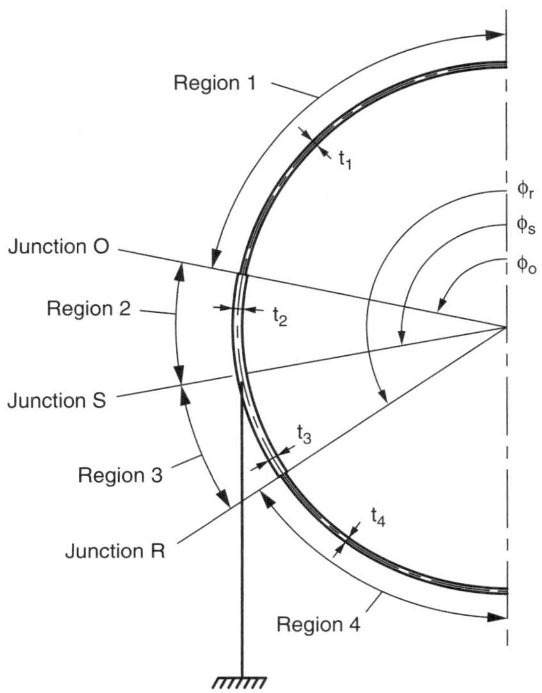

and Junction R respectively. The circle of support will be referred to as Junction S. Figure 8.5 shows these three junctions, and the associated geometric parameters. The parameters ϕ_o, ϕ_r and ϕ_s denote the meridional angles corresponding to Junctions O, R and S respectively. In general, subscripts o, r and s will be used to denote variables associated with Junctions O, R and S respectively.

Evidently, there are four regions of the vessel demarcated by the three junctions. Let us define these as follows (refer to Figure 8.5): Region 1: $0 \leq \phi \leq \phi_o$; Region 2: $\phi_o \leq \phi \leq \phi_s$; Region 3: $\phi_s \leq \phi \leq \phi_r$; Region 4: $\phi_r \leq \phi \leq \pi$. The thickness of the shell will be assumed to be constant over a given region, being denoted by t_1, t_2, t_3 and t_4 for Regions 1, 2, 3 and 4 respectively. Usually $t_2 = t_3$, but this need not be the case all the time. Numeral subscripts 1, 2, 3 and 4 will generally be used to denote variables for Regions 1, 2, 3 and 4 respectively.

The aim of this study is to investigate the amount of local thickening (in terms of the ratios t_1 / t_2 or t_4 / t_3) that may be required in order to bring the level of local stresses in the vicinity of the supports to the same order of magnitude as the membrane stresses elsewhere, and the extent from the support to which such thickening should be carried. In other words, we seek to *rationalise* the provision of local thickening and provide guidance to the designer, where this measure is needed.

Since the basic computational procedure is essentially the same as that already illustrated in the foregoing studies, it will not be necessary to give all the derivational details. Only the most important results, and the boundary conditions forming the basis of these, will be presented.

8.1.4.2 Membrane solution

The relevant membrane-solution results for the various regions and junctions are as follows:

8.1.4.2.1 HYDROSTATIC LIQUID PRESSURE

Interior stress resultants for Regions 1 and 2

$$N_\phi^m = \frac{\gamma a^2}{6}\left(\frac{1}{\sin^2\phi}\right)\left(1 - 3\cos^2\phi + 2\cos^3\phi\right) \tag{8.57a}$$

$$N_\theta^m = \frac{\gamma a^2}{6}\left(\frac{1}{\sin^2\phi}\right)\left(5 - 6\cos\phi - 3\cos^2\phi + 4\cos^3\phi\right) \tag{8.57b}$$

Interior stress resultants for Regions 3 and 4

$$N_\phi^m = \frac{\gamma a^2}{6}\left(\frac{1}{\sin^2\phi}\right)\left(5 - 3\cos^2\phi + 2\cos^3\phi\right) \tag{8.58a}$$

$$N_\theta^m = \frac{\gamma a^2}{6}\left(\frac{1}{\sin^2\phi}\right)\left(1 - 6\cos\phi - 3\cos^2\phi + 4\cos^3\phi\right) \tag{8.58b}$$

Net horizontal forces at junctions

$$H_o^m = H_r^m = 0 \tag{8.59}$$

$$H_s^m = -\frac{\gamma a^2}{6}\left(\frac{4\cos\phi_s}{\sin^2\phi_s}\right) \tag{8.60}$$

Deformations at edges of regions

$$V_{1o}^m = -\frac{1}{t_1}\frac{\gamma a^2}{E}\sin\phi_o \tag{8.61a}$$

$$V_{2o}^m = -\frac{1}{t_2}\frac{\gamma a^2}{E}\sin\phi_o \tag{8.61b}$$

$$V_{2s}^m = -\frac{1}{t_2}\frac{\gamma a^2}{E}\sin\phi_s \tag{8.62a}$$

$$V_{3s}^m = -\frac{1}{t_3}\frac{\gamma a^2}{E}\sin\phi_s \tag{8.62b}$$

$$V_{3r}^m = -\frac{1}{t_3}\frac{\gamma a^2}{E}\sin\phi_r \tag{8.63a}$$

$$V_{4r}^m = -\frac{1}{t_4}\frac{\gamma a^2}{E}\sin\phi_r \tag{8.63b}$$

$$\delta_{1o}^m = \frac{1}{t_1}\frac{\gamma a^3}{6E}\left(\frac{1}{\sin\phi_o}\right)\left\{\left(5-6\cos\phi_o-3\cos^2\phi_o+4\cos^3\phi_o\right)-\nu\left(1-3\cos^2\phi_o+2\cos^3\phi_o\right)\right\} \tag{8.64a}$$

$$\delta_{2o}^m = \frac{1}{t_2}\frac{\gamma a^3}{6E}\left(\frac{1}{\sin\phi_o}\right)\left\{\left(5-6\cos\phi_o-3\cos^2\phi_o+4\cos^3\phi_o\right)-\nu\left(1-3\cos^2\phi_o+2\cos^3\phi_o\right)\right\} \tag{8.64b}$$

$$\delta_{2s}^m = \frac{1}{t_2}\frac{\gamma a^3}{6E}\left(\frac{1}{\sin\phi_s}\right)\left\{\left(5-6\cos\phi_s-3\cos^2\phi_s+4\cos^3\phi_s\right)-\nu\left(1-3\cos^2\phi_s+2\cos^3\phi_s\right)\right\} \tag{8.65a}$$

$$\delta_{3s}^{m} = \frac{1}{t_3}\frac{\gamma a^3}{6E}\left(\frac{1}{\sin\phi_s}\right)\left\{\left(1-6\cos\phi_s-3\cos^2\phi_s+4\cos^3\phi_s\right)-\nu\left(5-3\cos^2\phi_s+2\cos^3\phi_s\right)\right\} \tag{8.65b}$$

$$\delta_{3r}^{m} = \frac{1}{t_3}\frac{\gamma a^3}{6E}\left(\frac{1}{\sin\phi_r}\right)\left\{\left(1-6\cos\phi_r-3\cos^2\phi_r+4\cos^3\phi_r\right)-\nu\left(5-3\cos^2\phi_r+2\cos^3\phi_r\right)\right\} \tag{8.66a}$$

$$\delta_{4r}^{m} = \frac{1}{t_4}\frac{\gamma a^3}{6E}\left(\frac{1}{\sin\phi_r}\right)\left\{\left(1-6\cos\phi_r-3\cos^2\phi_r+4\cos^3\phi_r\right)-\nu\left(5-3\cos^2\phi_r+2\cos^3\phi_r\right)\right\} \tag{8.66b}$$

8.1.4.2.2 UNIFORM SURCHARGE PRESSURE

Interior stress resultants for all regions

$$N_{\phi}^{m} = N_{\theta}^{m} = \frac{pa}{2} \tag{8.67}$$

Net horizontal forces at junctions

$$H_{o}^{m} = H_{r}^{m} = H_{s}^{m} = 0 \tag{8.68}$$

Deformations at edges of regions

$$V_{1o}^{m} = V_{2o}^{m} = V_{2s}^{m} = V_{3s}^{m} = V_{3r}^{m} = V_{4r}^{m} = 0 \tag{8.69}$$

$$\delta_{1o}^{m} = \frac{1}{t_1}\frac{pa^2}{2E}(1-\nu)\sin\phi_o \tag{8.70a}$$

$$\delta_{2o}^{m} = \frac{1}{t_2}\frac{pa^2}{2E}(1-\nu)\sin\phi_o \tag{8.70b}$$

$$\delta_{2s}^{m} = \frac{1}{t_2}\frac{pa^2}{2E}(1-\nu)\sin\phi_s \tag{8.71a}$$

$$\delta_{3s}^{m} = \frac{1}{t_3}\frac{pa^2}{2E}(1-\nu)\sin\phi_s \tag{8.71b}$$

$$\delta_{3r}^{m} = \frac{1}{t_3}\frac{pa^2}{2E}(1-\nu)\sin\phi_r \tag{8.72a}$$

$$\delta_{4r}^{m} = \frac{1}{t_4}\frac{pa^2}{2E}(1-\nu)\sin\phi_r \tag{8.72b}$$

8.1.4.3 Bending effects

8.1.4.3.1 ASSUMPTIONS

It is here assumed that Regions 2 and 3 bordering the support are sufficiently wide to permit bending effects associated with one edge of a region to be *decoupled* from those associated with the other edge of the region. In other words, edge-disturbance interaction between the bounding edges of either of these regions is assumed to be negligible.

According to the results of Section 6.3.8, the error incurred by making this assumption is of the order of only 5% if the locally thickened zone around the supports is continued for an angular distance of $4.25/\lambda$ on either side of the support circle. Now, most steel vessels designed for liquid storage at moderate pressures fall in the range $150 \le a/t \le 600$.[1] The lower limit of this range (i.e. $a/t = 150$) gives a value for λ of 15.74 (based on $\nu = 0.3$). For this value of λ, the 5% criterion for decoupling of edge effects would require shell thickening to extend for an angular distance of $\approx 15°$ on either side of the support circle. Many of the larger steel vessels actually lie in the range $300 \le a/t \le 400$. Taking the lower limit of this range (i.e. $a/t = 300$), the 5% decoupling criterion would require $\phi_s - \phi_o$ and $\phi_r - \phi_s$ to be at least $\approx 10°$, which is usually satisfied in practice. Thus we are quite justified in neglecting interaction of edge effects in this study. This greatly simplifies the mathematical formulation of the problem and the actual calculation of edge effects.

Figure 8.6(a) shows the positive sign conventions for the corrective bending moments and horizontal shear forces assumed as shell redundants at the edges of the various regions, while Figure 8.6(b) shows the positive sign conventions for the deformations (rotations and lateral movements) associated with the edges of these regions.

8.1.4.3.2 BOUNDARY CONDITIONS AND SOLUTIONS FOR SHELL REDUNDANTS

8.1.4.3.2.1 Thickness-discontinuity junctions

In order to preserve continuity of deformations across Junctions O and R, as well as the equilibrium of a shell element lying at either of these junctions, the following boundary conditions must be imposed:

Junction O

$$V_{1o}^{m} + V_{1o}^{b} = V_{2o}^{m} + V_{2o}^{b} \tag{8.73a}$$

$$\delta_{1o}^{m} + \delta_{1o}^{b} = \delta_{2o}^{m} + \delta_{2o}^{b} \tag{8.73b}$$

$$M_{2o} - M_{1o} = 0 \tag{8.73c}$$

$$H_{o}^{m} + H_{2o}^{b} - H_{1o}^{b} = 0 \tag{8.73d}$$

Junction R

$$V_{3r}^m + V_{3r}^b = V_{4r}^m + V_{4r}^b \tag{8.74a}$$

$$\delta_{3r}^m + \delta_{3r}^b = \delta_{4r}^m + \delta_{4r}^b \tag{8.74b}$$

$$M_{4r} - M_{3r} = 0 \tag{8.74c}$$

$$H_r^m + H_{4r}^b - H_{3r}^b = 0 \tag{8.74d}$$

In the above equations, the membrane-solution parameters V^m, δ^m, H^m for the various locations have already been given in Section 8.1.4.2 for the loadings of present relevance. The bending-related edge deformations V^b, δ^b in these equations can readily be written down in terms of the edge redundants M, H^b on the basis of relations (6.62), with due regard to the positive directions adopted in Figure 8.6.

From equations (8.73c) and (8.74c) respectively, let $M_{1o} = M_{2o} = M_o$ and $M_{3r} = M_{4r} = M_r$. Furthermore, since $H_o^m = H_r^m = 0$ (equations (8.59) and (8.68)), and in the light of conditions (8.73d) and (8.74d) respectively, let $H_{1o}^b = H_{2o}^b = H_o$ and $H_{3r}^b = H_{4r}^b = H_r$. Let us also define the following geometric parameters:

$$\rho_o = \frac{t_1}{t_2} \tag{8.75a}$$

$$\varsigma_o = 2\left(1+\rho_o^{5/2}\right)\left(1+\rho_o^{3/2}\right) - \left(1-\rho_o^2\right)^2 \tag{8.75b}$$

Figure 8.6 Positive convention for edge parameters: (a) redundant actions; (b) deformations.

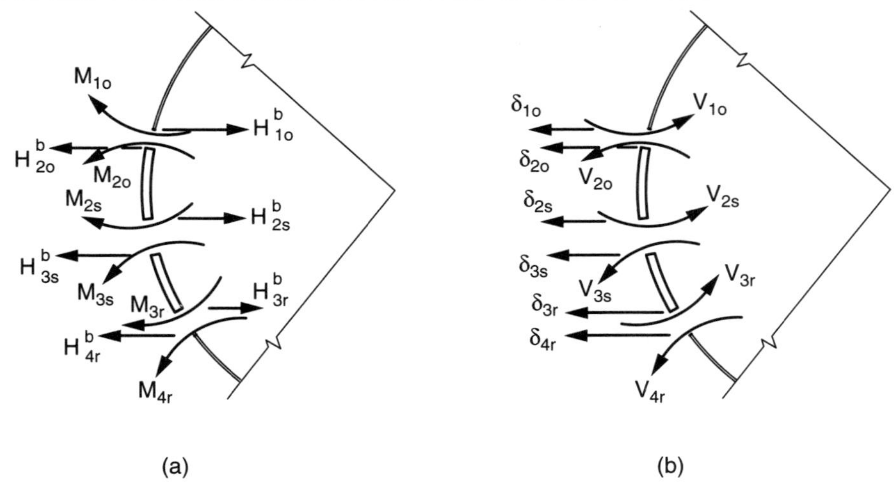

$$d_o^m = \left(5 - 6\cos\phi_o - 3\cos^2\phi_o + 4\cos^3\phi_o\right) - \nu\left(1 - 3\cos^2\phi_o + 2\cos^3\phi_o\right) \tag{8.75c}$$

$$\rho_r = \frac{t_4}{t_3} \tag{8.76a}$$

$$\varsigma_r = 2\left(1+\rho_r^{5/2}\right)\left(1+\rho_r^{3/2}\right) - \left(1-\rho_r^2\right)^2 \tag{8.76b}$$

$$d_r^m = \left(1 - 6\cos\phi_r - 3\cos^2\phi_r + 4\cos^3\phi_r\right) - \nu\left(5 - 3\cos^2\phi_r + 2\cos^3\phi_r\right) \tag{8.76c}$$

Upon solving the simultaneous equations ensuing from the sets of conditions in equations (8.73) and (8.74), and making use of the geometric parameters in equations (8.75) and (8.76), we obtain the following closed-form results for the shell-edge redundants:

Hydrostatic liquid pressure

Junction O

$$M_o = \frac{\gamma a^3}{2}\left(1-\rho_o\right)\frac{1}{\varsigma_o}\left\{\frac{1}{\lambda_1^2}\left(\frac{d_o^m}{6\sin^2\phi_o}\right)\left(1-\rho_o^2\right) - \frac{1}{\lambda_1^3}\left(1+\rho_o^{3/2}\right)\sin\phi_o\right\} \tag{8.77a}$$

$$H_o = \frac{\gamma a^2}{2}\left(\frac{1}{1+\rho_o}\right)\left[\frac{1}{\lambda_1^2} + 2\left(1+\rho_o^{5/2}\right)\frac{1}{\varsigma_o}\left\{\frac{1}{\lambda_1}\left(\frac{d_o^m}{6\sin^3\phi_o}\right)\left(1-\rho_o^2\right) - \frac{1}{\lambda_1^2}\left(1+\rho_o^{3/2}\right)\right\}\right] \tag{8.77b}$$

Junction R

$$M_r = \frac{\gamma a^3}{2}\left(1-\rho_r\right)\frac{1}{\varsigma_r}\left\{\frac{1}{\lambda_4^2}\left(\frac{d_r^m}{6\sin^2\phi_r}\right)\left(1-\rho_r^2\right) + \frac{1}{\lambda_4^3}\left(1+\rho_r^{3/2}\right)\sin\phi_r\right\} \tag{8.78a}$$

$$H_r = \frac{\gamma a^2}{2}\left(\frac{1}{1+\rho_r}\right)\left[\frac{1}{\lambda_4^2} - 2\left(1+\rho_r^{5/2}\right)\frac{1}{\varsigma_r}\left\{\frac{1}{\lambda_4}\left(\frac{d_r^m}{6\sin^3\phi_r}\right)\left(1-\rho_r^2\right) + \frac{1}{\lambda_4^2}\left(1+\rho_r^{3/2}\right)\right\}\right] \tag{8.78b}$$

Uniform surcharge pressure

Junction O

$$M_o = \frac{pa^2}{4}\frac{1}{\varsigma_o}\frac{1}{\lambda_1^2}\left(1-\nu\right)\left(1-\rho_o\right)\left(1-\rho_o^2\right) \tag{8.79a}$$

$$H_o = \frac{pa}{2}\frac{1}{\varsigma_o}\frac{1}{\lambda_1}\frac{1}{\sin\phi_o}(1-\nu)(1-\rho_o)(1+\rho_o^{5/2}) \tag{8.79b}$$

Junction R

$$M_r = \frac{pa^2}{4}\frac{1}{\varsigma_r}\frac{1}{\lambda_4^2}(1-\nu)(1-\rho_r)(1-\rho_r^2) \tag{8.80a}$$

$$H_r = -\frac{pa}{2}\frac{1}{\varsigma_r}\frac{1}{\lambda_4}\frac{1}{\sin\phi_r}(1-\nu)(1-\rho_r)(1+\rho_r^{5/2}) \tag{8.80b}$$

8.1.4.3.2.2 Support junction

For the support junction, we will consider two representative limiting cases of ring-beam behaviour: (*i*) ring beam of zero extensional and zero axi-rotational rigidities (i.e. ZAZT case); (*ii*) ring beam of infinite extensional and infinite axi-rotational rigidities (i.e. IAIT case). The boundary conditions for these support cases have already been encountered in Section 8.1.2, but are restated below specifically in terms of the notation for Regions 2 and 3.

ZAZT support conditions

$$V_{2s}^m + V_{2s}^b = V_{3s}^m + V_{3s}^b \tag{8.81a}$$

$$\delta_{2s}^m + \delta_{2s}^b = \delta_{3s}^m + \delta_{3s}^b \tag{8.81b}$$

$$M_{3s} - M_{2s} = 0 \tag{8.81c}$$

$$H_s^m + H_{3s}^b - H_{2s}^b = 0 \tag{8.81d}$$

IAIT support conditions

$$V_{2s}^m + V_{2s}^b = 0 \tag{8.82a}$$

$$\delta_{2s}^m + \delta_{2s}^b = 0 \tag{8.82b}$$

$$V_{3s}^m + V_{3s}^b = 0 \tag{8.82c}$$

$$\delta_{3s}^m + \delta_{3s}^b = 0 \tag{8.82d}$$

As was the case with the other two junctions, the membrane-solution parameters V^m, δ^m, H^m occurring in the above equations have already been evaluated for the loadings of interest, and were given in Section 8.1.4.2, while the bending-related edge deformations V^b, δ^b in these equations can also be written down in terms of the edge redundants M, H^b by reference to relations (6.62), and with due regard to the positive directions adopted in Figure 8.6. Let us assume that $t_2 = t_3$, as is usually the case. Denoting this thickness around the support by t_s (so that $\lambda_2 = \lambda_3 = \lambda_s$), we obtain, upon solving the simultaneous equations ensuing from the above sets of conditions, the results for shell-edge redundants as follows:

Hydrostatic liquid pressure

ZAZT support conditions

$$M_{2s} = M_{3s} = -\frac{\gamma a^3}{6}\frac{1}{\lambda_s}\cot\phi_s \tag{8.83a}$$

$$H^b_{2s} = \frac{\gamma a^2}{6}\left\{-2\left(\frac{\cos\phi_s}{\sin^2\phi_s}\right) + \frac{1}{\lambda_s}\left(\frac{1+\nu}{\sin^3\phi_s}\right)\right\} \tag{8.83b}$$

$$H^b_{3s} = \frac{\gamma a^2}{6}\left\{2\left(\frac{\cos\phi_s}{\sin^2\phi_s}\right) + \frac{1}{\lambda_s}\left(\frac{1+\nu}{\sin^3\phi_s}\right)\right\} \tag{8.83c}$$

IAIT support conditions

$$M_{2s} = \frac{\gamma a^3}{2}\frac{1}{\lambda_s^2}\left\{\left(\frac{d^m_{2s}}{6\sin^2\phi_s}\right) - \frac{1}{\lambda_s}\sin\phi_s\right\} \tag{8.84a}$$

$$M_{3s} = \frac{\gamma a^3}{2}\frac{1}{\lambda_s^2}\left\{\left(\frac{d^m_{3s}}{6\sin^2\phi_s}\right) + \frac{1}{\lambda_s}\sin\phi_s\right\} \tag{8.84b}$$

$$H^b_{2s} = \frac{\gamma a^2}{2}\frac{1}{\lambda_s}\left\{\left(\frac{d^m_{2s}}{3\sin^3\phi_s}\right) - \frac{1}{\lambda_s}\right\} \tag{8.84c}$$

$$H^b_{3s} = -\frac{\gamma a^2}{2}\frac{1}{\lambda_s}\left\{\left(\frac{d^m_{3s}}{3\sin^3\phi_s}\right) + \frac{1}{\lambda_s}\right\} \tag{8.84d}$$

where

$$d^m_{2s} = (5 - 6\cos\phi_s - 3\cos^2\phi_s + 4\cos^3\phi_s) - \nu(1 - 3\cos^2\phi_s + 2\cos^3\phi_s) \tag{8.85a}$$

$$d_{3s}^{m} = (1 - 6\cos\phi_s - 3\cos^2\phi_s + 4\cos^3\phi_s) - \nu(5 - 3\cos^2\phi_s + 2\cos^3\phi_s) \tag{8.85b}$$

Uniform surcharge pressure

ZAZT support conditions

$$M_{2s} = M_{3s} = 0 \tag{8.86a}$$

$$H_{2s}^{b} = H_{3s}^{b} = 0 \tag{8.86b}$$

IAIT support conditions

$$M_{2s} = M_{3s} = \frac{pa^2}{4}\frac{1}{\lambda_s^2}(1-\nu) \tag{8.87a}$$

$$H_{2s}^{b} = -H_{3s}^{b} = \frac{pa}{2}\frac{1}{\lambda_s}\left(\frac{1-\nu}{\sin\phi_s}\right) \tag{8.87b}$$

8.1.4.4 Findings of a parametric study

Let us consider a numerical example of a medium-sized steel vessel with radius $a = 6750$mm, and with vertical supports located at $\phi_s = 120°$.[10,11] Results for a steel vessel of this size are shown in Figure 8.7 (vessel of constant thickness $t = 15$mm) and Figure 8.8 (locally thickened vessel with $t_1 = t_4 = 15$mm and $t_2 = t_3 = 30$ mm). The plots cover both ZAZT and IAIT support cases, and show variations of membrane stresses (σ_ϕ^m and σ_θ^m) and total stresses (σ_ϕ^T and σ_θ^T) around the support as well as around the thickness-discontinuity locations. In showing the total stresses, a distinction has been made between stresses on the inner surface of the shell (denoted by subscript i) and stresses on the outer surface (denoted by subscript o). The membrane stresses are, of course, uniform across the shell thickness, while total stresses vary across the shell thickness owing to the flexural components.

The total stresses around the support are, of course, dependent on the type of support conditions (ZAZT or IAIT). Those around the thickness-discontinuity junctions, as shown in Figure 8.8, are only dependent on the discontinuity parameters; they are independent of support conditions, since they were derived by ignoring possible interaction between adjacent edges.

The findings and recommendations of a detailed parametric study,[10,11] which also took into account edge-disturbance interaction between the support and the thickness discontinuities above and below the support, are as follows:

(*a*) Peak shell stresses around the supports are, for all practical purposes, inversely proportional to shell thickness. Strictly, they tend to decrease somewhat more rapidly with increase of shell thickness for combinations of low support-ring location (i.e. ϕ_s approaching 120°) and low support-ring hoop stiffness (i.e. hoop stiffness approaching ZAZT conditions), but this trend is insignificant. Thus, for any steel vessel, the magnitude of the required shell thickening at the supports can be estimated on the basis of the inverse law relating peak stresses to shell thickness.

Figure 8.7 Stresses around the support for the constant-thickness vessel with t = 15 mm: (a) meridional stresses; (b) hoop stresses.

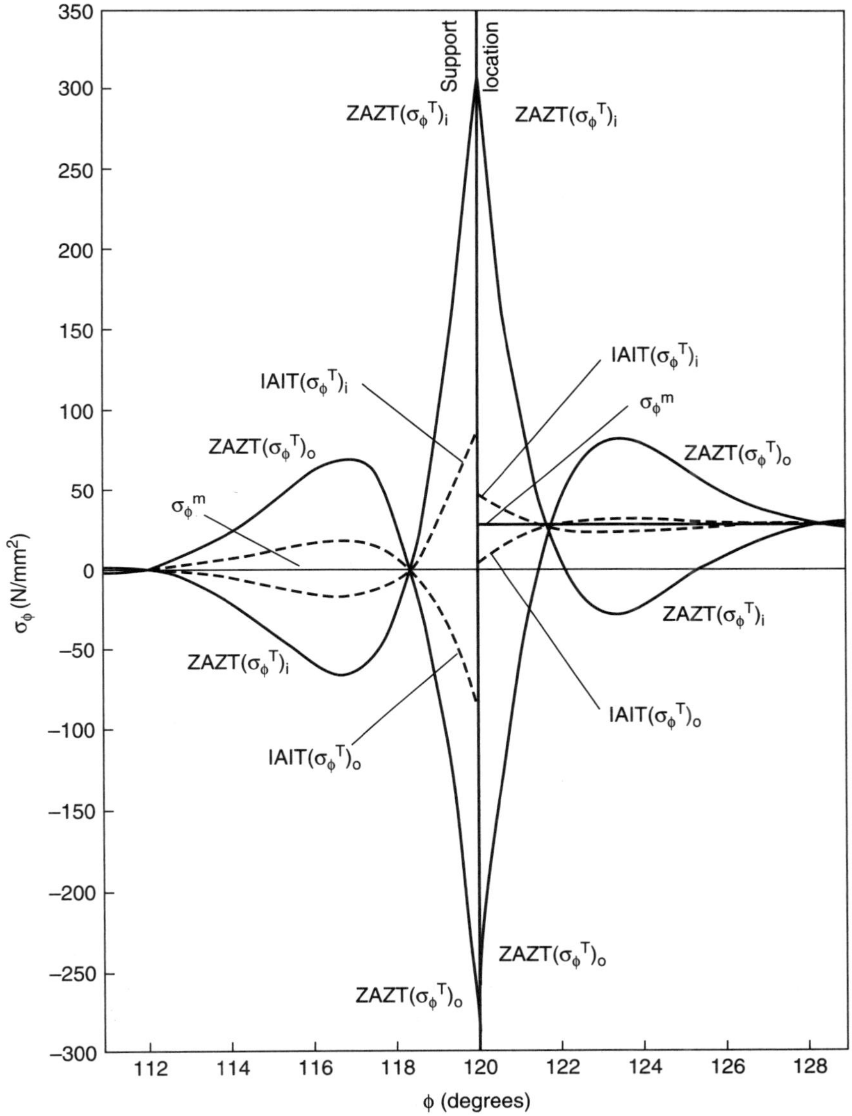

(a)

Figure 8.7 (*Continued*)

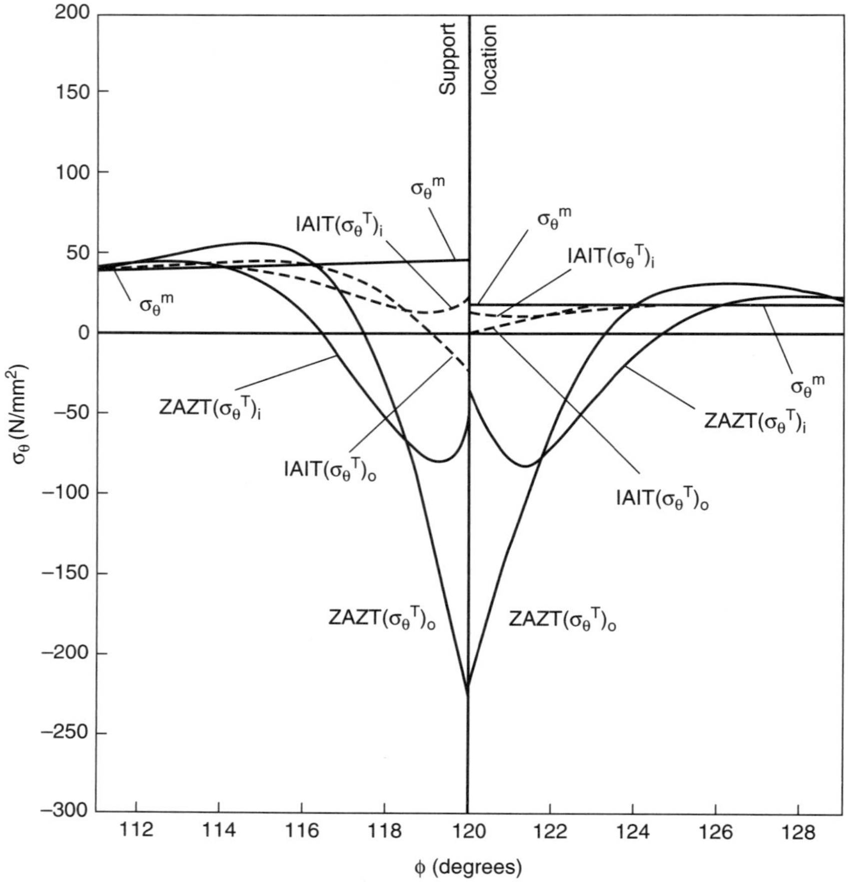

(b)

(*b*) For the range of thickening ratios likely to be adopted in practice (i.e. $0.3 \leq t_1 / t_2 \leq 0.7$ and $0.3 \leq t_4 / t_3 \leq 0.7$), discontinuity stresses at the edges of the thickened portion of the shell need not be a cause for concern, once the local stresses at the support itself (assumed to be located at or towards 120°) have been lowered to acceptable levels by thickness enhancement.

(*c*) However, for a support location $\phi_s = 120°$, if the hoop stiffness of the support ring closely approaches IAIT conditions, and the ratio t_4 / t_3 is close to the lower limit of 0.3, the largest stress in the entire vessel, after shell thickening, may shift from the support vicinity to just within the thinner side of the lower thickness discontinuity. It is useful to note that it is the *meridional* discontinuity stresses that always dictate thickening requirements at the support, and that may require to be checked at the lower thickness discontinuity.

(*d*) Now, for constant-thickness spherical shells, the effective range ψ_r of the bending disturbance from the edge of the shell may arbitrarily be taken as the half-wavelength of its oscillations. Alternatively, the distance over which the function $e^{-\lambda\psi}$ (controlling the amplitude of these oscillations) decays from unity at the edge ($\psi = 0$) to 0.05 (i.e. 5% of initial value) may be adopted as the effective range. These two different criteria correspond to $\lambda\psi = \pi$ and $\lambda\psi = 3$, respectively, and therefore lead to essentially the same result, since $\pi \approx 3$. It is preferred to define the effective range of the bending disturbance as $\psi_r = \pi / \lambda$, or $180° / \lambda$.

(*e*) However, local shell thickening around the supports need not extend the full effective range of the support bending disturbance. A narrow band of local thickening centred about the support circle, of total width no more than 10°, is recommended. This width of 10° is smaller than the effective bandwidth of the two-sided bending disturbance around the support (i.e. $360° / \lambda$), for thickening in the range

Figure 8.8 Stresses around isolated junctions for the locally thickened vessel with $t_1 = t_4 = 15$ mm and $t_2 = t_3 = 30$ mm: (a) meridional stresses; (b) hoop stresses.

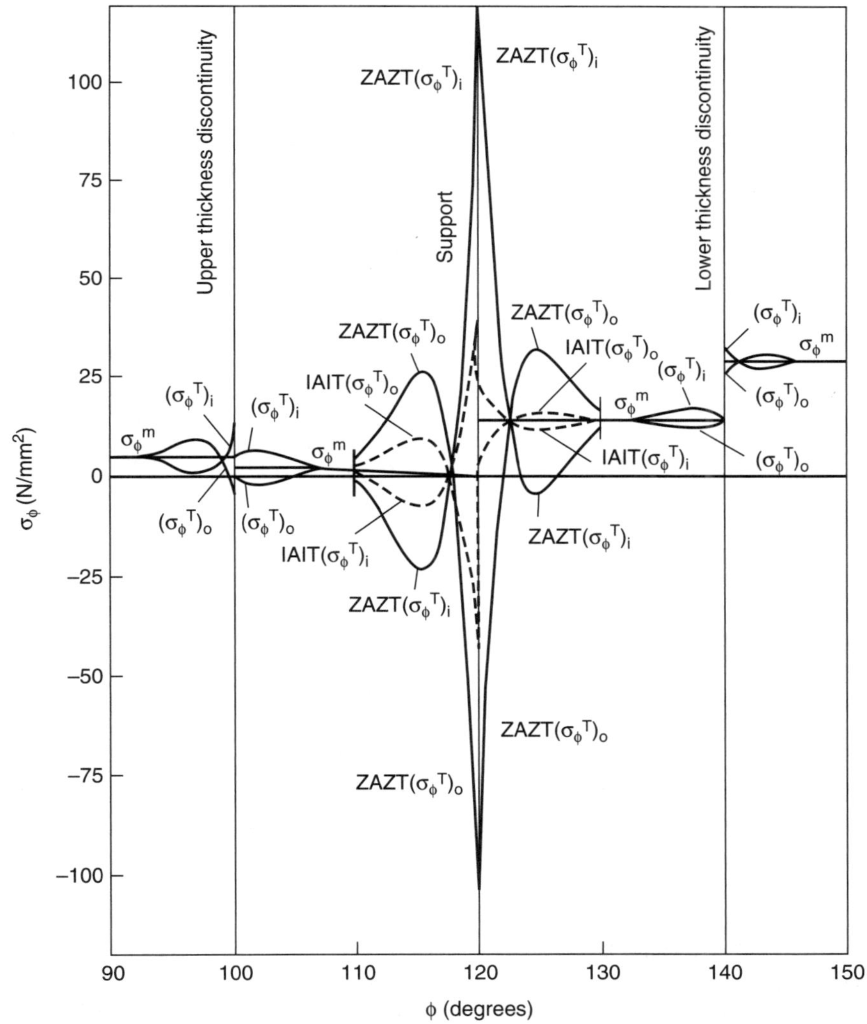

(a)

Figure 8.8 (*Continued*)

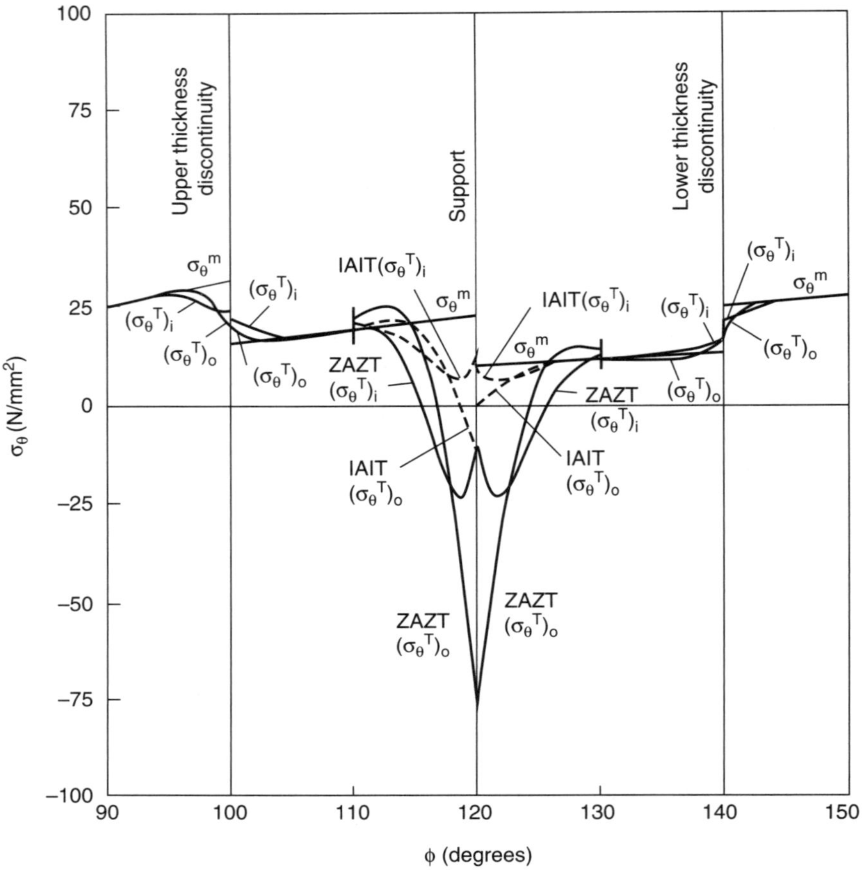

(b)

$100 \leq a/t \leq 400$. Such thickening is effective in cutting down the stress peaks in the immediate vicinity of the supports, which dictate shell-thickness requirements.

(*f*) The limitation on the extent of local thickening is also clearly desirable from the point of view of minimising material costs. However, it then becomes necessary, when calculating the stresses in a vessel of given dimensions, to take into account edge interaction. The simplified approach of decoupling edge effects, while generally leading to a conservative estimate of stresses at the thickness discontinuities, tends to slightly underestimate the peaks of the stresses at the support, which, unfortunately, are usually the most critical in the entire vessel.

8.2. Egg-shaped sludge digesters

8.2.1 Introduction

Sludge digesters at wastewater treatment plants are designed typically as short and wide cylindrical tanks with dome roofs. Digesters of this type, while relatively easy and inexpensive to construct, tend to have

Figure 8.9 Typical mixing patterns in sludge digesters: (a) conventional cylindrical digester; (b) egg-shaped digester.

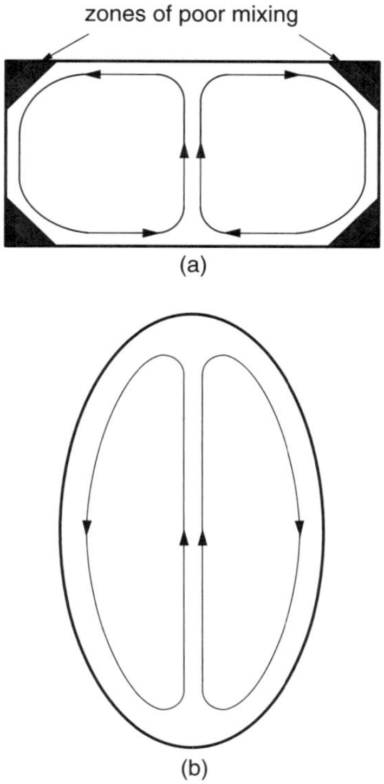

higher maintenance costs in the long term. This is due to the existence of zones of poor circulation of the sludge, resulting in solid deposits settling down at the bottom of the digester. These need to be cleared out at relatively frequent intervals.

In order to improve the mixing efficiency of sludge digesters, egg-shaped containment shells with a smooth surface-of-revolution profile have been proposed and, indeed, many such digesters have already been constructed in countries such as the USA, Germany, Australia, Taiwan and Japan.[14–17] Because of the smooth profile of the containment vessel, there are no zones of poor circulation of the sludge, and the accumulation of deposits at the bottom of the digester is greatly reduced in comparison with the conventional arrangement, resulting in lower operational costs over the long term.

Figure 8.9 compares one typical mixing pattern in the egg-shaped digester versus that in a conventional cylindrical digester. In the egg-shaped digester, the deposits that do settle to the bottom are easy to remove as they all collect in one relatively small area at the tapered bottom of the digester, and the removal of these deposits may be carried out on a continuous basis. Similarly, the tapered shape of the top of the egg-shaped digester allows the crust that forms on the surface of the sludge to be removed more conveniently than in the case of the conventional wide cylindrical tank, with its larger liquid-surface area. Other advantages are lower heat losses (owing to the smaller surface-to-volume ratio of the egg shape in comparison with the cylindrical shape), and more attractive appearance (despite the higher profile of the egg shape).

Figure 8.10 Geometric parameters of the egg-shaped digester.

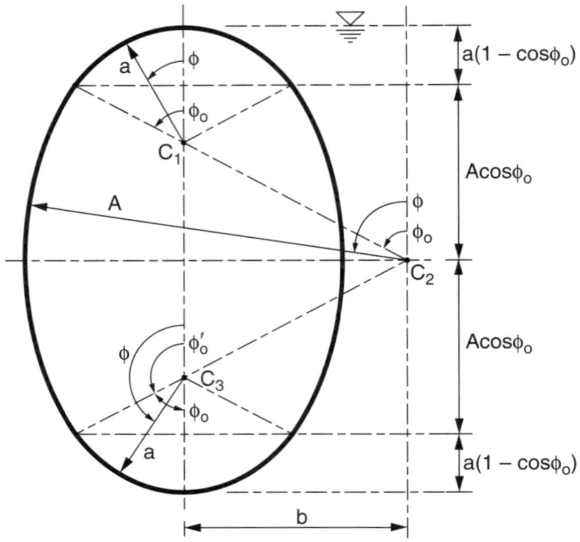

Obviously, the more complex geometry of the egg-shaped shell implies higher construction costs at the beginning, but the lower maintenance costs usually result in lower overall costs over the long term. Despite these advantages, egg-shaped digesters are not as widespread as they should be, owing to a lack of detailed and systematic information on their structural analysis and design. The treatment that follows is adopted from previous work of the author.[18,19]

Let us consider an egg-shaped digester comprising spherical top and bottom closures of radius a and half-angle of opening ϕ_o (this angle being typically 50° to 70°), connected by a middle portion of meridional radius of curvature A, as depicted in Figure 8.10. The upper and lower spherical closures have centres of curvature C_1 and C_3 on the axis of revolution of the assembly, and the middle portion has a centre of curvature C_2 that lies at a distance b to the right of the axis of revolution.

The middle portion is obtained by rotating a circular arc of radius A about a vertical axis that lies at a distance b to the left of the vertical axis through C_2, where $0 < b < A$. This gives an *ogival* shell of revolution – that is, one with pointed ends at the top and bottom. The pointed ends are then replaced by top and bottom spherical caps of radius a, to give a closed vessel that has the shape of an egg (Figure 8.10). The term *ogival* will be used to refer to the middle portion of radius A, although this no longer has pointed ends.

The spherical ends meet the middle portion tangentially. Thus the shell junctions exhibit no discontinuities in slope. However, the meridional radius of curvature jumps from a value of a in the spherical ends to a value of A in the central ogival region, as we move across either junction. We therefore expect discontinuity stresses at the junctions.

It should be pointed out that the upper and lower closures of the egg shell are often conical in shape[14–17] instead of being spherical caps. However, the applicability of the results that will be obtained for discontinuity effects at the shell junctions is not diminished by adopting spherical closures in place of conical closures, or vice versa, since for non-shallow shells of the present type (where the sides of the shell are relatively steep at the junctions of the middle part with the end closures), the localised edge responses of the spherical and the conical shell are essentially identical. The stresses calculated over the apex of the egg shell (regardless of whether this is conical or spherical) will not be of much significance anyway, since these will generally be relatively small and, in any case, the material around the apex is actually replaced by an opening in practice in order to allow access into the digester. Here, the spherical form for the end closures has been chosen for its better aesthetics, and considerable potential as a practical solution.

Without significant loss of generality from a practical point of view, the thickness of the upper and lower spherical closures will be assumed to be constant to begin with, while that of the central ogival portion will be allowed to vary in accordance with a certain law. For all points on the egg-shaped shell, the thickness t of the shell is relatively small in comparison with the least radius of curvature a occurring in the shell, so that thin-shell theory is applicable.

The principal loading condition in sludge-filled digesters is usually hydrostatic pressure, which, of course, varies linearly with depth. There may be some constant surcharge pressure due to gases inside the digester, but the analysis for constant pressure loading is relatively trivial, and will not be pursued here. As will be seen in due course, axisymmetric loading conditions stemming from the hydrostatic pressure are amenable to analytical solution procedures for stresses and deformations in the egg shell. Other loading conditions such as wind and earthquake accelerations are also possible, but in areas of little seismic activity, the pressure of the contained sludge will always be the dominant loading condition upon the walls of the digester. For the purposes of the analysis, the sludge may be assumed to have the same density as water $\left(1000\ \text{kg/m}^3\right)$, and obtained stresses may simply be scaled at the end by the actual specific gravity of the sludge mixture. A typical value for the specific gravity of sludge is 1.05.

The shell is assumed to be supported axisymmetrically around its lowest point, either over a contact interactive surface of finite area, or around a circular circumference of finite diameter. The precise support conditions around the bottom of the egg shell are not important in the present study, since the main concern is with the stresses and deformations that occur throughout the shell, *but not in the immediate vicinity of the contact interactive surface or line of the shell with the ground/foundation.*

Away from the supported zone, as one moves upwards towards the junction of the lower spherical closure with the ogival sides, the state of stress in the shell eventually becomes practically independent of the precise conditions of support at the bottom, and this observation assumes even greater validity in the rest of the egg shell over the middle and upper regions.

In the analysis that follows, and as depicted in Figure 8.10, the angular coordinate ϕ, being the angle between the axis of revolution of the shell and the normal to the shell middle surface at the point in question, will be used to locate points on the shell meridian. The values of ϕ at the upper and lower junctions of the egg shell are denoted by ϕ_o and ϕ_o' respectively, where $\phi_o' = \pi - \phi_o$. The equatorial level of the egg shell corresponds to $\phi = 90°$. As already pointed out, ϕ_o is assumed to lie in the range $50°$ to $70°$, which implies that ϕ_o' lies in the range $110°$ to $130°$. The position of the edge of the shell-soil contact surface, or of the line of support in the case of the egg shell being supported at a specific level, will be denoted by ϕ_s, and it is initially assumed that $\phi_s \geq 150°$.

Both the loading and shell geometry vary smoothly and not too rapidly over the entire surface of the digester shell except at the two junctions (corresponding to $\phi = \phi_o$ and $\phi = \phi_o'$), where discontinuities in meridional radius of curvature have already been noted. Therefore the membrane theory should yield good estimates of stresses and deformations throughout the entire shell except in the vicinity of the two junctions and of the support ($\phi = \phi_s$), where localised bending will occur.

In this study, we will derive closed-form results for membrane stress resultants in all the three regions of the egg shell, and for deformations at the edges of these regions. We will evaluate discontinuity effects at the shell junctions, and on the basis of the final stresses observed, propose design recommendations for egg-shaped sludge digesters.

8.2.2 Geometric preliminaries and loading components

Denoting the principal radii of curvature of the shell middle surface at any given point by r_1 (in the meridional plane) and r_2 (in the perpendicular plane containing the normal to the shell middle surface), we may write (by reference to Figure 8.10),

$$r_1 = r_2 = a \tag{8.88}$$

for the upper and lower spherical closures, and

$$r_1 = A \tag{8.89a}$$

$$r_2 = A - \frac{b}{\sin\phi} \tag{8.89b}$$

for the middle ogival portion of the digester.

At the junction of the ogival part with the spherical end closures, r_2 as given by expression (8.89b) must equal the radius a in the spherical end closures. That is,

$$(r_2)_{\phi=\phi_o} = (r_2)_{\phi=\phi'_o} = A - \frac{b}{\sin\phi_o} = a \tag{8.90a}$$

leading to the result

$$b = (A-a)\sin\phi_o \tag{8.90b}$$

Assuming, for design purposes, that the digester shell is completely filled with liquid of weight γ per unit volume, the hydrostatic pressure p_r at any given point defined by the angular coordinate ϕ is given by the expressions

$$p_r = \gamma a(1-\cos\phi) \qquad [0 \le \phi \le \phi_o] \tag{8.91a}$$

$$p_r = \gamma\{a + (A-a)\cos\phi_o - A\cos\phi\} \qquad [\phi_o \le \phi \le \phi'_o] \tag{8.91b}$$

$$p_r = \gamma\{2(A-a)\cos\phi_o + a(1-\cos\phi)\} \qquad [\phi_o' \le \phi \le \pi] \tag{8.91c}$$

corresponding to the upper spherical region, the middle ogival region and the lower spherical region, respectively.

8.2.3 Membrane stress resultants

For the upper spherical closure, substituting the values of r_1 and r_2 (equation (8.88)) into the general solution for the meridional stress resultant (equation (2.15)), and applying the condition that N_ϕ should remain finite at $\phi = 0$ (the presence of the small access opening at the top is ignored, since this does not significantly affect the state of stress in the rest of the shell), we obtain the well-known result

$$N_\phi = \frac{\gamma a^2}{6}\left(\frac{1-\cos\phi}{1+\cos\phi}\right)(1+2\cos\phi) \tag{8.92a}$$

The hoop stress resultant follows from equation (2.12). The result is also well known:

$$N_\theta = \frac{\gamma a^2}{6}\left(\frac{1-\cos\phi}{1+\cos\phi}\right)(5+4\cos\phi) \tag{8.92b}$$

For the middle ogival part, with r_1 and r_2 given by equations (8.89), evaluation of the integral in equation (2.15) leads to the result

$$\begin{aligned} N_\phi = \frac{1}{A\sin^2\phi - b\sin\phi}\Bigg[& \gamma A^2\left[A\frac{\cos^3\phi}{3} - \{a+(A-a)\cos\phi_o\}\frac{\cos^2\phi}{2}\right] \\ & -\gamma Ab\left[\{a+(A-a)\cos\phi_o\}(\sin\phi) - A\left(\frac{\phi}{2}+\frac{\sin 2\phi}{4}\right)\right] + k_1\Bigg] \end{aligned} \tag{8.93a}$$

The constant of integration k_1 is obtained from the condition that N_ϕ in the upper spherical shell and in the ogival shell must have the same values at the junction $\phi = \phi_o$ of these two regions. This condition yields the result

$$\begin{aligned} k_1 = & \frac{\gamma a^2}{6}\left(\frac{1-\cos\phi_o}{1+\cos\phi_o}\right)(1+2\cos\phi_o)(A\sin^2\phi_o - b\sin\phi_o) \\ & -\gamma A^2\left[A\frac{\cos^3\phi_o}{3} - \{a+(A-a)\cos\phi_o\}\frac{\cos^2\phi_o}{2}\right] \\ & +\gamma Ab\left[\{a+(A-a)\cos\phi_o\}(\sin\phi_o) - A\left(\frac{\phi_o}{2}+\frac{\sin 2\phi_o}{4}\right)\right] \end{aligned} \tag{8.93b}$$

The hoop stress resultant follows from equation (2.12):

$$N_\theta = \left(\frac{A\sin\phi - b}{\sin\phi}\right)\left[\gamma\{a+(A-a)\cos\phi_o - A\cos\phi\} - \frac{N_\phi}{A}\right] \tag{8.93c}$$

For the lower spherical closure, evaluation of the integral in equation (2.15) yields

$$N_\phi = \frac{1}{a\sin^2\phi}\left[-\gamma a^2\left\{2(A-a)(\cos\phi_o)+a\right\}\frac{\cos^2\phi}{2}+\gamma a^3\frac{\cos^3\phi}{3}+k_2\right] \tag{8.94a}$$

Applying the condition that N_ϕ in the ogival shell and in the lower spherical shell must have the same values at the junction $\phi=\phi_o'$ of these two regions, we obtain

$$\begin{aligned} k_2 &= \gamma a^2\left[(A-a)\cos^3\phi_o + a\frac{\cos^2\phi_o}{2} + a\frac{\cos^3\phi_o}{3}\right] \\ &-\gamma A^2\left[\left\{a+(A-a)\cos\phi_o\right\}\frac{\cos^2\phi_o}{2}+A\frac{\cos^3\phi_o}{3}\right] \\ &-\gamma Ab\left[\left\{a+(A-a)\cos\phi_o\right\}(\sin\phi_o)-A\left(\frac{\pi-\phi_o}{2}-\frac{\sin 2\phi_o}{4}\right)\right]+k_1 \end{aligned} \tag{8.94b}$$

The hoop stress resultant follows as

$$N_\theta = \gamma a\left\{2(A-a)\cos\phi_o + a(1-\cos\phi)\right\} - N_\phi \tag{8.94c}$$

Results in equations (8.94) are not valid over the supported portion of the egg shell, assumed to be a finite region around the bottom pole and bounded by the circle $\phi=\phi_s=150°$. They are also not adequate in the unsupported zone of the egg shell that is in the immediate vicinity of the edge $\phi=\phi_s$ of the supported region, and in the immediate vicinity of the junction $\phi=\phi_o'$ of the lower spherical shell with the middle ogival shell. Similarly, and as already pointed out, results (8.92) and (8.93) are not sufficient in the immediate vicinities of the shell junctions $\phi=\phi_o$ and $\phi=\phi_o'$. In the vicinities of the shell junctions, the membrane solution must be augmented with a bending solution that accounts for flexural and shearing effects in these regions. The flexibility formulation of Chapter 7 will be used to evaluate the edge effects. The calculations of these will require enforcement of the conditions of compatibility of deformations at the shell junctions. We will first calculate the membrane deformations.

8.2.4 Membrane deformations

Deformations of interest are the horizontal or lateral displacement of a point on the shell meridian, which will be denoted by δ and considered positive when the point moves away from the axis of revolution of the shell, and the rotation of the tangent to the shell meridian at that point as seen in the plane of the meridian, which will be denoted by V and considered positive when the tangent rotates in an *anticlockwise* sense on the side of the shell cross-section to the left of the axis of revolution (Figure 8.10). For axisymmetrically loaded shells of revolution, these deformations are related to the stress resultants N_ϕ and N_θ via equations (2.67) and (2.72).

Let us assume for now that the shell thickness t is constant and equal to h in the upper and lower spherical closures. In the middle ogival part, we allow the shell thickness to vary in such a way that $r_2 t$ =constant= ah. This implies that

$$t = \frac{ah\sin\phi}{A\sin\phi - b} \tag{8.95}$$

and allows us to reduce the shell thickness progressively as we move away from the locations $\phi = \phi_o$ and $\phi = \phi_o'$ of anticipated bending disturbances.

The membrane stress resultants N_ϕ and N_θ for the three regions of the egg shell were obtained as expressions (8.92), (8.93) and (8.94). Substituting these results into equations (2.67) and (2.72), and adopting the respective laws of shell-thickness variation as indicated above, we obtain closed-form explicit results for the membrane deformations δ and V as follows:

Upper spherical shell

$$\delta = \frac{\gamma a^3}{6Eh}(\sin\phi)\left(\frac{1-\cos\phi}{1+\cos\phi}\right)\{(5+4\cos\phi)-\nu(1+2\cos\phi)\} \tag{8.96a}$$

$$V = -\frac{\gamma a^2}{Eh}\sin\phi \tag{8.96b}$$

Lower spherical shell

$$\begin{aligned}\delta = \frac{a}{Eh}(\sin\phi)\Bigg[\gamma a\{2(A-a)\cos\phi_o + a(1-\cos\phi)\} - \left(\frac{1+\nu}{a\sin^2\phi}\right)\times \\ \times\left[k_2 - \gamma a^2\left\{2(A-a)(\cos\phi_o)\frac{\cos^2\phi}{2} + a\frac{\cos^2\phi}{2} - a\frac{\cos^3\phi}{3}\right\}\right]\Bigg]\end{aligned} \tag{8.97a}$$

$$V = -\frac{\gamma a^2}{Eh}\sin\phi \tag{8.97b}$$

Middle ogival shell

$$\delta = \frac{1}{EahA}\left(\frac{A\sin\phi - b}{\sin\phi}\right)^2\left[\gamma A(A\sin\phi - b)\{a+(A-a)\cos\phi_o - A\cos\phi\} - (A\sin\phi)f_1 f_2\right] \tag{8.98a}$$

$$V = \frac{1}{A}(f_3 f_4 - f_5 f_6 - f_7 f_8) \tag{8.98b}$$

where

$$f_1 = \frac{(A\sin\phi)(1+\nu) - b}{A(A\sin\phi - b)(\sin^2\phi)} \tag{8.99a}$$

$$f_2 = k_1 + \gamma A^2\left[A\frac{\cos^3\phi}{3} - \left\{a+(A-a)\cos\phi_o\right\}\frac{\cos^2\phi}{2}\right]$$
$$-\gamma Ab\left[\left\{a+(A-a)\cos\phi_o\right\}(\sin\phi) - A\left(\frac{\phi}{2}+\frac{\sin 2\phi}{4}\right)\right] \quad (8.99b)$$

$$f_3 = \frac{(\cos\phi)(A\sin\phi - b)}{Eah\sin^4\phi} \quad (8.99c)$$

$$f_4 = \left[\frac{(A\sin\phi)\left\{(A\sin\phi)(1+\mathrm{v})-\mathrm{v}b\right\}+(A\sin\phi-b)\left\{(A\sin\phi)(1+\mathrm{v})-b\right\}}{A(A\sin\phi-b)\sin\phi}\right]f_2$$
$$-\gamma(A\sin\phi-b)\left\{(A\sin\phi)(1+\mathrm{v})-b\right\}\left\{a+(A-a)\cos\phi_o - A\cos\phi\right\} \quad (8.99d)$$

$$f_5 = \frac{2b}{Eah}\frac{(A\sin\phi-b)(\cos\phi)}{\sin^3\phi} \quad (8.99e)$$

$$f_6 = \gamma\left(\frac{A\sin\phi-b}{\sin\phi}\right)\left\{a+(A-a)\cos\phi_o - A\cos\phi\right\} - f_1 f_2 \quad (8.99f)$$

$$f_7 = \frac{1}{Eah}\left(\frac{A\sin\phi-b}{\sin\phi}\right)^2 \quad (8.99g)$$

$$f_8 = \gamma\left[A(A\sin\phi-b)+b\left(\frac{\cos\phi}{\sin^2\phi}\right)\left\{a+(A-a)\cos\phi_o - A\cos\phi\right\}\right]$$
$$-\left[\frac{(\cos\phi)\left\{A(1+v)(\sin\phi)(b-2A\sin\phi)+b(3A\sin\phi-2b)\right\}}{A(A\sin\phi-b)^2\sin^3\phi}\right]f_2$$
$$-\left[\gamma A^2\cos\phi\sin\phi\left\{a+(A-a)\cos\phi_o - A\cos\phi\right\}\right.$$
$$\left.-\gamma Ab\left[\left\{a+(A-a)\cos\phi_o\right\}(\cos\phi) - \frac{A}{2}(1+\cos 2\phi)\right]\right]f_1 \quad (8.99h)$$

8.2.5 Numerical example

This example is taken from Ref. 18. Let us consider a relatively small egg-shaped digester with $a = 5$m, $A = 15$m and $\phi_o = 60°$. From equation (8.90b), the parameter b follows as 8.660 m. The total height of the egg shell is 20 m, and the contained volume is 781 m^3. It will be assumed that $\gamma = 9.810$kN/m^3 (the unit weight of water); if the actual weight density of the sludge is greater than this assumed value by a factor of η, all that will be necessary at the end of the calculations is to scale up all obtained values for stresses and deformations by a factor of η. Furthermore, the Young modulus E and Poisson's ratio v will be taken

Figure 8.11 Variation of membrane stresses σ_ϕ^m and σ_θ^m over the three regions of the egg shell, from $\phi = 0$ to $\phi = 135°$.

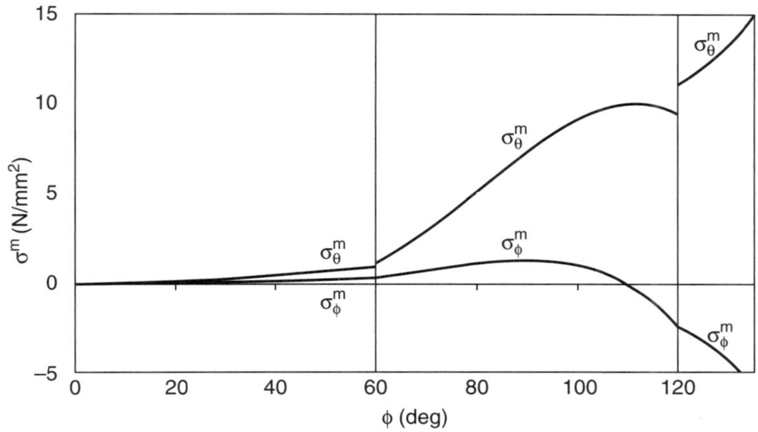

as 28kN/mm^2 and 0.15 respectively, which are typical values for structural concrete. The thickness t of the shell will be taken as constant at 100 mm in the spherical portions of the egg shell, and varying from a maximum value of 100 mm at the edges to a minimum value of 79 mm at the equator, in the ogival central part of the egg shell, the variation being in accordance with equation (8.95).

Figure 8.11 shows the variation of actual membrane stresses σ_ϕ^m and σ_θ^m in the three portions of the egg shell, from the apex $(\phi = 0)$ towards the bottom $(\phi = 180°)$. Values beyond $\phi = 135°$ are not shown, as the digester is assumed to be continuously supported over the lowest region $(150° \leq \phi \leq 180°)$.

From these results, it is evident that meridional compression will be a problem in the lower regions of the digester, where the shell thickness will almost certainly require to be enhanced to safeguard against possible buckling of the shell. In digesters of bigger size (which are more likely to be encountered in actual practice), this possibility becomes higher, and greater caution is certainly required. On the other hand, considerable prestressing will be required in the lower regions of the digester, to cater for the relatively large hoop tensions noted there.

Deformations δ_e and V_e at the edges of the three shell portions are obtained as follows (the subscripts 1, 2, 3 and 4 distinguish the four edges in question):

Edge of upper spherical shell

$$\delta_1^m = 0.1412\text{mm}$$

$$V_1^m = -7.5855 \times 10^{-5}$$

Upper edge of middle ogival shell

$$\delta_2^m = 0.1693\text{mm}$$

$$V_2^m = -1.1840 \times 10^{-4}$$

Lower edge of middle ogival shell

$$\delta_3^m = 1.5121\,\text{mm}$$

$$V_3^m = 3.5109 \times 10^{-4}$$

Edge of lower spherical shell

$$\delta_4^m = 1.7660\,\text{mm}$$

$$V_4^m = -7.5855 \times 10^{-5}$$

It is noted that all edges move laterally outwards as a result of the internal hydrostatic pressure, this movement being more noticeable in the lower regions of the digester. Rotations are generally very small. Evidently, $\delta_1 \neq \delta_2$, $V_1 \neq V_2$, $\delta_3 \neq \delta_4$ and $V_3 \neq V_4$, confirming that bending will occur around the shell junctions, as expected.

8.2.6 Bending solution

For the bending solution, we may adopt the Geckeler approximation for general shells of revolution (see Section 6.5). In the spherical ends of the egg-shaped digester shell, $r_1 = r_2 = a$ (constant) and the shell thickness t is also constant. Therefore, the parameter λ will be constant in these regions. In the middle ogival part of the digester shell, $r_1 = A$ (constant) and $r_2 t = ah$ (constant). Therefore, the parameter λ will also be constant. In cases where λ does vary, such as when the circular ogival shell is provided with constant thickness, the small variation of λ *over the relatively narrow extent of the bending-disturbance zone* allows the adoption of a constant *average* value of λ without incurring significant additional errors over those already inherent in the Geckeler approximation.

For constant λ, we may write the solution to equation (6.141) in the form

$$Q_\phi = Ce^{-\lambda\psi}\sin(\lambda\psi + \beta) \tag{8.100}$$

where C and β are constants of integration, and the coordinate ψ is the angle measured from the shell edge (Figure 8.12). Denoting the meridional angle corresponding to the shell edge by ϕ_e, the angle ψ is given by $\psi = \phi_e - \phi$ for shells with edges facing the direction of increasing ϕ, and by $\psi = \phi - \phi_e$ for shells with edges facing the direction of decreasing ϕ, as illustrated in Figure 8.12.

Stress resultants N_ϕ^b and N_θ^b associated with the bending disturbance or edge effect, bending moments M_ϕ and M_θ, as well as deformations δ^b and V^b of the edge effect, are obtained on the basis of equation (8.100), using the relationships of Chapter 6 in combination with Geckeler-type simplifications. For shells with edges facing the direction of increasing ϕ (i.e. shells for which $\psi = \phi_e - \phi$, so that $d\phi = -d\psi$), we have

$$\begin{aligned} N_\phi^b &= -Q_\phi \cot\phi \\ &= -\cot(\phi_e - \psi)Ce^{-\lambda\psi}\sin(\lambda\psi + \beta) \end{aligned} \tag{8.101a}$$

Figure 8.12 Shell-edge parameters ϕ_e, ψ, M_e and H_e for shell edges facing the direction of increasing ϕ (refer to left side of centreline) and shell edges facing the direction of decreasing ϕ (refer to right side of centreline).

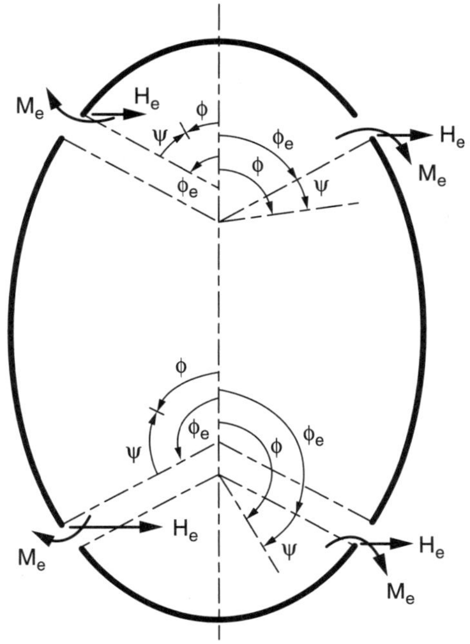

$$\begin{aligned} N_\theta^b &= -\left(\frac{r_2}{r_1}\frac{dQ_\phi}{d\phi} + \frac{Q_\phi}{r_1}\frac{dr_2}{d\phi}\right) \\ &\approx -\frac{r_2}{r_1}\frac{dQ_\phi}{d\phi} \\ &= -\frac{r_2}{r_1}\lambda\sqrt{2}Ce^{-\lambda\psi}\sin\left(\lambda\psi + \beta - \frac{\pi}{4}\right) \end{aligned} \tag{8.101b}$$

$$\begin{aligned} V^b &\approx \frac{1}{Et}\left(\frac{r_2^2}{r_1^2}\right)\frac{d^2Q_\phi}{d\phi^2} \\ &= -\frac{2\lambda^2}{Et}\left(\frac{r_2^2}{r_1^2}\right)Ce^{-\lambda\psi}\cos(\lambda\psi + \beta) \end{aligned} \tag{8.101c}$$

$$\begin{aligned} M_\phi &= -D\left(\frac{1}{r_1}\frac{dV}{d\phi} + \frac{\nu}{r_2}V\cot\phi\right) \\ &\approx -\frac{D}{r_1}\frac{dV}{d\phi} \\ &\approx -\frac{D}{Etr_1}\left(\frac{r_2^2}{r_1^2}\right)\frac{d^3Q_\phi}{d\phi^3} \\ &= \frac{1}{\lambda\sqrt{2}}r_1Ce^{-\lambda\psi}\sin\left(\lambda\psi + \beta + \frac{\pi}{4}\right) \end{aligned} \tag{8.101d}$$

$$M_\theta = -D\left(\frac{1}{r_2}V\cot\phi + \frac{\nu}{r_1}\frac{dV}{d\phi}\right) \approx -\nu\frac{D}{r_1}\frac{dV}{d\phi} = \nu M_\phi \tag{8.101e}$$

$$\delta^b = \frac{1}{Et}(r_2\sin\phi)(N_\theta - \nu N_\phi) \approx \frac{1}{Et}(r_2\sin\phi)N_\theta = -\left(\frac{r_2}{r_1}\right)\frac{\lambda\sqrt{2}}{Et}(r_2\sin\phi)Ce^{-\lambda\psi}\sin\left(\lambda\psi+\beta-\frac{\pi}{4}\right) \tag{8.101f}$$

For shell edges facing the direction of *increasing* ϕ (i.e. edges for which $\phi = \phi_e - \psi$; see the portion to the left of the centreline in Figure 8.12), the application of an edge bending moment M_e is associated with the boundary conditions

$$\left(M_\phi\right)_{\phi=\phi_e} = M_e \tag{8.102a}$$

$$\left(N_\phi^b\right)_{\phi=\phi_e} = 0 \tag{8.102b}$$

yielding the constants

$$C = \frac{2\lambda}{r_1}M_e \tag{8.103a}$$

$$\beta = 0 \tag{8.103b}$$

while the application of a horizontal shear force H_e at the edge is associated with the boundary conditions

$$\left(M_\phi\right)_{\phi=\phi_e} = 0 \tag{8.104a}$$

$$\left(N_\phi^b\right)_{\phi=\phi_e} = -H_e\cos\phi_e \tag{8.104b}$$

yielding the constants

$$C = -\sqrt{2}H_e\sin\phi_e \tag{8.105a}$$

$$\beta = -\frac{\pi}{4} \tag{8.105b}$$

Using the results for the constants C and β in equations (8.101c) and (8.101f), for the application of M_e and H_e taken separately and then superimposing the contributions of M_e and H_e, we obtain the edge deformations

$$\begin{bmatrix} V_e^b \\ \delta_e^b \end{bmatrix} = \begin{bmatrix} -\dfrac{4\lambda^3}{Etr_1}\left(\dfrac{r_2^2}{r_1^2}\right) & \dfrac{2\lambda^2}{Et}\left(\dfrac{r_2^2}{r_1^2}\right)\sin\phi_e \\ \dfrac{2\lambda^2}{Et}\left(\dfrac{r_2^2}{r_1^2}\right)\sin\phi_e & -\dfrac{2\lambda}{Et}\left(\dfrac{r_2^2}{r_1}\right)\sin^2\phi_e \end{bmatrix} \begin{bmatrix} M_e \\ H_e \end{bmatrix} \tag{8.106}$$

The interior actions due to the simultaneous application of the edge actions M_e and H_e are obtained as

$$N_\phi^b = -\{\cot(\phi_e - \psi)\} e^{-\lambda\psi}\left[\frac{2\lambda}{r_1} M_e \sin\lambda\psi - H_e(\sin\phi_e)(\sin\lambda\psi - \cos\lambda\psi)\right] \tag{8.107a}$$

$$N_\theta^b = -2\left(\frac{r_2}{r_1}\right)\lambda e^{-\lambda\psi}\left[\frac{\lambda}{r_1} M_e(\sin\lambda\psi - \cos\lambda\psi) + H_e(\sin\phi_e)\cos\lambda\psi\right] \tag{8.107b}$$

$$M_\phi = e^{-\lambda\psi}\left[M_e(\sin\lambda\psi + \cos\lambda\psi) - \frac{r_1}{\lambda} H_e(\sin\phi_e)\sin\lambda\psi\right] \tag{8.107c}$$

$$M_\theta = \nu M_\phi \tag{8.107d}$$

For the shell edges facing the direction of *decreasing* ϕ (i.e. $\phi = \phi_e + \psi$; see the portion to the right of the centreline in Figure 8.12), the simultaneous application of the edge actions M_e and H_e, directed as shown on the right-hand side of the diagram in Figure 8.12, leads to the edge deformations

$$\begin{bmatrix} V_e^b \\ \delta_e^b \end{bmatrix} = \begin{bmatrix} \dfrac{4\lambda^3}{Etr_1}\left(\dfrac{r_2^2}{r_1^2}\right) & \dfrac{2\lambda^2}{Et}\left(\dfrac{r_2^2}{r_1^2}\right)\sin\phi_e \\ \dfrac{2\lambda^2}{Et}\left(\dfrac{r_2^2}{r_1^2}\right)\sin\phi_e & \dfrac{2\lambda}{Et}\left(\dfrac{r_2^2}{r_1}\right)\sin^2\phi_e \end{bmatrix} \begin{bmatrix} M_e \\ H_e \end{bmatrix} \tag{8.108}$$

and the interior actions

$$N_\phi^b = -\{\cot(\phi_e + \psi)\} e^{-\lambda\psi}\left[\frac{2\lambda}{r_1} M_e \sin\lambda\psi + H_e(\sin\phi_e)(\sin\lambda\psi - \cos\lambda\psi)\right] \tag{8.109a}$$

$$N_\theta^b = -2\left(\frac{r_2}{r_1}\right)\lambda e^{-\lambda\psi}\left[\frac{\lambda}{r_1}M_e\left(\sin\lambda\psi - \cos\lambda\psi\right) - H_e\left(\sin\phi_e\right)\cos\lambda\psi\right] \tag{8.109b}$$

$$M_\phi = e^{-\lambda\psi}\left[M_e\left(\sin\lambda\psi + \cos\lambda\psi\right) + \frac{r_1}{\lambda}H_e\left(\sin\phi_e\right)\sin\lambda\psi\right] \tag{8.109c}$$

$$M_\theta = \nu M_\phi \tag{8.109d}$$

In evaluating the above quantities, it must be noted that $r_1 = r_2 = a$ for the two spherical ends, while $r_1 = A$ and $r_2 = A - (b/\sin\phi)$ for the middle ogival part.

For each of the three shell regions, once the redundant edge actions $\{M_e, H_e\}$ have been evaluated, the interior actions N_ϕ^b, N_θ^b, M_ϕ and M_θ (associated with the edge bending disturbance) become known explicitly via the above sets of equations. The net shell stresses are obtained by superimposing membrane stresses with edge effects in the usual manner.

8.2.7 Solutions for the edge redundants

In order to correct the noted mismatch in the membrane deformations δ^m and V^m between adjacent sides of the junctions $\phi = \phi_o$ and $\phi = \phi_o'$, it is assumed that axisymmetric bending moments and horizontal shearing forces act at the four edges of the separated shell regions, as shown in Figure 8.13(a). Thus the bending moment M_1 (per unit length) and horizontal shear force H_1 (per unit length) act over the (downward-facing) edge of the upper spherical cap, while M_4 and H_4 act over the (upward-facing) edge of the lower spherical cap; the action pairs $\{M_2, H_2\}$ and $\{M_3, H_3\}$ act over the upper and lower edges, respectively, of the middle ogival shell. The positive directions of these four edge actions are as shown, and these are consistent with the directions in Figure 8.12, in which the same parameters $\{M_e, H_e\}$ were then applied at each edge as *arbitrary actions*, and the ensuing effects noted. It is now considered that M_e and H_e have specific values that in general differ for each edge, and which are dependent on the actual edge conditions (or, rather, inter-edge equilibrium and compatibility conditions). Figure 8.13(b) depicts the positive directions for the edge deformations $\{\delta_1, V_1\}$, $\{\delta_2, V_2\}$, $\{\delta_3, V_3\}$ and $\{\delta_4, V_4\}$ at the four edges of the separated shell regions; these deformations may be due to membrane actions, bending actions, or both.

In order to obtain results that are applicable to both the upper and the lower junctions of the three shell portions, consider a generalised junction (Figure 8.14), for which $\{M_u, H_u\}$ and $\{M_l, H_l\}$ denote upper-edge and lower-edge corrective actions, respectively, at the junction. Thus, for the upper junction (located at $\phi = \phi_o$), $\{M_u, H_u\}$ and $\{M_l, H_l\}$ represent $\{M_1, H_1\}$ and $\{M_2, H_2\}$ respectively; for the lower junction (located at $\phi = \phi_o'$), the same parameters represent $\{M_3, H_3\}$ and $\{M_4, H_4\}$ respectively.

An element of the shell lying at the junction (refer to Figure 8.14) would be kept in equilibrium by the reactive actions $\{M_u, H_u, M_l, H_l\}$ and an additional action H_e^m, which is the resultant towards the axis of revolution of the shell, of the membrane meridional stress resultants occurring at the shell edges – that is,

$$H_e^m = \left\{\left(N_\phi^m\right)_u - \left(N_\phi^m\right)_l\right\}\cos\phi_e \tag{8.110}$$

Figure 8.13 Actions and deformations at the four edges of the separated shell regions: (a) edge actions $\{M_1, H_1\}$, $\{M_2, H_2\}$, $\{M_3, H_3\}$ and $\{M_4, H_4\}$; (b) edge deformations $\{\delta_1, V_1\}$, $\{\delta_2, V_2\}$, $\{\delta_3, V_3\}$ and $\{\delta_4, V_4\}$.

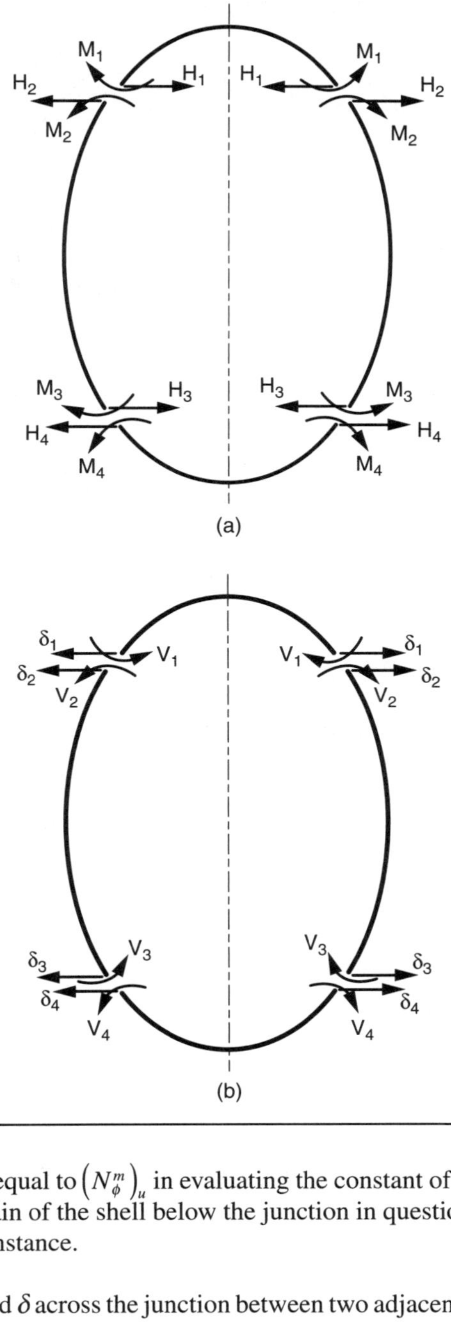

Since $\left(N_\phi^m\right)_l$ was previously set equal to $\left(N_\phi^m\right)_u$ in evaluating the constant of integration occurring in the general solution for N_ϕ^m in the domain of the shell below the junction in question, it follows that H_e^m in equation (8.110) is equal to zero in this instance.

Continuity of deformations V and δ across the junction between two adjacent regions of the shell requires that

$$V_u^T \left(= V_u^b + V_u^m\right) = V_l^T \left(= V_l^b + V_l^m\right) \tag{8.111a}$$

Figure 8.14 Generalised junction showing actions $\{M_u, H_u\}$ and $\{M_l, H_l\}$ on shell edges and on an inter-edge element of the shell.

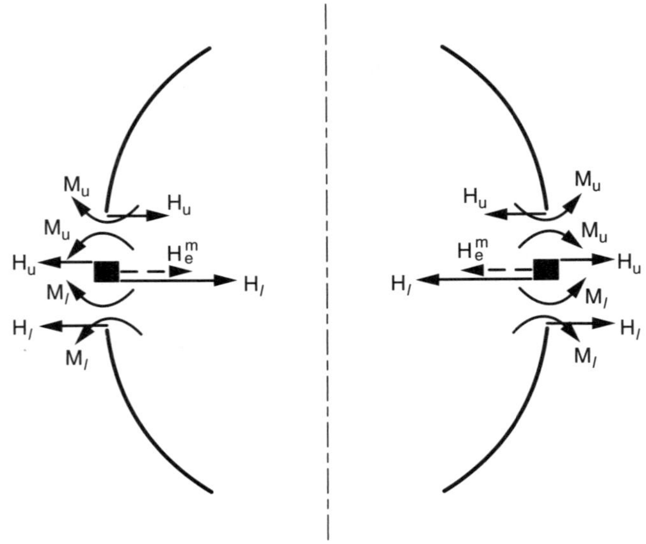

$$\delta_u^T \left(= \delta_u^b + \delta_u^m\right) = \delta_l^T \left(= \delta_l^b + \delta_l^m\right) \tag{8.111b}$$

where superscripts m, b and T denote membrane, bending-disturbance and total effects, respectively, while as before, subscripts u and l denote the upper and lower edges, respectively, of the junction in question.

For the bending-disturbance components of the net deformations, we may write

$$\begin{bmatrix} V_u^b \\ \delta_u^b \end{bmatrix} = \begin{bmatrix} I_{11} & I_{12} \\ I_{21} & I_{22} \end{bmatrix} \begin{bmatrix} M_u \\ H_u \end{bmatrix} \tag{8.112a}$$

$$\begin{bmatrix} V_l^b \\ \delta_l^b \end{bmatrix} = \begin{bmatrix} J_{11} & J_{12} \\ J_{21} & J_{22} \end{bmatrix} \begin{bmatrix} M_l \\ H_l \end{bmatrix} \tag{8.112b}$$

where $[I]$ and $[J]$ are the 2×2 matrices of influence coefficients applicable to the upper and lower edges, respectively, of the junction in question, as obtained earlier (equations (8.106) and (8.108)). This permits conditions (8.112) to be rewritten as

$$I_{11}M_u + I_{12}H_u + V_u^m = J_{11}M_l + J_{12}H_l + V_l^m \tag{8.113a}$$

$$I_{21}M_u + I_{22}H_u + \delta_u^m = J_{21}M_l + J_{22}H_l + \delta_l^m \tag{8.113b}$$

Equilibrium of the element of the shell lying at the junction of adjacent shell regions (refer to Figure 8.14) requires that

$$M_u - M_l = 0 \tag{8.114a}$$

$$H_u - H_l - H_e^m = 0 \tag{8.114b}$$

From the above conditions, and noting that $H_e^m = 0$ as already explained, the relationships

$$M_l = M_u \tag{8.115a}$$

$$H_l = H_u \tag{8.115b}$$

are obtained. Substituting for M_l and H_l in equations (8.113), we obtain

$$I_{11}M_u + I_{12}H_u + V_u^m = J_{11}M_u + J_{12}H_u + V_l^m \tag{8.116a}$$

$$I_{21}M_u + I_{22}H_u + \delta_u^m = J_{21}M_u + J_{22}H_u + \delta_l^m \tag{8.116b}$$

which, upon rearranging, become

$$(I_{11} - J_{11})M_u + (I_{12} - J_{12})H_u = V_l^m - V_u^m \tag{8.117a}$$

$$(I_{21} - J_{21})M_u + (I_{22} - J_{22})H_u = \delta_l^m - \delta_u^m \tag{8.117b}$$

These simultaneous equations lead to the closed-form results

$$M_u = \frac{(V_l^m - V_u^m)(I_{22} - J_{22}) - (\delta_l^m - \delta_u^m)(I_{12} - J_{12})}{(I_{22} - J_{22})(I_{11} - J_{11}) - (I_{21} - J_{21})(I_{12} - J_{12})} \tag{8.118a}$$

$$H_u = \frac{(\delta_l^m - \delta_u^m)(I_{11} - J_{11}) - (V_l^m - V_u^m)(I_{21} - J_{21})}{(I_{22} - J_{22})(I_{11} - J_{11}) - (I_{21} - J_{21})(I_{12} - J_{12})} \tag{8.118b}$$

Let us introduce the following parameters (noting that $I_{21} = I_{12}$ and $J_{21} = J_{12}$):

$$\xi = V_l^m - V_u^m \tag{8.119a}$$

$$\mu = \delta_l^m - \delta_u^m \tag{8.119b}$$

$$K_1 = I_{11} - J_{11} \tag{8.119c}$$

$$K_2 = I_{12} - J_{12} \tag{8.119d}$$

$$K_3 = I_{22} - J_{22} \tag{8.119e}$$

$$K_4 = (I_{11} - J_{11})(I_{22} - J_{22}) - (I_{12} - J_{12})^2 = K_1 K_3 - K_2^2 \tag{8.119f}$$

This permits us to express the results more compactly as follows:

$$M_u = \frac{\xi K_3 - \mu K_2}{K_4} (= M_l) \tag{8.120a}$$

$$H_u = \frac{\mu K_1 - \xi K_2}{K_4} (= H_l) \tag{8.120b}$$

8.2.8 Numerical results

Let us revisit our earlier example of a small egg-shaped digester with a = 5m, A = 15m and $\phi_o = 60°$. For the slenderness parameter, we obtain $\lambda = 9.2533$ for the spherical shell closures (with $\nu = 0.15$, $r_1 = r_2 = a = 5$m, $t = 0.1$m) and $\lambda = 27.7598$ for the ogival middle shell (with $\nu = 0.15$, $r_1 = A = 15$m, $r_2 t = ah = (5\text{m})(0.1\text{m})$). With $E = 28 \times 10^9$ N/m^2, equations(8.106) and (8.108) yield the following values for the influence coefficients:

(i) Upper junction ($\phi = \phi_o = 60°$)

Upper edge of upper junction (spherical top closure)

$$I_{11} = -2.2637 \times 10^{-7}\,\text{N}^{-1}$$

$$I_{12} = I_{21} = 5.2966 \times 10^{-8}\,\text{mN}^{-1}$$

$$I_{22} = -2.4786 \times 10^{-8}\,\text{m}^2\text{N}^{-1}$$

Lower edge of upper junction (upper end of ogival shell)

$$J_{11} = 2.2637 \times 10^{-7}\,\text{N}^{-1}$$

$$J_{12} = J_{21} = 5.2966 \times 10^{-8}\,\text{mN}^{-1}$$

$$J_{22} = 2.4786 \times 10^{-8}\,\text{m}^2\text{N}^{-1}$$

(ii) Lower junction ($\phi = \phi_o' = 120°$)

Upper edge of lower junction (lower end of ogival shell)

$I_{11} = -2.2637 \times 10^{-7}\,\text{N}^{-1}$

$I_{12} = I_{21} = 5.2966 \times 10^{-8}\,\text{mN}^{-1}$

$I_{22} = -2.4786 \times 10^{-8}\,\text{m}^2\text{N}^{-1}$

Lower edge of lower junction (spherical bottom closure)

$J_{11} = 2.2637 \times 10^{-7}\,\text{N}^{-1}$

$J_{12} = J_{21} = 5.2966 \times 10^{-8}\,\text{mN}^{-1}$

$J_{22} = 2.4786 \times 10^{-8}\,\text{m}^2\text{N}^{-1}$

Making use of the results for the membrane deformations at the edges of the three shell portions, the parameters ξ and μ may readily be evaluated on the basis of equations (8.119a) and (8.119b). The parameters K_1, K_2, K_3 and K_4 are evaluated on the basis of equations (8.119c)–(8.119f), using the now-known values of the influence coefficients. The redundant corrective actions M_u, H_u, M_l and H_l readily follow from equations (8.120). For the two junctions of the digester, the obtained numerical values are as follows:

(i) Upper junction ($\phi = \phi_o = 60°$)

$\xi = -4.2545 \times 10^{-5}$

$\mu = 2.814 \times 10^{-5}\,\text{m}$

$K_1 = -4.5274 \times 10^{-7}\,\text{N}^{-1}$

$K_2 = 0$

$K_3 = -4.9572 \times 10^{-8}\,\text{m}^2\text{N}^{-1}$

$K_4 = 2.2443 \times 10^{-14}\,\text{m}^2\text{N}^{-2}$

$M_u = M_l = 93.9732\text{Nm/m}$

$H_u = H_l = -567.6649\text{N/m}$

(ii) Lower junction ($\phi = \phi_o' = 120°$)

$\xi = -4.2695 \times 10^{-4}$

$$\mu = 2.539 \times 10^{-4}\,\text{m}$$

$$K_1 = -4.5274 \times 10^{-7}\,\text{N}^{-1}$$

$$K_2 = 0$$

$$K_3 = -4.9572 \times 10^{-8}\,\text{m}^2\text{N}^{-1}$$

$$K_4 = 2.2443 \times 10^{-14}\,\text{m}^2\text{N}^{-2}$$

$$M_u = M_l = 943.0342\text{Nm/m}$$

$$H_u = H_l = -5121.8948\text{N/m}$$

Interior shell actions stemming from the now-known redundant edge actions are evaluated on the basis of relations (8.107) for the edge zone of the upper spherical closure and the lower edge zone of the ogival portion, and relations (8.109) for the upper zone of the ogival portion and the edge zone of the lower spherical closure, using the applicable values of the edge actions as calculated above. Combining the membrane-solution stresses with those of the edge disturbance then yields the net stresses.

Figures 8.15 to 8.18 are plots of meridional (σ_ϕ) and hoop (σ_θ) stresses for (*a*) the edge zone of the upper spherical closure, (*b*) the upper edge zone of the ogival shell, (*c*) the lower edge zone of the ogival shell, and (*d*) the edge zone of the lower spherical closure. Each set of plots comprises the membrane component σ^m (constant across the shell thickness), and the total-stress values $\sigma^T(i)$ (for the inner surface of the shell) and $\sigma^T(o)$ (for the outer surface of the shell).

8.2.9 Conclusions and recommendations

In conclusion, it may be stated that, for the sizes and configurations of egg-shaped digester shells that have been considered, junction edge effects are generally not as significant as membrane effects, and the design of the shell will be governed primarily by calculated membrane stresses. The discontinuity effects do alter the *local* meridional stresses significantly, but the net stresses that ensue are considerably lower than membrane stresses elsewhere, so that the stress-raising tendency of the discontinuity effects has little bearing on the design of the shell. These conclusions may tentatively be extrapolated to larger egg-shaped digester shells of the same basic configuration, but in order to properly investigate the effects of the scale of the structure on the stress levels in the shell, a parametric study should be undertaken.

In order to control the steeply increasing membrane meridional compression (which may give rise to buckling problems in the thin shells of the type in question) and membrane hoop tension (which may give rise to cracking problems in the concrete) as one moves downwards from the lower junction of the digester through the lower spherical closure towards the bottom, a number of measures are recommended.

First, the thickness of the lower spherical closure should be increased rapidly with distance s from the shell edge. This thickness enhancement may, for instance, be in accordance with a parabolic law $(t = t_o + ks^2)$,

Figure 8.15 Stress plots for the edge zone of the upper spherical closure: (a) meridional stresses; (b) hoop stresses.

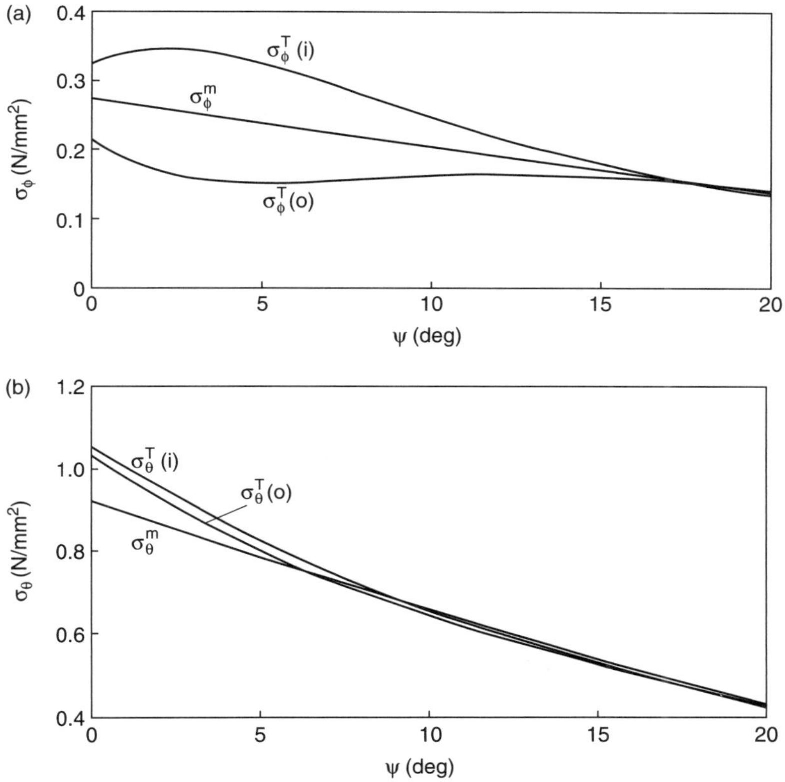

which would decrease the stresses almost as rapidly as the stress resultants N_ϕ^m and N_θ^m are rising. Above the lower junction $(\phi = \phi_o')$, and moving upwards, the shell thickness may continue to be decreased within the ogival shell in accordance with the law $r_2 t = at_o$ proposed earlier, until $\phi = 90°$ is reached, beyond which the shell thickness may be kept constant in the upper half of the ogival shell and throughout the upper spherical closure (since stress levels in these regions are generally small to moderate).

Alongside the shell thickening recommended in the lower half of the vessel, tensile reinforcement or pre-stressing must be provided, and the levels of these progressively increased as one moves from the equatorial level of the digester towards the bottom.

As the final measure for controlling the steeply increasing membrane stresses in the lower spherical closure, it is recommended that the support ring of the digester, or the edge of the shell-soil interactive surface in the case of the soil-supported digester, be raised to 45° from the downward direction of the axis of revolution of the digester shell (i.e. $\phi_s = 135°$). This raising of the support level will cut off the very excessive membrane stresses that would otherwise occur towards the bottom of the tank.

Figure 8.16 Stress plots for the upper edge zone of the ogival shell: (a) meridional stresses; (b) hoop stresses.

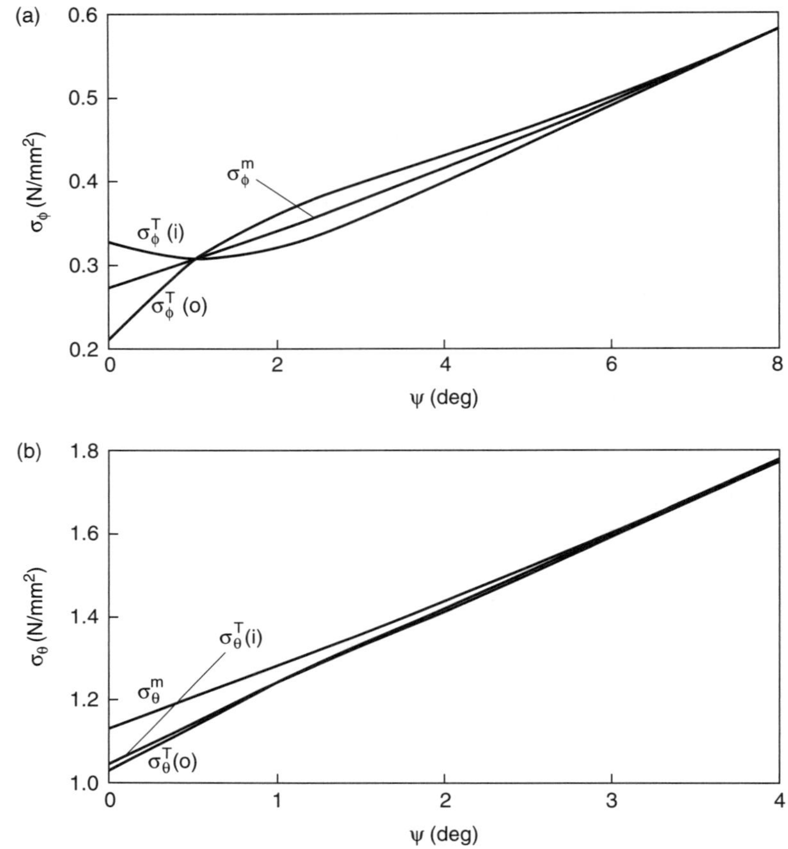

8.3. Spherical domes and caps

8.3.1 Introduction

In this last part of the present chapter, we turn our attention to spherical domes and caps. Some parametric studies are outlined, and the main findings and recommendations summarised. Details have previously been reported in the journal literature.[20,21]

8.3.2 Domes and caps of constant thickness

8.3.2.1 Introduction

We would like to investigate bending-disturbance stresses in roof domes and dome-shaped portions of containment shell structures, and assess the degree of effectiveness achieved in reducing the maximum stresses when recourse is made to thickening. The types of shells in question are single-edged, non-shallow spherical shells often used in dome-roof construction, or as end closures for cylindrical and conical containment vessels. In these applications, the main loading conditions are usually axisymmetric: self-weight and other gravity loadings for roof domes, and uniform pressure for end caps of gas-filled cylinders and bottom closures of deep Intze tanks. The latter type of container is regarded as *deep* if the head of liquid above the level of the crown of the spherical-shell bottom is large in comparison with the rise of the spherical shell, so that the pressure distribution over the shell is essentially uniform.

Specifically, we would like to investigate the laws that govern the relationship between maximum shell stresses (which usually occur in the zone bordering the supports) and geometric shell parameters such as a, t, a/t and ϕ_s. The domes (or shell caps) are assumed to be relatively thin ($a/t \geq 30$) and not too shallow ($\phi_s \geq 30°$). Thus we may use the Geckeler approximation to evaluate the edge effects. Of special interest is the effect on peak shell stresses of shell-thickness modifications, particularly the effectiveness of increasing shell thickness in reducing maximum shell stresses during the course of design.

Four limiting-type boundary conditions are of interest, namely (*i*) fully fixed (encastré) support; (*ii*) fully pinned (hinged) support; (*iii*) pinned support on rollers with vertical reactions; (*iv*) pinned support on rollers with tangential reactions. These four types of supports were illustrated in Figure 6.7. The last, being 'membrane-type', is completely free of edge effects, and will not be pursued any further.

Figure 8.17 Stress plots for the lower edge zone of the ogival shell: (a) meridional stresses; (b) hoop stresses.

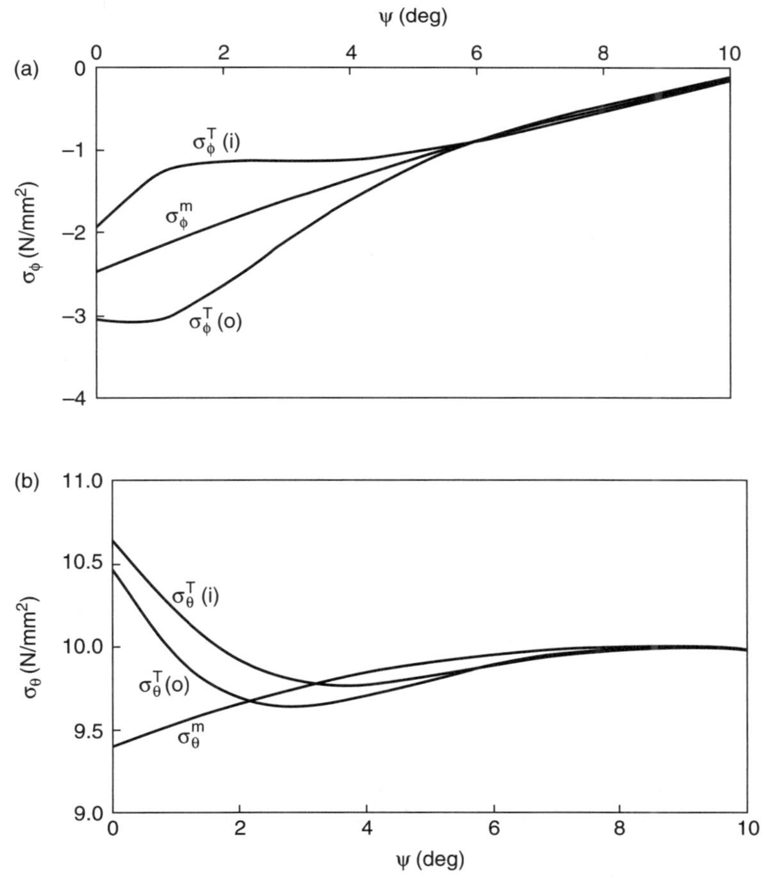

Figure 8.18 Stress plots for the edge zone of the lower spherical closure: (a) meridional stresses; (b) hoop stresses.

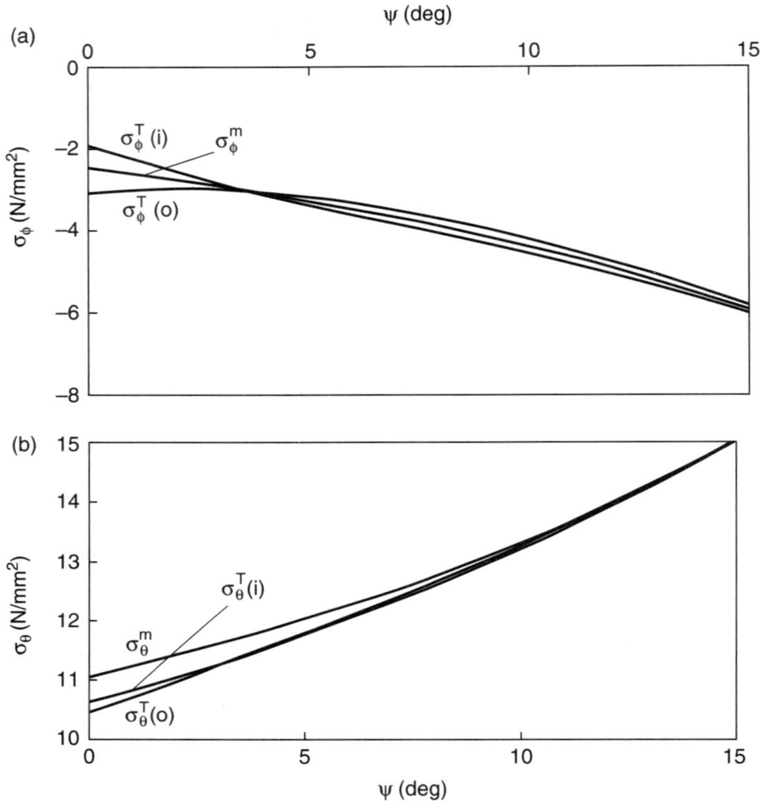

As regards loading, we will consider a uniform pressure p normal to the surface of the shell (tending to 'stretch' or 'compress' the shell, depending on whether the shell forms part of a pressure vessel or is the concave bottom of a deep Intze tank), and a uniformly distributed vertical load q per unit area of the shell surface.

For the gravitational loading, we will make a distinction between self-weight q_w ($= t\gamma$, where γ is the weight per unit volume of shell material) and superimposed dead load q_i. The positive direction for p is as shown in Figure 8.19(a), while that for q is as shown in Figure 8.19(b).

8.3.2.2 Results for net stresses

Membrane-solution results for spherical shells under a uniform *internal* pressure p were given by equations (3.12) and (3.13), while those for spherical domes under uniform loading q were given by equations (3.216). The same results were also given in Chapter 6 as equations (6.72) for p and (6.78) for q. Geckeler-approximation edge redundants are calculated in the usual way, as illustrated in Section 6.3.6. Bending-related interior actions follow from relations (6.68), and total inner- and outer-surface shell stresses are obtained from expressions (6.70).

Figure 8.19 Two types of axisymmetric loadings for domes: (a) uniform external pressure *p* normal to the surface of the shell; (b) uniformly distributed vertical load *q* per unit area of shell surface.

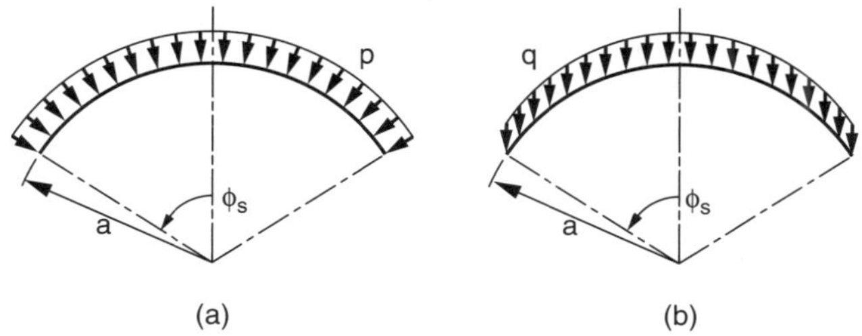

The dimensionless variations of σ_ϕ^T and σ_θ^T with ψ are given below in terms of the basic shell parameters a, a/t, ν, λ and ϕ_s. As stated earlier, a distinction has been made between self-weight q_w and imposed loading q_i. In order to bring out the effects of the geometric parameters a, a/t and ϕ_s (λ is not mentioned because it is essentially a function of a/t), especially as regards the contribution of the bending disturbance to total stresses, the results are given in a form that permits a parametric study to be conducted almost by inspection. Note that p is presently considered to act *towards the centre of the shell*, as shown in Figure 8.19. In the $\pm$ or μ symbols in front of the flexural terms, and consistent with our usual convention, the upper sign refers to the inner shell surface, while the lower sign refers to the outer surface.

8.3.2.2.1 UNIFORM EXTERNAL-PRESSURE LOADING

Encastré edge

$$\frac{\sigma_\phi^T}{p} = \frac{a}{t}\left\{-\frac{1}{2} + \left(e^{-\lambda\psi}\left[\left(\frac{1-\nu}{2\lambda}\right)\cot(\phi_s - \psi)\cos\lambda\psi \pm \frac{\sqrt{3}}{2}\left(\frac{1-\nu}{1+\nu}\right)^{1/2}(\sin\lambda\psi - \cos\lambda\psi)\right]\right)\right\}$$

$$\frac{\sigma_\theta^T}{p} = \frac{a}{t}\left\{-\frac{1}{2} + \left(e^{-\lambda\psi}\left[\left(\frac{1-\nu}{2}\right)(\sin\lambda\psi + \cos\lambda\psi) \pm \nu\frac{\sqrt{3}}{2}\left(\frac{1-\nu}{1+\nu}\right)^{1/2}(\sin\lambda\psi - \cos\lambda\psi)\right]\right)\right\} \qquad (8.121a, b)$$

Fully pinned edge

$$\frac{\sigma_\phi^T}{p} = \frac{a}{t}\left\{-\frac{1}{2} + \left(e^{-\lambda\psi}\left[-\left(\frac{1-\nu}{4\lambda}\right)\cot(\phi_s - \psi)(\sin\lambda\psi - \cos\lambda\psi) \pm \frac{\sqrt{3}}{2}\left(\frac{1-\nu}{1+\nu}\right)^{1/2}\sin\lambda\psi\right]\right)\right\}$$

$$\frac{\sigma_\theta^T}{p} = \frac{a}{t}\left\{-\frac{1}{2} + \left(e^{-\lambda\psi}\left[\left(\frac{1-\nu}{2}\right)\cos\lambda\psi \pm \nu\frac{\sqrt{3}}{2}\left(\frac{1-\nu}{1+\nu}\right)^{1/2}\sin\lambda\psi\right]\right)\right\} \qquad (8.122a, b)$$

Pinned edge on vertically reacting supports

$$\frac{\sigma_\phi^T}{p} = \frac{a}{t}\left\{-\frac{1}{2}+\left(\frac{a}{t}\right)^{1/2}\left(e^{-\lambda\psi}\sin\phi_s\cos\phi_s\left[-\frac{1}{2}\left(\frac{a}{t}\right)^{-1/2}\cot(\phi_s-\psi)(\sin\lambda\psi-\cos\lambda\psi)\pm\left(\frac{27}{1-\nu^2}\right)^{1/4}\sin\lambda\psi\right]\right)\right\}$$

$$\frac{\sigma_\theta^T}{p} = \frac{a}{t}\left\{-\frac{1}{2}+\left(\frac{a}{t}\right)^{1/2}\left(e^{-\lambda\psi}\sin\phi_s\cos\phi_s\left[\{3(1-\nu^2)\}^{1/4}\cos\lambda\psi\pm\nu\left(\frac{27}{1-\nu^2}\right)^{1/4}\sin\lambda\psi\right]\right)\right\} \quad (8.123\text{a, b})$$

8.3.2.2.2 UNIFORM GRAVITY LOADING

$$\frac{\sigma_\phi^T}{q} = \frac{a}{t}(f_\phi + d_\phi) \quad (8.124\text{a})$$

$$\frac{\sigma_\theta^T}{q} = \frac{a}{t}(f_\theta + d_\theta) \quad (8.124\text{b})$$

where, for self-weight, $q = q_w = \gamma t$, and for uniform superimposed dead load, $q = q_i$.

The non-dimensional parameters f_ϕ and f_θ are membrane-solution functions, and are thus independent of the boundary conditions at the edge of the shell:

$$f_\phi = -\left\{\frac{1}{1+\cos(\phi_s-\psi)}\right\} \quad (8.125\text{a})$$

$$f_\theta = \left\{\frac{1}{1+\cos(\phi_s-\psi)}\right\} - \cos(\phi_s-\psi) \quad (8.125\text{b})$$

On the other hand, the non-dimensional parameters d_ϕ and d_θ are bending-disturbance functions that depend on the boundary conditions at the edge of the shell. For the three types of supports of present interest, these functions are defined as follows:

Encastré edge

$$d_\phi = g_\phi = e^{-\lambda\psi}\left\{\frac{1}{\lambda}\left(\left(\frac{2+\nu}{2\lambda}\right)\sin\phi_s\sin\lambda\psi - \left[\left\{\frac{1+\nu}{1+\cos\phi_s} - \cos\phi_s\right\} - \left(\frac{2+\nu}{2\lambda}\right)\sin\phi_s\right]\cos\lambda\psi\right)\right.$$

$$\left.\times\cot(\phi_s-\psi)\pm 2\left(\frac{3}{1-\nu^2}\right)^{1/2}\left(-\frac{1}{2}\left\{\frac{1+\nu}{1+\cos\phi_s} - \cos\phi_s\right\}\sin\lambda\psi\right.\right.$$

$$+\left[\frac{1}{2}\left\{\frac{1+\nu}{1+\cos\phi_s}-\cos\phi_s\right\}-\left(\frac{2+\nu}{2\lambda}\right)\sin\phi_s\right]\cos\lambda\psi\Bigg)\Bigg\} \tag{8.126a}$$

$$d_\theta = g_\theta = e^{-\lambda\psi}\Bigg\{\Bigg(\left[-\left\{\frac{1+\nu}{1+\cos\phi_s}-\cos\phi_s\right\}+2\left(\frac{2+\nu}{2\lambda}\right)\sin\phi_s\right]\sin\lambda\psi-\left\{\frac{1+\nu}{1+\cos\phi_s}-\cos\phi_s\right\}\cos\lambda\psi\Bigg)$$

$$\pm 2\nu\left(\frac{3}{1-\nu^2}\right)^{1/2}\Bigg(-\frac{1}{2}\left\{\frac{1+\nu}{1+\cos\phi_s}-\cos\phi_s\right\}\sin\lambda\psi+$$

$$+\left[\frac{1}{2}\left\{\frac{1+\nu}{1+\cos\phi_s}-\cos\phi_s\right\}-\left(\frac{2+\nu}{2\lambda}\right)\sin\phi_s\right]\cos\lambda\psi\Bigg)\Bigg\} \tag{8.126b}$$

Fully pinned edge

$$d_\phi = h_\phi = e^{-\lambda\psi}\left(\frac{1+\nu}{1+\cos\phi_s}-\cos\phi_s\right)\left\{\frac{1}{2\lambda}(\sin\lambda\psi-\cos\lambda\psi)\cot(\phi_s-\psi)\mu\left(\frac{3}{1-\nu^2}\right)^{1/2}\sin\lambda\psi\right\}$$

$$d_\theta = h_\theta = e^{-\lambda\psi}\left(\frac{1+\nu}{1+\cos\phi_s}-\cos\phi_s\right)\left\{-\cos\lambda\psi\mu\nu\left(\frac{3}{1-\nu^2}\right)^{1/2}\sin\lambda\psi\right\} \tag{8.127a, b}$$

Pinned edge on vertically reacting supports

$$d_\phi = y_\phi = e^{-\lambda\psi}\left(\frac{\sin\phi_s\cos\phi_s}{1+\cos\phi_s}\right)\left(\frac{a}{t}\right)^{1/2}\left\{-\left(\frac{a}{t}\right)^{-1/2}(\sin\lambda\psi-\cos\lambda\psi)\cot(\phi_s-\psi)\pm\left(\frac{432}{1-\nu^2}\right)^{1/4}\sin\lambda\psi\right\}$$

$$d_\theta = y_\theta = e^{-\lambda\psi}\left(\frac{\sin\phi_s\cos\phi_s}{1+\cos\phi_s}\right)\left(\frac{a}{t}\right)^{1/2}\left\{2\{3(1-\nu^2)\}^{1/4}\cos\lambda\psi\pm\nu\left(\frac{432}{1-\nu^2}\right)^{1/4}\sin\lambda\psi\right\} \tag{8.128a, b}$$

In the expressions relating to the total meridional stress σ_ϕ^T, the term corresponding to the meridional stress resultant N_ϕ^b (i.e. the term in $\cot(\phi_s-\psi)$) may be neglected in comparison with the flexural term corresponding to the bending moment M_ϕ (i.e. the term preceded by the $\pm$ or μ symbol), as it is generally very small in comparison with the latter, particularly as a/t is increased and/or ϕ_s approaches 90°. Let us try to understand this from a more general perspective.

An examination of the general expressions for N_ϕ^b and M_ϕ (expressions (6.68)), and hence of the corresponding stresses (N_ϕ^b/t and $\pm 6M_\phi/t^2$), shows that the ratio of the *stresses* stemming from M_ϕ to those associated with N_ϕ^b is of the order of $\{3/(1-\nu^2)\}^{1/2}\lambda\tan\phi_s$. Assuming, for instance, a Poisson's ratio of 0.15, the neglecting of the N_ϕ^b term in comparison with the M_ϕ term would introduce an error in the total-stress values of only ≈ 14% for the least favourable combination of $a/t = 30$ (corresponding to $\lambda \approx 7$) and $\phi_s = 30°$; for $\phi_s \geq 60°$ and any λ (≥ 7), or for $\lambda \geq 20$ and any ϕ_s ($\geq 30°$), the error does not exceed the order of 5%.

From Section 6.3.7, where it was shown that the general error of the Geckeler approximation relative to the asymptotic approximation of Hetényi is of the order of $(1+2\nu)(\cot\phi_s)/(2\lambda)$, it may be concluded that, for meridional stresses, any neglecting of the stress term in $\cot(\phi_s - \psi)$ in relation to the flexural term merely involves errors of the same magnitude as those already inherent in the Geckeler approximation. It follows, therefore, that whether we include these small terms for completeness, or decide to neglect them for convenience, is really of negligible consequence upon the overall accuracy of the results.

Expressions (8.121) and (8.122) for uniform external-pressure loading, corresponding to the two cases of laterally restrained shell edges, can readily be applied to the situation of a uniform rise in temperature ΔT throughout the entire dome. All we need to do is to disregard the first term $(-1/2)$ corresponding to the membrane solution, and then replace $p(a/t)$ with $-2E\alpha(\Delta T)/(1-\nu)$, where α is the coefficient of thermal expansion of the shell material, and $\{E, \nu\}$ are the usual isotropic material properties. This conversion may be readily derived on the basis of equivalent strains or displacements. The negative sign simply means that external pressurisation of the shell (as assumed in Figure 8.19(a)) tends to cause movement of the shell inwards, whereas a uniform rise in temperature would cause the shell to move outwards.

It is clear that such a uniform rise in temperature does not induce a membrane state of stress, as the shell would simply expand freely outwards were it provided with membrane-type boundary conditions. Similarly, temperature-induced edge stresses would not occur where the dome is supported on vertically reacting rollers, since the lack of lateral restraint would allow the dome to move freely inwards or outwards in response to any changes in temperature.

A useful feature of the results in equations (8.121)–(8.128) is their wide range of applicability, as all the shell parameters (ϕ_s, a, t and ν (hence also a/t and λ)) appear in arbitrary form for a given loading and support type. If more than one loading applies (e.g. self-weight and imposed loading), we can obtain the combined effects by superposition. The only restriction on the use of the derived sets of equations is that $\phi_s \geq 30°$ and $\lambda \geq 7$, since we have made use of the Geckeler approximation, which requires that these criteria are satisfied for reasonable accuracy. In accordance with the error estimate of Section 6.3.7, the expressions become more accurate as ϕ_s approaches 90° and/or as λ attains higher values, becoming practically exact for $\phi_s = 90°$ regardless of the value λ, provided this is at least 7.

8.3.2.3 Effect of shell geometry on shell stresses

For the two support conditions where lateral movement of the shell edge is completely restrained (i.e. the encastré edge and the fully pinned edge), peak bending-disturbance stresses caused by either uniform-pressure loading p or imposed loading q_i are essentially proportional to the ratio a/t, which is the same law as that which governs membrane stresses. However, for self-weight loading q_w, the peak values of both bending-disturbance stresses and membrane stresses are directly proportional to shell radius a and independent of shell thickness t. Excessive stresses due to self-weight cannot be reduced by shell thickening.

For a dome with a fixed or fully pinned edge, a uniform temperature rise ΔT throughout the whole shell does not induce any membrane stresses, but generates bending-disturbance stresses in the edge zone whose peaks depend only on the material parameters E (Young's modulus), ν (Poisson's ratio) and α (coefficient of thermal expansion); the geometric parameters t, a, a/t and ϕ_s have no influence at all on the magnitude of the stresses induced by such a temperature rise.

In the case where lateral restraint is lacking and the reactions of the supports are vertical (i.e. the shell edge is pinned onto rollers that move on a horizontal plane), peak stresses are dominated by the edge effect, and are essentially proportional to $(a/t)^{3/2}$ for loading p or q_i, and to $a(a/t)^{1/2}$ for loading q_w, provided that the dome is not near hemispherical. If the dome is near hemispherical (ϕ_s approaching 90°), lateral restraint becomes less important, since the lateral thrust transmitted to the supports by membrane action diminishes; the support conditions approach the requirements of the membrane hypothesis, the bending disturbance vanishes, and membrane action (for which the laws governing peak stresses have already been discussed) dominates.

For this type of roller support, the least favourable depth of shell (with regard to the severity of the edge effect) corresponds to $\phi_s = 45°$ for uniform pressure loading p, and $\phi_s = 51.8°$ for gravity loading (both q_w and q_i). With regard to bending-disturbance effects, this type of support generally results in a shell response to gravity loading q that lies somewhere in between that of a pure membrane shell and a plate under bending, noting that for (*a*) a pure membrane shell, (*b*) a shell with the support type in question, and (*c*) a plate under bending, stresses are proportional to (in that order) t^0, $t^{-1/2}$ and t^{-1} for q_w, and t^{-1}, $t^{-3/2}$ and t^{-2} for q_i.

These findings are based on limiting-type support conditions. In static and kinematic terms, these limiting conditions may be stated as (*a*) {zero lateral movement, zero meridional rotation} for the encastré edge, (*b*) {zero lateral movement, zero meridional moment} for the fully pinned edge, and (*c*) {zero lateral force, zero meridional moment} for the pinned edge on horizontal rollers. Should the shell edge in a real dome be essentially supported in one of these ways, then the assumed boundary conditions are exact, and the results are directly applicable.

However, where lateral and rotational constraints at the supports are finite and non-zero, as is sometimes the case, then the results provide upper and lower bounds for shell response, which may be used as a basis for conservative shell design in the region of the support, particularly where the actual stiffness of the support is difficult to quantify. Conversely, the designer can specify, within the limits imposed by practical considerations, fixity details at the edge of the shell that best approximate limiting support conditions that are most favourable to shell stresses.

Quite often a ring beam is provided at the edge of the dome. The limiting-type support conditions of present considerations may be regarded as special cases of ring-beam behaviour. An encastré support is equivalent to the provision of a ring beam with infinite extensional and infinite axi-rotational stiffnesses (IAIT type of Section 8.1.2). A fully pinned support is equivalent to the provision of a ring beam with infinite extensional but zero axi-rotational stiffness (IAZT type). A support pinned on rollers that move on a horizontal plane is equivalent to providing a ring beam of zero extensional and zero axi-rotational stiffnesses (ZAZT type), with the reactions on the ring soffit being in the vertical direction.

For ring beams of real proportions and in real materials (i.e. having finite stiffness properties), and as already implied above, the results in equations (8.121)–(8.128) may serve as upper- and lower-bound estimates for the design of the dome, but if needed, a more exact assessment of stresses in the edge zone of the dome may be made on the basis of the shell-ring considerations of Chapter 7.

8.3.2.4 Design rules for concrete domes

The closed-form results for total stresses presented in Section 8.3.2.2 may be used as a basis for a detailed parametric study of bending-disturbance effects in non-shallow concrete domes.[20,21] Figures 8.20–8.25 show non-dimensional stress variations for the loadings and support types that have been dealt with. In these plots, unbroken curves labelled *m* denote the membrane solution; the remaining unbroken curves refer to the total solution for $a / t = 30$, while the broken curves refer to the total solution for $a / t = 300$. Following the usual notation, for the total-solution plots, labels i and o refer to the inner and outer shell surfaces, respectively.

The following general rules have been established for the design of unreinforced or lightly reinforced concrete domes. In formulating these rules, we do not neglect the tensile strength of concrete, but we make use of the fact that this is considerably lower than the compressive strength.

Figure 8.20 Non-dimensional stress variations for a dome subjected to uniform external pressure and with the edge fixed. The plots are applicable for all values of ϕ_s within the non-shallow range.

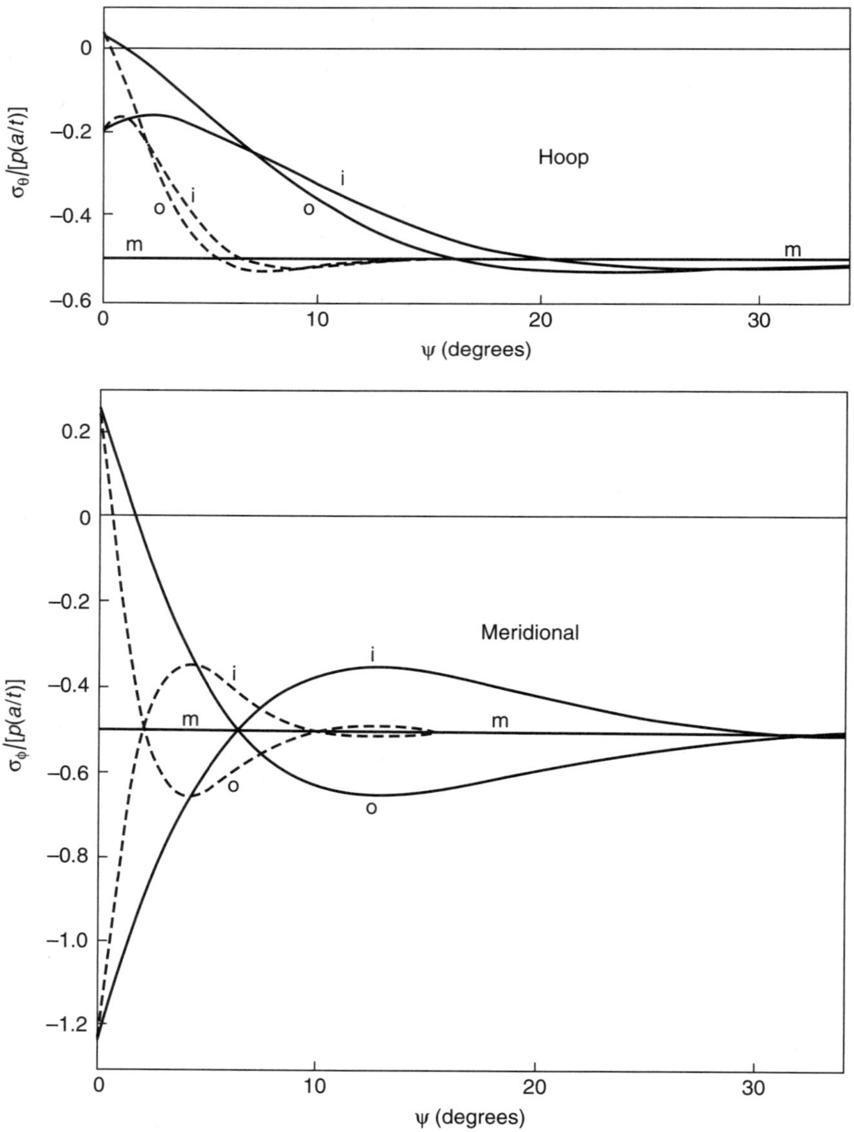

Figure 8.21 Non-dimensional stress variations for a dome subjected to uniform external pressure and with the edge fully pinned. The plots are applicable for all values of ϕ_s within the non-shallow range.

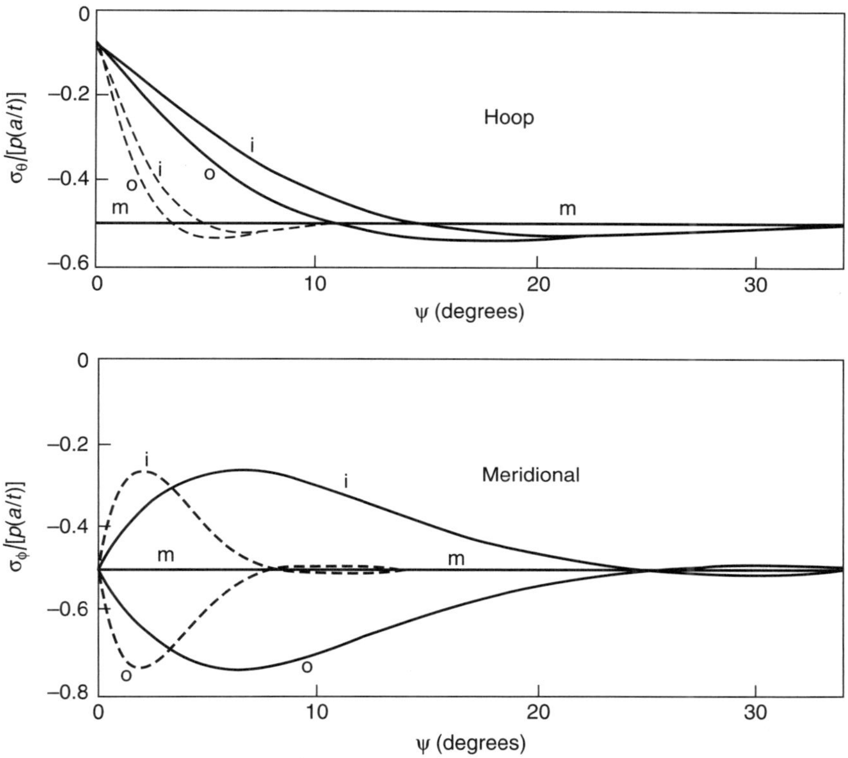

When lateral movement of the shell edge is completely restrained (as in the cases of the fixed and the fully pinned edges), the presence of the edge disturbance is beneficial in concrete domes subjected to uniform gravity loading, including self-weight and superimposed dead load, provided that the springing angle ϕ_s is greater than 45° – 52°.

When the springing angle exceeds 60°, the edge disturbance results in maximum tensile values of the hoop total-stress function ($(f_\theta + g_\theta)$ or $(f_\theta + h_\theta)$) in the vicinity of the support of somewhat smaller magnitude than the tensile value of the hoop membrane-solution function (f_θ) at the edge, permitting the shell thickness to be selected, rather conservatively, on the basis of membrane calculations only. This rule assumes that the compressive strength of concrete is at least six times larger than its tensile strength, which is almost invariably the case in practice.

On the other hand, when ϕ_s is less than 45° – 52°, the compressive membrane stress field throughout the dome ensures that total-stress functions remain essentially negative (i.e. compressive) throughout, or that they attain relatively insignificant tensile values in the immediate vicinity of the springing. This means that the thickness of the dome can be selected on the basis of the compressive strength of the concrete rather than on its considerably smaller tensile strength, resulting in greater efficiency (buckling considerations aside).

Figure 8.22 Non-dimensional stress variations for a dome subjected to uniform external pressure and with the edge pinned on vertically reacting rollers, for $\phi_s = 45°$.

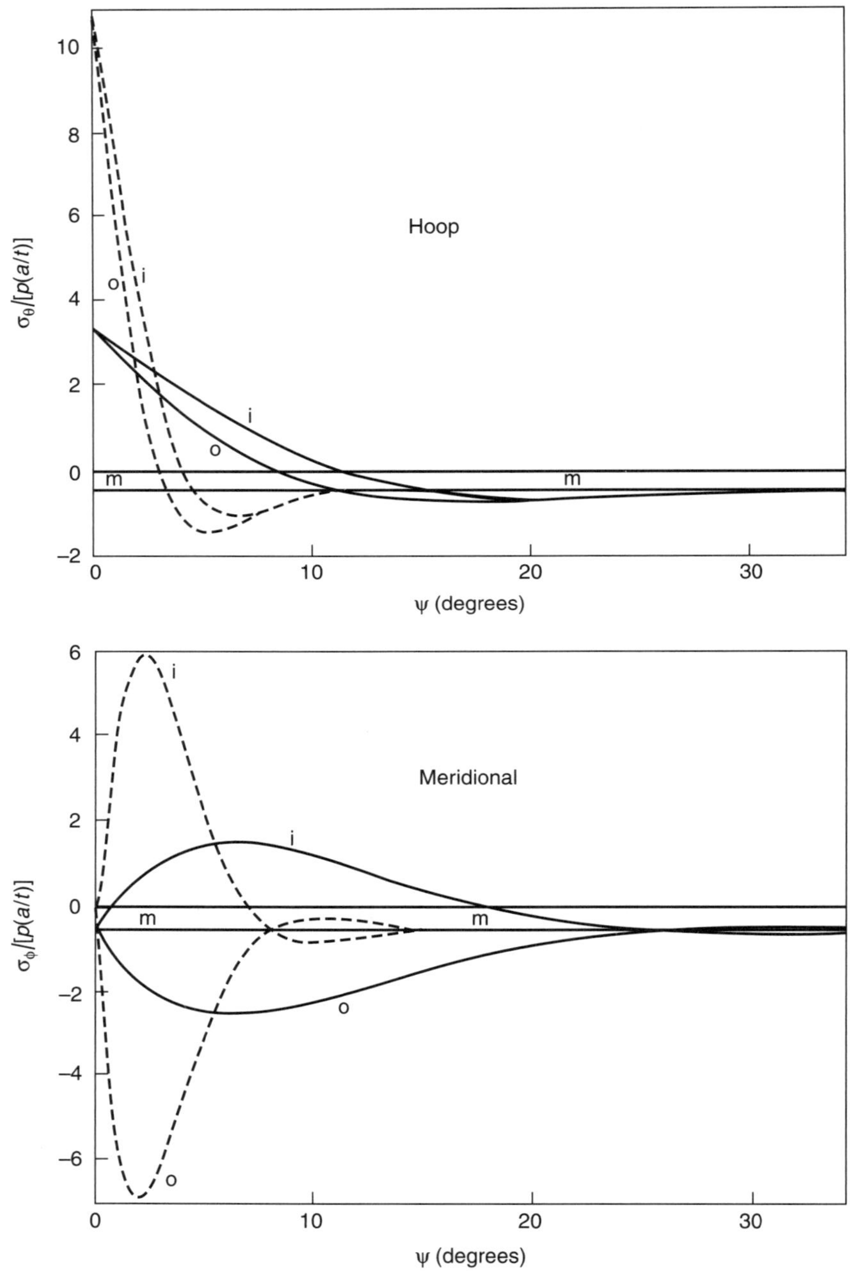

Figure 8.23 Non-dimensional stress variations for a dome subjected to uniform gravity loading and with the edge fixed, for $\phi_s = 90°$.

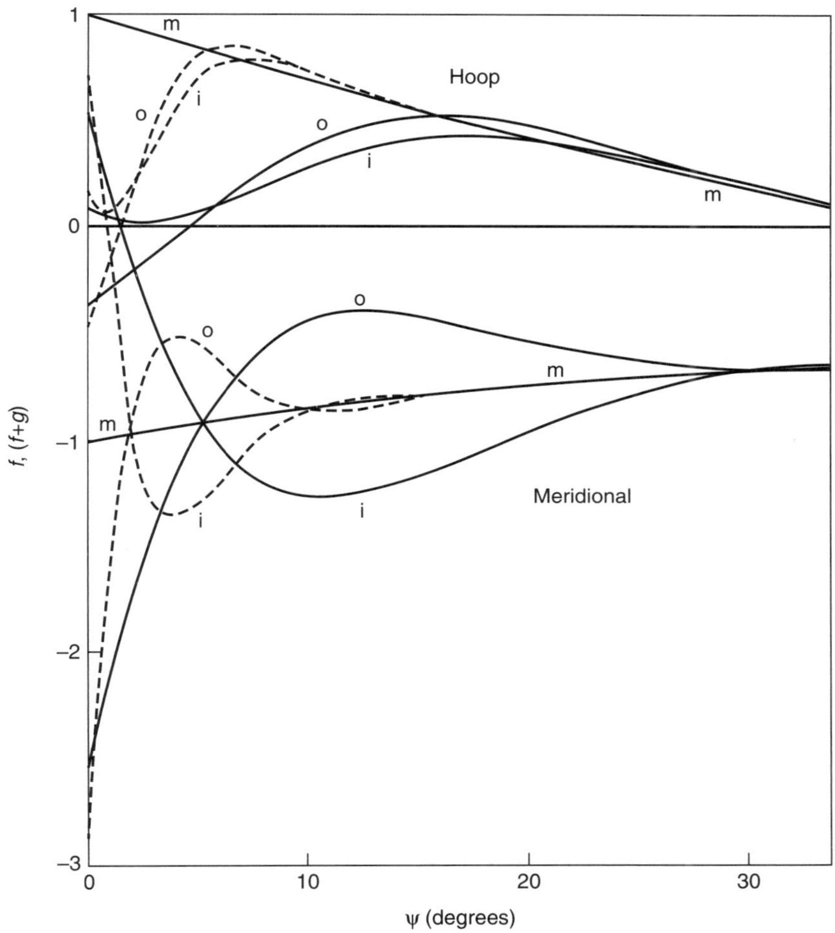

The following two general rules may be stated for concrete domes restrained against lateral movement at the base and subjected to gravity loading:

(*i*) *The peak stresses are essentially proportional to* qat^{-1} *so that, in the case of self-weight* ($q = \gamma t \;\Rightarrow\; qat^{-1} = \gamma a$), *thickening is ineffectual in reducing such stresses.*

This rule, well known for membrane calculations, and earlier shown to apply also to edge-disturbance stresses, may now be seen to hold for the superimposed effects.

(*ii*) *If the magnitude of the loading and the shell dimensions are such that* qat^{-1} *(or, in the case of self-weight,* γa*) does not exceed the tensile strength of the concrete, the design is adequate.*

Figure 8.24 Non-dimensional stress variations for a dome subjected to uniform gravity loading and with the edge fully pinned, for $\phi_s = 90°$.

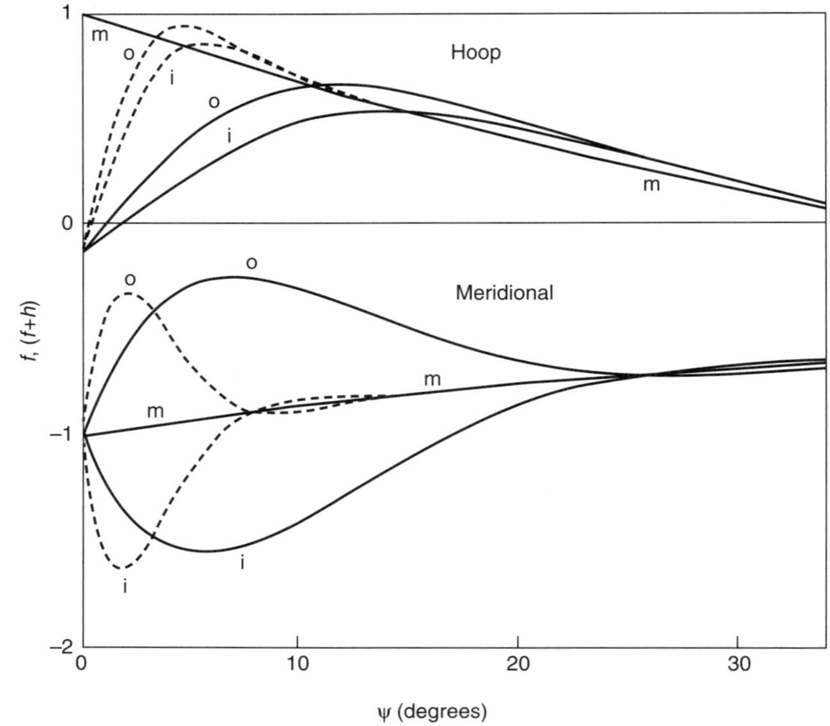

This is based on the fact that the largest possible tensile value of the membrane solution, for all values of ϕ_s, occurs when $\phi_s = 90°$, and is equal to qat^{-1}. This value is therefore an upper bound to any possible tensile stresses in all shallower domes. Compressive stresses, though attaining larger magnitudes than this value, are not a problem, since concrete has relatively large compressive strength.

In the case of uniform external-pressure loading, no net tensile stresses would develop anywhere on the shell when the edge is fully pinned, enabling the design of the shell to be based on a compressive stress simply 1.5 times larger than the constant membrane value of $-0.5pat^{-1}$. This holds for all values of ϕ_s and a/t within the usual limits ($\phi_s \geq 30°$; $a/t \geq 30$).

If, on the other hand, the edge is clamped, the dome thickness may be chosen on the basis of a tensile stress of $0.25pat^{-1}$ (i.e. 50% of the magnitude of the compressive membrane value) for all values of ϕ_s and a/t. Alternatively, the dome thickness may be selected on the basis of a compressive stress that is simply 2.5 times larger than the constant membrane value, for all values of ϕ_s and a/t, provided the dome is meridionally reinforced on the outer surface against the tensile stress of $0.25pat^{-1}$ for an angular distance of 2° from the springing.

Effects of a uniform rise or drop in temperature for the laterally restrained cases of support may be estimated directly from the uniform-pressure results through the equivalence $pat^{-1} \equiv 2E\alpha(\Delta T)/(1-\nu)$, disregarding

Figure 8.25 Non-dimensional stress variations for a dome subjected to uniform gravity loading and with the edge pinned on vertically reacting rollers, for $\phi_s = 60°$.

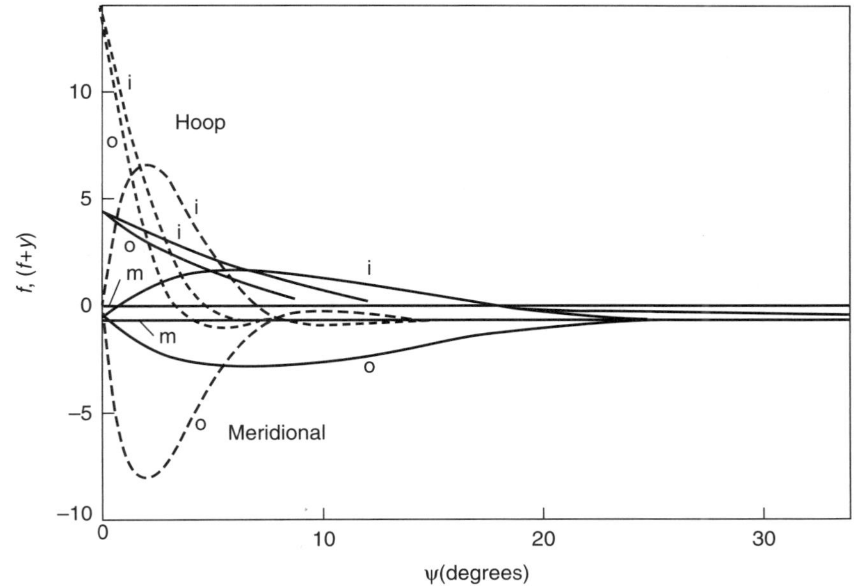

the membrane-solution component of the pressure-loading results, since only bending disturbances occur as a result of uniform temperature changes.

For the encastré edge, the maximum tensile stress is meridional, with a value of $\approx 1.5E\alpha(\Delta T)/(1-\nu)$, typically 0.42 N/mm^2 per °C; it occurs irrespective of whether the temperature rises or falls. For the fully pinned edge, the peak tensile stress is a circumferential one, at $\approx 0.85E\alpha(\Delta T)/(1-\nu)$ – typically 0.24 N/mm^2 per °C – for a temperature drop and $\approx 0.5E\alpha(\Delta T)/(1-\nu)$ – typically 0.14 N/mm^2 per °C – for a temperature rise.

Clearly, these peaks of the temperature-induced stresses are independent of the geometric parameters of the dome (t, a, ϕ_s), and are only affected by the material properties of the concrete (α, E, ν). As the latter generally do not vary widely from one concrete to another, excessive stresses due to temperature changes in the course of design calculations may not be easy to reduce in domes that are restrained against lateral movement at their edges.

For loadings with a net vertical resultant (for instance, applied pressure and gravity loadings), lack of lateral restraint at the edge is generally very detrimental, and leads to severe bending in the edge zone (hence an unnecessarily excessive requirement for shell thickness) if stress values are to be kept within permissible levels. The worst situation occurs when $\phi_s = 45°$ for uniform pressure loading, and when $\phi_s = 51.8°$ for uniform gravity loading. For these two loadings, the peak tensile stresses work out at $\approx 0.65pat^{-1}(a/t)^{1/2}$ and $\approx 0.80qat^{-1}(a/t)^{1/2}$, respectively.

Taking $a/t = 300$, for example, these estimates give $\approx 11pat^{-1}$ and $\approx 14qat^{-1}$, values that are much higher than the tensile-stress bounds of $0.25pat^{-1}$ and qat^{-1} established for laterally restrained cases for all values of a/t and ϕ_s. Thus, unless ϕ_s is close to 90° (the special instance when the vertical reactions also become tangential to the shell midsurface and the edge disturbance vanishes), adequate lateral restraint must be provided at the edge of the shell.

8.3.3 Domes of gradually varied local thickening

8.3.3.1 Introduction

In the previous section, we saw how excessive flexural stresses can develop at the base of the dome, particularly if no lateral restraint is provided. Even if a dome is provided with adequate lateral restraint at the base (via connection to a ring beam, another structure or a sufficiently rigid foundation), it is often desirable to increase the thickness of the shell in the approaches to the springing, in order to accommodate any edge effects there. Since concrete can easily be moulded into a variety of shapes, this local thickening can be effected in a gradual manner, to avoid thickness-discontinuity stresses of the type that we dealt with in Section 8.1.4.

In this final section, we seek to obtain a reasonable estimate of bending-disturbance stresses in the edge zone of a spherical dome whose thickness is gradually increased in the approaches to the support, the increase being in accordance with a particular law. We will also seek to establish the effectiveness of such tapered local thickening as a design option in reducing the peak stresses in the edge zones of spherical domes (including spherical closures of vessels).

Starting from the apex of the shell, where $\phi = 0$, and following a meridian, the shell thickness is assumed to remain constant at a value $t = t_o$, down to a certain location denoted by ϕ_e, beyond which the thickness is gradually increased to a maximum value of t_s at the support (where $\phi = \phi_s$). We will use the notation $\Delta\phi$ to denote the angular bandwidth of the tapered lower portion of the shell (i.e. $\Delta\phi = \phi_s - \phi_e$). The thickening of the shell is assumed to be symmetrical about the spherical midsurface of the unthickened dome (i.e. the midsurface of the thickened dome coincides with that of the unthickened dome).

In the considerations of constant-thickness shells of Section 8.1.4.4, we defined the effective range of the bending disturbance, from the edge of the shell, as $\psi_r = \pi/\lambda$. For shells with a thickness that gradually decreases from the edge, the effective range of the bending disturbance is more difficult to define, since λ now varies over this distance. However, for the purposes of estimating a suitable width over which to gradually change the thickness of the shell, we will take the effective range of the edge effect as $\psi_r = \pi/\lambda_s$, where λ_s is the value of λ at the support. This tends to overestimate ψ_r, and hence is on the conservative side.

From an enhanced value of t_s at the edge of the dome, the thickness of the shell is assumed to be gradually decreased to a constant interior value t_o over an angular distance given by this upper-bound estimate of the effective range of the bending disturbance. Thus, $\Delta\phi = \left(\phi_s - \phi_e\right) = \pi/\lambda_s$, so that $\phi_e = \phi_s - \left(\pi/\lambda_s\right)$. Once we have decided on t_s, this simple formula gives us a rationale way of deciding on a suitable width of the thickened zone.

8.3.3.2 Assumed thickness variation in the edge zone

Over the thickened edge zone ($\Delta\phi = \pi/\lambda_s$), let us assume that the variation of λ follows the law

$$\lambda = \frac{\alpha}{\beta + \phi} \tag{8.129}$$

where α and β are constants, and ϕ is, of course, the meridional-angle coordinate. As will be shown in due course, equation (8.129) represents a quadratic variation of thickness with the angle ϕ, of the form $t = \mu(\beta + \phi)^2$, where μ is a constant. This variation was considered in Section 3.3.3 with regard to membrane effects.

The above variation has been adopted for two reasons. First, within the context of the axisymmetric bending of thin shells of revolution, this law is one of the few for which a closed-form solution of the equation

$$\frac{d^4 Q_\phi}{d\phi^4} + 4\lambda^4 Q_\phi = 0 \tag{8.130}$$

exists, if λ is not constant (owing to the variation with respect to ϕ of one or more of the geometrical parameters r_1, r_2 or t). As shown in Section 6.5, equation (8.130) stems from Geckeler-type simplifications of the Reissner–Meissner pair of second-order ordinary differential equations in the variables Q_ϕ and V. For the variation of λ given by equation (8.129), the solution of equation (8.130) has been presented by Spotts,[22] who considered spherical shells, and by Rygol,[23] who extended the solution to general shells of revolution. Flügge[13] also treated the same variation, while Nielsen[24] suggested a slight modification to the ensuing relations for internal actions in order to improve the accuracy of the results.

Secondly, even if the actual variation of thickness in the edge zone does not follow the above law, the parameters α and β can usually be chosen such that the ensuing variation of λ as given by equation (8.129) reasonably approximates the actual variation, *at least in the narrow edge zone experiencing bending disturbances*. Nielsen[24] has proposed choosing the constants α and β to fit the actual λ values at two specific locations, namely the points within the effective bending-disturbance range that lie at angular distances of $0.2(\Delta\phi)$ from either end of the effective bending-disturbance range, $\Delta\phi$ being the effective range of the bending disturbance from the support location ϕ_s, taken by Nielsen as $0.75\pi / \lambda_s$.

Since suitable values of α and β can always be chosen to suit any arbitrary variation of thickness likely to be adopted in practice, we will henceforth assume equation (8.129) is satisfied *exactly*. We also want to investigate the effectiveness of the exact law of equation (8.129) in its own right. When plotted, thickness profiles based on equation (8.129) look quite reasonable from a practical point of view; if the structural performance turns out to be good, we could then specifically *choose* this law of thickness variation in the edge zone of the dome.

It may be noted that a closed-form solution also exists for the case of shell thickness t varying linearly with distance along a meridian. This variation, whose solution has been given by Basuli,[25] can be written in the form $\lambda^2 = \alpha / (\beta + \phi)$, or $t = A + B\phi$, where α, β, A and B are constants. However, we prefer to use the quadratic law represented by equation (8.129), since this is better than a linear law in modelling thickness variations that are encountered in practice, where shell thickness is usually made to increase more rapidly as the edge of the dome is approached.

8.3.3.3 Calculation of parameters α and β

If local thickening is required to vary from a value of t_o somewhere near the edge of the dome to a value of t_s at the edge, we may substitute into equation (8.129) the known values of λ at the upper and lower ends of the zone of tapered thickening (i.e. λ_s at $\phi = \phi_s$ and λ_e at $\phi = \phi_e$), to obtain two simultaneous equations in

α and β. If we then make use of the fact that the value of λ at the upper end (i.e. λ_e) is equal to the value of λ in the constant-thickness portion of the dome (i.e. λ_o), we may write the solutions for α and β as follows:

$$\alpha = \left(\frac{\lambda_s \lambda_o}{\lambda_o - \lambda_s}\right)(\phi_s - \phi_e)$$

$$= \left(\frac{\lambda_o}{\lambda_o - \lambda_s}\right)\pi \qquad (8.131a)$$

$$\beta = \left(\frac{\lambda_s \phi_s - \lambda_o \phi_e}{\lambda_o - \lambda_s}\right)$$

$$= \left(\frac{\lambda_o}{\lambda_o - \lambda_s}\right)\frac{\pi}{\lambda_s} - \phi_s \qquad (8.131b)$$

The first lines of the above expressions allow ϕ_e to be assigned any preferred arbitrary value, but the second lines assume that $\phi_e = \phi_s - (\pi / \lambda_s)$ as earlier proposed.

As already pointed out, where the variation of thickness does not follow the law of equation (8.129), the constants α and β must be chosen such that the ensuing variation fits the actual variation as closely as possible. In this case, the 'best-fit' values of α and β may need to be calculated not from the end values $(\{\phi_s, \lambda_s\}, \{\phi_e, \lambda_o\})$ of the tapered portion as given by the first lines of equations (8.131), but from the values corresponding to locations a little distance from each end (*pivot* points). In this case, we have to replace $\{\phi_s, \lambda_s\}$ and $\{\phi_e, \lambda_o\}$ in the first lines of expressions (8.131) by the pair of values for the pivotal location nearer the end in question.

Now, for a dome of constant radius a and variable thickness t, the parameter λ at any point is given by

$$\lambda = \{3(1-\nu^2)\}^{1/4}\left(\frac{a}{t}\right)^{1/2} \qquad (8.132)$$

Rearranging this equation to express t in terms of λ, and then making use of equation (8.129) to eliminate λ, we obtain

$$t = \frac{a}{\alpha^2}\{3(1-\nu^2)\}^{1/2}(\beta + \phi)^2 = \mu(\beta + \phi)^2 \qquad (8.133)$$

where μ is a constant defined as follows:

$$\mu = \frac{a}{\alpha^2}\{3(1-\nu^2)\}^{1/2} \qquad (8.134)$$

and α and β are as defined in equations (8.131). Thus the variation of λ given by equation (8.129) represents a quadratic variation of shell thickness with the angular coordinate ϕ.

In a typical design procedure, ϕ_s, t_s (hence λ_s) and t_o (hence λ_o) are first decided. Then ϕ_e is determined as earlier indicated (i.e. $\phi_e = \phi_s - (\pi / \lambda_s)$). The parameters α and β are next calculated from the second lines of expressions (8.131). The ensuing thickness profile over the range $\phi_e \leq \phi \leq \phi_s$ is evaluated from expressions (8.133) and (8.134), and plotted to check that the profile is reasonable and practicable.

8.3.3.4 Development of general formulae for edge effects and net stresses

Substituting for λ (as given by equation (8.129)) in equation (8.130), we obtain the differential equation

$$\frac{d^4 Q_\phi}{d\phi^4} + \frac{4\alpha^4}{(\beta + \phi)^4} Q_\phi = 0 \tag{8.135}$$

A very close solution to the above equation is

$$Q_\phi = C_1 (\beta + \phi)^{-(\alpha - 0.5)} \sin\left\{\alpha \ln\left(\frac{\beta + \phi}{\beta + \phi_1}\right) - \eta_1\right\} + C_2 (\beta + \phi)^{(\alpha + 0.5)} \sin\left\{\alpha \ln\left(\frac{\beta + \phi}{\beta + \phi_2}\right) - \eta_2\right\} \tag{8.136}$$

where, for the general case of a shell with two edges, ϕ_1 is the value of ϕ at the upper edge, while ϕ_2 is the value of ϕ at the lower edge; C_1, C_2, η_1 and η_2 are constants to be determined from the boundary conditions at the two edges. Equation (8.136) is an alternative form of the solution of equation (8.135) as given by Flügge[13] and Nielsen.[24]

Expression (8.136) is not an exact solution to equation (8.135), since in the derivation leading to the solution, a term of order unity has to be dropped in comparison with another of order $8\alpha^2$. However, the error incurred by this simplification is insignificant for all practical purposes. From equation (8.131a), $\alpha \geq 5.68$ for all likely values of the ratio t_o / t_s (i.e. $t_o / t_s \geq 0.2$). This means that the parameter $8\alpha^2$ exceeds 258, a value that is clearly much greater than unity.

For a shell with no opening at the top, or one whose upper edge is very remote from the lower edge, the local stress distribution at the lower edge is represented by the second term of equation (8.136), which for a dome supported at $\phi = \phi_s$ (i.e. $\phi_2 = \phi_s$) becomes

$$Q_\phi = C (\beta + \phi)^{(\alpha + 0.5)} \sin\left\{\alpha \ln\left(\frac{\beta + \phi}{\beta + \phi_s}\right) - \eta\right\} \tag{8.137}$$

(The subscript 2 has now been dropped from C and η, as this is no longer necessary.)

On the basis of equation (8.137), general expressions for interior actions and deformations of interest are derived as follows:

$$\begin{aligned} N_\phi^b &= -Q_\phi \cot\phi \\ &= -(\cot\phi)\, C (\beta + \phi)^{(\alpha + 0.5)} \sin\left\{\alpha \ln\left(\frac{\beta + \phi}{\beta + \phi_s}\right) - \eta\right\} \end{aligned} \tag{8.138}$$

$$N_\theta^b = -\frac{dQ_\phi}{d\phi}$$

$$= -C\,\alpha(\beta+\phi)^{(\alpha-0.5)}\left[\left(\frac{\alpha+0.5}{\alpha}\right)\sin\left\{\alpha\ln\left(\frac{\beta+\phi}{\beta+\phi_s}\right)-\eta\right\}+\cos\left\{\alpha\ln\left(\frac{\beta+\phi}{\beta+\phi_s}\right)-\eta\right\}\right]$$

$$\approx -\sqrt{2}\;C\,\alpha(\beta+\phi)^{(\alpha-0.5)}\sin\left\{\alpha\ln\left(\frac{\beta+\phi}{\beta+\phi_s}\right)-\eta+\frac{\pi}{4}\right\} \tag{8.139}$$

$$V^b \approx \frac{1}{Et}\frac{d^2Q_\phi}{d\phi^2}$$

$$= \frac{1}{Et}C\,(\beta+\phi)^{(\alpha-1.5)}\left[-\frac{1}{4}\sin\left\{\alpha\ln\left(\frac{\beta+\phi}{\beta+\phi_s}\right)-\eta\right\}+2\alpha^2\cos\left\{\alpha\ln\left(\frac{\beta+\phi}{\beta+\phi_s}\right)-\eta\right\}\right]$$

$$\approx \frac{2\alpha^2}{Et}C\,(\beta+\phi)^{(\alpha-1.5)}\cos\left\{\alpha\ln\left(\frac{\beta+\phi}{\beta+\phi_s}\right)-\eta\right\} \tag{8.140}$$

$$M_\phi \approx -\frac{D}{a}\frac{dV^b}{d\phi}$$

$$= \frac{a}{4\alpha^4}(\beta+\phi)^{(\alpha+1.5)}\,C\times$$

$$\left[(\alpha-3.5)\left[\frac{1}{4}\sin\left\{\alpha\ln\left(\frac{\beta+\phi}{\beta+\phi_s}\right)-\eta\right\}-2\alpha^2\cos\left\{\alpha\ln\left(\frac{\beta+\phi}{\beta+\phi_s}\right)-\eta\right\}\right]\right.$$

$$\left.+\alpha\left[\frac{1}{4}\cos\left\{\alpha\ln\left(\frac{\beta+\phi}{\beta+\phi_s}\right)-\eta\right\}+2\alpha^2\sin\left\{\alpha\ln\left(\frac{\beta+\phi}{\beta+\phi_s}\right)-\eta\right\}\right]\right]$$

$$\approx \frac{a}{2\alpha}(\beta+\phi)^{(\alpha+1.5)}\,C\left[\sin\left\{\alpha\ln\left(\frac{\beta+\phi}{\beta+\phi_s}\right)-\eta\right\}-\left(\frac{\alpha-3.5}{\alpha}\right)\cos\left\{\alpha\ln\left(\frac{\beta+\phi}{\beta+\phi_s}\right)-\eta\right\}\right]$$

$$\approx \frac{a}{\alpha\sqrt{2}}C\,(\beta+\phi)^{(\alpha+1.5)}\sin\left\{\alpha\ln\left(\frac{\beta+\phi}{\beta+\phi_s}\right)-\eta-\frac{\pi}{4}\right\} \tag{8.141}$$

$$M_\theta \approx \nu M_\phi \tag{8.142}$$

$$\delta^b \approx \frac{a}{Et}(\sin\phi)\, N_\theta^b$$

$$= -\frac{a}{Et}(\sin\phi)\sqrt{2}\;C\,\alpha\,(\beta+\phi)^{(\alpha-0.5)}\sin\left\{\alpha\ln\left(\frac{\beta+\phi}{\beta+\phi_s}\right)-\eta+\frac{\pi}{4}\right\} \tag{8.143}$$

It is important to note that the final expression of equation (8.139) is obtained by assuming that $\alpha >> 0.5$. On the other hand, the final expression of equation (8.140) and the third expression of equation (8.141) are both obtained by assuming $8\alpha^2 >> 1$, a simplification of negligible consequence, and one that is already inherent in our basic solution (equation 8.136), as already pointed out. However, the simplification that is likely to have the greatest bearing on the overall accuracy of the results is the assumption that $\alpha >> 3.5$ made in arriving at the final expression of equation (8.141). Such an assumption (that $\alpha >> 3.5$) is obviously quite restrictive on the range of validity of the results, but it is a desirable approximation, since it ensures that the results will be simple enough for practical use. We will assume that this approximation is valid, but should this not be the case, then the second last expression for M_ϕ must be used.

The constants of integration C and η corresponding to the application of bending moments M_s and horizontal shear forces H_s at the edge $\phi = \phi_s$ (acting as shown in Figure 6.6, with the arbitrary edge variables $\{\phi_e, M_e, H_e\}$ now replaced with $\{\phi_s, M_s, H_s\}$ for the support) are obtained by considering the static boundary conditions for the respective edge loadings as given by equations (6.50) and (6.56). The results for the present problem are as follows:

Application of M_s

$$C = -\frac{2\lambda_s}{a}(\beta+\phi_s)^{-(\alpha+0.5)}\,M_s \tag{8.144a}$$

$$\eta = 0 \tag{8.144b}$$

Application of H_s

$$C = \sqrt{2}\;(\beta+\phi_s)^{-(\alpha+0.5)}\,H_s\sin\phi_s \tag{8.145a}$$

$$\eta = -\frac{\pi}{4} \tag{8.145b}$$

Had we not made the assumption that $\alpha >> 3.5$ in arriving at the final result of equation (8.141), the expression for C corresponding to the application of M_s, and the expressions for both C and η in the case of the application of H_s, would have been considerably less convenient.

If we substitute expressions (8.144) for C and η into equations (8.138)–(8.143), we obtain the interior actions and deformations due to M_s. Similarly, if we make use of expressions (8.145) to eliminate C and η from equations (8.138)–(8.143), we obtain the interior actions and deformations due to H_s. Superposition then gives us the interior stress resultants $\{N_\phi^b, N_\theta^b\}$, bending moments $\{M_\phi, M_\theta\}$ and deformations $\{V^b, \delta^b\}$ due to the actions M_s and H_s applied simultaneously at the edge of the shell. For the deformations, it is, of course, their values at the edge of the shell (i.e. V_s^b and δ_s^b) that are of interest. Introducing the notation

$$Y = \frac{\beta + \phi}{\beta + \phi_s} = \left(\frac{t}{t_s}\right)^{1/2} \tag{8.146}$$

enables us to write the results more conveniently, as follows:

$$N_\phi^b = (\cot\phi)\, Y^{(\alpha+0.5)} \left[\frac{2\lambda_s}{a}\sin(\alpha \ln Y)\, M_s - (\sin\phi_s)\{\sin(\alpha\ln Y) + \cos(\alpha \ln Y)\}\, H_s\right] \tag{8.147}$$

$$N_\theta^b = 2\lambda_s\, Y^{(\alpha-0.5)} \left[\frac{\lambda_s}{a}\{\sin(\alpha\ln Y) + \cos(\alpha\ln Y)\}\, M_s - \{(\sin\phi_s)\cos(\alpha\ln Y)\}\, H_s\right] \tag{8.148}$$

$$M_\phi = -Y^{(\alpha+1.5)} \left[\{\sin(\alpha\ln Y) - \cos(\alpha\ln Y)\}\, M_s - \frac{a}{\lambda_s}\{(\sin\phi_s)\sin(\alpha\ln Y)\}\, H_s\right] \tag{8.149}$$

$$M_\theta = \nu M_\phi \tag{8.150}$$

$$V_s^b = -\frac{4\lambda_s^3}{Eat_s} M_s + \frac{2\lambda_s^2}{Et_s}(\sin\phi_s)\, H_s \tag{8.151}$$

$$\delta_s^b = \frac{2\lambda_s^2}{Et_s}(\sin\phi_s)\, M_s - \frac{2a\lambda_s}{Et_s}(\sin^2\phi_s)\, H_s \tag{8.152}$$

By applying boundary conditions (6.71) for the fixed edge, (6.84) for the fully pinned edge, and (6.82) for the pinned edge on vertically reacting rollers, we can easily obtain general solutions for the edge redundants for any axisymmetric distribution of surface loads in terms of the edge parameters of the membrane solution. The results, which are equally applicable to shells of constant thickness and those of variable thickness, are as follows:

Fixed edge

$$M_s = \frac{Eat_s}{2\lambda_s^3} V_s^m + \frac{Et_s}{2\lambda_s^2}\left(\frac{1}{\sin\phi_s}\right)\delta_s^m \tag{8.153a}$$

$$H_s = \frac{Et_s}{2\lambda_s^2}\left(\frac{1}{\sin\phi_s}\right)V_s^m + \frac{Et_s}{a\lambda_s}\left(\frac{1}{\sin^2\phi_s}\right)\delta_s^m \tag{8.153b}$$

Fully pinned edge

$$M_s = 0 \tag{8.154a}$$

$$H_s = \frac{Et_s}{2a\lambda_s}\left(\frac{1}{\sin^2\phi_s}\right)\delta_s^m \tag{8.154b}$$

Pinned edge on vertically reacting rollers

$$M_s = 0 \tag{8.155a}$$

$$H_s = \left(N_\phi^m\right)_s \cos\phi_s \tag{8.155b}$$

The last case is, of course, statically determinate: $\left(N_\phi^m\right)_s$ is simply the value of the membrane meridional stress resultant at the edge of the shell.

For the dome with local thickening of the form given by equation (8.129) or (8.133), $\left\{\left(N_\phi^m\right)_s, V_s^m, \delta_s^m\right\}$ were evaluated in Section 3.3.3, and are given by expressions {3.237, 3.238} and {3.239, 3.240} for uniform superimposed dead load and self-weight loading, respectively. Expressions (8.153)–(8.155) for shell-edge redundants are therefore closed-form results, since all quantities are fully known.

With the shell-edge redundants $\{M_s, H_s\}$ now known in explicit terms, bending-related interior actions follow from expressions (8.147)–(8.150). Membrane stress resultants in the constant-thickness upper part of the dome ($0 \le \phi \le \phi_e$) are given by expressions (3.216) for both uniform superimposed dead load and self-weight. For the gradually thickened lower part of the dome ($\phi_e \le \phi \le \phi_s$), expressions (3.216) still give the membrane stress resultants due to uniform superimposed dead load over the entire surface of the dome, but for self-weight loading, the membrane stress resultants have to be calculated from expressions (3.235). Net stresses in the dome are obtained on the basis of relations (6.70), using the values for shell actions and shell thickness applicable to the location in question.

8.3.3.5 Conclusions of a parametric study

Let us consider the case of the laterally unrestrained edge, which is the support type for which local shell thickening is most likely to be required. In climates of large diurnal temperature variation, it may sometimes be desirable to leave the edges of a dome essentially unrestrained in order to prevent the occurrence of severe temperature-induced stresses.

Figure 8.26 Self-weight stresses for a dome with a = 30m and ϕ_s = 52°: (a) meridional stresses; (b) hoop stresses. Each set of plots compares three cases of shell thickness: (i) t = 100mm throughout; (ii) t = 150mm throughout; (iii) t gradually varied from 100mm to 150mm in the edge zone ($\Delta\phi$ = 9.72°). The edge of the dome is pinned and free to translate on the horizontal plane.

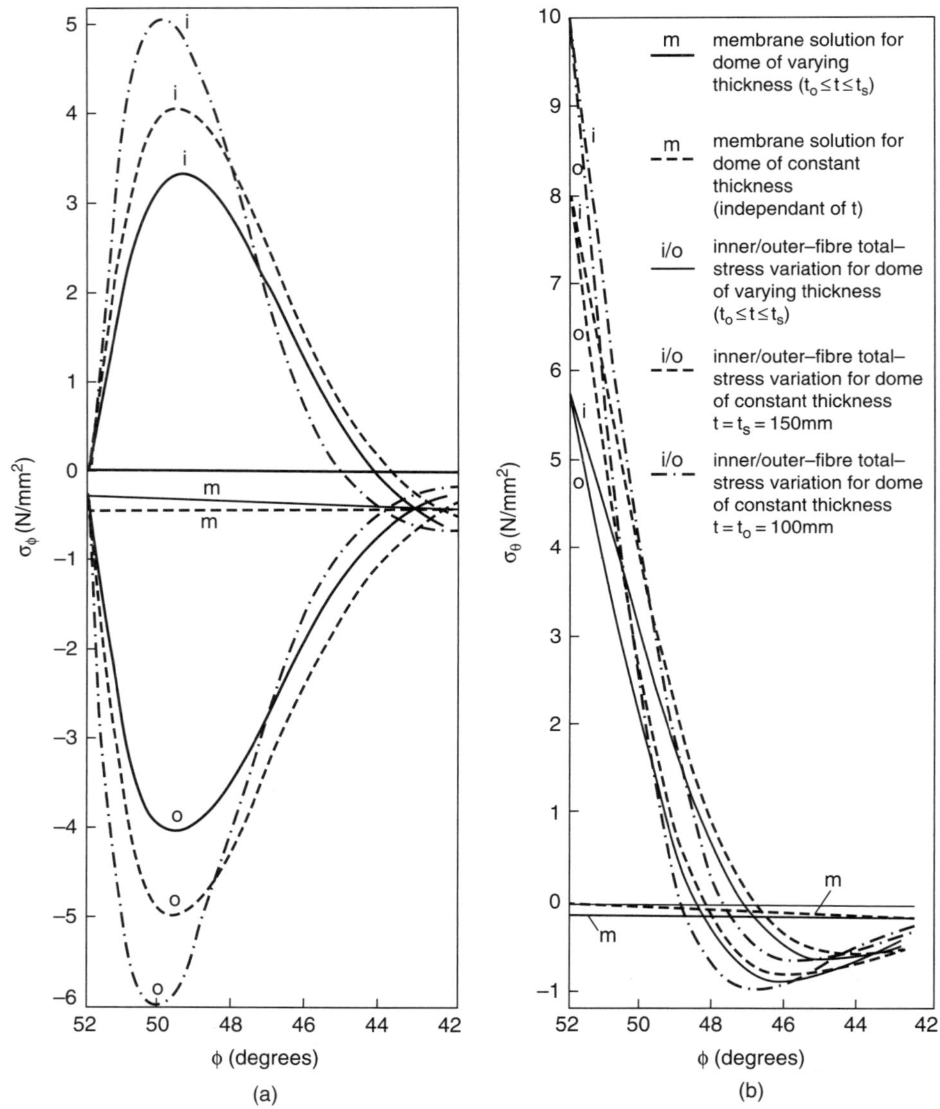

We assume the shell is gradually thickened from $t = t_o$ at the location $\phi_e \left(= \phi_s - \pi / \lambda_s\right)$ to $t = t_s$ at the support ($\phi = \phi_s$) in accordance with the law $t = \mu(\beta + \phi)^2$. A concrete dome with a = 30m and ϕ_s = 52° (the springing angle associated with the severest self-weight bending disturbance) is considered. The thickness parameters are t_o = 100 mm and t_s = 150 mm. Taking Poisson's ratio as $\nu = 0.15$, we obtain the values of other relevant parameters as $\phi_e \approx 42°$, $\alpha = 17.12$, $\beta = 0.018$ and $\mu = 175.2$. For the edge zone of the dome, Figure 8.26 shows stresses σ_ϕ and σ_θ due to self-weight. The findings of a parametric study are as follows.

For spherical domes in the intermediate-depth range $30° \leq \phi_s \leq 70°$ (typified by the case $\phi_s = 52°$), uniform thickening of the shell from t_o to t_s $(t_s > t_o)$ throughout the entire shell $(0 \leq \phi \leq \phi_s)$ is not very effective in reducing peak values of *self-weight* stresses occurring in the edge zones of laterally unrestrained domes.

On the other hand, gradually reducing the shell thickness in accordance with the proposed law is not only more desirable from the point of view of reducing material usage, but is also considerably more effective in reducing the peak magnitudes of the self-weight stresses in the edge zones of domes that are not provided with adequate lateral restraint. This is evident in Figure 8.26, when we compare the peaks for the three cases of the dome with: (*i*) constant thickness $t = 100$mm throughout; (*ii*) constant thickness $t = 150$mm throughout; (*iii*) constant thickness $t = 100$mm over most of the interior, gradually varied from $t = 100$mm to $t = 150$ mm in the edge zone.

As expected, the extent of the zone of significant bending disturbances for gradually varied local thickening $(t_o \leq t \leq t_s)$ lies in between that for the constant-thickness cases of (*i*) $t = t_o$ throughout and (*ii*) $t = t_s$ throughout. This is clearly evident in Figure 8.26.

Tapering the shell thickness from t_s at the support to a smaller interior value t_o over the recommended angle of π / λ_s results in considerable lowering of peak self-weight stresses in the vicinity of the dome edge, compared with the peak stresses that occur if uniform thickening from t_o to t_s is applied over the entire dome. Such tapering also results in a smaller penetration of bending disturbances into the interior of the shell, compared with a shell of uniform thickness t_s.

Stated slightly differently, applying an increased shell thickness over the entire dome, just to cater for the large stresses observed in the vicinity of the support, not only wastes material, but also achieves a smaller reduction in peak stresses and results in a more extensive bending zone, compared with a localised increase of shell thickness over the narrow edge zone only.

References

1. Tooth A.S. Storage vessels. In *Developments in Thin-Walled Structures* (Vol. 1). Applied Science Publishers, London, 1982, 1–52.
2. CBI. *Hortonsphere pressure vessels*. Chicago Bridge & Iron Co., Oak Brook (Illinois), 1980.
3. Zick L.P. and Carlson C.E. Stress analysis of a hortonsphere. *Welding Research Supplement*, 1949, May, 205s–214s.
4. Hicks R. Theoretical analysis of the stresses induced in a spherical pressure vessel due to constraining effect of a cylindrical skirt. *Proceedings of the Institution of Mechanical Engineers*, 1958, **172**, 707–718.
5. Bailey R.W. and Hicks R. Stress analysis problems associated with the design of reactor pressure vessels. In *Nuclear Reactor Containment Buildings and Pressure Vessels*. Butterworths, London, 1960, 134–149.
6. Houghton D.S. Discontinuity problems in shell structures. In *Nuclear Reactor Containment Buildings and Pressure Vessels*. Butterworths, London, 1960, 191–220.
7. Zingoni A. Bending-disturbance considerations in the axisymmetric response of non-shallow spherical shell structures. PhD Thesis, University of London, 1991.

8. Zingoni A. and Pavlovic M.N. Effect of support conditions in liquid-filled spherical vessels. Part 1: Limiting ring-beam stiffnesses. *Proceedings of the Institution of Civil Engineers*, 1991, **91** (2), 323–346.
9. Zingoni A. and Pavlovic M.N. Effect of support conditions in liquid-filled spherical vessels. Part 2: Inclined supports. *Proceedings of the Institution of Civil Engineers*, 1991, **91** (2), 347–363.
10. Zingoni A. and Pavlovic M.N. Discontinuity phenomena around the supports of stepwise-thickened spherical steel tanks. Part 1: Theoretical considerations and parametric results. *International Journal for Pressure Vessels and Piping*, 1993, **53**, 405–435.
11. Zingoni A. and Pavlovic M.N. Discontinuity phenomena around the supports of stepwise-thickened spherical steel tanks. Part 2: Numerical examples and design recommendations. *International Journal for Pressure Vessels and Piping*, 1993, **53**, 437–456.
12. Kitching R. and Olsen B. Pressure stresses at discrete supports on spherical shells. *Journal of Strain Analysis*, 1967, **2**, 298–306.
13. Flügge W. *Stresses in Shells*. Springer-Verlag, Berlin, 1973.
14. Nojiri, Y. Egg-shaped prestressed concrete digestor tanks, Yokohama (Japan). *IABSE Periodica*, 1989, **2**, 30–31.
15. Aeberhard, H.U. Egg-shaped digestor tanks in Matsumoto City (Japan). *IABSE Periodica*, 1989, **2**, 32–33.
16. Sutter, G. and Jager, P. Egg-shaped anaerobic digestors, Taiwan. *Structural Engineering International*, 1994, **4** (3), 160–163.
17. Jager, P. Two egg-shaped digestors, Australia. *Structural Engineering International*, 1997, **7** (4), 271–274.
18. Zingoni A. Stresses and deformations in egg-shaped sludge digesters: Membrane effects. *Engineering Structures*, 2001, **23** (11), 1365–1372.
19. Zingoni A. Stresses and deformations in egg-shaped sludge digesters: Discontinuity effects. *Engineering Structures*, 2001, **23** (11), 1373–1382.
20. Pavlovic M.N. and Zingoni A. Edge disturbances in spherical shells with varying geometric parameters and support types, with particular reference to thickening on peak stresses. *Proceedings of the Institution of Civil Engineers*, 1991, **91** (2), 495–516.
21. Pavlovic M.N. and Zingoni A. On the elastic design of concrete domes. *Structural Engineering Review*, 1994, **6**, 57–69.
22. Spotts M.F. Analysis of spherical shells of variable wall thickness. *Transactions of the ASME: Journal of Applied Mechanics*, 1939, **6**, A97–A102.
23. Rygol J. The calculation of thin shells of revolution of variable thickness. *Civil Engineering and Public Works Review*, 1960, **55**, 1015–1020.
24. Nielsen T.B. *Axisymmetric Bending of Shells*. Danish Technical Press, The Institution of Danish Civil Engineers, Copenhagen, 1962.
25. Basuli S. Bending of thin shells of revolution of variable thickness. *Archiwum Mechaniki Stosowanej*, 1963, **15**, 183–195.

Shell Structures in Civil and Mechanical Engineering
ISBN 978-0-7277-6028-9

http://dx.doi.org/10.1680/ssicame.60289.327

Chapter 9
Membrane theory and solutions for general cylinders

9.1. Definition

We define a general cylinder as a shell of translation for which the midsurface is generated by translating a straight line around a closed or an open smooth curve while keeping the straight line always parallel to its initial orientation and perpendicular to the plane of the curve. We will refer to the straight line as the *generator*. Thus, the midsurface of a general cylinder is a developable surface, and the shell geometry belongs to the class of shells of zero Gaussian curvature.

Depending on whether the curve is circular, elliptical, parabolic or some other shape, the cylinder is described as circular, elliptic, parabolic, and so forth. The circumferentially closed circular cylinder is a special case in the sense that it is also a shell of revolution. For circumferentially closed circular cylinders under axisymmetric loading conditions, the relevant membrane theory was developed in Chapters 2 and 3; where the loading is non-axisymmetric (e.g. in horizontal circular cylindrical tanks), the theory of general cylinders may be used.

9.2. Governing equations and general solution

To define points on a general cylinder, let us adopt a 2-dimensional coordinate system (ϕ, x) as shown in Figure 9.1(a). The coordinate x, measured along the direction of translation of the general cylinder, defines the position of any given cross-section of the general cylinder relative to some datum cross-section. The coordinate ϕ, as seen in a cross-section of the general cylinder, is analogous to the ϕ coordinate in the meridional section of a shell of revolution: the tangent of ϕ gives the slope of the profile at any given point. The radius of curvature of the cross-section $x = \text{constant}$, at any point of the cross-section defined by the coordinate ϕ, will be denoted by r_1. The second principal radius of curvature (r_2) is, of course, infinite for a general cylinder.

The shaded area in Figure 9.1(a) shows an element of the general cylinder bounded by two adjacent cross-sections $x = \text{constant}$ and $(x + dx) = \text{constant}$, and two adjacent generators $\phi = \text{constant}$ and $(\phi + d\phi) = \text{constant}$ The same element is shown separately in Figure 9.1(b), where it is now acted upon by external surface-loading components p_ϕ (in the tangential ϕ direction), p_x (in the x direction) and p_r (in the direction normal to the shell midsurface), as well as in-plane stress resultants N_ϕ, N_x, $N_{\phi x}$ and $N_{x\phi}$ acting over the edges of the element. The shear stress resultants $N_{\phi x}$ and $N_{x\phi}$ are, of course, equal to each other.

Let us consider the equilibrium of all the forces acting on the element. Summing up forces in the ϕ direction (tangent to the surface of the element at its centre), and equating the resultant force to zero, we have

$$\left(N_\phi + \frac{\partial N_\phi}{\partial \phi} d\phi\right) dx - N_\phi\, dx + \left(N_{x\phi} + \frac{\partial N_{x\phi}}{\partial x} dx\right) r_1\, d\phi - N_{x\phi}\, r_1\, d\phi + p_\phi\, r_1\, d\phi dx = 0$$

from which we can write

$$\frac{\partial N_\phi}{\partial \phi} + r_1 \frac{\partial N_{x\phi}}{\partial x} = -r_1 p_\phi \tag{9.1}$$

Resolving forces in the x direction, we have

$$\left(N_x + \frac{\partial N_x}{\partial x} dx\right) r_1 \, d\phi - N_x r_1 \, d\phi + \left(N_{\phi x} + \frac{\partial N_{\phi x}}{\partial \phi} d\phi\right) dx - N_{\phi x} \, dx + p_x r_1 \, d\phi dx = 0$$

which reduces to the equation

$$\frac{\partial N_x}{\partial x} + \frac{1}{r_1} \frac{\partial N_{\phi x}}{\partial \phi} = -p_x \tag{9.2}$$

Figure 9.1 General cylinder: (a) coordinate system $\{\phi, x\}$ showing an element bound by the curves $\{x = \text{constant}; x + dx = \text{constant}\}$ and the straight lines $\{\phi = \text{constant}; \phi + d\phi = \text{constant}\}$; (b) in-plane stress resultants and external load components on the element.

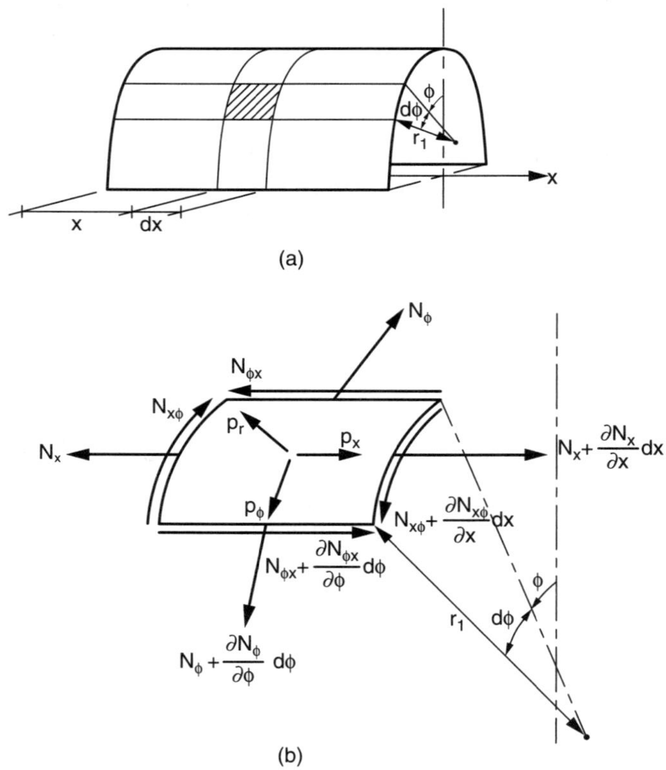

Finally, considering equilibrium in the direction normal to the shell midsurface, we obtain

$$\left(N_\phi + \frac{\partial N_\phi}{\partial \phi} d\phi\right) dx \sin\frac{d\phi}{2} + N_\phi \, dx \sin\frac{d\phi}{2} - p_r \, r_1 \, d\phi dx = 0$$

Applying the small-angle approximation $\sin(d\phi/2) \approx (d\phi/2)$, and ignoring small terms, we arrive at the relation

$$N_\phi = r_1 p_r \tag{9.3}$$

Equations (9.1)–(9.3) are the three equations of equilibrium for a general cylindrical shell under external surface loading and interior membrane action. The circumferential stress resultant N_ϕ follows immediately from the third equation of equilibrium. While this is very convenient from the point of view of solving the equations of equilibrium, the fact that we cannot freely prescribe arbitrary values of N_ϕ at the longitudinal boundaries $\phi = constant$ of a circumferentially open shell (such as the one depicted in Figure 9.1(a)) is the first indication that the membrane solution will generally be inadequate in the vicinity of longitudinal edges.

Knowing the expression for N_ϕ, we integrate equation (9.1) with respect to x to obtain $N_{x\phi}$:

$$N_{x\phi} = -\int \left(p_\phi + \frac{1}{r_1}\frac{\partial N_\phi}{\partial \phi}\right) dx + f_1(\phi) \tag{9.4}$$

With the expression for $N_{x\phi} \left(= N_{\phi x}\right)$ now known from (9.4), we integrate equation (9.2) with respect to x to obtain N_x:

$$N_x = -\int \left(p_x + \frac{1}{r_1}\frac{\partial N_{x\phi}}{\partial \phi}\right) dx + f_2(\phi) \tag{9.5}$$

The parameters $f_1(\phi)$ and $f_2(\phi)$, which are the constants of integration with respect to x, are functions of ϕ only. They may be obtained by specifying $N_{x\phi}$ and N_x at an edge $x = constant$. In shells for which the variation of $N_{x\phi}$ and N_x with respect to ϕ is different at the two curved ends, we require four constants of integration with respect to x, say arbitrary functions $f_1(\phi)$, $f_2(\phi)$, $f_3(\phi)$ and $f_4(\phi)$, to be able to satisfy the four boundary conditions for $N_{x\phi}$ and N_x at the edges $x = x_1$ (constant) and $x = x_2$ (constant). However, our general solutions (equations (9.4) and (9.5)) have only one constant of integration each. This means that, where the boundary conditions for $\left\{N_{x\phi}, N_x\right\}$ at one end are different to those at the other end, we have to choose which are the more important boundary conditions to satisfy, and ignore the others. This is yet another inherent shortcoming of the membrane solution.

In many important practical problems, we have $p_x = 0$, with p_ϕ and p_r being both independent of x. From relation (9.3), N_ϕ would also be independent of x, permitting the integral in equation (9.4) to be evaluated to give

$$\begin{aligned} N_{x\phi} &= -\left(p_\phi + \frac{1}{r_1}\frac{dN_\phi}{d\phi}\right) x + f_1(\phi) \\ &= -x\, g(\phi) + f_1(\phi) \end{aligned} \tag{9.6}$$

where

$$g(\phi) = p_\phi + \frac{1}{r_1}\frac{dN_\phi}{d\phi} \tag{9.7}$$

Using this result in equation (9.5) with p_x set to zero, and performing the integration, we obtain

$$N_x = \frac{1}{r_1}\left(\frac{x^2}{2}\right)\frac{dg(\phi)}{d\phi} - \frac{1}{r_1}(x)\frac{df_1(\phi)}{d\phi} + f_2(\phi) \tag{9.8}$$

9.3. Boundary conditions

Let us first consider a cantilever cylindrical shell with the edge $x = 0$ free and the edge $x = l$ built-in (Figure 9.2a). At $x = 0$, we have the boundary conditions

$$\left(N_{x\phi}\right)_{x=0} = 0 \tag{9.9a}$$

$$\left(N_x\right)_{x=0} = 0 \tag{9.9b}$$

Applying these conditions to equations (9.6) and (9.8), we obtain $f_1(\phi) = 0$ and $f_2(\phi) = 0$, so that equations (9.6) and (9.8) reduce to

$$N_{x\phi} = -x\,g(\phi) \tag{9.10a}$$

$$N_x = \frac{x^2}{2r_1}\frac{dg(\phi)}{d\phi} \tag{9.10b}$$

Now the variations of transverse shear force $SF(x)$ and bending moment $BM(x)$ along a cantilever beam carrying a uniformly distributed transverse loading of intensity q per unit length (refer to Figure 9.3(a)) are given by

$$SF(x) = -x\,q \tag{9.11a}$$

$$BM(x) = x^2\left(-\frac{q}{2}\right) \tag{9.11b}$$

Figure 9.2 Simple boundary conditions at the ends $\{x = 0;\ x = l\}$ of cylindrical shells: (a) cantilever cylinder; (b) cylinder with diaphragm-type end supports.

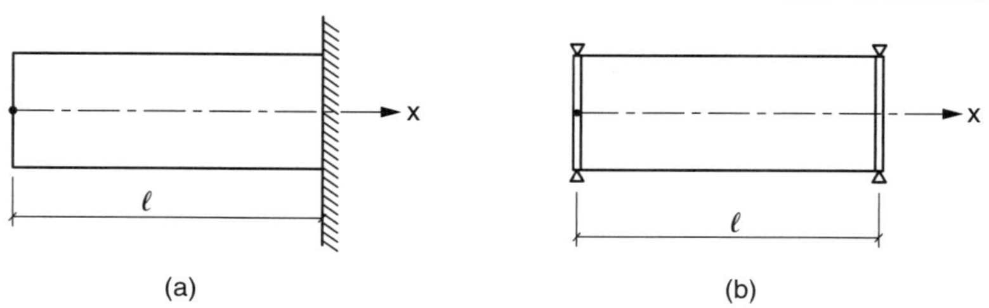

Thus, along a straight generator of the cantilever cylinder ($\phi = constant$), the variations of $N_{x\phi}$ and N_x, being proportional to x and x^2 respectively, resemble the variations of $SF(x)$ and $BM(x)$ along a cantilever beam carrying a uniformly distributed loading q (bearing in mind that the bending moment at a given cross-section of the beam is simply the resultant of axial-force couples acting across the section).

Let us next consider a horizontal cylindrical shell with diaphragm supports at both ends. In the present context, a diaphragm is a plane stiffening member such as a solid plate, a ring beam or an arch, which can resist forces acting in its plane, but has negligible resistance to forces transverse to its plane. Such supports may accept $N_{x\phi}$ forces from the cylindrical shell, but not the N_x forces. Figure 9.2(b) shows a horizontal cylindrical shell provided with simply supported diaphragms at its two ends. The relevant boundary conditions are

$$(N_x)_{x=0} = 0 \tag{9.12a}$$

$$(N_x)_{x=l} = 0 \tag{9.12b}$$

Applying these conditions to expression (9.8), we obtain the results

$$f_2(\phi) = 0 \tag{9.13a}$$

from the first condition, and

$$\frac{l^2}{2r_1}\frac{dg(\phi)}{d\phi} - \frac{l}{r_1}\frac{df_1(\phi)}{d\phi} = 0 \tag{9.13b}$$

from the second condition, when we also make use of result (9.13a).

Integrating expression (9.13b), we obtain

$$\frac{l}{2}g(\phi) = f_1(\phi) + k'$$

where k' is a constant of integration. If we let $k = -k'$, we may write

$$f_1(\phi) = \frac{l}{2}g(\phi) + k \tag{9.14}$$

Figure 9.3 Beam analogues of the cylindrical-shell cases of Figure 9.2: (a) cantilever beam; (b) simply supported beam. Both beams are carrying a uniformly distributed vertical downward load of intensity q per unit length.

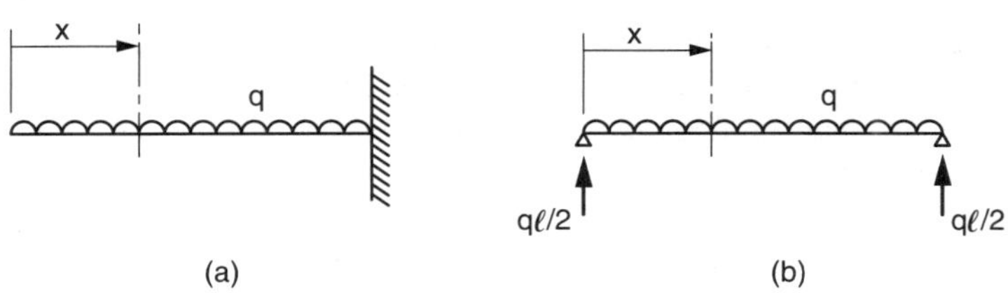

Substituting this result into expression (9.6), we obtain

$$N_{x\phi} = \left(\frac{l}{2} - x\right) g(\phi) + k \tag{9.15}$$

When $x = 0$, we have

$$\left(N_{x\phi}\right)_{x=0} = \frac{l}{2}\, g(\phi) + k \tag{9.16a}$$

When $x = l$, we have

$$\left(N_{x\phi}\right)_{x=l} = -\frac{l}{2}\, g(\phi) + k \tag{9.16b}$$

These results must be applicable to all possible x-wise distributions of loading, symmetric and non-symmetric alike. Let us consider the cases when the loading distribution is symmetric about $x = l/2$, for which symmetry requires that the value of $N_{x\phi}$ at $x = 0$ must be equal in magnitude but of opposite sign to the value of $N_{x\phi}$ at $x = l$. This condition gives us a way of deducing what the value of k must be. From expressions (9.16), k must be equal to zero for this condition to hold. Equations (9.14) and (9.15), with $k = 0$, reduce to

$$f_1(\phi) = \frac{l}{2}\, g(\phi) \tag{9.17}$$

$$N_{x\phi} = \left(\frac{l}{2} - x\right) g(\phi) \tag{9.18a}$$

Substituting results (9.13a) and (9.17) into expression (9.8), we obtain

$$N_x = -\frac{1}{2r_1}\, x(l - x)\, \frac{dg(\phi)}{d\phi} \tag{9.18b}$$

Consider a simply supported beam under a uniformly distributed transverse loading q per unit length, as shown in Figure 9.3(b). The lengthwise distributions of transverse shear force and bending moment are given by

$$SF(x) = \left(\frac{l}{2} - x\right) q \tag{9.19a}$$

$$BM(x) = \frac{q}{2}\, x(l - x) \tag{9.19b}$$

We conclude that, as in the case of the cantilever cylinder, the lengthwise variations of $N_{x\phi}$ and N_x in a simply supported cylindrical shell provided with diaphragm-type end conditions are of the same form as those for shear force and bending moment along a simply supported beam subjected to a uniformly distributed loading q over its entire span.

In the problems that follow, we will prescribe simply supported diaphragm-type conditions at both ends of the shell, and obtain some useful solutions for particular applications.

9.4. Solutions for horizontal circumferentially closed cylindrical vessels filled with liquid

Horizontal circumferentially closed cylindrical shells find widespread application as liquid-storage tanks and pressure vessels. Let us consider the general situation of hydrostatic liquid pressure γh (where γ is the weight per unit volume of the liquid and h is the depth of liquid at a given point) co-existing with a constant surcharge pressure p. The loading components are

$$p_x = 0; \;\; p_\phi = 0; \;\; p_r = p + \gamma h \tag{9.20}$$

Vessels of this type usually have a circular or an elliptical cross-section, as shown in Figure 9.4, where the tank is assumed to be completely filled with liquid. The relevant geometrical parameters of these two types of cross-section are as shown.

9.4.1 Circular cylinders

For the liquid-filled horizontal cylinder of circular cross-section (Figure 9.4b), we may express the depth h in terms of the coordinate ϕ, so that

$$p_r = p + \gamma a(1 - \cos\phi) \tag{9.21}$$

Using this expression for p_r in equation (9.3), with $r_1 = a$ for a circular cylinder, we obtain

$$N_\phi = \{p + \gamma a(1 - \cos\phi)\}\, a \tag{9.22}$$

From equation (9.7), with $p_\phi = 0$, we have

$$g(\phi) = \frac{1}{a}\frac{dN_\phi}{d\phi} = \gamma a \sin\phi \tag{9.23}$$

Figure 9.4 Simply supported horizontal cylindrical tank filled with liquid: (a) longitudinal view; (b) circular cross-section; (c) elliptic cross-section.

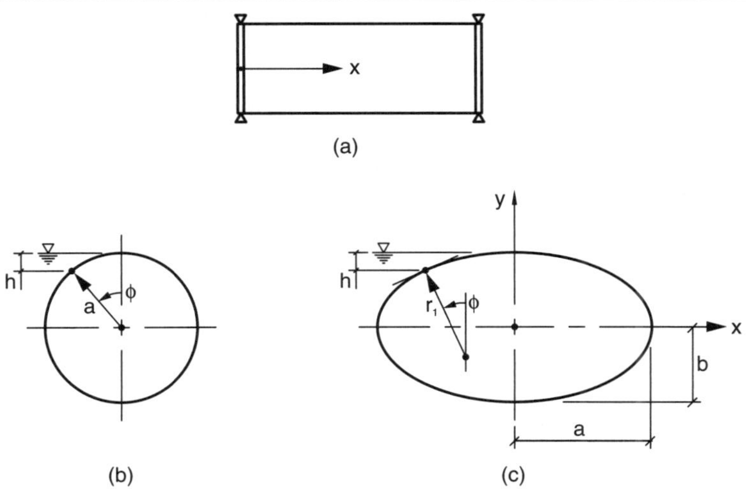

Expressions (9.18) therefore become

$$N_{x\phi} = \frac{\gamma a}{2}(l - 2x)\sin\phi \tag{9.24a}$$

$$N_x = -\frac{\gamma}{2}x(l - x)\cos\phi \tag{9.24b}$$

Since $h = a(1 - \cos\phi)$, we may replace $\cos\phi$ in the expression for N_x by $(a - h)/a$ to obtain

$$N_x = -\frac{\gamma}{2}x(l - x)\left(\frac{a - h}{a}\right) = K\left(\frac{a - h}{a}\right) \tag{9.25}$$

Equation (9.25) shows that, for a given cross-section ($x = constant$) of the cylinder, the distribution of N_x across the depth of the liquid-filled cross-section varies linearly from K at the top ($h = 0$) to $-K$ at the bottom ($h = 2a$). This variation resembles the distribution of bending stresses across the depth of a beam.

Clearly, the solution for N_x is not valid if the cylinder has end closures (flat or curved), since the resultant of the pressure forces over the surface of the closure is transmitted to the cylinder walls. Thus, the boundary conditions $N_x = 0$ at $\{x = 0; x = l\}$ are no longer correct. The boundary conditions of vanishing N_x at the ends of the cylinder would only be valid if we do not have end closures, and the cylinder ends are allowed to move freely in the axial direction. An example of such a situation is a pipeline with stiffening rings at regular intervals, where a long interval between two adjacent stiffeners constitutes a cylinder, and each cylinder is separated from adjacent cylinders by short connecting portions carrying expansion joints.

Liquid-filled horizontal circular cylindrical tanks have end closures, and require to be treated differently. Consider the case of hemispherical end closures. We may analyse each end closure as a hemispherical shell with a horizontal axis of revolution, and subjected to hydrostatic liquid pressure that varies linearly with the vertical distance coordinate, the line of support of the hemisphere being in the vertical plane of its junction to the cylinder. The ensuing non-axisymmetric distribution of the N_ϕ stress resultants around the circular edge of the hemisphere then gives the required N_x distribution around the edge of the circular cylinder, to be used as a boundary condition for determining N_x throughout the cylinder. End closures in the form of semi-ellipsoids of revolution may be treated in a similar manner.

In the case of flat plate closures, and as a convenient approximation, we may assume the depthwise distribution of edge reactions transferred to the cylinder to be of the same form as the variation with depth of the pressure acting on the surface of the plate. Thus, at a point on the circular edge of the plate defined by the angular coordinate ϕ (which is the same coordinate as that defining the straight generators of the cylinder), we may write the force transferred from the plate to the cylinder in the form

$$N_x = C_1 + C_2\cos\phi \tag{9.26}$$

where C_1 and C_2 are constants. This variation is of the same form as the pressure variation in equation (9.21).

Figure 9.5 shows a narrow surface strip of the plate (hatched area) at a depth defined by the coordinate ϕ. The strip of plate is attached at its ends to the cylinder via the dark-shaded interface areas. For the strip, and making use of the relationship in equation (9.26), we may write the elemental axial force (in the x direction of the cylinder) that is transferred through the dark-shaded contact areas as

$$dF = 2N_x\, a\, d\phi = 2(C_1 + C_2 \cos\phi)\, a\, d\phi$$

Therefore, the total axial force transferred through the entire contact circumference of the plate and the cylinder is

$$F = \int_0^{\pi} 2(C_1 + C_2 \cos\phi)\, a\, d\phi = 2a\left[C_1\phi + C_2 \sin\phi\right]_0^{\pi} = 2\pi a C_1$$

The resultant horizontal force over the hatched surface area of the strip of the plate, due to the pressure acting over this surface, is given by

$$dP = \left[p + \gamma a(1-\cos\phi)\right](a\, d\phi \sin\phi)(2a\sin\phi) = 2a^2\left[p + \gamma a(1-\cos\phi)\right]\sin^2\phi\, d\phi$$

noting that the strip of plate is of depth $(a\, d\phi)\sin\phi$ and overall width $2a\sin\phi$, while the pressure over the surface of the strip is $\left[p + \gamma a(1-\cos\phi)\right]$. The total horizontal force on the entire surface of the plate, due to liquid and surcharge pressure, is therefore

$$P = \int_0^{\pi} 2a^2\left[p + \gamma a(1-\cos\phi)\right]\sin^2\phi\, d\phi$$

$$= 2a^2\left[(p+\gamma a)\left(\frac{\phi}{2} - \frac{1}{4}\sin 2\phi\right) - \frac{\gamma a}{3}\sin^3\phi\right]_0^{\pi}$$

$$= \pi a^2(p + \gamma a)$$

Figure 9.5 Transfer of pressure acting on the end closure to the walls of the cylinder in accordance with equation (9.26).

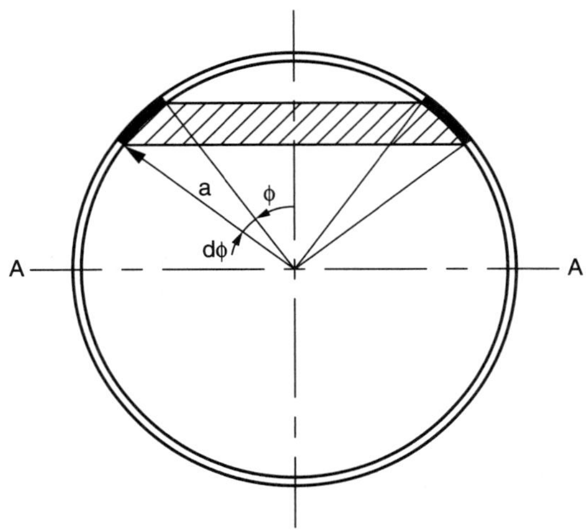

Force equilibrium of the plate closure requires that F and P are equal, which gives

$$C_1 = \frac{a}{2}(p + \gamma a) \tag{9.27a}$$

For the strip of plate, the moment dM_F of the elemental axial force dF about the horizontal plane AA (refer to Figure 9.5) is given by

$$dM_F = (2N_x\, a\, d\phi)\, a\cos\phi = 2(C_1 + C_2\cos\phi)\, a\, d\phi\, a\cos\phi$$

Therefore, the resultant moment M_F of all the dF elemental forces about the horizontal plane AA is given by

$$M_F = \int_0^\pi 2a^2(C_1 + C_2\cos\phi)\cos\phi\, d\phi = 2a^2\left[C_1\sin\phi + C_2\left(\frac{\phi}{2} + \frac{1}{4}\sin 2\phi\right)\right]_0^\pi = \pi a^2 C_2$$

The moment dM_P of the elemental surface force dP about the plane AA is given by

$$dM_P = \left\{2a^2\left[p + \gamma a(1 - \cos\phi)\right]\sin^2\phi\, d\phi\right\} a\cos\phi$$

Therefore, the resultant moment M_P due to the pressure acting over the entire surface of the end plate is given by

$$\begin{aligned} M_P &= \int_0^\pi 2a^3\left[p + \gamma a(1 - \cos\phi)\right](\sin^2\phi)\cos\phi\, d\phi \\ &= 2a^3\left[(p + \gamma a)\left(\frac{\sin^3\phi}{3}\right) - \gamma a\left(\frac{\phi}{8} - \frac{1}{32}\sin 4\phi\right)\right]_0^\pi \\ &= -\frac{\pi}{4}\gamma a^4 \end{aligned}$$

Moment equilibrium of the plate closure requires that M_F and M_P be equal to each other, which gives the result

$$C_2 = -\frac{\gamma a^2}{4} \tag{9.27b}$$

Substituting into expressions (9.26) the values of the constants C_1 and C_2 as given by equations (9.27), we may finally write down the distribution of N_x around the circular ends of the cylinder (which are defined by the coordinates $x = 0$ and $x = l$) as

$$N_x = \frac{a}{2}(p + \gamma a) - \frac{\gamma a^2}{4}\cos\phi \tag{9.28}$$

Applying this boundary condition to expression (9.8), we have

$$(N_x)_{x=0} = f_2(\phi) = \frac{a}{2}(p + \gamma a) - \frac{\gamma a^2}{4}\cos\phi \tag{9.29a}$$

$$(N_x)_{x=l} = \frac{l^2}{2a}\frac{dg(\phi)}{d\phi} - \frac{l}{a}\frac{df_1(\phi)}{d\phi} + f_2(\phi) = \frac{a}{2}(p+\gamma a) - \frac{\gamma a^2}{4}\cos\phi \tag{9.29b}$$

From equations (9.29), we deduce that

$$\frac{l^2}{2a}\frac{dg(\phi)}{d\phi} - \frac{l}{a}\frac{df_1(\phi)}{d\phi} = 0 \tag{9.30}$$

Integrating this expression, we obtain

$$\frac{l}{2}g(\phi) = f_1(\phi) + k'$$

where k' is a constant. If we let $k = -k'$, we may write

$$f_1(\phi) = \frac{l}{2}g(\phi) + k \tag{9.31}$$

Substituting this result into expression (9.6), we may write the shear stress resultant as

$$N_{x\phi} = \left(\frac{l}{2} - x\right)g(\phi) + k \tag{9.32}$$

Noting that expressions (9.31) and (9.32) are identical to expressions (9.14) and (9.15) respectively, we deduce, by the same argument as given earlier, that k must be zero, so that $f_1(\phi)$ and $N_{x\phi}$ reduce to the same values as given by expressions (9.17) and (9.18a). Thus, for a circumferentially closed horizontal circular cylinder filled with liquid, N_ϕ and $N_{x\phi}$ are given by the same expressions, for the case of the cylinder with flat end closures and the case of the cylinder without end closures. On the other hand, and as expected, N_x will be different for these two cases. Substituting into expression (9.8) the results for $f_1(\phi)$ and $f_2(\phi)$ as given by equations (9.17) and (9.29a) respectively, we obtain

$$N_x = \left(\frac{x^2}{2a} - \frac{x}{a}\frac{l}{2}\right)\frac{dg(\phi)}{d\phi} + \frac{a}{2}(p+\gamma a) - \frac{\gamma a^2}{4}\cos\phi$$

which, upon substituting the result for $g(\phi)$ as given by equation (9.23), finally becomes

$$N_x = \frac{a}{2}(p+\gamma a) - \frac{\gamma}{4}\left\{2x(l-x) + a^2\right\}\cos\phi \tag{9.33}$$

9.4.2 Elliptic cylinders

For the elliptic cross-section shown in Figure 9.4(c), we note that $h = b - y$. Using equation (3.24) to eliminate y, we may write the expression for p_r (see equations (9.20)) as

$$p_r = p + \gamma(b-y)$$

$$= (p+\gamma b) - \gamma\frac{b^2\cos\phi}{\left(a^2\sin^2\phi + b^2\cos^2\phi\right)^{1/2}} \tag{9.34}$$

Using this expression for p_r, and the expression for r_1 as derived in Chapter 3 (see equation (3.27)), we obtain the circumferential stress resultant from equation (9.3):

$$N_\phi = \frac{(p+\gamma b)\, a^2 b^2}{\left(a^2 \sin^2\phi + b^2 \cos^2\phi\right)^{3/2}} - \frac{\gamma\, a^2 b^4 \cos\phi}{\left(a^2 \sin^2\phi + b^2 \cos^2\phi\right)^2} \tag{9.35}$$

As before, let us first assume the boundary conditions $N_x = 0$ at $\{x = 0;\ x = l\}$, as in the case of a length of pipeline provided with expansion joints on either side. To obtain the stress resultants $N_{x\phi}$ and N_x, we first evaluate the function $g(\phi)$. From equation (9.7), this is obtained as follows:

$$g(\phi) = \frac{1}{r_1}\frac{dN_\phi}{d\phi} = -3(p+\gamma b)\left(a^2 - b^2\right)\left\{\frac{\cos\phi \sin\phi}{a^2 \sin^2\phi + b^2 \cos^2\phi}\right\} + \gamma b^2 \left\{\frac{a^2 + 3\left(a^2 - b^2\right)\cos^2\phi}{\left(a^2 \sin^2\phi + b^2 \cos^2\phi\right)^{3/2}}\right\}\sin\phi \tag{9.36}$$

Substituting this expression for $g(\phi)$ into equations (9.18), we obtain the results

$$N_{x\phi} = \left(\frac{l-2x}{2}\right)\left[-3(p+\gamma b)\left(a^2 - b^2\right)\left\{\frac{\cos\phi \sin\phi}{a^2 \sin^2\phi + b^2 \cos^2\phi}\right\} + \gamma b^2 \left\{\frac{a^2 + 3\left(a^2 - b^2\right)\cos^2\phi}{\left(a^2 \sin^2\phi + b^2 \cos^2\phi\right)^{3/2}}\right\}\sin\phi\right] \tag{9.37a}$$

$$N_x = \left(\frac{x(l-x)}{2a^2}\right)\left[-3(p+\gamma b)\left(\frac{a^2 - b^2}{b^2}\right)\left\{\frac{a^2 \sin^2\phi - b^2 \cos^2\phi}{\left(a^2 \sin^2\phi + b^2 \cos^2\phi\right)^{1/2}}\right\} + \gamma \left\{\frac{8a^4 \sin^2\phi - a^2 b^2\left(4 + 5\sin^2\phi\right) + 3b^4 \cos^2\phi}{a^2 \sin^2\phi + b^2 \cos^2\phi}\right\}\cos\phi\right] \tag{9.37b}$$

Tanks of elliptic section will have end closures in the form of flat elliptical plates, or doubly curved triaxial ellipsoids. (A triaxial ellipsoid is a quadratic surface having the equation $\left(x^2/a^2\right) + \left(y^2/b^2\right) + \left(z^2/c^2\right) = 1$ in the Cartesian x, y, z space, where the semi-axes a, b, c of the ellipsoid are generally all different.) For such tanks, the stress resultants N_ϕ and $N_{x\phi}$ in the elliptic cylinder remain the same as for the case without end closures, but the determination of N_x in the walls of the cylinder must now take into account the boundary condition that the circumferential distribution of N_x at the ends of the cylinder ($x = 0$; $x = l$) must match the distribution of horizontal forces around the edge of the elliptical plate or ellipsoidal shell cap.

In the case of an elliptic plate closure, the problem of finding the N_x distribution at the ends of the cylinder is identical to finding the distribution of vertical reactions around the circumference of a horizontal elliptic plate simply supported continuously along its edge, and subjected to a transverse pressure over its surface that varies linearly in the direction parallel to one semi-axis but is constant in the direction parallel to the other semi-axis of the elliptic plate. This is not a trivial plate problem.

In the case of a closure in the form of a triaxial semi-ellipsoid, we may find the distribution of axial force transmitted to the elliptic cylinder by directly tackling the problem of the hydrostatically loaded triaxial semi-ellipsoid (which is not so easy to solve), or via the method of affine transformation,[1] where the problem of the triaxial semi-ellipsoid is transformed to an equivalently loaded hemisphere, which may be a little easier to solve.

9.5. Solutions for horizontal troughs filled with liquid

Liquid-filled troughs in the form of general cylinders are encountered as open tanks, aquariums, suspended pools, elevated open channels, and the like. Three types of cross-sections for general cylindrical troughs (semi-circular, semi-elliptic and parabolic) are illustrated in Figure 9.6. If the trough has flat or curved end closures, then N_x in the cylindrical walls of the trough may be determined, as before, from the boundary condition that the variation of N_x along an end profile ($x = 0$ or $x = l$) of the cylinder must match the distribution of axial reactions around the edge of the closure, which arise as a result of the outward hydrostatic pressure acting over the surface of the closure.

The exact distribution of these interface reactions is difficult to obtain analytically, regardless of whether the closure is a plate or a shell, owing to the complex nature of the boundary conditions of the closure: the curved part of the boundary is supported, by virtue of its attachment to the cylinder wall, while the horizontal edge is a free boundary. The non-uniform distribution of pressure over the surface of the closure further complicates the problem.

We will therefore limit our considerations to the simpler boundary conditions $N_x = 0$ at $\{x = 0;\, x = l\}$. These boundary conditions apply when the cylinder is not provided with end closures (as in the case of a length of elevated open channel with expansion joints on either side), or where the cylinder has end closures, but the net outward push on the end closures (due to the liquid pressure) is not transferred to the cylindrical walls. The latter conditions may easily be realised by providing rigid external supports (props) to the end closures.

Even in the more typical cases where end closures are directly attached to the cylinder (permitting transfer of end loads to the cylinder walls), adopting the boundary conditions $N_x = 0$ at $\{x = 0;\, x = l\}$ will still give the correct results for N_ϕ and $N_{x\phi}$. It is only in the computation of N_x that we need to take into account the transfer of end loads from the closure into the cylinder walls. It is easy to see why this conclusion holds for N_ϕ: from equation (9.3), N_ϕ is clearly independent of end conditions at $\{x = 0;\, x = l\}$. In the case of $N_{x\phi}$, the conclusion holds if the boundary conditions at $x = 0$ are identical to those at $x = l$, which is usually the case in practice.

Figure 9.6 Simply supported horizontal cylindrical trough filled with liquid: (a) longitudinal view; (b) semi-circular cross-section; (c) semi-elliptic cross-section; (d) parabolic cross-section.

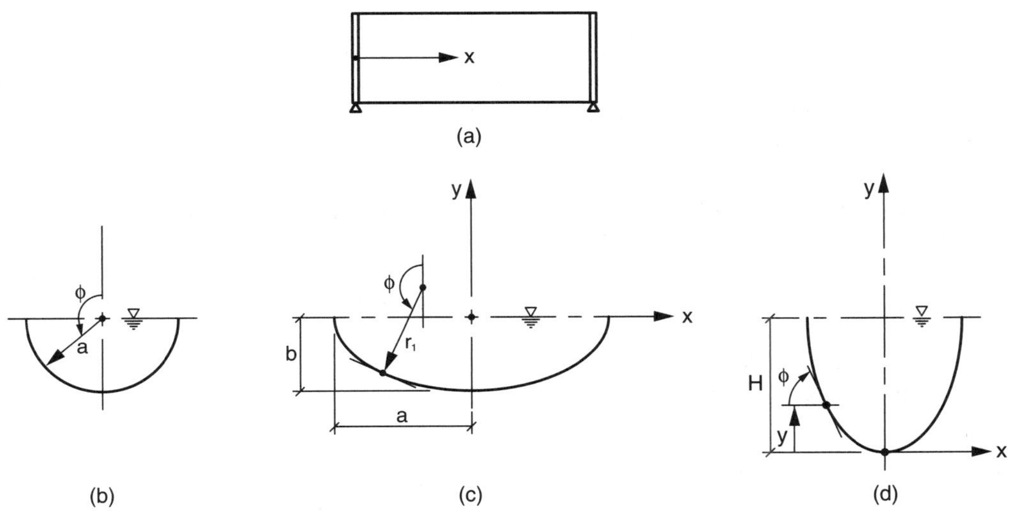

To see how the above conclusion holds for $N_{x\phi}$, we note, from expression (9.8), that if the circumferential variation of N_x at $x = 0$ is identical to that at $x = l$, then result (9.14) holds. For symmetrical support conditions, $N_{x\phi}$ at $x = 0$ is of equal magnitude but of opposite sign to $N_{x\phi}$ at $x = l$, which leads to relations (9.17) and (9.18a). From expression (9.18a), $N_{x\phi}$ will be independent of end conditions if the function $g(\phi)$ is independent of end conditions. Now, and by reference to equation (9.7), $g(\phi)$ is equal to $(1/r_1)(dN_\phi/d\phi)$ for all loadings for which $p_\phi = 0$, and is therefore independent of end conditions, since we have already shown that N_ϕ is independent of end conditions. It follows, therefore, that $N_{x\phi}$ is independent of end conditions.

9.5.1 Semi-circular trough

With reference to Figure 9.6(b), we have $r_1 = a$ and

$$p_r = -\gamma a \cos\phi \quad (90^\circ \le \phi \le 180^\circ) \tag{9.38}$$

From expression (9.3), we obtain

$$N_\phi = -\gamma a^2 \cos\phi \tag{9.39}$$

The parameter $g(\phi)$ follows from equation (9.7) with p_ϕ taken as zero:

$$g(\phi) = \frac{1}{a}\frac{dN_\phi}{d\phi} = \gamma a \sin\phi \tag{9.40}$$

Comparing expressions (9.23) and (9.40), we observe that $g(\phi)$ is the same for circumferentially closed circular-cylindrical liquid-filled horizontal vessels as for circumferentially open vessels (i.e. troughs). From relations (9.18), it follows that the $N_{x\phi}$ and N_x expressions are also the same – that is,

$$N_{x\phi} = \frac{\gamma a}{2}(l - 2x)\sin\phi \tag{9.41a}$$

$$N_x = -\frac{\gamma}{2}x(l - x)\cos\phi \tag{9.41b}$$

At the horizontal free edges of the trough (i.e. the horizontal top edges defined by the generators $\phi = \pm 90^\circ$), we expect the shear stress resultant $N_{x\phi}$ to be zero. However, expression (9.41a) shows that $N_{x\phi}$ does not vanish at these edges, implying that we require edge beams along the longitudinal boundaries of the trough to take the $N_{x\phi}$ from the shell.

Following a longitudinal edge of the cylinder, $N_{x\phi}$ varies linearly from $\gamma al/2$ at $x = 0$, through zero at $x = l/2$, to $-\gamma al/2$ at $x = l$, as represented by the arrows of varying length in Figure 9.7. The shaded shell element with arrows along its edges reminds us of the positive directions for $N_{x\phi}$. The adjacent edge beam will experience an equal but opposite distribution of axial force, which induces compression in the beam.

The magnitude of the compressive force C at any given cross-section of the beam is given by the integral of the $N_{x\phi}$ forces on one side of the section – that is,

$$C = \int_0^x \frac{\gamma a}{2}(l - 2x)\,dx = \frac{\gamma a}{2}x(l - x) \tag{9.42}$$

Figure 9.7 $N_{x\phi}$ actions in shell edge and adjacent edge beam for a semi-circular horizontal trough. The edge beam is in compression.

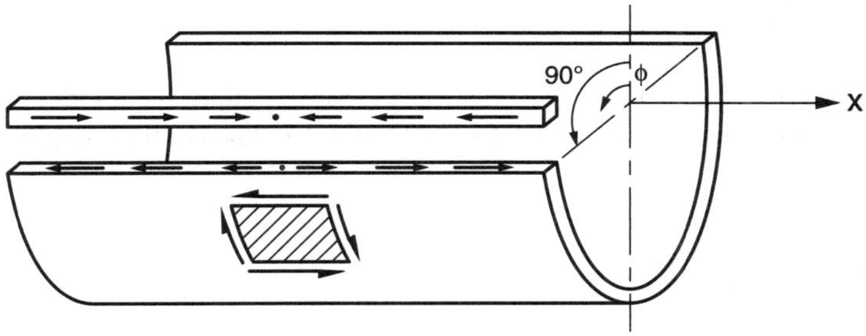

Clearly this compressive force is zero at $x = 0$ and $x = l$, implying that the ends of the beam can be free. The compressive force attains a maximum of $\gamma a l^2 / 8$ at $x = l/2$.

Some bending will occur in the vicinity of the longitudinal edges of the shell, due to incompatibilities of deformations between the shell edge and the longitudinal beam.

9.5.2 Semi-elliptic trough

For the semi-elliptic section shown in Figure 9.6(c), we have

$$r_1 = \frac{a^2 b^2}{\left(a^2 \sin^2 \phi + b^2 \cos^2 \phi\right)^{3/2}} \tag{9.43}$$

$$p_r = -\gamma y = -\frac{\gamma b^2 \cos \phi}{\left(a^2 \sin^2 \phi + b^2 \cos^2 \phi\right)^{1/2}} \quad \left(90° \le \phi \le 180°\right) \tag{9.44}$$

Using these values in equation (9.3), we obtain

$$N_\phi = -\frac{\gamma a^2 b^4 \cos \phi}{\left(a^2 \sin^2 \phi + b^2 \cos^2 \phi\right)^2} \tag{9.45}$$

Also,

$$g(\phi) = \frac{1}{r_1} \frac{dN_\phi}{d\phi} = \gamma b^2 \left\{ \frac{a^2 + 3\left(a^2 - b^2\right)\cos^2 \phi}{\left(a^2 \sin^2 \phi + b^2 \cos^2 \phi\right)^{3/2}} \right\} \sin \phi \tag{9.46}$$

and from relations (9.18),

$$N_{x\phi} = \gamma b^2 \left(\frac{l - 2x}{2}\right) \left\{ \frac{a^2 + 3\left(a^2 - b^2\right)\cos^2 \phi}{\left(a^2 \sin^2 \phi + b^2 \cos^2 \phi\right)^{3/2}} \right\} \sin \phi \tag{9.47a}$$

$$N_x = \frac{\gamma}{2a^2} x\left(l - x\right) \left\{ \frac{8a^4 \sin^2 \phi - a^2 b^2 \left(4 + 5\sin^2 \phi\right) + 3b^4 \cos^2 \phi}{a^2 \sin^2 \phi + b^2 \cos^2 \phi} \right\} \cos \phi \tag{9.47b}$$

When $\phi = 90°$, we have

$$\left(N_{x\phi}\right)_{\phi=90°} = \frac{\gamma b^2}{2a}(l-2x) \neq 0$$

Along the longitudinal edges $\phi = \pm 90°$, $N_{x\phi}$ varies in the same pattern as for the circular trough, from $\gamma b^2 l / 2a$ at $x = 0$, through zero at $x = l/2$, to $-\gamma b^2 l / 2a$ at $x = l$. The compressive force C induced in the edge beams is given by

$$C = \int_0^x \frac{\gamma b^2}{2a}(l-2x)\,dx = \frac{\gamma b^2}{2a} x(l-x) \tag{9.48}$$

This is zero at $x = \{0, l\}$ and a maximum of $\gamma b^2 l^2 / (8a)$ at $x = l/2$.

9.5.3 Parabolic trough

Consider a parabolic trough whose cross-section is given by the equation $y = kx^2$, as shown in Figure 9.6(d). For this profile, we have

$$\frac{dy}{dx} = 2kx$$

By definition, the first derivative of y with respect to x is equal to $\tan\phi$. We may therefore write

$$x = \frac{\tan\phi}{2k}$$

so that

$$y = kx^2 = \frac{\tan^2\phi}{4k}$$

Differentiating the expression for dy/dx with respect to x, we obtain

$$\frac{d^2y}{dx^2} = 2k$$

The radius of curvature at a given point follows as

$$r_1 = \frac{\left[1 + \left(\frac{dy}{dx}\right)^2\right]^{3/2}}{\frac{d^2y}{dx^2}} = \frac{\left[1 + 4k^2x^2\right]^{3/2}}{2k} = \frac{1}{2k\cos^3\phi} \tag{9.49}$$

If H is the overall depth of the trough, the hydrostatic pressure at the coordinate ϕ is given by

$$p_r = \gamma(H-y) = \gamma\left(H - \frac{\tan^2\phi}{4k}\right) = \frac{\gamma}{4k}\left(4kH - \tan^2\phi\right) \tag{9.50}$$

From equation (9.3), the circumferential stress resultant becomes

$$N_\phi = \frac{\gamma}{8k^2}\left(\frac{4kH - \tan^2\phi}{\cos^3\phi}\right) \tag{9.51}$$

The value of the coordinate ϕ at the top of the trough (i.e. at $y = H$) may be obtained by setting the expression for p_r equal to zero. This gives

$$(\phi)_{y=H} = \tan^{-1} 2\sqrt{kH}$$

For the parabolic trough, the domain of ϕ is therefore $0 \le \phi \le \tan^{-1} 2\sqrt{kH}$.

Note that N_ϕ always has the value zero at the longitudinal edges of the trough, regardless of the shape of the cross-section, since it is directly proportional to p_r (equation 9.3), and the latter is zero at the top of the trough (assumed to coincide with the surface of the liquid). However, and as will be seen shortly, the shear stress resultant $N_{x\phi}$ is not zero along the longitudinal edges of the trough, which means a free edge will not suffice.

Substituting the above expressions for r_1 and N_ϕ into equation (9.7), we obtain

$$g(\phi) = \frac{1}{r_1}\frac{dN_\phi}{d\phi} = \frac{\gamma}{4k}\left\{-\frac{2}{\cos^2\phi} + 3\left(4kH - \tan^2\phi\right)\right\}\tan\phi \tag{9.52}$$

Relations (9.18) then yield

$$N_{x\phi} = \frac{\gamma}{8k}(l - 2x)\left(\frac{12kH\cos^2\phi - 3\sin^2\phi - 2}{\cos^2\phi}\right)\tan\phi \tag{9.53a}$$

$$N_x = -\frac{\gamma}{4}x(l - x)\left(\frac{12kH\cos^2\phi - 13\sin^2\phi - 2}{\cos\phi}\right) \tag{9.53b}$$

At the longitudinal edges, $\tan\phi = 2\sqrt{kH}$ so that

$$\cos^2\phi = \frac{1}{1 + 4kH}$$

$$\sin^2\phi = \frac{4kH}{1 + 4kH}$$

Substituting these values for $\tan\phi$, $\cos^2\phi$ and $\sin^2\phi$ into expression (9.53a) for $N_{x\phi}$, we obtain the value of $N_{x\phi}$ along the longitudinal edges as

$$\left(N_{x\phi}\right)_{y=H} = -\left[\frac{\gamma}{2k}(1 + 4kH)\sqrt{kH}\right](l - 2x) = -K(l - 2x)$$

This varies from $-Kl$ at $x = 0$, through zero at $x = l/2$, to $+Kl$ at $x = l$. Unlike in the cases of the semi-circular and semi-elliptic troughs where compression was induced in the longitudinal edge beams, tension is induced in the longitudinal edge beams of the parabolic trough. Denoting the tensile force by T, we obtain this by integrating the $N_{x\phi}$ distribution:

$$T = \int_0^x K(l - 2x)\, dx = Kx(l - x) \tag{9.54}$$

This varies from zero at the ends ($x = \{0, l\}$), to a maximum of $Kl^2/4$ in the middle ($x = l/2$).

9.6. Solutions for barrel roofs

For concrete barrel-shell roofs, the principal loading is self-weight and weight of cladding. We will represent the combined effects by a uniformly distributed vertical force q per unit area of the shell surface. The loading components are thus

$$p_r = -q \cos\phi \tag{9.55a}$$

$$p_\phi = q \sin\phi \tag{9.55b}$$

We assume diaphragm-type end conditions, which may easily be achieved in the form of thin vertical end walls, arches or plane trusses. All these structural elements have substantial stiffness in their vertical plane, but offer relatively small resistance to any thrust or pull in the direction transverse to this plane. Thus, for barrel-roof problems, the conditions that $N_x = 0$ at $\{x = 0;\, x = l\}$ are more readily achievable than in the case of liquid-retaining horizontal cylinders, where the N_x stress resultants in the cylinder may be subject to the influence of the lateral surface forces transmitted from the end closures.

9.6.1 Barrel roof of circular section

With $r_1 = a$ and p_r given by equation (9.55a), we obtain, from equation (9.3),

$$N_\phi = -qa \cos\phi \tag{9.56}$$

The parameter $g(\phi)$ is evaluated on the basis of equation (9.7), with p_ϕ given by equation (9.55b). Thus

$$g(\phi) = p_\phi + \frac{1}{a}\frac{dN_\phi}{d\phi} = q \sin\phi + \frac{1}{a} qa \sin\phi = 2q \sin\phi \tag{9.57}$$

Expressions (9.18) then yield

$$N_{x\phi} = q(l - 2x)\sin\phi \tag{9.58a}$$

$$N_x = -\frac{q}{a} x(l - x)\cos\phi \tag{9.58b}$$

At the longitudinal edges ($\phi = \phi_s$), N_ϕ vanishes only if $\phi_s = 90°$ (i.e. only if the roof is semi-circular). Unlike for the liquid-filled semi-circular trough where the magnitude of $N_{x\phi}$ in the shell depends on both the length l and the radius a of the cylinder, the magnitude of $N_{x\phi}$ in the circular barrel roof depends only on the length l of the cylinder, and is independent of the radius a. Along an edge where $\phi = \phi_s = 90°$, $N_{x\phi} = q(l - 2x)$. This varies linearly from ql at $x = 0$, through zero at $x = l/2$, to $-ql$ at $x = l$, a pattern that is illustrated in Figure 9.8.

Figure 9.8 $N_{x\phi}$ actions in shell edge and adjacent edge beam for a circular barrel roof. The edge beam is in tension.

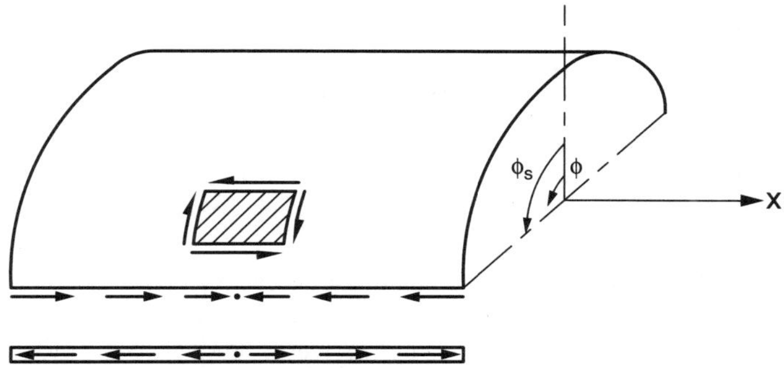

As before, the $N_{x\phi}$ forces are self-equilibrating, and may be resisted by an edge beam in tension. Taking the special case when the barrel shell roof is semi-circular, the tensile force T in the edge beam, at a section defined by the coordinate x, is given by

$$T = \int_0^x q(l-2x)\,dx = qx(l-x) \tag{9.59}$$

This tensile force is zero at the ends of the edge beam, attaining a maximum value of $ql^2/4$ at the centre of the beam.

As already pointed out, if $\phi_s < 90°$, N_ϕ does not vanish at the longitudinal edges. There is an outward thrust of $qa\cos\phi_s$ which has to be resisted by a system of internal roof ties, external buttress walls or inclined props. Longitudinal beams would still be required to absorb the $N_{x\phi}$ actions there, but as the cross-section of the barrel roof becomes shallower (i.e. for smaller values of ϕ_s), the values of the $N_{x\phi}$ actions at the longitudinal edges reduce (see equation (9.58a)), enabling smaller edge beams to be used. Irrespective of the value of ϕ_s, some bending of the shell will occur in the vicinity of the longitudinal edges, due to incompatibilities of deformations between the shell edge and the longitudinal beam.

9.6.2 Barrel roof of elliptic section

Substituting into equation (9.3) the expressions for r_1 and p_r given by equations (9.43) and (9.55a) respectively, we obtain the stress resultant N_ϕ as

$$N_\phi = -qa^2b^2\frac{\cos\phi}{\left(a^2\sin^2\phi + b^2\cos^2\phi\right)^{3/2}} \tag{9.60}$$

Therefore, from equation (9.7),

$$g(\phi) = p_\phi + \frac{1}{r_1}\frac{dN_\phi}{d\phi} = q\left[\frac{3a^2 - \left(a^2\sin^2\phi + b^2\cos^2\phi\right)}{a^2\sin^2\phi + b^2\cos^2\phi}\right]\sin\phi \tag{9.61}$$

so that, on the basis of expressions (9.18), the other two stress resultants become

$$N_{x\phi} = \frac{q}{2}(l-2x)\left[\frac{3a^2 - \left(a^2\sin^2\phi + b^2\cos^2\phi\right)}{a^2\sin^2\phi + b^2\cos^2\phi}\right]\sin\phi \tag{9.62a}$$

$$N_x = -\frac{q}{2a^2b^2}x(l-x)\left[\frac{\cos\phi}{\left(a^2\sin^2\phi + b^2\cos^2\phi\right)^{1/2}}\right]\times$$

$$\left[\left(a^2\sin^2\phi + b^2\cos^2\phi\right)\left\{3a^2 - \left(a^2\sin^2\phi + b^2\cos^2\phi\right)\right\} - 6a^2\left(a^2 - b^2\right)\sin^2\phi\right] \tag{9.62b}$$

If we have a semi-elliptic barrel roof ($\phi_s = 90°$), N_ϕ vanishes at the longitudinal edges $\phi = \phi_s$, as also observed for the semi-circular barrel roof. At the longitudinal edges of such a semi-elliptic barrel roof, $N_{x\phi} = q(l-2x)$. This is the same result as was obtained for the semi-circular barrel roof. We therefore conclude that, for semi-circular and semi-elliptic barrel roofs of the same plan dimensions, the tensile force T induced in the longitudinal edge beams will have the same variation $qx(l-x)$ along a beam, peaking at a value of $ql^2/4$ at the centre of the edge beam. In both cases, T is completely independent of the cross-sectional dimensions (a in the case of a circular barrel section; a and b in the case of an elliptic barrel section). Clearly if $\phi_s \neq 90°$, both the parameters a and b of the elliptic section will then have an influence on the magnitude of the tensile force T in the longitudinal beam, since $N_{x\phi}$ (which determines T) is generally dependent on both a and b (see equation (9.62a)).

9.6.3 Barrel roof of parabolic section

Owing to the shape of the parabola, barrel roofs of parabolic section are ideal for aircraft hangars, where a large headroom is desirable. With r_1 given by equation (9.49), we obtain

$$N_\phi = -\frac{q}{2k\cos^2\phi} \tag{9.63}$$

Thus,

$$g(\phi) = p_\phi + \frac{1}{r_1}\frac{dN_\phi}{d\phi} = -q\sin\phi \tag{9.64}$$

Use of this result for $g(\phi)$ in equation (9.18) leads to

$$N_{x\phi} = -\frac{q}{2}(l-2x)\sin\phi \tag{9.65a}$$

$$N_x = qkx(l-x)\cos^4\phi \tag{9.65b}$$

In a parabolic barrel roof, like in the circular and elliptic barrels, N_ϕ is compressive throughout. However, unlike in the semi-circular and semi-elliptic barrels, where N_ϕ is a maximum (in compression) at the crown (being $-qa$ for the circular barrel, and $-qa^2/b$ for the elliptic barrel) and decreases to zero at the longitudinal edge $\phi_s = 90°$, the depthwise distribution of N_ϕ in the parabolic barrel starts off with a magnitude $(q/2k)$ at the crown, and increases in compression with increasing ϕ, attaining its largest value at the longitudinal edges. Thus at the longitudinal edges, both N_ϕ and $N_{x\phi}$ are non-zero. The supports along the longitudinal edges of the parabolic barrel must therefore be designed to take two shell actions (N_ϕ and $N_{x\phi}$), instead of only one action ($N_{x\phi}$) as in the cases of the semi-circular and semi-elliptic barrel roofs.

9.7. Bending considerations for barrel roofs

In general, the membrane solution is not capable of satisfying all the boundary conditions at the edges. Where edge beams are provided, there will also be incompatibilities between the membrane deformations of the shell and those of the edge beam at their junction. All this implies that bending of the shell is inevitable in the boundary zones of barrel roofs. A rigorous derivation of the governing differential equations for the bending of barrel shells, as well as some rigorous and semi-rigorous solution approaches, may be seen in the work of Flügge.[1]

One of the simplest approximations for estimating edge effects in circular cylindrical barrel roofs is Schorer's theory (refer to the historical account in Section 1.4.5.2). This assumes that the bending moment M_x and transverse shear Q_x in the x direction, as well as the twisting moments $M_{x\phi}$ and $M_{\phi x}$, are all zero (i.e. $M_x = 0$; $Q_x = 0$; $M_{x\phi} = M_{\phi x} = 0$). Reference may be made to Figure 1.14 for all the forces and moments acting upon an element of a circular cylindrical shell. Furthermore, Schorer's theory assumes that the circumferential strain ε_ϕ and the shear strain $\gamma_{x\phi}$ are both small and may be neglected in relation to the longitudinal strain ε_x (i.e. $\varepsilon_\phi = 0$; $\gamma_{x\phi} = 0$).

Schorer's theory leads to an eighth-order partial differential equation of relatively simple form. The solution of this equation may readily be used to estimate the bending disturbances originating at the longitudinal edges of a barrel shell, by adopting a set of bending-related actions at the edges of the unloaded shell as the homogeneous solution, and the membrane solution for the barrel shell as the particular solution. This application of Schorer's theory to cylindrical shells is somewhat similar to the way Geckeler's approximation is used to estimate edge effects in spherical shells (refer to Chapter 6).

A full derivation of Schorer's theory and its application to the analysis of barrel roofs may readily be found in the literature (particularly in the older texts devoted to shell roofs alone or cylindrical shells alone), and will not be presented here. The approximation of Schorer gives reliable results only in the case of long shells – that is, those for which $l / a \geq \pi$, where l is the length of the shell in the x direction, and a is the radius of the cylinder.

In general, the bending analysis of cylindrical barrel roofs is generally very lengthy, even if use is made of a simplified analytical approach such as Schorer's theory. For this reason, various tables have been compiled for the convenience of designers. An authoritative and readily available source of such data is Manual 31 of the American Society of Civil Engineers.[2] Detailed considerations of the practical analysis and design of circular concrete barrel roofs, taking into account bending effects at the edges, may be seen in the work of Billington,[3] which also lists the tables of Manual 31 of the American Society of Civil Engineers.

References

1. Flügge W. *Stresses in Shells*. Springer-Verlag, Berlin, 1973.
2. ASCE. Design of cylindrical concrete shell roofs. *Manual of Engineering Practice No. 31*. American Society of Civil Engineers, New York, 1952.
3. Billington D.P. *Thin Shell Concrete Structures*. McGraw-Hill, New York, 1982.

Shell Structures in Civil and Mechanical Engineering
ISBN 978-0-7277-6028-9

http://dx.doi.org/10.1680/ssicame.60289.349

Chapter 10
Membrane theory and solutions for shells of arbitrary shape

10.1. Governing equations

We consider an element *ABCD* of a shell of arbitrary midsurface geometry, in a Cartesian *xyz* reference system, such that edges *AB* and *DC* lie in planes parallel to the *xz* plane, while edges *AD* and *BC* lie in planes parallel to the *yz* plane, edges *AB* and *AD* being assumed to actually coincide with the *xz* and *yz* planes, respectively (Figure 10.1). The projection of the element *ABCD* on the *xy* plane is a rectangle $\bar{A}\bar{B}\bar{C}\bar{D}$, with sides dx (in the x direction) and dy (in the y direction).

Let us denote by α the angle that edge *AB* makes with the x axis, and by β the angle that edge *AD* makes with the y axis; the angle between *AB* and *AD* will be denoted by γ. In terms of the projection dimensions, the edges *AB* and *AD* are therefore of length $dx / \cos\alpha$ and $dy / \cos\beta$, respectively; the area dA of the element *ABCD* is approximately twice the area of the triangle *ABD*, and is therefore equal to $(dx / \cos\alpha)(dy / \cos\beta)\sin\gamma$.

For the element *ABCD*, we adopt a system of curvelinear coordinates (x, y), the curves $x =$ constant and $y =$ constant on the element's midsurface being the coordinate curves. The stress resultants N_x, N_y, N_{xy} and N_{yx} shown in Figure 10.1 are forces per unit length of the coordinate curve of the element over which they act. Thus, the sum of the N_x forces over edge *AD* is $N_x \times$ (length of *AD*), that is

$$N_x \frac{dy}{\cos\beta}$$

with a component in the x direction of

$$N_x \frac{dy}{\cos\beta} \cos\alpha \tag{10.1}$$

since, as is clear from Figure 10.1, N_x is inclined at an angle α to the x direction.

Now, on the projection $\bar{A}\bar{B}\bar{C}\bar{D}$, we have a system of plan projections of the stress resultants, namely $\{\bar{N}_x, \bar{N}_y, \bar{N}_{xy}, \bar{N}_{yx}\}$, which are forces per unit length of the projection of the coordinate curves on the xy plane. Thus, the sum of the $\bar{N}_x$ forces over the projection $\bar{A}\bar{D}$ is $\bar{N}_x \times$ (length of $\bar{A}\bar{D}$), that is

$$\bar{N}_x \, dy \tag{10.2}$$

Equating terms (10.1) and (10.2), which refer to the same force, we obtain

$$\bar{N}_x = \frac{\cos\alpha}{\cos\beta} N_x \tag{10.3a}$$

Similarly, we have

$$\bar{N}_y = \frac{\cos\beta}{\cos\alpha} N_y \tag{10.3b}$$

If we also equate $\bar{N}_{xy}\, dy$ (net $\bar{N}_{xy}$ force over $\bar{A}\bar{D}$, which points in the negative y direction) to the component in the (negative) y direction of the resultant of the N_{xy} forces over AD, this component being

$$N_{xy} \frac{dy}{\cos\beta} \cos\beta = N_{xy}\, dy,$$

Figure 10.1 Element $ABCD$ of a shell of arbitrary shape and its projection $\bar{A}\bar{B}\bar{C}\bar{D}$ on the xy plane.

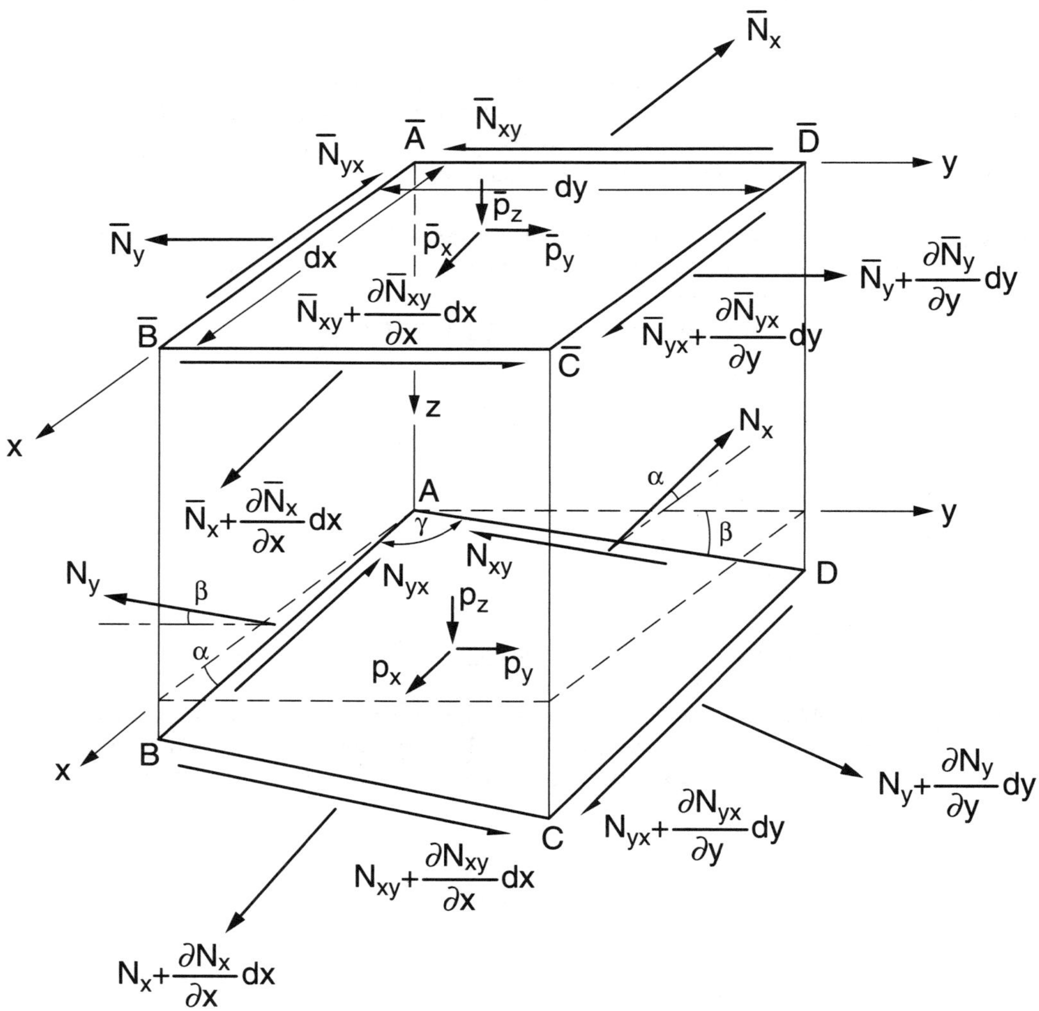

we obtain the relationship

$$\bar{N}_{xy} = N_{xy}$$

Similarly,

$$\bar{N}_{yx} = N_{yx}$$

Since $N_{xy} = N_{yx}$ (from moment equilibrium of the element $ABCD$ about the z axis), we may write

$$\bar{N}_{xy} = N_{xy} = N_{yx} = \bar{N}_{yx} \tag{10.3c}$$

The loading intensities per unit area of the shell element are p_x (in the x direction), p_y (in the y direction) and p_z (in the z direction). The loading intensities per unit area of the *projection* of the element on the xy plane are therefore

$$p_x \frac{dA}{dx\,dy}; \quad p_y \frac{dA}{dx\,dy}; \quad p_z \frac{dA}{dx\,dy}$$

in the x, y and z directions, respectively. By definition, these are the parameters $\bar{p}_x$, $\bar{p}_y$ and $\bar{p}_z$ associated with the projection $\bar{A}\bar{B}\bar{C}\bar{D}$ of the element. Replacing dA with $(dx/\cos\alpha)(dy/\cos\beta)\sin\gamma$, we may therefore write

$$\bar{p}_x = \left(\frac{\sin\gamma}{\cos\alpha\cos\beta}\right) p_x \tag{10.4a}$$

$$\bar{p}_y = \left(\frac{\sin\gamma}{\cos\alpha\cos\beta}\right) p_y \tag{10.4b}$$

$$\bar{p}_z = \left(\frac{\sin\gamma}{\cos\alpha\cos\beta}\right) p_z \tag{10.4c}$$

The equations of force equilibrium of the shell element in the x and y directions may be written down in terms of the variables of the projection element, by considering the force equilibrium of the latter. Summing up forces in the x and y directions in turn, we have

$$\left(\bar{N}_x + \frac{\partial \bar{N}_x}{\partial x}\,dx\right)dy - \bar{N}_x\,dy + \left(\bar{N}_{yx} + \frac{\partial \bar{N}_{yx}}{\partial y}\,dy\right)dx - \bar{N}_{yx}\,dx + \bar{p}_x\,dx\,dy = 0$$

$$\left(\bar{N}_y + \frac{\partial \bar{N}_y}{\partial y}\,dy\right)dx - \bar{N}_y\,dx + \left(\bar{N}_{xy} + \frac{\partial \bar{N}_{xy}}{\partial x}\,dx\right)dy - \bar{N}_{xy}\,dy + \bar{p}_y\,dx\,dy = 0$$

which reduce to

$$\frac{\partial \bar{N}_x}{\partial x} + \frac{\partial \bar{N}_{yx}}{\partial y} + \bar{p}_x = 0 \tag{10.5a}$$

$$\frac{\partial \bar{N}_y}{\partial y} + \frac{\partial \bar{N}_{xy}}{\partial x} + \bar{p}_y = 0 \tag{10.5b}$$

To obtain the equation of force equilibrium of the shell element in the z direction, we first write down the net resultants in the z direction of the N_x, N_y, N_{xy} and N_{yx} forces, by reference to the element itself rather than its projection. The component in the negative z direction of the resultant of the N_x forces over AD is

$$\begin{aligned} N_x \frac{dy}{\cos\beta} \sin\alpha &= \left(N_x \frac{\cos\alpha}{\cos\beta} \right) \frac{\sin\alpha}{\cos\alpha} \, dy \\ &= \bar{N}_x (\tan\alpha) \, dy \\ &= \left(\bar{N}_x \frac{\partial z}{\partial x} \right) dy \end{aligned} \tag{10.6}$$

making use of relation (10.3a) to eliminate N_x, and replacing $\tan\alpha$ with $(\partial z / \partial x)$, where $z = f(x, y)$ defines the shell midsurface. As we move along a coordinate curve y = constant (i.e. in the x direction), this force varies, becoming

$$\left(\bar{N}_x \frac{\partial z}{\partial x} \right) dy + \frac{\partial}{\partial x} \left\{ \left(\bar{N}_x \frac{\partial z}{\partial x} \right) dy \right\} dx \tag{10.7}$$

at the edge BC of the element, in the positive z direction. Thus, the net force in the positive z direction associated with the stress resultant N_x is the difference between (10.7) and the final term of (10.6), that is

$$\frac{\partial}{\partial x} \left(\bar{N}_x \frac{\partial z}{\partial x} \right) dx \, dy = \left(\frac{\partial \bar{N}_x}{\partial x} \frac{\partial z}{\partial x} + \bar{N}_x \frac{\partial^2 z}{\partial x^2} \right) dx \, dy \tag{10.8a}$$

Similarly, the net force in the positive z direction associated with the stress resultant N_y is

$$\left(\frac{\partial \bar{N}_y}{\partial y} \frac{\partial z}{\partial y} + \bar{N}_y \frac{\partial^2 z}{\partial y^2} \right) dx \, dy \tag{10.8b}$$

The component in the negative z direction of the resultant of the N_{xy} forces over AD is

$$\begin{aligned} N_{xy} \frac{dy}{\cos\beta} \sin\beta &= \bar{N}_{xy} (\tan\beta) \, dy \\ &= \left(\bar{N}_{xy} \frac{\partial z}{\partial y} \right) dy \end{aligned}$$

making use of equation (10.3c) to eliminate N_{xy}, and replacing $\tan\beta$ with $(\partial z / \partial y)$. Taking into account the variation of this force across the width of the element in the x direction, the net force in the positive z direction associated with the stress resultant N_{xy} is therefore

$$\frac{\partial}{\partial x}\left(\bar{N}_{xy}\frac{\partial z}{\partial y}\right)dx\,dy = \left(\frac{\partial \bar{N}_{xy}}{\partial x}\frac{\partial z}{\partial y} + \bar{N}_{xy}\frac{\partial^2 z}{\partial x\,\partial y}\right)dx\,dy \tag{10.8c}$$

Similarly, the net force in the positive z direction associated with N_{yx} is

$$\left(\frac{\partial \bar{N}_{yx}}{\partial y}\frac{\partial z}{\partial x} + \bar{N}_{yx}\frac{\partial^2 z}{\partial x\,\partial y}\right)dx\,dy$$

which, from relation (10.3c), may be written in terms of $\bar{N}_{xy}$ as

$$\left(\frac{\partial \bar{N}_{xy}}{\partial y}\frac{\partial z}{\partial x} + \bar{N}_{xy}\frac{\partial^2 z}{\partial x\,\partial y}\right)dx\,dy \tag{10.8d}$$

The net force in the positive z direction associated with the loading is, of course,

$$p_z\,dA = \bar{p}_z\,dx\,dy \tag{10.8e}$$

Summing up the five components of force in the positive z direction, represented by expressions (10.8a–e), and dividing throughout by $dx\,dy$, we obtain

$$\bar{N}_x\frac{\partial^2 z}{\partial x^2} + \bar{N}_y\frac{\partial^2 z}{\partial y^2} + 2\bar{N}_{xy}\frac{\partial^2 z}{\partial x\,\partial y} + \left(\frac{\partial \bar{N}_x}{\partial x} + \frac{\partial \bar{N}_{xy}}{\partial y}\right)\frac{\partial z}{\partial x} + \left(\frac{\partial \bar{N}_y}{\partial y} + \frac{\partial \bar{N}_{xy}}{\partial x}\right)\frac{\partial z}{\partial y} + \bar{p}_z = 0$$

From relations (10.5) the first and second terms in brackets are equal to $-\bar{p}_x$ and $-\bar{p}_y$ respectively. We may therefore write the above equation in the following final form:

$$\bar{N}_x\frac{\partial^2 z}{\partial x^2} + 2\bar{N}_{xy}\frac{\partial^2 z}{\partial x\,\partial y} + \bar{N}_y\frac{\partial^2 z}{\partial y^2} = \bar{p}_x\frac{\partial z}{\partial x} + \bar{p}_y\frac{\partial z}{\partial y} - \bar{p}_z \tag{10.9}$$

Equation (10.9), together with the pair (10.5), constitute the governing equations for the membrane theory of shells of arbitrary shape.

10.2. Solution approach based on a stress function

The form of equations (10.5), which resembles that of the equations expressing the equilibrium of a plane-stress system,[1] suggests a solution approach based on the introduction of a stress function. If we assume the existence of a stress function Φ, such that

$$\bar{N}_x = \frac{\partial^2 \Phi}{\partial y^2} - \int \bar{p}_x\,dx \tag{10.10a}$$

$$\bar{N}_y = \frac{\partial^2 \Phi}{\partial x^2} - \int \bar{p}_y\,dy \tag{10.10b}$$

$$\bar{N}_{xy} = -\frac{\partial^2 \Phi}{\partial x\, \partial y} \tag{10.10c}$$

then equations (10.5) are automatically satisfied, while equation (10.9) becomes

$$\frac{\partial^2 \Phi}{\partial x^2}\frac{\partial^2 z}{\partial y^2} - 2\frac{\partial^2 \Phi}{\partial x\, \partial y}\frac{\partial^2 z}{\partial x\, \partial y} + \frac{\partial^2 \Phi}{\partial y^2}\frac{\partial^2 z}{\partial x^2}$$

$$= -\bar{p}_z + \bar{p}_x\frac{\partial z}{\partial x} + \bar{p}_y\frac{\partial z}{\partial y} + \left(\int \bar{p}_x\, dx\right)\frac{\partial^2 z}{\partial x^2} + \left(\int \bar{p}_y\, dy\right)\frac{\partial^2 z}{\partial y^2} \tag{10.11}$$

10.3. Application to shell-roof problems

10.3.1 Hyperbolic paraboloid with edges parallel to straight-line generators

For a hyperbolic paraboloidal shell roof with edges parallel to the straight-line generators of the surface, the equation of the surface is

$$z = \frac{x}{a}\frac{y}{b}h \tag{10.12}$$

where the x and y coordinate directions are taken parallel to the straight-line generators as seen in plan, a and b are the plan dimensions in the x and y directions respectively, and h is the rise of the shell roof (Figure 10.2).

Let us assume, for simplicity, that the loading components p_x and p_y are zero, and the vertical load p_z (per unit area of the shell midsurface) is constant and practically equal to $\bar{p}_z$ (per unit area of the horizontal projection of the shell midsurface), the approximation $\bar{p}_z \approx p_z$ being valid for relatively shallow shells (i.e. when $h \ll a$, $h \ll b$). Many shell roofs encountered in practice conform very closely to these assumptions, since self-weight and uniform superimposed dead load are usually the dominant loading conditions, and the rise of the roof is usually small in comparison with the plan dimensions of the roof. We may therefore denote $\bar{p}_z$ as simply q (constant).

With z as given by equation (10.12), the first and third terms of the left-hand side of equation (10.11) vanish, so that, with all terms of the right-hand side of equation (10.11) also zero except the first, equation (10.11) reduces to

Figure 10.2 Hyperbolic paraboloid with edges parallel to straight generators.

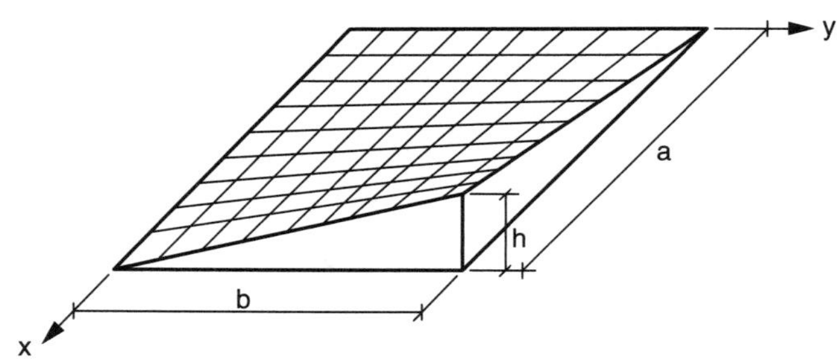

$$-2\frac{\partial^2\Phi}{\partial x\,\partial y}\frac{h}{ab}=-q$$

– that is,

$$\frac{\partial^2\Phi}{\partial x\,\partial y}=\frac{qab}{2h} \qquad (10.13)$$

Using this result in equation (10.10c), we immediately obtain

$$\bar{N}_{xy}=N_{xy}=-\frac{qab}{2h} \qquad (10.14)$$

Integrating equation (10.13) once with respect to x and once with respect to y, we obtain

$$\Phi=\frac{qab}{2h}xy+f(x)+g(y) \qquad (10.15)$$

where $f(x)$ and $g(y)$ are arbitrary functions of x and y respectively.

From equations (10.10a) and (10.10b), with Φ as given by expression (10.15) above, we obtain

$$\bar{N}_x=\frac{\partial^2\Phi}{\partial y^2}=\frac{d^2g(y)}{dy^2} \qquad (10.16a)$$

$$\bar{N}_y=\frac{\partial^2\Phi}{\partial x^2}=\frac{d^2f(x)}{dx^2} \qquad (10.16b)$$

From relations (10.3a) and (10.3b), and using the above results, we may write

$$N_x=\left(\frac{\cos\beta}{\cos\alpha}\right)\bar{N}_x=\left(\frac{\cos\beta}{\cos\alpha}\right)\frac{d^2g(y)}{dy^2} \qquad (10.17a)$$

$$N_y=\left(\frac{\cos\alpha}{\cos\beta}\right)\bar{N}_y=\left(\frac{\cos\alpha}{\cos\beta}\right)\frac{d^2f(x)}{dx^2} \qquad (10.17b)$$

Now, by reference to Figure 10.2, $\tan\alpha$ varies linearly from zero along $y=0$ to h/a along $y=b$. That is, the slope of the generator $y=$constant is given by

$$\tan\alpha=\frac{y}{b}\frac{h}{a}=\frac{yh}{ab} \qquad (10.18)$$

implying that

$$\cos\alpha=\frac{ab}{\sqrt{a^2b^2+y^2h^2}} \qquad (10.19a)$$

Similarly,

$$\cos\beta \;=\; \frac{ab}{\sqrt{a^2b^2 + x^2h^2}} \tag{10.19b}$$

Using these results for $\cos\alpha$ and $\cos\beta$ in expressions (10.17), we may finally write

$$N_x = \left(\frac{a^2b^2 + y^2h^2}{a^2b^2 + x^2h^2}\right)^{1/2} \frac{d^2g(y)}{dy^2} \tag{10.20a}$$

$$N_y = \left(\frac{a^2b^2 + x^2h^2}{a^2b^2 + y^2h^2}\right)^{1/2} \frac{d^2f(x)}{dx^2} \tag{10.20b}$$

Along a generator y = constant, $\bar{N}_x$ is constant; along a generator x = constant, $\bar{N}_y$ is constant (see expressions (10.16)). Thus, having prescribed the distribution of $\bar{N}_x$ across the edge $x = 0$ or $x = a$, the solution requires that $\bar{N}_x$ remains constant in moving across the shell in the x direction along a given generator y = constant; therefore, we may not be able to simultaneously satisfy the $\bar{N}_x$ boundary conditions at the opposite edges $x = 0$ and $x = a$ of the shell. The same applies to $\bar{N}_y$, whose boundary conditions at the opposite edges $y = 0$ and $y = b$ may be different, and therefore impossible to satisfy simultaneously.

From result (10.14), N_{xy} is constant throughout the shell. If the edges of the shell cannot resist lateral forces, then $\bar{N}_x = \bar{N}_y = 0$ at the edges, and therefore throughout, which, from relations (10.17), implies that N_x and N_y are zero throughout. Under such conditions, N_{xy} is the only membrane stress resultant in the shell, constant throughout the shell and therefore non-zero at the edges. The non-zero values of N_{xy} at the edges are resisted by edge beams (or side walls), which receive tensile or compressive actions in directions opposite to the N_{xy} forces in the adjacent shell edge.

In practical shell-roof construction, the basis hyperbolic paraboloidal unit of Figure 10.2 is often combined with other similar units to form a wide variety of architectural shapes, such as those shown in Figure 10.3. Along the junction of adjacent units, the effects of the N_{xy} forces from the adjacent panels may be such that they reinforce or counter each other, so that we may or may not require inter-panel beams, depending on the net effect of the superimposed N_{xy} forces.

10.3.2 Hyperbolic paraboloid with edges along parabolic generators

The 'saddle' roof shown in Figure 10.4 is formed by translating a 'hogging' parabola over a 'sagging' parabola fixed in a vertical plane, while keeping the plane of the moving parabola vertical and at right angles to the plane of the fixed parabola. In plan, the roof is a rectangle of dimensions $2a$ (in the x direction) and $2b$ (in the y direction), h_x being the rise of the central hogging parabola (in the xz plane) above its ends, and h_y the fall of the central sagging parabola (in the yz plane) below its ends. The origin of the xyz coordinate system is on the surface of the saddle, at its centre in plan. The locus of all points on the surface of the shell such that z =constant is a hyperbola, hence the name *hyperbolic paraboloid.*

We may write the equation of the surface as

$$z = \frac{x^2}{A} - \frac{y^2}{B} \tag{10.21}$$

where

$$A = \frac{a^2}{h_x}; \quad B = \frac{b^2}{h_y} \tag{10.22}$$

Again assuming $p_x = p_y = 0$ and $\overline{p}_z \approx p_z = q$ (constant), equation (10.11), with z as given by expression (10.21), becomes

$$-\frac{2}{B}\frac{\partial^2 \Phi}{\partial x^2} + \frac{2}{A}\frac{\partial^2 \Phi}{\partial y^2} = -q$$

Figure 10.3 Some combinations of the basic hyperbolic paraboloidal unit of Figure 10.2: (a) umbrella roof (outer perimeter is horizontal and lowermost); (b) gable-ended roof (meeting edges are horizontal and uppermost).

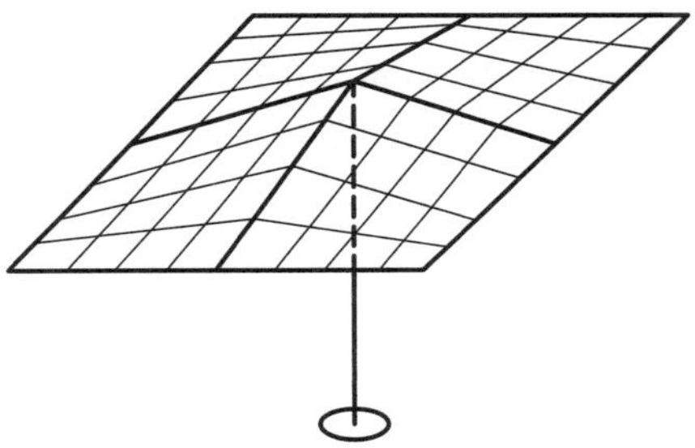

(a)

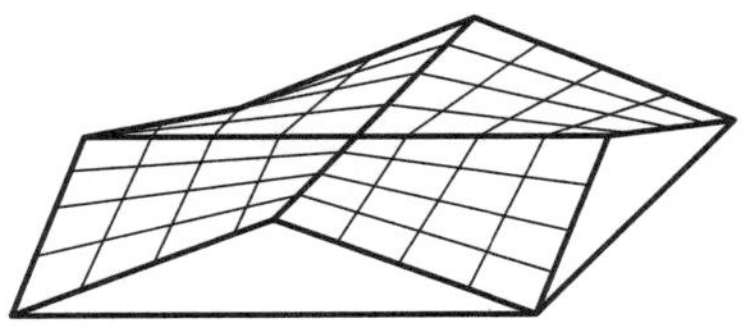

(b)

Figure 10.4 Hyperbolic paraboloid with edges along parabolic generators.

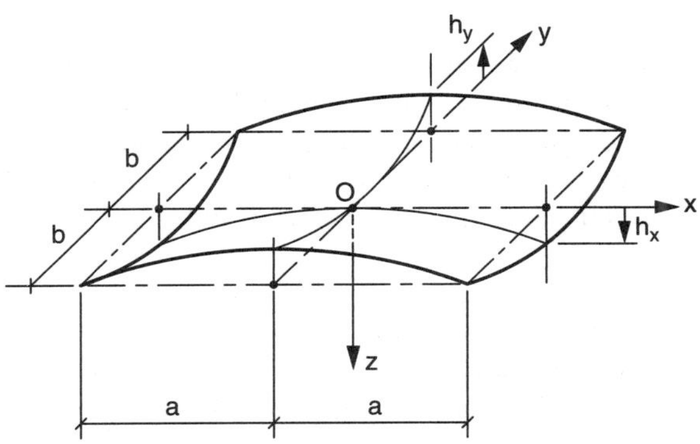

– that is,

$$\frac{2}{B}\frac{\partial^2\Phi}{\partial x^2} - \frac{2}{A}\frac{\partial^2\Phi}{\partial y^2} = q \tag{10.23}$$

Particular solutions of equation (10.23) may be taken as follows:

$$\Phi_1 = \frac{qBx^2}{4} \tag{10.24a}$$

$$\Phi_2 = -\frac{qAy^2}{4} \tag{10.24b}$$

$$\Phi_3 = \frac{q}{8}\left(Bx^2 - Ay^2\right) \tag{10.24c}$$

Taking the first solution, we obtain, from expressions (10.10),

$$\bar{N}_x = 0; \quad \bar{N}_y = \frac{qB}{2}; \quad N_{xy} = 0 \tag{10.25}$$

The second solution yields

$$\bar{N}_x = -\frac{qA}{2}; \quad \bar{N}_y = 0; \quad N_{xy} = 0 \tag{10.26}$$

while the third solution gives the results

$$\bar{N}_x = -\frac{qA}{4}; \quad \bar{N}_y = \frac{qB}{4}; \quad N_{xy} = 0 \tag{10.27}$$

In the first case (results (10.25)), a system of hanging parabolas x = constant (in planes parallel to the yz plane) carries all the loading in tension (N_y is positive). This load-carrying mechanism is possible if we can provide laterally braced arches along the hogging parabolic end profiles $y = \pm b$, to support the load-carrying hanging parabolas x = constant, at the same time ensuring that the sagging parabolic end profiles $x = \pm a$ are without lateral bracing. We would therefore normally adopt the first stress function if the edges of the hyperbolic paraboloid are approximately of this nature.

In the second case (results (10.26)), a system of parabolic arches y =constant (in planes parallel to the xz plane) carries all the loading in compression (N_x is negative). This mode is possible if the sagging parabolic end profiles $x = \pm a$ are restrained against lateral movement (i.e. rigid abutments are provided to support the load-carrying parabolic arches y = constant), with the hogging parabolic end profiles $y = \pm b$ left unsupported. Actual edge conditions approaching this idealisation would point to our adoption of the second stress function.

In the third case (results (10.27)), the load is carried by both a system of hanging parabolas in tension and a system of parabolic arches in compression. The choice of the third stress function would be valid if the edges $y = \pm b$ were of comparable rigidity with the edges $x = \pm a$. Clearly, provision of lateral restrain to all the four edges of the shell is desirable, since this allows the loading to be resisted by a two-way load-sharing mechanism, leading to greater efficiency in the use of shell material.

10.3.3 Elliptic paraboloid with edges along parabolic generators

The surface depicted in Figure 10.5 is formed by translating a hogging parabola over another hogging parabola fixed in a vertical plane, while keeping the plane of the moving parabola vertical and at right angles to the plane of the fixed parabola. The plan view of the roof is a rectangle of dimensions $2a$ (in the x direction) and $2b$ (in the y direction), h_x and h_y being the rises of the central parabolic profiles in the xz plane and the yz plane, respectively, relative to their respective ends. The origin of the xyz coordinate system is at the highest point on the surface of the shell, which in plan coincides with the centre of the rectangular area. Contours of equal height (i.e. z = constant) on the surface of the shell are ellipses, hence the name *elliptic paraboloid*.

The equation of the surface may be written as

$$z = \frac{x^2}{A} + \frac{y^2}{B} \tag{10.28}$$

where A and B remain as defined by equation (10.22). Equation (10.28) is similar to (10.21) for the hyperbolic paraboloid, except for the sign of the second term.

As before, we assume $p_x = p_y = 0$ and $\bar{p}_z \approx p_z = q$ (constant); using the definition of z given by equation (10.28) in the governing equation (10.11), we obtain

$$\frac{2}{B}\frac{\partial^2\Phi}{\partial x^2} + \frac{2}{A}\frac{\partial^2\Phi}{\partial y^2} = -q \tag{10.29}$$

Particular solutions of equation (10.29) may be taken as

$$\Phi_1 = -\frac{qBx^2}{4} \tag{10.30a}$$

Figure 10.5 Elliptic paraboloid with edges along parabolic generators.

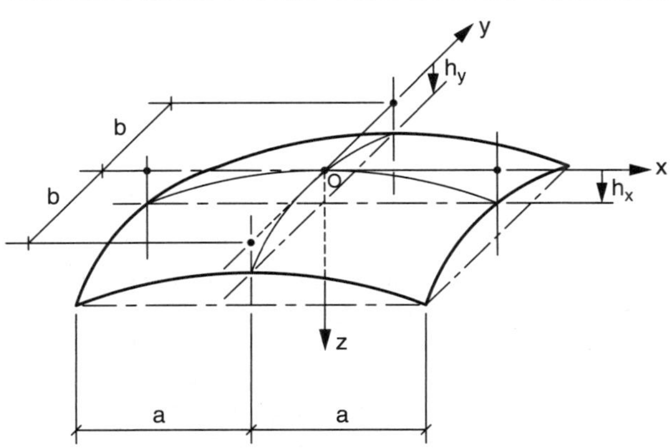

$$\Phi_2 = -\frac{qAy^2}{4} \tag{10.30b}$$

$$\Phi_3 = -\frac{q}{8}\left(Bx^2 + Ay^2\right) \tag{10.30c}$$

which, of course, are similar to those in (10.24) but for the sign of the terms in x^2. These three solutions yield

$$\bar{N}_x = 0; \ \ \bar{N}_y = -\frac{qB}{2}; \ \ N_{xy} = 0 \tag{10.31}$$

$$\bar{N}_x = -\frac{qA}{2}; \ \ \bar{N}_y = 0; \ \ N_{xy} = 0 \tag{10.32}$$

$$\bar{N}_x = -\frac{qA}{4}; \ \ \bar{N}_y = -\frac{qB}{4}; \ \ N_{xy} = 0 \tag{10.33}$$

respectively.

In the first case, the load is carried by parabolic arches x = constant, in compression, so that edges $y = \pm b$ have to be laterally restrained while edges $x = \pm a$ are unsupported. In the second case, the load is carried by parabolic arches y = constant, also in compression, so that the edges $x = \pm a$ have to be laterally braced, with edges $y = \pm b$ left unsupported. The third case represents a mechanism where the loading is carried by two orthogonal systems of arches (the x = constant arches, and the y = constant arches) in compression, so that all four edges of the shell must be capable of resisting lateral thrust.

In practice, the shell is usually supported by relatively thin vertical edge walls that can resist forces in the vertical plane, but have negligible resistance to lateral thrust from the roof. Alternatively, the shell may be supported only at the four corners, with edge beams in the form of vertical parabolic arches transferring shear forces from the shell to the supports. Such vertical arches also have high load-resisting capacity in their (vertical) planes, but negligible resistance to out-of-plane (i.e. lateral) forces, except at the corners. For such thin-wall or arch diaphragm-type supports, we therefore require that $N_x = 0$ (so that $\bar{N}_x = 0$) at $x = \pm a$, and $N_y = 0$ (so that $\bar{N}_y = 0$) at $y = \pm b$. From relations (10.10), with $\bar{p}_x = \bar{p}_y = 0$, we may therefore write

$$\left(\frac{\partial^2\Phi}{\partial y^2}\right)_{x=\pm a} = 0 \tag{10.34a}$$

$$\left(\frac{\partial^2\Phi}{\partial x^2}\right)_{y=\pm b} = 0 \tag{10.34b}$$

The first condition implies that Φ varies linearly with y (i.e. $\Phi = k_1 + k_2 y$, with k_1 and k_2 being constants) along the edges $x = \pm a$. Similarly, from the second condition, we deduce that Φ varies linearly with x (i.e. $\Phi = k_3 + k_4 x$, where k_3 and k_4 are constants) along the edges $y = \pm b$. Moreover, from the symmetry of the shell configuration, the applied loading, and the boundary conditions about both the x and y axes, the variations of Φ along the edges $x = \pm a$ and $y = \pm b$ must be even functions of y and x respectively, implying that $k_2 = 0$ and $k_4 = 0$. Thus, $\Phi = k_1$ (constant) along edges $x = \pm a$, and $\Phi = k_3$ (constant) along edges $y = \pm b$. Since all three stress resultants $\bar{N}_x$, $\bar{N}_y$ and $\bar{N}_{xy}$ are second derivatives of Φ (refer to expressions (10.10)), the actual values of the constants k_1 and k_3 do not affect the results, and hence we may set these to zero. We have therefore reduced boundary conditions (10.34) to the more convenient form

$$(\Phi)_{x=\pm a} = 0 \tag{10.35a}$$

$$(\Phi)_{y=\pm b} = 0 \tag{10.35b}$$

The particular solutions (10.30a) and (10.30b) were capable of satisfying the boundary conditions $N_x = 0$ at $x = \pm a$ (see results (10.31)), or $N_y = 0$ at $y = \pm b$ (see results (10.32)), but not both. Let us choose a particular solution that automatically satisfies the condition $N_y = 0$ at $y = \pm b$ (i.e. condition (10.34b), which is equivalent to condition (10.35b)). Such a particular solution is given by expression (10.30b), modified to vanish at the edges $y = \pm b$ by the incorporation of the constant term b^2:

$$\Phi_P = \frac{qA}{4}\left(b^2 - y^2\right) \tag{10.36}$$

Let us also assume a homogeneous solution of the form

$$\Phi_H = \sum_n^{\infty} H_n \cosh\left(\frac{n\pi}{2b}\right)kx \; \cos\left(\frac{n\pi}{2b}\right)y \tag{10.37}$$

where $n = 1, 2, 3, \ldots, \infty$. For the nth term, we have

$$\left(\Phi_H\right)_n = H_n \cosh\left(\frac{n\pi}{2b}\right)kx \; \cos\left(\frac{n\pi}{2b}\right)y \tag{10.38}$$

At $y = \pm b$, $\left(\Phi_H\right)_n$ vanishes if $\cos\left(n\pi/2\right) = 0$, that is, if $n = 1, 3, 5, \ldots$ Thus, if we take only the odd terms of n in the summation of the series for Φ_H, then Φ_H, like Φ_P, automatically satisfies boundary condition (10.35b).

For Φ_H to satisfy the homogeneous equation

$$\frac{2}{B}\frac{\partial^2\Phi}{\partial x^2} + \frac{2}{A}\frac{\partial^2\Phi}{\partial y^2} = 0 \tag{10.39}$$

we must have, for each nth term of the series,

$$\left\{H_n\left(\frac{n\pi}{2b}\right)^2 \cosh\left(\frac{n\pi}{2b}\right)kx \; \cos\left(\frac{n\pi}{2b}\right)y\right\}\left\{\frac{2}{B}k^2 - \frac{2}{A}\right\} = 0$$

– that is,

$$k = \left(\frac{B}{A}\right)^{1/2} = \frac{b}{a}\left(\frac{h_x}{h_y}\right)^{1/2} \tag{10.40}$$

Since we have seen that both the particular solution (10.36) and the assumed homogeneous solution (10.37) individually satisfy boundary condition (10.35b), it follows that the net solution

$$\Phi_T = \Phi_P + \Phi_H \tag{10.41}$$

also satisfies (10.35b). It now remains to choose the constant H_n in such a way that Φ_T also satisfies boundary condition (10.35a).

Before we can impose this boundary condition (10.35a), it is necessary to first express Φ_P in the same form as Φ_H; that is, as a Fourier-series expansion in the y direction, with terms corresponding to those of expression (10.38). We obtain

$$\Phi_P = \frac{qA}{4}\left(b^2 - y^2\right) = \frac{8qAb^2}{\pi^3} \sum_{n=1,3,5,...}^{\infty} (-1)^{(n-1)/2} \frac{1}{n^3} \cos\left(\frac{n\pi}{2b}\right) y \tag{10.42}$$

which, of course, satisfies condition (10.35b).

For the nth term of Φ_T, we have

$$\left(\Phi_T\right)_n = \left(\Phi_P\right)_n + \left(\Phi_H\right)_n = \frac{8qAb^2}{\pi^3 n^3} (-1)^{(n-1)/2} \cos\left(\frac{n\pi}{2b}\right) y \;+\; H_n \cosh\left(\frac{n\pi}{2b}\right) kx \;\cos\left(\frac{n\pi}{2b}\right) y \tag{10.43}$$

Imposing boundary condition (10.35a) for $\left(\Phi_T\right)_n$, we obtain

$$H_n \;=\; -\frac{8qAb^2}{\pi^3 n^3}\left[\frac{(-1)^{(n-1)/2}}{\cosh\left\{\left(\frac{n\pi}{2b}\right) ka\right\}}\right] \;=\; -\frac{8qa^2 b^2}{\pi^3 n^3 h_x}\left[\frac{(-1)^{(n-1)/2}}{\cosh\left\{\left(\frac{n\pi}{2}\right)\left(\frac{h_x}{h_y}\right)^{1/2}\right\}}\right] \tag{10.44}$$

The net solution may finally be written as follows:

$$\Phi_T \;=\; \Phi_P + \Phi_H \;=\; \frac{qa^2}{4h_x}\left(b^2 - y^2\right) - \frac{8qa^2 b^2}{\pi^3 h_x} \sum_{n=1,3,5,..}^{\infty} (-1)^{(n-1)/2} \frac{1}{n^3} \left[\frac{\cosh\left\{\left(\frac{n\pi x}{2a}\right)\left(\frac{h_x}{h_y}\right)^{1/2}\right\}}{\cosh\left\{\left(\frac{n\pi}{2}\right)\left(\frac{h_x}{h_y}\right)^{1/2}\right\}}\right] \cos\frac{n\pi y}{2b} \tag{10.45}$$

The results for $\bar{N}_x$, $\bar{N}_y$ and $\bar{N}_{xy}$ follow as

$$\bar{N}_x \;=\; \frac{\partial^2 \Phi_T}{\partial y^2} \;=\; \frac{qa^2}{h_x}\left[-\frac{1}{2} + \frac{2}{\pi} \sum_{n=1,3,5,..}^{\infty} (-1)^{(n-1)/2} \frac{1}{n} \left[\frac{\cosh\left\{\left(\frac{n\pi x}{2a}\right)\left(\frac{h_x}{h_y}\right)^{1/2}\right\}}{\cosh\left\{\left(\frac{n\pi}{2}\right)\left(\frac{h_x}{h_y}\right)^{1/2}\right\}}\right] \cos\frac{n\pi y}{2b}\right] \tag{10.46a}$$

$$\bar{N}_y = \frac{\partial^2 \Phi_T}{\partial x^2} = -\frac{2qb^2}{\pi h_y} \sum_{n=1,3,5,..}^{\infty} (-1)^{(n-1)/2} \frac{1}{n} \left[\frac{\cosh\left\{ \left(\frac{n\pi x}{2a}\right)\left(\frac{h_x}{h_y}\right)^{1/2} \right\}}{\cosh\left\{ \left(\frac{n\pi}{2}\right)\left(\frac{h_x}{h_y}\right)^{1/2} \right\}} \right] \cos\frac{n\pi y}{2b} \tag{10.46b}$$

$$\bar{N}_{xy} = -\frac{\partial^2 \Phi_T}{\partial x\, \partial y} = -\frac{2qab}{\pi\sqrt{h_x h_y}} \sum_{n=1,3,5,..}^{\infty} (-1)^{(n-1)/2} \frac{1}{n} \left[\frac{\sinh\left\{ \left(\frac{n\pi x}{2a}\right)\left(\frac{h_x}{h_y}\right)^{1/2} \right\}}{\cosh\left\{ \left(\frac{n\pi}{2}\right)\left(\frac{h_x}{h_y}\right)^{1/2} \right\}} \right] \sin\frac{n\pi y}{2b} \tag{10.46c}$$

Now, for the surface given by equation (10.28), we have

$$\tan\alpha = \frac{\partial z}{\partial x} = \frac{2x}{A} \tag{10.47a}$$

$$\tan\beta = \frac{\partial z}{\partial y} = \frac{2y}{B} \tag{10.47b}$$

and therefore

$$\cos\alpha = \frac{A}{\sqrt{A^2 + 4x^2}} \tag{10.48a}$$

$$\cos\beta = \frac{B}{\sqrt{B^2 + 4y^2}} \tag{10.48b}$$

so that the actual stress resultants in the shell are

$$N_x = \frac{\cos\beta}{\cos\alpha} \bar{N}_x = \frac{B}{A}\sqrt{\frac{A^2 + 4x^2}{B^2 + 4y^2}}\; \bar{N}_x \tag{10.49a}$$

$$N_y = \frac{\cos\alpha}{\cos\beta} \bar{N}_y = \frac{A}{B}\sqrt{\frac{B^2 + 4y^2}{A^2 + 4x^2}}\; \bar{N}_y \tag{10.49b}$$

$$N_{xy} = \bar{N}_{xy} \tag{10.49c}$$

The series for N_x and N_y are convergent; that for N_{xy} is also convergent, except towards the corners $(x, y) = (a, b)$, $(a, -b)$, $(-a, b)$, $(-a, -b)$. The vanishing of both N_x and N_y at the corners means that only N_{xy} remains available there to equilibrate the external loading q. However, the elliptic paraboloid, being a surface without twist (i.e. $\partial^2 z/(\partial x\, \partial y) = 0$), has shell elements whose opposite edges are exactly parallel to each other, so that the equal N_{xy} forces at opposite edges are exactly parallel and in opposite directions, and therefore without a resultant in the direction of the external loading. That is why the N_{xy} stress resultant tends to infinity as the corners are approached. In reality, N_{xy} is, of course, finite at the corners, hence considerable bending must occur around the corners in order to equilibrate the loading.

If we write the stress resultants as

$$N_x = -\frac{B}{A}\sqrt{\frac{A^2+4x^2}{B^2+4y^2}}\,\frac{qa^2}{h_x}\,C_1 \tag{10.50a}$$

$$N_y = -\frac{A}{B}\sqrt{\frac{B^2+4y^2}{A^2+4x^2}}\,\frac{qb^2}{h_y}\,C_2 \tag{10.50b}$$

$$N_{xy} = -\frac{qab}{\sqrt{h_x h_y}}\,C_3 \tag{10.50c}$$

then we may use Parme's tables[2] to read off (or interpolate) the coefficients C_1, C_2 and C_3 for most practical combinations of $\left\{(x/a),\ (y/b),\ (h_x/h_y)\right\}$, which greatly facilitates evaluation of the stress resultants.

As is often the case, the edges of the shell may be supported by parabolic arches, which must be designed to carry the shear forces N_{xy} shed to them by the shell.

References

1. Timoshenko S.P. and Goodier J.N. *Theory of Elasticity*. McGraw-Hill, New York, 1982.
2. Parme A.L. Shells of double curvature. *Transactions of the American Society of Civil Engineers*, 1958, **123**, 989.

Shell Structures in Civil and Mechanical Engineering
ISBN 978-0-7277-6028-9

http://dx.doi.org/10.1680/ssicame.60289.365

Chapter 11
Buckling of shells

11.1. Introduction

When thin shells experience excessive levels of compression, failure may occur by buckling. Buckling is a structural-mechanics phenomenon characterised by sudden and relatively large displacements transverse to the axis (straight or curved) of a 1-dimensional member (e.g. a column, a beam or an arch) or transverse to the midsurface (plane or curved) of a 2-dimensional member (e.g. a plate or a shell), typically as a result of compressive stresses acting along the axis of a 1-dimensional member or in the midsurface of a 2-dimensional member. For a shell, as the level of compression is raised from zero, deflections remain relatively small (almost unnoticeable) until a certain point is reached when, all of a sudden, one or more very visible buckles appear in the walls of the shell.

Depending on the type of shell, buckling may result in the immediate collapse of the structure, or it may merely deform the shell very visibly without necessarily causing immediate collapse. In the latter cases, the shell will still be able to carry a further increase in the loading, but the amplitude of the wall buckles will increase, or other buckles will emerge and propagate, until collapse eventually occurs (usually after severe deformations).

The stress analysis of the preceding chapters would have revealed zones of the shell that have significant compressive stresses, and hence may be vulnerable to buckling. However, to be able to predict the stress levels at which buckling will occur, it is necessary to carry out a buckling analysis of the shell for the loading in question. The buckling strength of a shell depends on several factors, namely: (*a*) the shape of the shell (i.e. midsurface geometry); (*b*) the manner in which the edges are supported (i.e. boundary conditions); (*c*) the nature of the applied loading; (*d*) the geometric proportions of the shell (e.g. radius-to-thickness ratio, radius-to-length ratio, angle of opening, height-to-diameter ratio); (*e*) the existence of imperfections. With regard to the last factor, geometric deviations of the shell midsurface from the intended shape can have a very significant detrimental effect on the buckling strength of the shell.

In this chapter, we will study the buckling of circular cylindrical shells in some detail, from the theoretical formulation of the problem, to the derivation of critical buckling loads for specific loading configurations. We will restrict our attention to linear elastic buckling theory for perfect shell geometries, and will not be concerned with non-linear and post-buckling behaviour. The theoretical treatment of other shell geometries (such as conical, spherical, ellipsoidal and toroidal shells) involves cumbersome solution procedures, hence we will simply present the key results for these and point the reader to relevant literature if full derivational details are required. From a practical point of view, the buckling analysis of these other shell types is best performed using numerical methods (e.g. the Finite Element Method, which is the subject of Chapter 12).

11.2. Cylindrical shells

11.2.1 Introduction

Cylindrical shells are the commonest of all shell forms, and their stability behaviour has been studied the most. Applications in which buckling may need to be taken into account include steel tanks, silos, chimneys, pipelines, tubular masts, hollow members of large steel structures, submarine hulls and aerospace structures. The findings of various theoretical, numerical and experimental studies on the stability of liquid-containment shells of revolution (including cylindrical tanks) have been summarised in a recent literature survey.[1]

In Chapter 1, reference was made to the Donnell theory[2] for cylindrical shells. Although this is less exact than the theory of Flügge,[3] Donnell's theory has the advantage of simplicity, and leads to a less complicated formulation of the buckling problem. The treatment that follows is based on the Donnell theory, with occasional reference to the results of Flügge's theory for comparison purposes.

11.2.2 Governing equation

Figure 11.1 shows an element of a circular cylindrical shell of radius a and lying in the $\{x, \theta\}$ coordinate system. The element is bound between two adjacent straight generators $\{\theta, \theta+d\theta\}$ and two adjacent hoop circles $\{x, x+dx\}$, and is subjected to a general system of direct forces $\{N_x, N_\theta\}$, in-plane shear forces $\{N_{x\theta}, N_{\theta x}\}$, transverse shear forces $\{Q_x, Q_\theta\}$, bending moments $\{M_x, M_\theta\}$ and twisting moments $\{M_{x\theta}, M_{\theta x}\}$, all being per unit length of the relevant edge of the element. As will be justified shortly, this system may be taken as the set of internal actions that counter the additional transverse forces due to buckling.

For the element, let us resolve forces: (*i*) parallel to the axis of the cylinder (in the direction of increasing x), (*ii*) along the central tangent to the central hoop circle of the element (in the direction of increasing θ), and (*iii*) in the direction normal to the shell midsurface. Let us also take moments: (*iv*) about the centroidal axis of the element parallel to the x axis, and (*v*) about the central tangent to the central hoop circle of the element. These five operations result in five equations of equilibrium:

$$a\frac{\partial N_x}{\partial x} + \frac{\partial N_{\theta x}}{\partial \theta} = 0 \tag{11.1a}$$

$$a\frac{\partial N_{x\theta}}{\partial x} + \frac{\partial N_\theta}{\partial \theta} - Q_\theta = 0 \tag{11.1b}$$

$$a\frac{\partial Q_x}{\partial x} + \frac{\partial Q_\theta}{\partial \theta} + N_\theta = 0 \tag{11.1c}$$

$$\frac{\partial M_\theta}{\partial \theta} - a\frac{\partial M_{x\theta}}{\partial x} - aQ_\theta = 0 \tag{11.2a}$$

$$\frac{\partial M_{\theta x}}{\partial \theta} - a\frac{\partial M_x}{\partial x} + aQ_x = 0 \tag{11.2b}$$

From equations (11.2a) and (11.2b), we get

$$Q_\theta = \frac{1}{a}\frac{\partial M_\theta}{\partial \theta} - \frac{\partial M_{x\theta}}{\partial x} \tag{11.3a}$$

Figure 11.1 Element of a circular cylindrical shell of radius *a*, subjected to a general system of: (a) direct forces and shearing forces; (b) bending moments and twisting moments. A prime against a quantity denotes a derivative of the quantity with respect to *x*; a dot above a quantity denotes a derivative of the quantity with respect to θ.

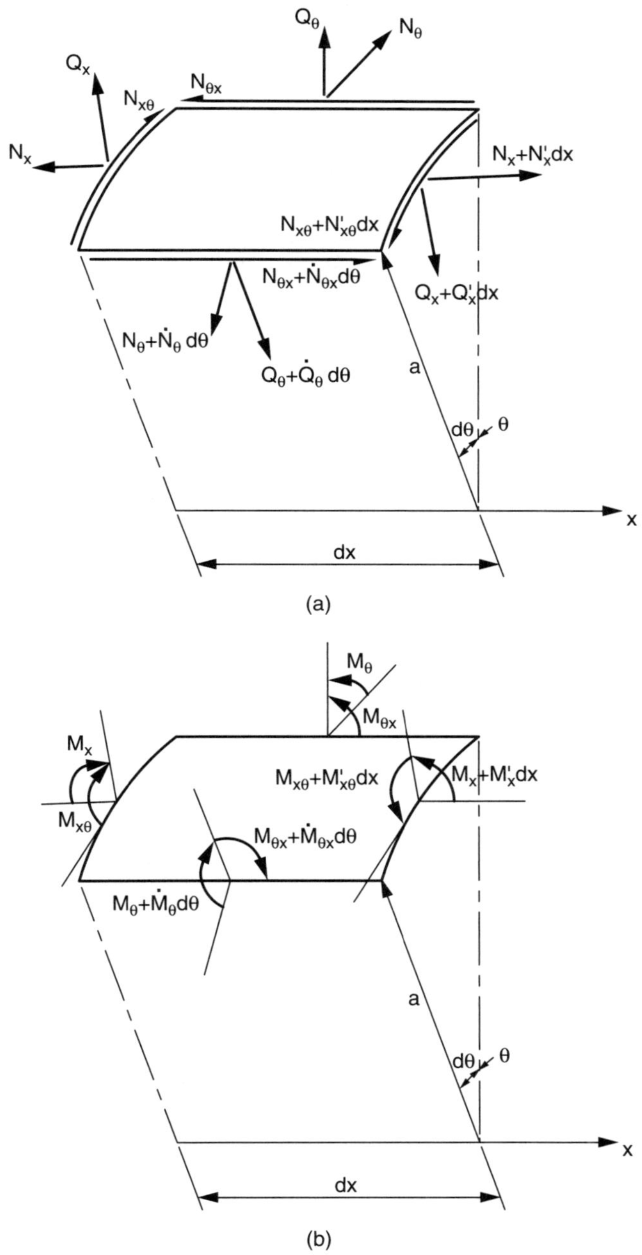

$$Q_x = \frac{\partial M_x}{\partial x} - \frac{1}{a}\frac{\partial M_{\theta x}}{\partial \theta} \tag{11.3b}$$

Substituting these results into equation (11.1c), with $M_{\theta x} = M_{x\theta}$, we obtain

$$\frac{\partial^2 M_x}{\partial x^2} - \frac{2}{a}\frac{\partial^2 M_{x\theta}}{\partial x \partial \theta} + \frac{1}{a^2}\frac{\partial^2 M_\theta}{\partial \theta^2} + \frac{N_\theta}{a} = 0 \tag{11.4}$$

The moment-curvature relationships, written in terms of the displacement w, are as follows

$$M_x = D\left[\frac{\partial^2 w}{\partial x^2} + \frac{v}{a^2}\frac{\partial^2 w}{\partial \theta^2}\right] \tag{11.5a}$$

$$M_\theta = D\left[\frac{1}{a^2}\frac{\partial^2 w}{\partial \theta^2} + v\frac{\partial^2 w}{\partial x^2}\right] \tag{11.5b}$$

$$M_{x\theta} = -\frac{D}{a}(1-v)\frac{\partial^2 w}{\partial x\,\partial \theta} \tag{11.5c}$$

Substituting these results into equation (11.4), we obtain

$$D\left[\frac{\partial^4 w}{\partial x^4} + \frac{2}{a^2}\frac{\partial^4 w}{\partial x^2\,\partial \theta^2} + \frac{1}{a^4}\frac{\partial^4 w}{\partial \theta^4}\right] + \frac{N_\theta}{a} = 0 \tag{11.6a}$$

or

$$D\nabla^4 w + \frac{N_\theta}{a} = 0 \tag{11.6b}$$

where

$$\nabla^2 f = \frac{\partial^2 f}{\partial x^2} + \frac{1}{a^2}\frac{\partial^2 f}{\partial \theta^2} \tag{11.7a}$$

$$\nabla^4 f = \nabla^2\left(\nabla^2 f\right) = \frac{\partial^2\left(\nabla^2 f\right)}{\partial x^2} + \frac{1}{a^2}\frac{\partial^2\left(\nabla^2 f\right)}{\partial \theta^2} \tag{11.7b}$$

Equation (11.6a) features plate-bending action (the biharmonic first term) and the radial component of the membrane hoop stress resultant (N_θ / a).

At the onset of buckling, let the shell midsurface move to an adjacent position while the existing membrane stress resultants N_x, N_θ and $N_{x\theta}$ remain unchanged. If the additional transverse displacements due to buckling are denoted by w, the displaced membrane stress resultants along the edges of the shell element will have components of force in the transverse direction. These components are similar to the resultants in the z direction of the forces N_x, N_y and N_{xy} acting along the edges of the element of Figure 10.1 (see Chapter 10). Replacing the z in equation (10.9) with w, the dimension dy by $a\,d\theta$ (so that $\partial/\partial y = a^{-1}\partial/\partial\theta$), and the parameters N_x, N_y and N_{xy} by N_x, N_θ and $N_{x\theta}$ of present considerations, we obtain the net transverse force Δp as

$$\Delta p = N_x \frac{\partial^2 w}{\partial x^2} + \frac{2}{a} N_{x\theta} \frac{\partial^2 w}{\partial x\, \partial\theta} + \frac{1}{a^2} N_\theta \frac{\partial^2 w}{\partial\theta^2} \tag{11.8}$$

Incorporating this resultant of in-plane membrane action into equation (11.6b), we obtain

$$D\nabla^4 w + \frac{N_\theta}{a} = \Delta p$$

– that is,

$$D\nabla^4 w + \frac{N_\theta}{a} = N_x \frac{\partial^2 w}{\partial x^2} + \frac{2}{a} N_{x\theta} \frac{\partial^2 w}{\partial x\, \partial\theta} + \frac{1}{a^2} N_\theta \frac{\partial^2 w}{\partial\theta^2} \tag{11.9}$$

Now N_θ on the left-hand side is a function of all three displacement components: u and v in the plane of the shell midsurface (u being tangent to a hoop circle, and v being parallel to the x axis), and w in the direction normal to the shell midsurface. We may obtain an expression for derivatives of u in terms of derivatives of w, and another for derivatives of v in terms of derivatives of w. Upon substitution of these relationships into equation (11.9), we end up with the well-known Donnell equation in uncoupled form:

$$D\nabla^8 w + \frac{Et}{a^2}\frac{\partial^4 w}{\partial x^4} = \nabla^4\left(N_x \frac{\partial^2 w}{\partial x^2} + \frac{2}{a} N_{x\theta} \frac{\partial^2 w}{\partial x\, \partial\theta} + \frac{1}{a^2} N_\theta \frac{\partial^2 w}{\partial\theta^2}\right) \tag{11.10}$$

Denoting the pre-buckling membrane stress resultants (which we assume do not change during buckling) by $\overline{N}_x$, $\overline{N}_\theta$ and $\overline{N}_{x\theta}$, we may write equation (11.10) in terms of only one unknown w, as follows:

$$D\nabla^8 w + \frac{Et}{a^2}\frac{\partial^4 w}{\partial x^4} = \nabla^4\left(\overline{N}_x \frac{\partial^2 w}{\partial x^2} + \frac{2}{a} \overline{N}_{x\theta} \frac{\partial^2 w}{\partial x\, \partial\theta} + \frac{1}{a^2} \overline{N}_\theta \frac{\partial^2 w}{\partial\theta^2}\right) \tag{11.11}$$

11.2.3 Cylinder subjected to axial compression

11.2.3.1 General solution

Consider a cylinder subjected to a uniformly distributed axial compressive force of P per unit length of the shell circumference, as shown in Figure 11.2. The pre-buckling membrane stress resultants are therefore:

$$\overline{N}_x = -P; \quad \overline{N}_\theta = 0; \quad \overline{N}_{x\theta} = 0 \tag{11.12}$$

Substituting these values into equation (11.11), we obtain

$$D\nabla^8 w + \frac{Et}{a^2}\frac{\partial^4 w}{\partial x^4} + P\nabla^4\left(\frac{\partial^2 w}{\partial x^2}\right) = 0 \tag{11.13}$$

Figure 11.2 Cylindrical shell under axial compression *P* per unit length of shell circumference.

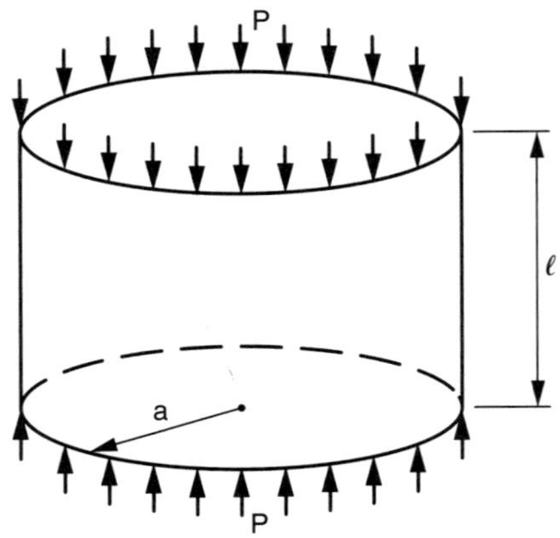

Let us assume a buckled shape of the form

$$w = A \sin \frac{n\pi x}{l} \sin m\theta \tag{11.14}$$

where n and $2m$ are the numbers of half-waves along the length of the cylinder and around the circumference respectively. The case when $n = 2$ and $m = 4$ is illustrated in Figure 11.3.

At $x = 0$ and $x = l$, expression (11.14) gives

$$w = 0\,; \quad \frac{\partial^2 w}{\partial x^2} = 0 \tag{11.15}$$

which implies that both the deflection and the bending moment M_x are zero. Thus the assumed solution (equation (11.14)) satisfies simply-supported end conditions. In practice, the actual boundary conditions at the ends of the cylinder generally differ from this assumption, but if the cylinder is long enough, we would expect the buckling load not to be significantly influenced by the precise conditions at the ends of the cylinder.

Let us introduce a parameter λ, defined as

$$\lambda = \frac{n\pi a}{l} \tag{11.16}$$

We may rewrite equation (11.14) as

$$w = A \sin \frac{\lambda x}{a} \sin m\theta \tag{11.17}$$

Figure 11.3 Buckling mode $m = 4$ and $n = 2$.

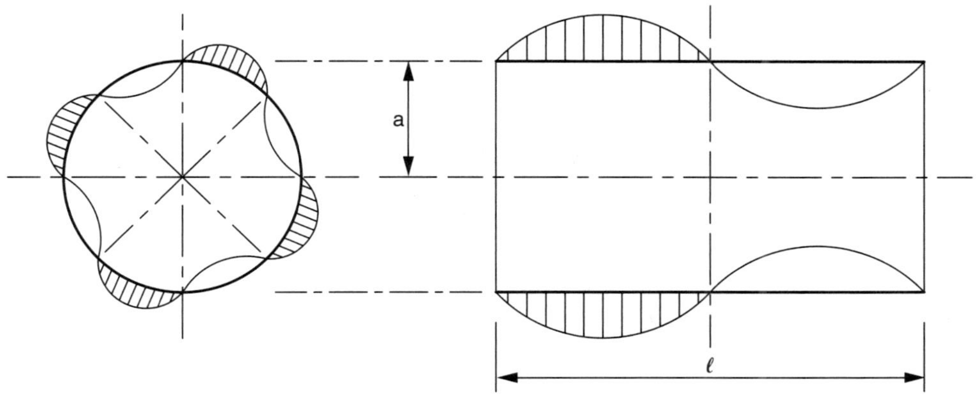

We want to substitute this expression for w into equation (11.13). In this process, the following relationships are useful:

$$\nabla^4 w = \nabla^2\left(\nabla^2 w\right) = \frac{\partial^2}{\partial x^2}\left(\frac{\partial^2 w}{\partial x^2} + \frac{1}{a^2}\frac{\partial^2 w}{\partial \theta^2}\right) + \frac{1}{a^2}\frac{\partial^2}{\partial \theta^2}\left(\frac{\partial^2 w}{\partial x^2} + \frac{1}{a^2}\frac{\partial^2 w}{\partial \theta^2}\right)$$

$$= \frac{\partial^4 w}{\partial x^4} + \frac{2}{a^2}\frac{\partial^4 w}{\partial x^2 \partial \theta^2} + \frac{1}{a^4}\frac{\partial^4 w}{\partial \theta^4} \tag{11.18a}$$

$$\nabla^8 w = \nabla^4\left(\nabla^4 w\right) = \left[\frac{\partial^4}{\partial x^4} + \frac{2}{a^2}\frac{\partial^4}{\partial x^2 \partial \theta^2} + \frac{1}{a^4}\frac{\partial^4}{\partial \theta^4}\right]\left[\frac{\partial^4 w}{\partial x^4} + \frac{2}{a^2}\frac{\partial^4 w}{\partial x^2 \partial \theta^2} + \frac{1}{a^4}\frac{\partial^4 w}{\partial \theta^4}\right]$$

$$= \frac{\partial^8 w}{\partial x^8} + \frac{4}{a^2}\frac{\partial^8 w}{\partial x^6 \partial \theta^2} + \frac{6}{a^4}\frac{\partial^8 w}{\partial x^4 \partial \theta^4} + \frac{4}{a^6}\frac{\partial^8 w}{\partial x^2 \partial \theta^6} + \frac{1}{a^8}\frac{\partial^8 w}{\partial \theta^8} \tag{11.18b}$$

$$\nabla^4\left(\frac{\partial^2 w}{\partial x^2}\right) = \frac{\partial^4}{\partial x^4}\left(\frac{\partial^2 w}{\partial x^2}\right) + \frac{2}{a^2}\frac{\partial^4}{\partial x^2 \partial \theta^2}\left(\frac{\partial^2 w}{\partial x^2}\right) + \frac{1}{a^4}\frac{\partial^4}{\partial \theta^4}\left(\frac{\partial^2 w}{\partial x^2}\right)$$

$$= \frac{\partial^6 w}{\partial x^6} + \frac{2}{a^2}\frac{\partial^6 w}{\partial x^4 \partial \theta^2} + \frac{1}{a^4}\frac{\partial^6 w}{\partial x^2 \partial \theta^4} \tag{11.18c}$$

Substituting the assumed solution for w as given by equation (11.17) into equation (11.13), with the help of the above relationships, and simplifying the ensuing expression, we obtain

$$D\left[\left(\frac{\lambda}{a}\right)^8 + \frac{4}{a^2}\left(\frac{\lambda}{a}\right)^6 m^2 + \frac{6}{a^4}\left(\frac{\lambda}{a}\right)^4 m^4 + \frac{4}{a^6}\left(\frac{\lambda}{a}\right)^2 m^6 + \frac{1}{a^8} m^8\right]$$

$$+\frac{Et}{a^2}\left(\frac{\lambda}{a}\right)^4 - P\left[\left(\frac{\lambda}{a}\right)^6 + \frac{2}{a^2}\left(\frac{\lambda}{a}\right)^4 m^2 + \frac{1}{a^4}\left(\frac{\lambda}{a}\right)^2 m^4\right] = 0$$

or

$$\frac{D}{a^2}\left[\lambda^8 + 4\lambda^6 m^2 + 6\lambda^4 m^4 + 4\lambda^2 m^6 + m^8\right] + Et\lambda^4 - P\left[\lambda^6 + 2\lambda^4 m^2 + \lambda^2 m^4\right] = 0$$

– that is,

$$\frac{D}{a^2}\left(\lambda^2 + m^2\right)^4 + Et\lambda^4 - P\lambda^2\left(\lambda^2 + m^2\right)^2 = 0 \tag{11.19}$$

For the critical buckling load, this gives the solution

$$P_{cr} = \frac{D}{a^2}\frac{\left(\lambda^2 + m^2\right)^2}{\lambda^2} + Et\frac{\lambda^2}{\left(\lambda^2 + m^2\right)^2} \tag{11.20}$$

If we let

$$\beta = \frac{\left(\lambda^2 + m^2\right)^2}{\lambda^2} \tag{11.21}$$

we may write

$$P_{cr} = \frac{D}{a^2}\beta + Et\frac{1}{\beta} \tag{11.22}$$

The value of β at which P_{cr} is a minimum is obtained from the condition:

$$\frac{dP_{cr}}{d\beta} = \frac{D}{a^2} - \frac{Et}{\beta^2} = 0 \tag{11.23}$$

This gives the solution

$$\beta = \sqrt{\frac{Eta^2}{D}} = 2\frac{a}{t}\sqrt{3\left(1 - \nu^2\right)} \tag{11.24}$$

Substituting this value of β into equation (11.22), we obtain

$$\left(P_{cr}\right)_{\min} = \frac{D}{a^2}2\frac{a}{t}\sqrt{3\left(1-\nu^2\right)} + Et\frac{t}{2a\sqrt{3\left(1-\nu^2\right)}}$$

$$= \frac{E}{\sqrt{3\left(1-\nu^2\right)}}\frac{t^2}{a} \tag{11.25}$$

Table 11.1 Mode combinations {*n*, *m*} for the numerical example: $a/t = 500$; $l/a = 3.0$; $\nu = 0.3$.

n (no. of longitudinal half-waves)	λ ($\lambda = n\pi/3$ for $l/a = 3.0$)	m (no. of circumferential full waves)
1	1.047	6.44 (6)
2	2.094	8.98 (9)
5	5.236	13.62 (14)

The corresponding buckling stress is given by

$$\sigma_{cr\min} = \frac{(P_{cr})_{\min}}{t} = \frac{E}{\sqrt{3(1-\nu^2)}}\frac{t}{a} \tag{11.26}$$

For steel ($\nu = 0.3$), this gives $\sigma_{cr\min} = 0.605\, Et/a$. Notice that we did not need to know λ and m in order to evaluate the minimum buckling load. The minimum buckling load is the same for an infinite range of combinations of λ and m. Consider the example of a steel cylinder with $a/t = 500$ and $l/a = 3.0$. From equation (11.24), $\beta = 2(500)(1.652) = 1652$. From equation (11.16), $\lambda = n\pi/3$. For a given longitudinal mode n, the parameter λ is therefore known, and m follows from equation (11.21). The calculated values of m for $n = 1$, $n = 2$ and $n = 5$ are shown in Table 11.1. These calculated values are rounded off to the nearest integer (shown in brackets) as m is an integer.

11.2.3.2 Special case of axisymmetric buckling

For buckling modes that are axisymmetric, we put $m = 0$ in equation (11.20) to obtain

$$P_{cr} = \frac{D}{a^2}\lambda^2 + Et\frac{1}{\lambda^2} \tag{11.27}$$

Substituting the expression for λ as given by equation (11.16), we may write equation (11.27) in the alternative form

$$P_{cr} = D\left[\frac{n^2\pi^2}{l^2} + \frac{Et}{Da^2}\frac{l^2}{n^2\pi^2}\right] \tag{11.28}$$

The buckling load is minimum with respect to λ^2 when $\lambda^2 = \sqrt{Eta^2/D}$, giving

$$(P_{cr})_{\min} = \frac{E}{\sqrt{3(1-\nu^2)}}\frac{t^2}{a} \quad ; \quad \sigma_{cr\min} = \frac{E}{\sqrt{3(1-\nu^2)}}\frac{t}{a}$$

as before. The condition $\lambda^2 = \sqrt{Eta^2/D}$ may be written in terms of the longitudinal mode n (see equation (11.16)) as follows

$$\frac{n^2\pi^2 a^2}{l^2} = \sqrt{\frac{Eta^2}{D}} = \sqrt{12\left(1-\nu^2\right)\frac{a^2}{t^2}} \quad \Rightarrow \quad n = \frac{\sqrt{2}}{\pi}\left[3\left(1-\nu^2\right)\right]^{1/4}\frac{l}{\sqrt{at}} \tag{11.29}$$

For steel, taking $\nu = 0.3$, we get

$$n = 0.58\frac{l}{\sqrt{at}} \tag{11.30a}$$

while for concrete, taking $\nu = 0.15$, we get

$$n = 0.59\frac{l}{\sqrt{at}} \tag{11.30b}$$

Considering a cylindrical shell with $l = a$, and taking typical values of $a/t = 500$ for a steel shell and $a/t = 100$ for a concrete shell, we obtain $n \approx 13$ for the steel shell and $n \approx 6$ for the concrete shell. This illustrates the fact that n (number of longitudinal half-waves) is a relatively large number for thin shells, particularly where l exceeds the diameter of the cylinder ($l \geq 2a$). Our earlier example of a steel cylinder with $l = 3a$ and $a/t = 500$ would axisymmetrically buckle with ≈ 39 longitudinal half-waves!

11.2.3.3 Limiting cases

11.2.3.3.1 VERY SHORT CYLINDERS

For very short cylinders ($a >> l$), it is reasonable to assume the cylinder buckles with only one half-wave in the axial direction – that is, $n = 1$. Thus, from equation (11.16), $\lambda = \pi(a/l)$, which is a large number since a/l is large. Using the approximation that $\lambda >> m$, equation (11.20) becomes

$$P_{cr} \approx \frac{D}{a^2}\lambda^2 + Et\frac{1}{\lambda^2} \approx \frac{D}{a^2}\lambda^2 = \frac{D}{a^2}\left(\frac{\pi^2 a^2}{l^2}\right) = \frac{\pi^2 D}{l^2} = \frac{\pi^2 E}{l^2}\left[\frac{t^3}{12\left(1-\nu^2\right)}\right] \tag{11.31}$$

The final result is equivalent to the buckling load of an Euler strip of unit width, with the term in square brackets taking the place of I (second moment of area of the cross-section). Ignoring the second term of equation (11.20), we can see that, for a given value of λ, P_{cr} is smallest when m is smallest – that is, when $m = 0$ – indicating that the buckling of very short cylinders under axial compression is axisymmetric.

11.2.3.3.2 VERY LONG CYLINDERS

As l becomes very large in relation to a (i.e. $l >> a$), the parameter λ approaches zero, and P_{cr} (equation (11.20)) becomes very large, which clearly is contrary to expectation. The Donnell theory cannot account for the true behaviour. The shell behaves like a column, and therefore Euler buckling theory must be used: $P_{cr} = \pi^2 EI/l^2$ where $I = \pi a^3 t$ (for flexural bending).

Based on a more rigorous formulation (in which the three equations of force equilibrium of the element are expressed in terms of the basic displacements u, v and w taking into account the stretching of the shell midsurface and all rotations of the shell edges due to the displacements u, v and w), Flügge[3] came up with the following solution for the critical buckling load of an axially compressed cylinder:

Figure 11.4 Buckling curves for axial compression: plots of q_{cr} versus l/na for a steel shell with $a/t = 500$.

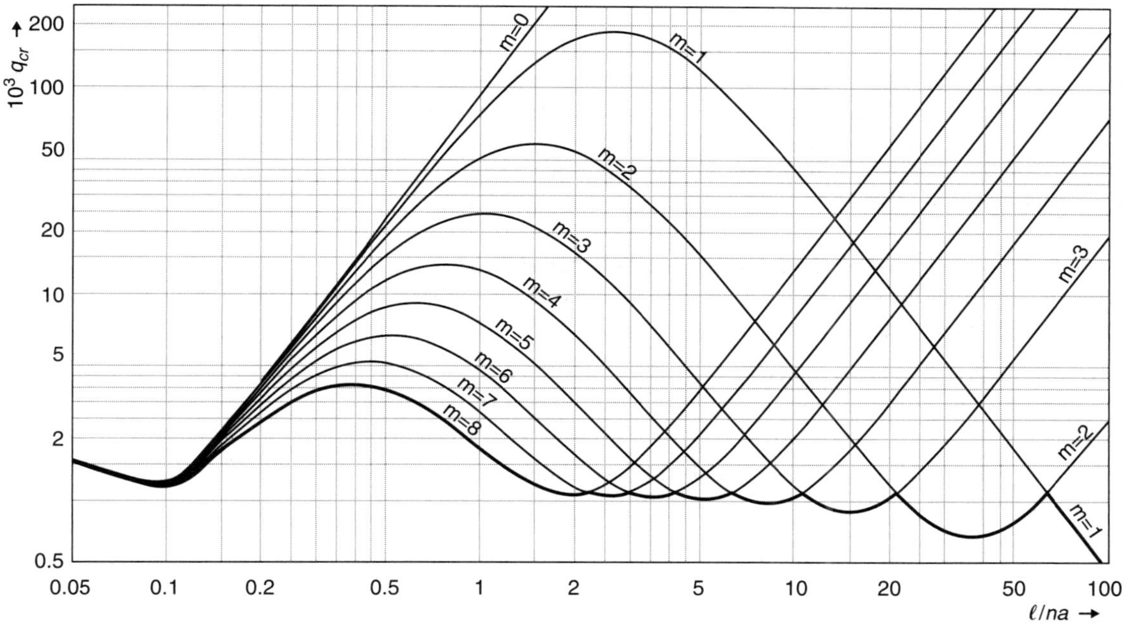

$$q_{cr} = \frac{P_{cr}}{Et}\left(1-\nu^2\right) = \frac{1}{\lambda^2\left(\lambda^2+m^2\right)^2+\lambda^2 m^2} \times$$

$$\left\{\left(1-\nu^2\right)\lambda^4+\xi\left[\left(\lambda^2+m^2\right)^4-2\left\{\nu\lambda^6+3\lambda^4m^2+\left(4-\nu\right)\lambda^2m^4+m^6\right\}+2\left(2-\nu\right)\lambda^2m^2+m^4\right]\right\} \qquad (11.32)$$

where

$$\xi = \frac{t^2}{12a^2} \qquad (11.33)$$

For every integer m, we may plot q_{cr} versus λ to generate a buckling curve. However, instead of plotting q_{cr} versus λ, let us follow Flügge[3] and plot q_{cr} versus l/na (which is equal to π/λ). It is also better to use logarithmic scales for both q_{cr} and l/na, in order to cover a much larger range of parameters than would be possible with linear scales. Figure 11.4 shows the plots for $m = 0$ up to $m = 8$ for a steel shell ($\nu = 0.3$) with $a/t = 500$ (i.e. $\xi = 0.333 \times 10^{-6}$). Note that:

(*a*) the curve for $m = 0$ exhibits a minimum, and then rises indefinitely
(*b*) the curve for $m = 1$ exhibits a minimum, rises to a maximum, and then descends towards zero in the range when l/na is very large

(*c*) all other curves exhibit a minimum, a maximum, then a second minimum, before rising indefinitely with further increase in l / na.

The lower-bound envelope of the curves is a festoon-shaped curve that comprises a portion of the curve $m = 0$ on the extreme left, a portion of the curve $m = 1$ on the extreme right, and arcs of curves $m = 2, 3, 4$ and so on in between. This lower-bound envelope defines the buckling condition, with all points below this curve being stable, and all points above the festoon being unstable. Notice that, unlike the theory of Donnell, Flügge's theory is able to account for the behaviour of a very long cylinder ($l >> a$). This is represented by the descending part of the curve $m = 1$ on the extreme right.

11.2.4 Cylinder subjected to uniform lateral pressure

When the cylinder is subjected to a uniform external lateral pressure p as shown in Figure 11.5, the pre-buckling membrane stress resultants are:

$$\overline{N}_x = 0; \quad \overline{N}_\theta = -pa; \quad \overline{N}_{x\theta} = 0 \tag{11.34}$$

Substituting these values into equation (11.11), we obtain

$$D\nabla^8 w + \frac{Et}{a^2}\frac{\partial^4 w}{\partial x^4} + \frac{p}{a}\nabla^4\left(\frac{\partial^2 w}{\partial \theta^2}\right) = 0 \tag{11.35}$$

We assume the same deflection function as given by equations (11.14) and (11.17). The expression for $\nabla^8 w$ was given by equation (11.18b). We also need the expanded form of $\nabla^4(\partial^2 w / \partial\theta^2)$:

Figure 11.5 Cylindrical shell under uniform external pressure p.

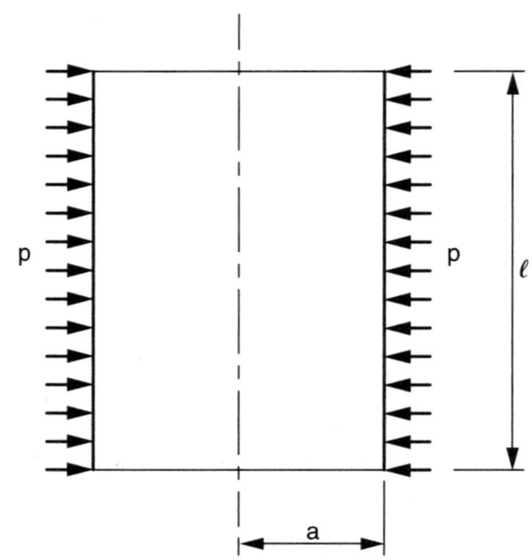

$$\nabla^4\left(\frac{\partial^2 w}{\partial \theta^2}\right) = \frac{\partial^4}{\partial x^4}\left(\frac{\partial^2 w}{\partial \theta^2}\right)+\frac{2}{a^2}\frac{\partial^4}{\partial x^2\,\partial \theta^2}\left(\frac{\partial^2 w}{\partial \theta^2}\right)+\frac{1}{a^4}\frac{\partial^4}{\partial \theta^4}\left(\frac{\partial^2 w}{\partial \theta^2}\right)$$

$$= \frac{\partial^6 w}{\partial x^4\,\partial \theta^2}+\frac{2}{a^2}\frac{\partial^6 w}{\partial x^2\,\partial \theta^4}+\frac{1}{a^4}\frac{\partial^6 w}{\partial \theta^6} \tag{11.36}$$

Substituting for w (as given by equation (11.17)) in equation (11.35), with the help of expansions (11.18b) and (11.36), and simplifying the ensuing expression, we obtain

$$D\left[\left(\frac{\lambda}{a}\right)^8+\frac{4}{a^2}\left(\frac{\lambda}{a}\right)^6 m^2+\frac{6}{a^4}\left(\frac{\lambda}{a}\right)^4 m^4+\frac{4}{a^6}\left(\frac{\lambda}{a}\right)^2 m^6+\frac{1}{a^8}m^8\right]$$

$$+\frac{Et}{a^2}\left(\frac{\lambda}{a}\right)^4-\frac{p}{a}\left[\left(\frac{\lambda}{a}\right)^4 m^2+\frac{2}{a^2}\left(\frac{\lambda}{a}\right)^2 m^4+\frac{1}{a^4}m^6\right]=0$$

or

$$\frac{D}{a^2}\left[\lambda^8+4\lambda^6 m^2+6\lambda^4 m^4+4\lambda^2 m^6+m^8\right]+Et\lambda^4-pa\left[\lambda^4 m^2+2\lambda^2 m^4+m^6\right]=0$$

– that is,

$$\frac{D}{a^2}\left(\lambda^2+m^2\right)^4+Et\lambda^4-pa\,m^2\left(\lambda^2+m^2\right)^2=0 \tag{11.37}$$

For the critical buckling load, this gives the solution

$$p_{cr}a = \frac{D}{a^2}\frac{\left(\lambda^2+m^2\right)^2}{m^2}+Et\frac{\lambda^4}{m^2\left(\lambda^2+m^2\right)^2} \tag{11.38}$$

One axial half-wave ($n = 1$) always gives the lowest buckling load. From equation (11.16), λ therefore becomes $\pi a / l$, or π / r where $r = l / a$. Putting this value of λ into equation (11.38), and also replacing D, we obtain for the critical buckling pressure

$$p_{cr} = \frac{E}{12\left(1-\nu^2\right)}\left(\frac{t}{a}\right)^3\frac{\left(\pi^2+r^2m^2\right)^2}{r^4m^2}+E\left(\frac{t}{a}\right)\frac{\pi^4}{m^2\left(\pi^2+r^2m^2\right)^2} \tag{11.39}$$

For $m = 0$, both terms are subject to division by zero, implying that p_{cr} is infinite. This may very well be a true indication that the shell cannot buckle axisymmetrically under uniform external pressure, but at the same time we cannot rule out the possibility that this is yet another manifestation of the shortcoming of Donnell's theory.

Let us consider the limiting case of a very long cylinder – that is, r very large. The second term of equation (11.39) may be neglected in comparison with the first, so that

$$p_{cr} \approx \frac{E}{12(1-\nu^2)}\left(\frac{t}{a}\right)^3 m^2 = \frac{Dm^2}{a^3} \tag{11.40}$$

This result is not valid for $m = 0$ and $m = 1$. Taking $m = 2$, we obtain

$$p_{cr} \approx \frac{Et^3}{3(1-\nu^2)a^3} \tag{11.41}$$

for a very long cylinder. For the same limiting case of a very long cylinder, the more exact theory of Timoshenko and Gere[4] reduces to

$$p_{cr} \approx \frac{E}{12(1-\nu^2)}\left(\frac{t}{a}\right)^3 (m^2-1) \tag{11.42}$$

which for $m = 2$ gives the result

$$p_{cr} \approx \frac{Et^3}{4(1-\nu^2)a^3} \tag{11.43}$$

The even more exact theory of Flügge[3] also yields this result for p_{cr} in the limit when l/a tends to infinity and m is assigned a value of 2. Thus, compared with more exact theories, the Donnell theory overestimates the critical buckling load of an externally pressurised very long cylinder by 33%. Clearly as m (the number of circumferential waves) becomes larger, the approximation given by equation (11.40) becomes closer to the result in equation (11.42). In general, the accuracy of Donnell's theory is not always very good for low values of m.

To obtain more insights, let us pursue the more exact approach of Flügge. For the problem of a cylindrical shell subjected to uniform external pressure, Flügge gives the following solution for the critical buckling pressure:[3]

$$q_{cr} = \frac{ap_{cr}}{Et}(1-\nu^2) = \frac{1}{m^2(\lambda^2+m^2)^2 - m^2(3\lambda^2+m^2)} \times$$

$$\left\{(1-\nu^2)\lambda^4 + \xi\left[(\lambda^2+m^2)^4 - 2\left\{\nu\lambda^6 + 3\lambda^4 m^2 + (4-\nu)\lambda^2 m^4 + m^6\right\} + 2(2-\nu)\lambda^2 m^2 + m^4\right]\right\} \tag{11.44}$$

where ξ is as defined by equation (11.33).

As before, we may plot q_{cr} versus l/na for integer values of m, and obtain the festoon curve as the lower-bound envelope of these plots. Experimental results show that the shell will always buckle with one axial half-wave, so n should be taken as 1.0. The horizontal coordinate therefore reduces to l/a, while λ becomes simply $\pi a/l$. From equation (11.44), we note that the curve for $m = 0$ cannot be plotted since the denominator becomes zero and q_{cr} goes to infinity. This suggests that the axisymmetric buckling mode is very difficult to maintain under uniform external pressure, an observation that was also made in discussing solution (11.39).

The case $m = 1$ reveals some interesting behaviour. If we put $m = 1$ into equation (11.44), we get, after some simplifications,

$$[q_{cr}]_{m=1} = \frac{(1-\nu^2)\lambda^2 + \xi\lambda^4\left[\lambda^2 + 2(2-\nu)\right]}{\lambda^2 - 1} \tag{11.45}$$

If it so happens that $\lambda = 1$ (i.e. $l = \pi a \approx 3a$), then q_{cr} is infinite, implying that, under uniform external pressure, the shell cannot buckle in this mode ($m = 1$). As we shorten the cylinder ($l < \pi a$), λ becomes greater than 1 and q_{cr} decreases to finite values, but these are still larger than those for $m = 2, 3, 4$ and so on, and so will not be part of the festoon curve. On the other hand, if we lengthen the cylinder ($l > \pi a$), λ becomes smaller than 1 and q_{cr} becomes *negative*. This suggests that for relatively long cylinders, buckling may occur under *internal* pressure, a result that is not particularly obvious. As λ approaches zero (very long cylinder), the above expression becomes

$$[q_{cr}]_{m=1;\,\lambda\to 0} \approx \frac{(1-\nu^2)\lambda^2}{-1} = -(1-\nu^2)\lambda^2 \tag{11.46}$$

If we write q_{cr} in terms of p_{cr} (refer to equation (11.44)), and write λ in terms of a and l, we obtain

$$\frac{ap_{cr}}{Et}(1-\nu^2) = -(1-\nu^2)\frac{\pi^2 a^2}{l^2}$$

which gives

$$p_{cr} = -\pi^2 E\left(\frac{at}{l^2}\right) \tag{11.47}$$

This is the buckling pressure of a pipeline (very long cylinder) under internal pressure.

For all ranges of l/a, buckling under external pressure will therefore only occur in the circumferential modes $m = 2, 3, 4$ and so on. Figure 11.6 shows the plots for $m = 2, 3, \ldots, 8$ for a steel shell ($\nu = 0.3$) with $a/t = 500$ (and therefore $\xi = 0.333 \times 10^{-6}$).

We note that, in the higher ranges of l/a, buckling occurs with two circumferential full waves ($m = 2$). Furthermore, the critical buckling value gradually approaches a lower limit as l/a approaches infinity. To obtain this lower limit, we set m equal to 2 in equation (11.44), and also set λ equal to zero. Thus,

$$[q_{cr}]_{m=2;\,\lambda=0} = \frac{1}{48}\times\left\{0 + \xi\left[256 - 2\{64\} + 16\right]\right\} = \frac{144\xi}{48} = 3\xi$$

– that is,

$$\frac{ap_{cr}}{Et}(1-\nu^2) = 3\left(\frac{t^2}{12a^2}\right) = \frac{t^2}{4a^2}$$

Figure 11.6 Buckling curves for uniform external pressure: plots of q_{cr} versus l / a for a steel shell with $a / t = 500$.

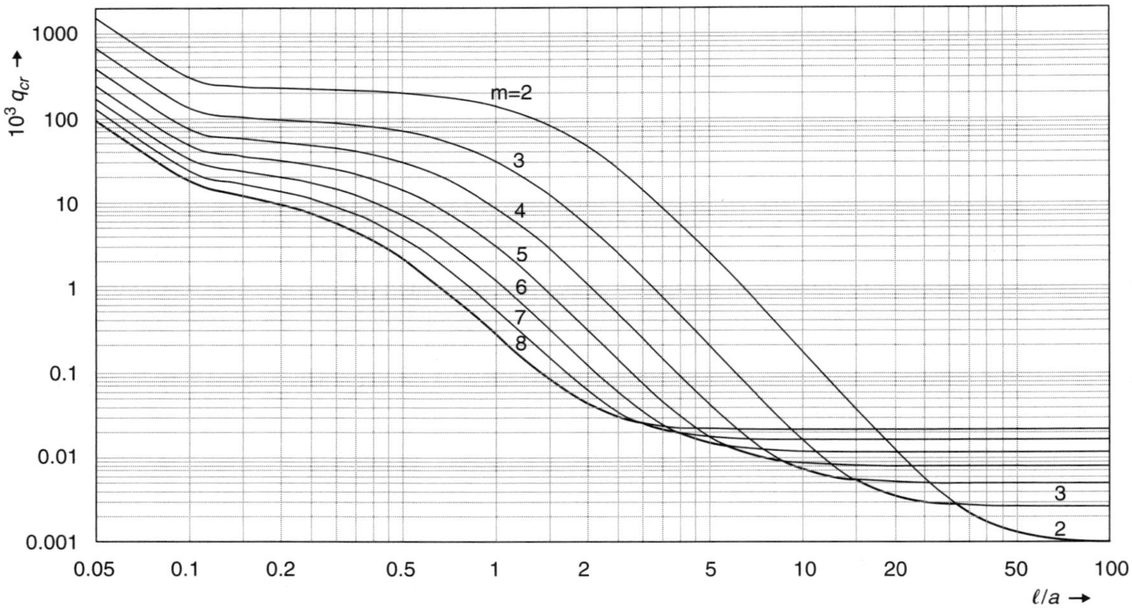

This gives

$$p_{cr} = \frac{Et^3}{4(1-\nu^2)a^3} \tag{11.48}$$

which is the result that was given earlier without proof.

Figure 11.6 also reveals that, as the cylinder becomes shorter (decreasing l / a), the shell will buckle with an increasing number of circumferential waves (m gets bigger), and at increasingly larger values of the external pressure.

It may be remarked here that theoretical values of the buckling stress are always considerably larger than the real values as obtained by experimental testing, owing to the presence of imperfections in real shells. For design purposes, knock-down factors have to be applied to theoretical values. It has also been observed that the discrepancy between theoretical and experimental values is smaller for external lateral pressure than for axial compression, pointing to the higher sensitivity of axial buckling load to the presence of shell imperfections.

11.2.5 Cylinder subjected to uniform torsion

Figure 11.7 shows a cylinder subjected to a total torque of T about its axis. Let us assume shear stress τ is constant over the annular area $2\pi at$ of the shell cross-section, so that

Figure 11.7 Cylindrical shell subjected to a torque T about the axis of revolution of the shell.

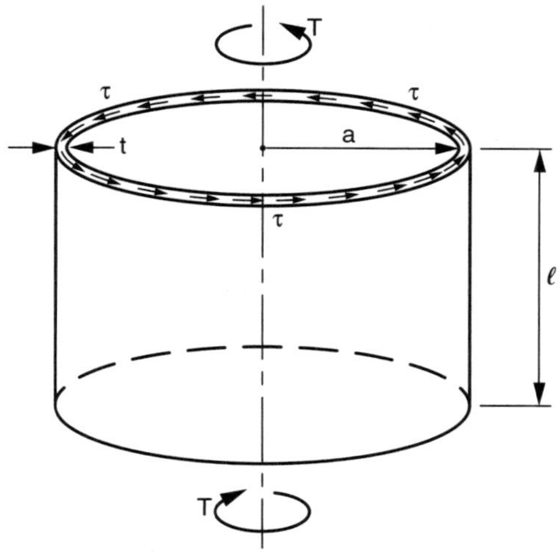

$$T=(2\pi a t\tau)\times a \quad\Rightarrow\quad \tau=\frac{T}{2\pi a^2 t} \tag{11.49}$$

The shear force per unit length, $N_{x\theta}$, is thus

$$N_{x\theta}=\tau\times(\text{area of unit length})=\frac{T}{2\pi a^2 t}(1\times t)=\frac{T}{2\pi a^2} \tag{11.50}$$

The pre-buckling membrane stress resultants are therefore

$$\overline{N}_x=0;\quad \overline{N}_\theta=0;\quad \overline{N}_{x\theta}=-\frac{T}{2\pi a^2} \tag{11.51}$$

Equation (11.11) becomes

$$D\nabla^8 w+\frac{Et}{a^2}\frac{\partial^4 w}{\partial x^4}+\frac{T}{\pi a^3}\nabla^4\left(\frac{\partial^2 w}{\partial x\,\partial\theta}\right)=0 \tag{11.52}$$

Here we cannot assume a solution of the form of equation (11.17), because during buckling lines of zero displacement (i.e. nodal lines) do not coincide with the generators of the cylindrical surface. Instead, these take a spiralling form, a suitable description of which is

$$w=A\sin\left(\frac{\lambda x}{a}-m\theta\right) \tag{11.53}$$

where λ is defined by equation (11.16). Clearly, normal support conditions at the edges $x=$ constant are unlikely to be satisfied by this function, but if the cylinder is long enough, the precise conditions at the edges will not significantly affect the magnitude of the buckling load. Thus, a solution based on equation (11.53) may be adopted for a cylinder with any arbitrary set of end conditions, provided the cylinder is very long. For such a cylinder, we substitute solution (11.53) into equation (11.52), leading to the result

$$T_{cr} = \pi D \frac{(\lambda^2 + m^2)^2}{\lambda m} + Et\pi a^2 \frac{\lambda^3}{m(\lambda^2 + m^2)^2}$$

$$= \frac{\pi E t^3}{12(1-\nu^2)} \left[\frac{(\lambda^2 + m^2)^2}{\lambda m} + 12(1-\nu^2)\left(\frac{a}{t}\right)^2 \frac{\lambda^3}{m(\lambda^2 + m^2)^2} \right]$$

$$= \pi D \left[\frac{(\lambda^2 + m^2)^2}{\lambda m} + 12(1-\nu^2)\left(\frac{a}{t}\right)^2 \frac{\lambda^3}{m(\lambda^2 + m^2)^2} \right] \tag{11.54}$$

For long cylinders, the shell buckles in four circumferential half-waves (i.e. $m = 2$). Putting $m = 2$ in the above expression, we obtain

$$T_{cr} = \pi D \left[\frac{(\lambda^2 + 4)^2}{2\lambda} + 6(1-\nu^2)\left(\frac{a}{t}\right)^2 \frac{\lambda^3}{(\lambda^2 + 4)^2} \right] \tag{11.55}$$

Since the shell is long, it may be assumed that $\lambda << 4$. We may thus discard λ^2 in relation to 4, so that the above expression simplifies to

$$T_{cr} = \pi D \left[\frac{8}{\lambda} + \frac{3}{8}(1-\nu^2)\left(\frac{a}{t}\right)^2 \lambda^3 \right] \tag{11.56}$$

It is evident that T_{cr} is a minimum with respect to λ when

$$\lambda = \left[\frac{64}{9}\left(\frac{1}{1-\nu^2}\right)\left(\frac{t}{a}\right)^2 \right]^{1/4} \tag{11.57}$$

Substituting this value of λ into equation (11.56), we obtain the minimum value of T_{cr} as

$$T_{cr\,min} = \frac{\pi D}{\lambda}\left[8 + \frac{3}{8}(1-\nu^2)\left(\frac{a}{t}\right)^2 \lambda^4 \right]$$

$$= \frac{32}{3}\frac{\pi D}{\lambda}$$

$$= \frac{32\pi}{3}\frac{E t^3}{12(1-\nu^2)}\left[\frac{9}{64}(1-\nu^2)\left(\frac{a}{t}\right)^2 \right]^{1/4}$$

$$= \quad 1.71 \frac{Et^3}{\left(1-\nu^2\right)^{3/4}} \left(\frac{a}{t}\right)^{1/2} \tag{11.58}$$

The corresponding critical buckling shear stress resultant is given by

$$\left(N_{x\theta}\right)_{cr\min} = \frac{T_{cr\min}}{2\pi a^2} = 0.272 \frac{Et}{\left(1-\nu^2\right)^{3/4}} \left(\frac{t}{a}\right)^{3/2} \tag{11.59}$$

and the critical buckling shear stress is given by

$$\tau_{cr\min} = \frac{\left(N_{x\theta}\right)_{cr\min}}{t} = 0.272 \frac{E}{\left(1-\nu^2\right)^{3/4}} \left(\frac{t}{a}\right)^{3/2} \tag{11.60}$$

The coefficient 0.272 appearing in equation (11.60) is, of course, based on Donnell's theory. When a more exact formulation is used,[4] the coefficient works out as 0.236. Thus Donnell's theory overestimates the value of the critical buckling stress for a long cylinder under torsion by 15%. As already pointed out, Donnell's theory is not always very accurate for low circumferential modes ($m \leq 2$). In considering axial compression (Section 11.2.3), we also observed that Donnell's theory cannot model the column-buckling mode. Despite these shortcomings, the theory generally provides satisfactory estimates of critical buckling stresses for a fairly wide range of cylindrical shells subjected to simple loading conditions.

Treatment of the buckling problems of cylindrical shells under combined axial compression and external pressure, and combined shear and axial compression, may be seen elsewhere.[3,4]

11.3. Results for other shells

Within the voluminous literature on the buckling of shells, there are many papers that have tackled the linear elastic buckling of various other shell geometries such as conical, spherical, ellipsoidal and toroidal shells. Compared with circular cylindrical shells, exact solutions for these geometries generally require more complex mathematical tools, and the final results are not always convenient for practical use. Here, rather than attempting to derive the theoretical buckling solutions for these shells (like we did for cylindrical shells), we will simply present practical results for typical loading configurations, for the convenience of the shell designer.

11.3.1 Conical shells

Conical shells experiencing compression are encountered in liquid-storage vessels, pressure vessels, pipelines, silos and aerospace vehicles, among others.

The theoretical critical buckling load for long conical frusta subjected to uniform axial compression has been obtained by Seide as follows:[5]

$$P_{cr} = \frac{2\pi E}{\left[3\left(1-\nu^2\right)\right]^{1/2}} t^2 \cos^2 \alpha \tag{11.61}$$

Figure 11.8 Conical shell frustum under compression: (a) shell subjected to a total axial load *P* uniformly distributed around shell opening; (b) shell subjected to uniform external pressure *p*.

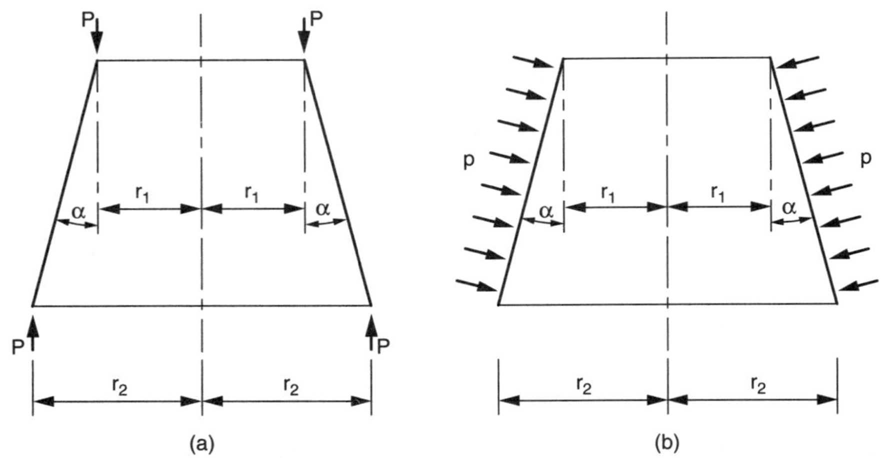

where, as illustrated in Figure 11.8(a), α denotes the vertical angle of the cone, r_1 the radius of the smaller opening, and r_2 the radius of the larger opening; P is the *total* load over the circumference of the opening (not the load per unit length of circumference).

Owing to imperfections, experimental test results are generally lower than the above prediction. A lower-bound to the experimental data may be obtained by applying a factor γ as follows:

$$P_{cr} = \gamma \frac{2\pi E}{\left[3\left(1-\nu^2\right)\right]^{1/2}} t^2 \cos^2\alpha \tag{11.62}$$

NASA[6] recommends a value of 0.33 for γ, provided $10° \leq \alpha \leq 75°$.

For conical frusta subjected to uniform external pressure p (see Figure 11.8(b)), the critical buckling pressure may be expressed as[7]

$$p_{cr} = \gamma \frac{0.92E}{\left(\frac{L}{R}\right)\left(\frac{R}{t}\right)^{5/2}} \tag{11.63}$$

where R is the average *second radius of curvature* of the conical shell, and L is the length of the shell measured along the inclined sides:

$$R = \frac{r_1 + r_2}{2\cos\alpha} \tag{11.64a}$$

$$L = \frac{r_2 - r_1}{\sin\alpha} \tag{11.64b}$$

The factor γ depends on the ratio r_1 / r_2 of the conical frustum. When r_1 / r_2 approaches 1.0, γ may be taken as 1.0; the p_{cr} value of the cone becomes practically identical to that of a circular cylindrical shell of radius R and length L. For lower ratios of r_1 / r_2, NASA[6] recommends a value for γ of 0.75 as a safe lower-bound for experimental data. Note that this γ value (0.75) is much higher than that recommended for axial compression (0.33), since the sensitivity of conical shells to geometric imperfections, like that of cylindrical shells, is higher for axial compression than for uniform external pressure.

11.3.2 Spherical shells

The buckling of a complete spherical shell under uniform external pressure seems to have been first studied by Zoelly in 1915.[4] The theoretical development of the solution for axisymmetric buckling may be seen in some of the early texts on shells.[3,4] For this classical problem, the result obtained for the critical buckling pressure p_{cr} is as follows:

$$p_{cr} = \frac{2E}{\left[3\left(1-\nu^2\right)\right]^{1/2}} \frac{t^2}{a^2} \tag{11.65a}$$

The corresponding critical buckling stress is

$$\sigma_{cr} = \frac{ap_{cr}}{2t} = \frac{E}{\left[3\left(1-\nu^2\right)\right]^{1/2}} \frac{t}{a} \tag{11.65b}$$

which, interestingly, is the same result as was obtained for the buckling stress of an axially loaded cylinder of radius a – see equation (11.26). It must be pointed out that the buckling capacity of a thin spherical shell is particularly sensitive to small deviations of the surface from a perfect sphere, with the result that experimental values will be much lower than theoretical predictions.

For shallow spherical shells under uniform external pressure, failure may occur via a non-linear phenomenon called snap-through buckling, which has been the subject of many studies.[8–10] The critical buckling pressure p_{cr} of a spherical cap (see Figure 11.9) may be expressed as a ratio of the critical buckling pressure p_{ss} of the complete spherical shell:

$$\frac{p_{cr}}{p_{ss}} = f\left(\mu\right) \tag{11.66}$$

where p_{ss} is as given by equation (11.65a), and the parameter μ is defined as follows:

$$\mu = 2\left[12\left(1-\nu^2\right)\right]^{1/4} \left(\frac{a}{t}\right)^{1/2} \sin\frac{\phi_s}{2} \tag{11.67}$$

The function $f\left(\mu\right)$ depends on the conditions at the edge of the shell. To allow for the large discrepancy between theory and experiment, and for the scatter in the test data (which depends on the actual support conditions and the exact nature of the shell imperfections), a lower-bound ratio for spherical caps with fixed edges has been recommended as follows:[11]

$$\frac{p_{cr}}{p_{ss}} = 0.14 + \frac{3.2}{\mu^2} \quad \left(\mu > 2.0\right) \tag{11.68}$$

Figure 11.9 Spherical shell cap under uniform external pressure *p*.

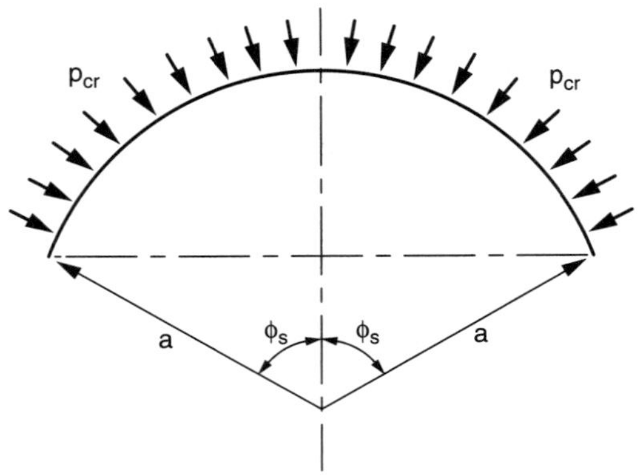

11.3.3 Ellipsoidal shells

Studies of the stability of ellipsoids of revolution have been conducted by many investigators over the past several decades.[12] Figure 11.10 shows two types of ellipsoidal shells of revolution, one being the case when the vertical semi-axis b is greater than the lateral semi-axis a (giving a *prolate* ellipsoid), and the other being the case when b is smaller than a (giving an *oblate* ellipsoid). The special case when $b = a$ gives a spherical shell, which we have already discussed.

The buckling behaviour of an ellipsoid of revolution very much depends on the ratio b / a. In the present context, we will denote this ratio by ρ. Clearly external pressure p acting on the ellipsoid will always induce compression in the shell (irrespective of the ratio ρ). Internal pressure generally tends to induce tension in the shell, but as we saw in Chapter 3, internal pressure will also induce compression in the hoop direction in zones adjacent to the equator if the ratio ρ is less than $1/\sqrt{2}$. For such oblate ellipsoids (with $\rho < 1/\sqrt{2}$), local buckling of the shell could occur if the internal pressure is high enough and the shell is thin enough.

In their well-known monograph, Mushtari and Galimov[13] treat the stability problem of the ellipsoid of revolution under uniform external pressure, and present a general solution for the critical buckling pressure as

$$p_{cr} = \frac{2Et^2}{\left[3\left(1-\nu^2\right)\right]^{1/2}}\left(\frac{1}{2r_1r_2 - r_2^2}\right) \tag{11.69}$$

where, similar to the notation generally used in this book, r_1 and r_2 are the two principal radii of curvature of the shell midsurface at the point in question. For the ellipsoid of revolution, r_1 and r_2 may be written in terms of the radius R of the circle of latitude at any given point (refer to Chapter 3) as follows:

$$r_1 = \frac{\left[\left(b^2 - a^2\right)R^2 + a^4\right]^{3/2}}{a^4 b} \tag{11.70a}$$

$$r_2 = \frac{\left[\left(b^2 - a^2\right)R^2 + a^4\right]^{1/2}}{b} \tag{11.70b}$$

Figure 11.10 Two types of ellipsoidal shells of revolution: (a) prolate ellipsoid ($b > a$); (b) oblate ellipsoid ($b < a$).

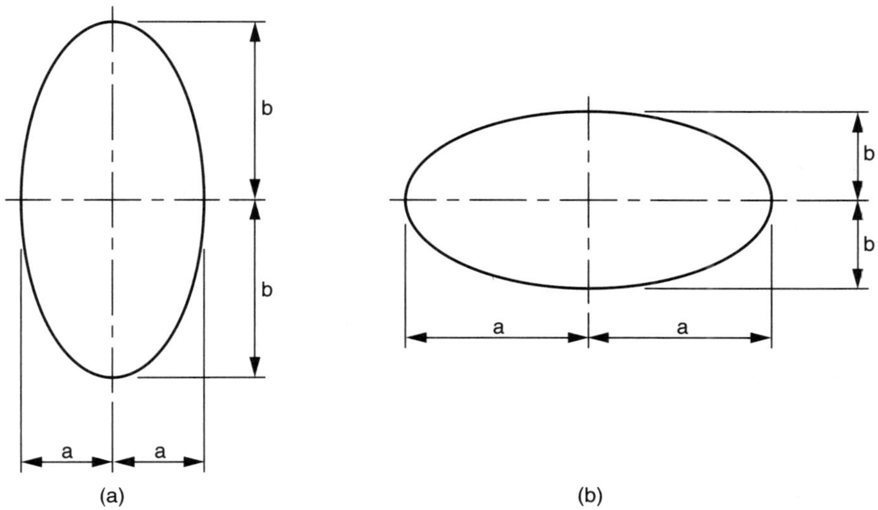

For the prolate ellipsoid, the weakest plane is the equatorial plane, where $R = a$. Evaluating the principal radii of curvature at this location, we get

$$r_1 = \frac{\left[\left(b^2 - a^2\right)a^2 + a^4\right]^{3/2}}{a^4 b} = \frac{b^3 a^3}{a^4 b} = \frac{b^2}{a} \tag{11.71a}$$

$$r_2 = \frac{\left[\left(b^2 - a^2\right)a^2 + a^4\right]^{1/2}}{b} = \frac{ba}{b} = a \tag{11.71b}$$

Substituting these values into equation (11.69), we obtain

$$p_{cr} = \frac{2Et^2}{\left[3\left(1-\nu^2\right)\right]^{1/2}} \left(\frac{1}{2b^2 - a^2}\right)$$

$$= \frac{2E}{\left[3\left(1-\nu^2\right)\right]^{1/2}} \frac{t^2}{a^2} \left(\frac{1}{2\rho^2 - 1}\right) \quad ; \quad \rho = \frac{b}{a} > 1.0 \tag{11.72}$$

If we write this relationship as

$$\frac{p_{cr}\left[3\left(1-\nu^2\right)\right]^{1/2} a^2}{2Et^2} = \frac{p_{cr}}{p_{ss}} = \frac{1}{2\rho^2 - 1} = f(\rho) \tag{11.73}$$

where

$$f(\rho) = \frac{1}{2\rho^2 - 1} \tag{11.74}$$

we may plot the function $f(\rho)$ versus ρ, and use the ensuing curve to read off the numerical values of the non-dimensional critical buckling pressure p_{cr} / p_{ss}, which is the factor by which the critical buckling stress of the ellipsoid is reduced relative to that of a sphere of radius a. This curve is nearly the same as that obtained by Danielson[14] when $a / t = 1000$.

For the oblate ellipsoid, the weakest regions are those around the axis of revolution (i.e. around the poles of the ellipsoid), where $R = 0$. Evaluating the principal radii of curvature at these locations, we obtain

$$r_1 = \frac{\left[(b^2 - a^2)0 + a^4\right]^{3/2}}{a^4 b} = \frac{a^6}{a^4 b} = \frac{a^2}{b} \tag{11.75a}$$

$$r_2 = \frac{\left[(b^2 - a^2)0 + a^4\right]^{1/2}}{b} = \frac{a^2}{b} \tag{11.75b}$$

As expected from the symmetry around the poles, r_1 and r_2 are equal. Substituting for r_1 and r_2 in equation (11.69), we obtain

$$\begin{aligned} p_{cr} &= \frac{2Et^2}{\left[3(1-\nu^2)\right]^{1/2}} \frac{b^2}{a^4} \\ &= \frac{2E}{\left[3(1-\nu^2)\right]^{1/2}} \frac{t^2}{a^2} \frac{b^2}{a^2} \\ &= \frac{2E}{\left[3(1-\nu^2)\right]^{1/2}} \frac{t^2}{a^2} \rho^2 \quad ; \quad \rho = \frac{b}{a} < 1.0 \end{aligned} \tag{11.76}$$

For the oblate ellipsoid, the radius $\bar{a}$ of the equivalent sphere (i.e. the sphere with the same buckling pressure) is therefore given by

$$\bar{a} = \frac{a^2}{b} \tag{11.77}$$

In conclusion, we note that the critical buckling pressure rapidly decreases (in relation to that of a sphere of radius a) as b increases above a in the case of prolate ellipsoids, or as b decreases below a in the case of oblate ellipsoids. As with the spherical shell (and all other shells for that matter), experimental results differ substantially from theoretical predictions, and knock-down factors are necessary in order to adjust theoretical values to more realistic levels.

Figure 11.11 Circular toroidal shell under uniform external pressure *p*.

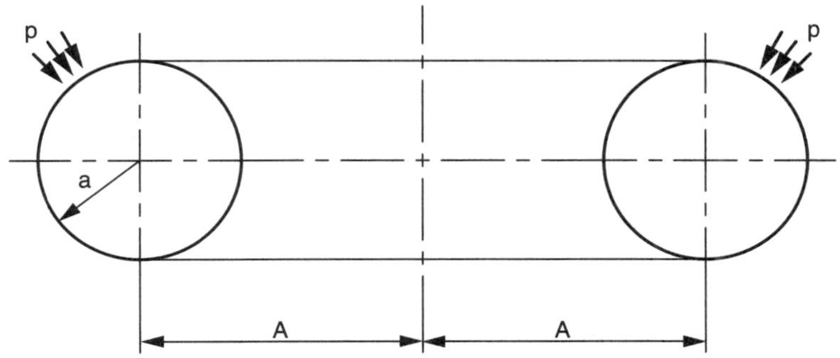

11.3.4 Toroidal shells

Studies of the stability of circular toroidal shells under uniform external pressure include the theoretical papers of Sobel and Flügge,[15] Jordan[16] and Panagiotopoulos,[17] and the numerical computations of Galletly and Blachut.[18] A simple but fairly accurate prediction of the theoretical buckling pressure of a circular toroidal shell under uniform external pressure was provided by Jordan:[16]

$$p_{cr} = \frac{0.1738\,E}{\left(1-\nu^2\right)^{2/3}}\left(\frac{t}{a}\right)^{7/3}\left(\frac{a}{A}\right)^{2/3} \tag{11.78}$$

where a is the radius of the circular cross-section, and A is the mean radius of the torus (Figure 11.11). If Poisson's ratio is taken as $\nu = 0.3$, the above expression simplifies to

$$p_{cr} = 0.185\,E\left(\frac{t}{a}\right)^{7/3}\left(\frac{a}{A}\right)^{2/3} \tag{11.79}$$

The equatorial bending of toroidal shells with prolate ellipsoidal cross-sections[19] shows similarities with that of cylindrical shells of equivalent radius. In the neighbourhood of the intrados of the toroid (inner shell wall), the equivalent cylindrical radius is $A - a$, while in the neighbourhood of the extrados (outer shell wall), the equivalent cylindrical radius is $A + a$. For externally pressurised toroidal shells of prolate ellipsoidal cross-section, and in the absence of better information, the results for cylindrical shells of equivalent radius and length equal to $2b$ (the vertical dimension of the elliptical cross-section as illustrated in Figure 11.10(a)) may serve as lower-bound estimates of the critical buckling pressure, particularly where $b / a > 3.0$.

References

1. Zingoni A. Liquid-containment shells of revolution: A review of recent studies on strength, stability and dynamics. *Thin-Walled Structures*, 2015, **87**, 102–114.
2. Donnell L.H. A new theory for the buckling of thin cylinders under axial compression and bending. *Transactions of the American Society of Mechanical Engineers*, 1934, **56**, 795–806.
3. Flügge W. *Stresses in Shells*. Springer-Verlag, Berlin, 1973.
4. Timoshenko S.P. and Gere J.M. *Theory of Elastic Stability*. McGraw Hill, New York, 1961.

5. Seide P. Axisymmetric buckling of circular cones under axial compression. *Journal of Applied Mechanics*, 1956, **23** (4), 625–628.
6. NASA SP-8019. *Buckling of Thin-Walled Truncated Cones*. National Aeronautics and Space Administration, Langley Research Centre, Hampton (Virginia), 1968.
7. Seide P. On the buckling of truncated conical shells under uniform hydrostatic pressure. *Proceedings of the IUTAM Symposium on the Theory of Thin Elastic Shells*. North-Holland Publishing Company, Amsterdam, 1960, 363–388.
8. Von Karman T. and Tsien H.S. The buckling of spherical shells by external pressure. *Journal of Aeronautical Science*, 1939, **7**, 43–50.
9. Weinitschke H. On the stability problem for shallow spherical shells. *Journal of Mathematics and Physics*, 1960, **38** (4), 209–231.
10. Budiansky B. Buckling of clamped shallow spherical shells. *Proceedings of the IUTAM Symposium on the Theory of Thin Elastic Shells*. North-Holland Publishing Company, Amsterdam, 1960, 64–94.
11. NASA SP-8032. *Buckling of Thin-Walled Doubly Curved Shells*. National Aeronautics and Space Administration, Langley Research Centre, Hampton (Virginia), 1969.
12. Krivoshapko S.N. Research on general and axisymmetric ellipsoidal shells used as domes, pressure vessels and tanks. *Applied Mechanics Reviews, ASME*, 2007, **60**, 336–355.
13. Mushtari H.M. and Galimov H.Z. *Nonlinear Theory of Elastic Shells*. Kazan, Tatknigoizdat, 1957.
14. Danielson D.A. Buckling and initial postbuckling behaviour of spheroidal shells under pressure. *American Institute of Aeronautics and Astronautics (AIAA) Journal*, 1969, **7** (5), 936–944.
15. Sobel L.H. and Flügge W. Stability of toroidal shells under uniform external pressure. *American Institute of Aeronautics and Astronautics (AIAA) Journal*, 1969, **5**, 425–431.
16. Jordan P.F. Buckling of toroidal shells under hydrostatic pressure. *American Institute of Aeronautics and Astronautics (AIAA) Journal*, 1973, **11**, 1439–1441.
17. Panagiotopoulos G.D. Stress and stability analysis of toroidal shells. *International Journal of Pressure Vessels and Piping*, 1985, **20**, 87–100.
18. Galletly G.D. and Blachut J. Stability of complete circular and non-circular toroidal shells. *Proceedings of the Institution of Mechanical Engineers*, 1995, **209**, 245–255.
19. Zingoni A., Enoma N. and Govender N. Equatorial bending of an elliptic toroidal shell. *Thin-Walled Structures*, 2015, **96**, 286–294.

Shell Structures in Civil and Mechanical Engineering
ISBN 978-0-7277-6028-9

http://dx.doi.org/10.1680/ssicame.60289.391

Chapter 12
Finite element analysis of shells

12.1. Introduction

Many problems of the analysis of shells are not amenable to solution by exact mathematical methods, either because the associated differential equations are too complex, or the boundary conditions at the edges of the shell are impossible to satisfy. Even where exact mathematical solutions exist, these may be too cumbersome for practical implementation, often involving complex functions and summations of series that do not always converge in a satisfactory manner. Fortunately, resort can be made to approximate but sufficiently accurate numerical solution procedures as a means of getting around these problems.

Two numerical methods, particularly useful in the context of plate analysis and shell analysis, are the Finite Difference (FD) method and the Finite Element (FE) method. The FD method approximates the differential equations with difference equations (i.e. derivative terms are replaced with equivalent finite-difference expressions), which are then solved numerically. On the other hand, the FE method subdivides the physical domain of the problem (e.g. the shell) into many smaller and much simpler pieces or elements (the properties of which can easily be determined), and then tries to predict the behaviour of the complete structure by combining (or assembling) the individual properties of the elements in a systematic way. Both methods ultimately involve solving (usually large) sets of simultaneous algebraic equations, a task that is very much suited to implementation via computer programs. The FE method is more versatile, and is certainly used more frequently. For these reasons, we will devote this chapter to the FE analysis of shells.

It must be pointed out at the outset that the FE method is a large area of study in computational mechanics, covering a wide diversity of topics ranging from constitutive modelling, choice of displacement fields and element formulations, to mesh generation, assembly methods and solution procedures, as well as related mathematical considerations such as convergence of solutions and error estimates. The literature on the subject is vast, and books devoted exclusively to the FE method or to FE analysis are readily available.[1–7] The aim of this chapter is to present the essence of FE analysis, and to illustrate how the concepts are implemented in the context of shell analysis.

Like many other structures, a shell may be modelled using very general 3-dimensional finite elements. However, and similar to the benefits gained by using shell theory instead of the general theory of elasticity, we may exploit the property that the thickness of a shell is much smaller than its other two dimensions, and adopt plate and shell elements in our modelling, instead of general 3-dimensional finite elements, which involve more computational effort.

From a physical perspective, finite elements are distinguished by their shape, and also by the number and arrangement of their nodes. Having a variety of possible geometrical shapes allows considerable flexibility in discretising the structure, to ensure a choice that models the physical domain of the structure (including its boundaries) as precisely as possible, and which delivers accurate results at reasonable cost. The nodes of the element are the locations at which we may specify the degrees of freedom of the element (i.e. the directions in which it can translate or the planes in which it can rotate).

Figure 12.1 Examples of shell finite elements: (a) flat triangular shell element with 3 nodes; (b) flat rectangular shell element with 4 nodes; (c) doubly curved triangular shell element with 6 nodes; (d) doubly curved quadrilateral shell element with 8 nodes; (e) doubly curved thick shell element with 20 nodes.

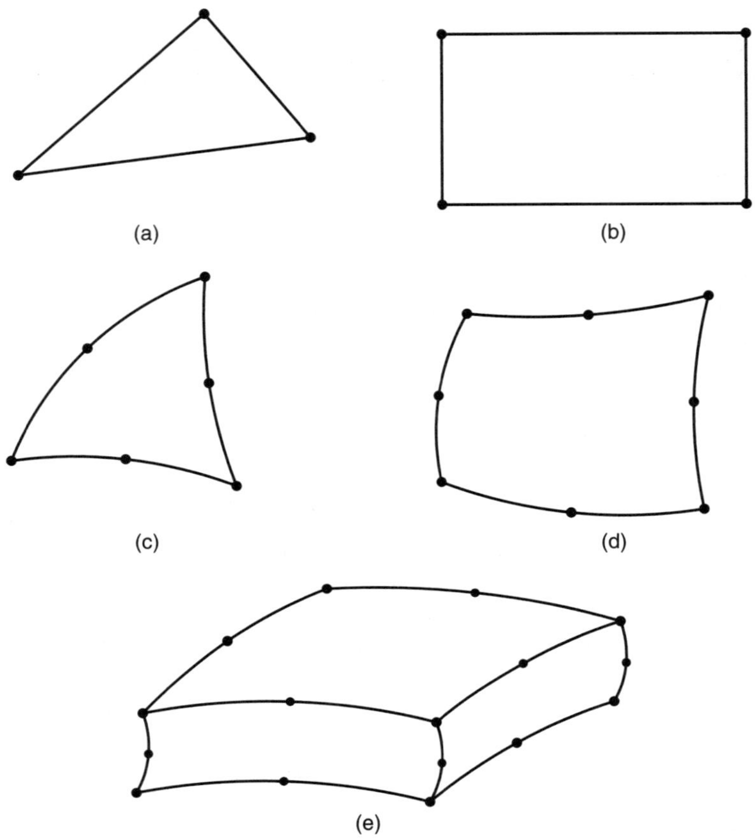

Typically, nodes are located at the ends of the element, at the corners, or along the edges. The nodes also serve as the points of connection between adjacent elements, across which conditions of continuity can be enforced, and overall assembly achieved.

In general, the greater the number of degrees of freedom an element has, the more accurately it can simulate the real behaviour of the structure, but also the more expensive the associated computations (typically numerical integrations) become. A common strategy is to use the simplest possible finite element capable of delivering an acceptable level of accuracy, but clearly other considerations need to be taken into account.

Shell elements may be flat (like plate elements), or they may be curved. Thin elements will suffice for thin shells, but where the shell is relatively thick and transverse shear deformations cannot be ignored, brick-like shell elements, with more than one layer of nodes in the thickness direction, may be necessary. Figure 12.1 shows examples of shell finite elements, while Figure 12.2 shows examples of shell structures discretised into finite-element meshes.

In the next sections, we will begin by illustrating basic properties of shell finite elements by reference to a simple 2-noded conical shell element. This will be followed by an outline of the general procedures for

Figure 12.2 Examples of finite-element meshes for shell structures: (a) cylindrical shell barrel roof; (b) conical shell space capsule.

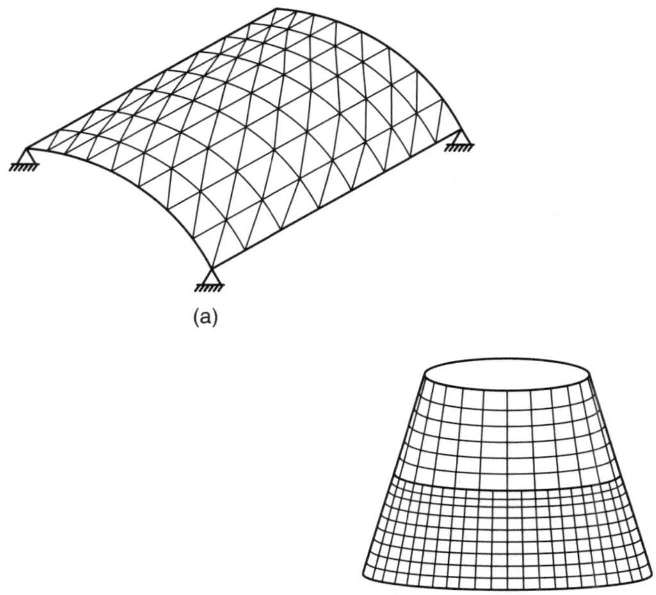

(a)

(b)

finite-element stress analysis, dynamic analysis and stability analysis (linear elastic buckling). Towards the end of the chapter, we will briefly discuss the use of flat elements in modelling shell structures, and then consider in more detail the concept of isoparametric finite elements as applied to general shells and shells of revolution.

12.2. A simple formulation for shells of revolution

12.2.1 Strain-displacement and constitutive relations

For arbitrary curved shells with axisymmetry, the simplest approach is to discretise the curved meridian into a series of short straight segments, so that the whole shell of revolution becomes an assembly of short conical frusta. A refinement of this modelling is to allow the segments to be curved, so that the profile of the meridian is more closely matched.

Figure 12.3 shows a curved element in the $\{s, \theta\}$ coordinate system, where s is the arc-length coordinate along the curved meridian, and θ is the angular coordinate along a hoop circle. For axisymmetric loading conditions, the actions of interest are the direct stress resultants $\{N_s, N_\theta\}$ and the bending moments $\{M_s, M_\theta\}$, all per unit length of the respective shell edge (see Figure 12.3(a)).

Figure 12.3(b) shows the displacements acting at a point on a meridian of the element: v in the direction of the tangent to the meridian, and w normal to the shell midsurface.

For this element experiencing axisymmetric deformations, we may readily write down the strains of the midsurface (i.e. direct strains and changes in curvature) in terms of the basic displacements v and w, using the relationships already established in Chapters 2 and 6. Taking equations (2.54), (2.57), (6.10) and (6.12), and replacing $d\phi$ with (ds / r_1) and r_2 with $(R / \sin\phi)$, we obtain

$$\boldsymbol{\varepsilon} = \begin{Bmatrix} \varepsilon_s \\ \varepsilon_\theta \\ \chi_s \\ \chi_\theta \end{Bmatrix} = \begin{Bmatrix} \dfrac{dv}{ds} + \dfrac{w}{r_1} \\ \dfrac{1}{R}(v\cos\phi + w\sin\phi) \\ -\dfrac{d}{ds}\left(\dfrac{v}{r_1}\right) + \dfrac{d^2w}{ds^2} \\ -\left(\dfrac{v}{r_1}\right)\dfrac{\cos\phi}{R} + \left(\dfrac{dw}{ds}\right)\dfrac{\cos\phi}{R} \end{Bmatrix} \tag{12.1}$$

Figure 12.3 A doubly curved shell element in the sθ coordinate system: (a) actions N_s, N_θ, M_s and M_θ; (b) displacements v and w.

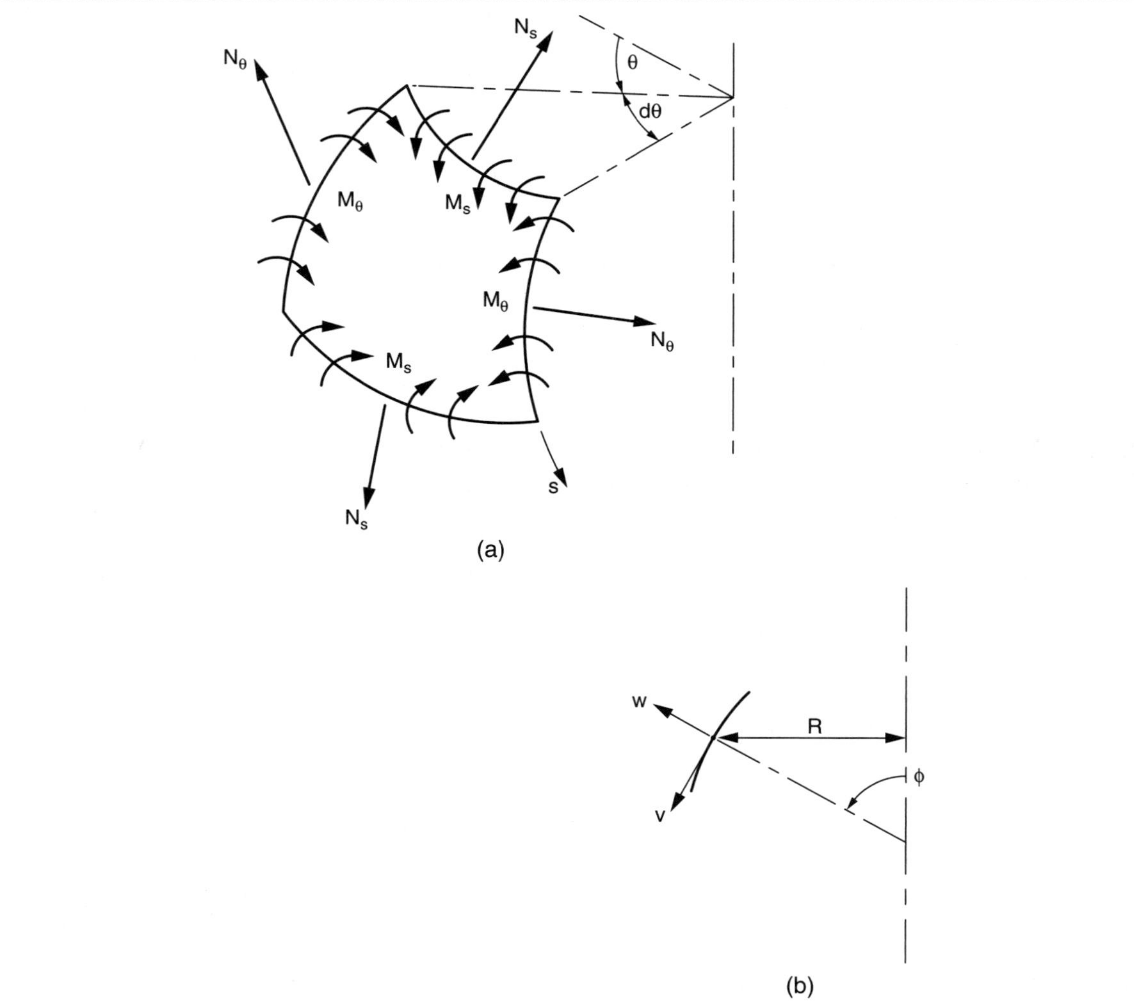

The actions in the shell (i.e. stress resultants and bending moments) may also be readily expressed in terms of the strains in the shell. From equations (6.6) and (6.16), we may write

$$\boldsymbol{\sigma} = \begin{Bmatrix} N_s \\ N_\theta \\ M_s \\ M_\theta \end{Bmatrix} = \begin{Bmatrix} \dfrac{Et}{1-\nu^2}\left(\varepsilon_s + \nu\varepsilon_\theta\right) \\ \dfrac{Et}{1-\nu^2}\left(\nu\varepsilon_s + \varepsilon_\theta\right) \\ \dfrac{Et^3}{12\left(1-\nu^2\right)}\left(\chi_s + \nu\chi_\theta\right) \\ \dfrac{Et^3}{12\left(1-\nu^2\right)}\left(\nu\chi_s + \chi_\theta\right) \end{Bmatrix} = \mathbf{D}\boldsymbol{\varepsilon} \tag{12.2}$$

where $\mathbf{D}$ is a matrix of material stiffnesses:

$$\mathbf{D} = \frac{Et}{1-\nu^2}\begin{bmatrix} 1 & \nu & 0 & 0 \\ \nu & 1 & 0 & 0 \\ 0 & 0 & \dfrac{t^2}{12} & \dfrac{\nu t^2}{12} \\ 0 & 0 & \dfrac{\nu t^2}{12} & \dfrac{t^2}{12} \end{bmatrix} \tag{12.3}$$

In the treatment of this chapter, bold notation (such as $\boldsymbol{\varepsilon}$, $\boldsymbol{\sigma}$ and $\mathbf{D}$) denotes column vectors and matrices. To explicitly show the elements of a column vector or a matrix, we will use curled brackets for a column vector and square brackets for a matrix. Note that here $\boldsymbol{\sigma}$ is used to denote the vector of actions $\{N_s, N_\theta, M_s, M_\theta\}$, and is not a vector of stresses.

12.2.2 Element shape functions

Let us pursue the simple example of the straight-sided conical shell element, and use this to illustrate fairly general concepts. Let the element have two nodes at the ends, denoted by the labels 1 and 2 as shown in Figure 12.4. The length of the element is denoted by l. We will use the distance coordinate s, measured from node 1 as shown, to define the position of any point on the element.

We seek to express the displacements at any arbitrary point within the element in terms of the displacements of the two nodes of the element. The displacements at each node will comprise v in the axial direction, w in the direction normal to the axis of the element, and the rotation φ (given by dw/ds). Thus our simple 2-noded element has a total of six degrees of freedom (three per node), which we may collectively represent by the column vector $\mathbf{d}$ as follows:

$$\mathbf{d} = \begin{Bmatrix} d_1 \\ d_2 \end{Bmatrix} = \begin{Bmatrix} v_1 \\ w_1 \\ \varphi_1 \\ v_2 \\ w_2 \\ \varphi_2 \end{Bmatrix} \quad ; \quad \varphi = \frac{dw}{ds} \tag{12.4}$$

where subscripts 1 and 2 refer to nodes 1 and 2 respectively.

Figure 12.4 A 2-noded conical shell element of a shell of revolution.

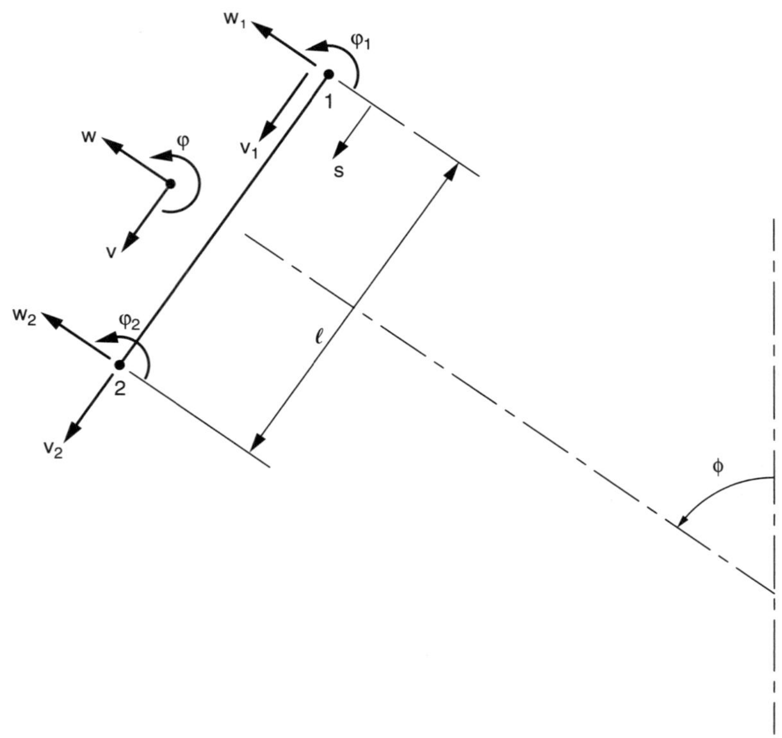

In the FE method, the usual approach is to assume a displacement field within the element in the form of polynomial functions or some other suitable functions. For our 2-noded element, let us take v as linearly varying with s, and w as a cubic function of s:

$$v(s) = a_1 + a_2 s \tag{12.5a}$$

$$w(s) = a_3 + a_4 s + a_5 s^2 + a_6 s^3 \tag{12.5b}$$

We will need an expression for φ. This follows from equation (12.5b) as

$$\varphi(s) = \frac{dw(s)}{ds} = a_4 + 2a_5 s + 3a_6 s^2 \tag{12.5c}$$

Note that such a displacement field allows for axial strain to be constant across the length of the element, and for meridional curvature to vary linearly across the element.

For this displacement field, the boundary conditions are

At node 1:

$$v(0) = v_1 \quad \Rightarrow \quad a_1 = v_1 \tag{12.6a}$$

$$w(0) = w_1 \quad \Rightarrow \quad a_3 = w_1 \tag{12.6b}$$

$$\varphi(0) = \varphi_1 \quad \Rightarrow \quad a_4 = \varphi_1 \tag{12.6c}$$

At node 2:

$$v(l) = v_2 \quad \Rightarrow \quad a_1 + a_2 l = v_2 \tag{12.6d}$$

$$w(l) = w_2 \quad \Rightarrow \quad a_3 + a_4 l + a_5 l^2 + a_6 l^3 = w_2 \tag{12.6e}$$

$$\varphi(l) = \varphi_2 \quad \Rightarrow \quad a_4 + 2a_5 l + 3a_6 l^2 = \varphi_2 \tag{12.6f}$$

Solving the above set of equations for the constants a_1 to a_6, we obtain the full set of solutions as

$$a_1 = v_1 \tag{12.7a}$$

$$a_2 = \frac{v_2 - v_1}{l} \tag{12.7b}$$

$$a_3 = w_1 \tag{12.7c}$$

$$a_4 = \varphi_1 \tag{12.7d}$$

$$a_5 = \frac{1}{l^2}\left(3w_2 - 3w_1 - 2l\varphi_1 - l\varphi_2\right) \tag{12.7e}$$

$$a_6 = \frac{1}{l^3}\left(2w_1 - 2w_2 + l\varphi_1 + l\varphi_2\right) \tag{12.7f}$$

Substituting these solutions into equations (12.5a) and (12.5b) and rearranging, we obtain the displacements at arbitrary interior points in terms of the displacements at the nodes as follows:

$$v(s) = \left(1 - \frac{s}{l}\right)v_1 + \left(\frac{s}{l}\right)v_2 = \left(1 - \bar{s}\right)v_1 + \bar{s}\,v_2 \tag{12.8a}$$

$$\begin{aligned} w(s) &= \left(1 - 3\frac{s^2}{l^2} + 2\frac{s^3}{l^3}\right)w_1 + l\left(\frac{s}{l} - 2\frac{s^2}{l^2} + \frac{s^3}{l^3}\right)\varphi_1 + \left(3\frac{s^2}{l^2} - 2\frac{s^3}{l^3}\right)w_2 + l\left(-\frac{s^2}{l^2} + \frac{s^3}{l^3}\right)\varphi_2 \\ &= \left(1 - 3\bar{s}^2 + 2\bar{s}^3\right)w_1 + l\left(\bar{s} - 2\bar{s}^2 + \bar{s}^3\right)\varphi_1 + \left(3\bar{s}^2 - 2\bar{s}^3\right)w_2 + l\left(-\bar{s}^2 + \bar{s}^3\right)\varphi_2 \end{aligned} \tag{12.8b}$$

where

$$\bar{s} = \frac{s}{l} \quad \left(0 \le \bar{s} \le 1.0\right) \tag{12.9}$$

In matrix form, we may write

$$\begin{Bmatrix} v \\ w \end{Bmatrix} = \begin{bmatrix} (1-\bar{s}) & 0 & 0 & \bar{s} & 0 & 0 \\ 0 & (1-3\bar{s}^2+2\bar{s}^3) & l(\bar{s}-2\bar{s}^2+\bar{s}^3) & 0 & (3\bar{s}^2-2\bar{s}^3) & l(-\bar{s}^2+\bar{s}^3) \end{bmatrix} \begin{Bmatrix} v_1 \\ w_1 \\ \varphi_1 \\ v_2 \\ w_2 \\ \varphi_2 \end{Bmatrix} \tag{12.10}$$

– that is,

$$\mathbf{u} = \begin{Bmatrix} v \\ w \end{Bmatrix} = \mathbf{Nd} \tag{12.11}$$

where **N** is the matrix of *element shape functions* (a 2×6 matrix in the present example).

12.2.3 Strain-displacement matrix

A straight-sided (conical) element has no meridional curvature, so $r_1 = \infty$ and the strain matrix simplifies to

$$\boldsymbol{\varepsilon} = \begin{Bmatrix} \varepsilon_s \\ \varepsilon_\theta \\ \chi_s \\ \chi_\theta \end{Bmatrix} = \begin{Bmatrix} \dfrac{dv}{ds} \\ \dfrac{1}{R}(v\cos\phi + w\sin\phi) \\ \dfrac{d^2w}{ds^2} \\ \dfrac{dw}{ds}\left(\dfrac{\cos\phi}{R}\right) \end{Bmatrix} \tag{12.12}$$

Evaluating the four components of the strain matrix of equation (12.12), using the expressions for the displacement field $\{v(s), w(s)\}$ in terms of nodal displacements (i.e. equations (12.8)), we obtain

$$\varepsilon_s = \frac{dv}{ds} = -\frac{1}{l}v_1 + \frac{1}{l}v_2 \tag{12.13a}$$

$$\begin{aligned} \varepsilon_\theta = \frac{1}{R}(v\cos\phi + w\sin\phi) &= \left(1-\frac{s}{l}\right)\left(\frac{\cos\phi}{R}\right)v_1 + \frac{s}{l}\left(\frac{\cos\phi}{R}\right)v_2 + \left(1-3\frac{s^2}{l^2}+2\frac{s^3}{l^3}\right)\left(\frac{\sin\phi}{R}\right)w_1 \\ &+ l\left(\frac{s}{l}-2\frac{s^2}{l^2}+\frac{s^3}{l^3}\right)\left(\frac{\sin\phi}{R}\right)\varphi_1 + \left(3\frac{s^2}{l^2}-2\frac{s^3}{l^3}\right)\left(\frac{\sin\phi}{R}\right)w_2 + l\left(-\frac{s^2}{l^2}+\frac{s^3}{l^3}\right)\left(\frac{\sin\phi}{R}\right)\varphi_2 \end{aligned} \tag{12.13b}$$

$$\chi_s = \frac{d^2w}{ds^2} = \left(-\frac{6}{l^2}+\frac{12s}{l^3}\right)w_1 + \left(-\frac{4}{l}+\frac{6s}{l^2}\right)\varphi_1 + \left(\frac{6}{l^2}-\frac{12s}{l^3}\right)w_2 + \left(-\frac{2}{l}+\frac{6s}{l^2}\right)\varphi_2 \tag{12.13c}$$

$$\chi_\theta = \frac{dw}{ds}\left(\frac{\cos\phi}{R}\right) = \left(-\frac{6s}{l^2}+\frac{6s^2}{l^3}\right)\left(\frac{\cos\phi}{R}\right)w_1 + \left(1-\frac{4s}{l}+\frac{3s^2}{l^2}\right)\left(\frac{\cos\phi}{R}\right)\varphi_1 + $$

$$+\left(\frac{6s}{l^2}-\frac{6s^2}{l^3}\right)\left(\frac{\cos\phi}{R}\right)w_2 + \left(-\frac{2s}{l}+\frac{3s^2}{l^2}\right)\left(\frac{\cos\phi}{R}\right)\varphi_2 \tag{12.13d}$$

Substituting expressions (12.13) into equation (12.12), we may finally write the strains within the element in terms of the nodal displacements of the element as follows:

$$\boldsymbol{\varepsilon} = \begin{Bmatrix} \varepsilon_s \\ \varepsilon_\theta \\ \chi_s \\ \chi_\theta \end{Bmatrix} = \mathbf{B}\begin{Bmatrix} v_1 \\ w_1 \\ \varphi_1 \\ v_2 \\ w_2 \\ \varphi_2 \end{Bmatrix} = \mathbf{B\,d} \tag{12.14}$$

where **B** is the *strain-displacement matrix*:

B =

$$\begin{bmatrix} -\frac{1}{l} & 0 & 0 & \frac{1}{l} & 0 & 0 \\ (1-\bar{s})\left(\frac{\cos\phi}{R}\right) & (1-3\bar{s}^2+2\bar{s}^3)\left(\frac{\sin\phi}{R}\right) & l(\bar{s}-2\bar{s}^2+\bar{s}^3)\left(\frac{\sin\phi}{R}\right) & \bar{s}\left(\frac{\cos\phi}{R}\right) & (3\bar{s}^2-2\bar{s}^3)\left(\frac{\sin\phi}{R}\right) & l(-\bar{s}^2+\bar{s}^3)\left(\frac{\sin\phi}{R}\right) \\ 0 & \frac{1}{l^2}(-6+12\bar{s}) & \frac{1}{l}(-4+6\bar{s}) & 0 & \frac{1}{l^2}(6-12\bar{s}) & \frac{1}{l}(-2+6\bar{s}) \\ 0 & (-6\bar{s}+6\bar{s}^2)\left(\frac{\cos\phi}{lR}\right) & (1-4\bar{s}+3\bar{s}^2)\left(\frac{\cos\phi}{R}\right) & 0 & (6\bar{s}-6\bar{s}^2)\left(\frac{\cos\phi}{lR}\right) & (-2\bar{s}+3\bar{s}^2)\left(\frac{\cos\phi}{R}\right) \end{bmatrix} \tag{12.15}$$

12.3. General procedure for stress analysis

12.3.1 Element stiffness matrix

Nodal forces **f** (in the generalised sense of being inclusive of forces and moments) may be thought of as actions that are statically equivalent to the system of loads actually acting over the element. They act in the same sense as the nodal displacements. Let **G** denote the distributed actual forces per unit volume of material of the element, and $\boldsymbol{\Phi}$ denote the distributed actual forces per unit area of the surface of the element. Examples of **G** are self-weight, inertial forces and thermal loads, while examples of $\boldsymbol{\Phi}$ are fluid pressure, friction and superimposed dead loads acting on the shell surface.

To obtain the equivalent nodal forces **f**, we will use the concept of virtual work. Let us impose a virtual nodal displacement $\delta\mathbf{d}$ at the nodes. This external virtual displacement will induce within the element displacements and strains given by equations (12.11) and (12.14) as follows:

$$\delta\mathbf{u} = \mathbf{N}\,\delta\mathbf{d} \tag{12.16a}$$

$$\delta\boldsymbol{\varepsilon} = \mathbf{B}\,\delta\mathbf{d} \tag{12.16b}$$

The virtual work done by the equivalent nodal actions is the summation of products of equivalent nodal actions (forces or moments) and corresponding virtual nodal displacements (translations or rotations):

$$\delta \mathbf{W}_e = \{\delta \mathbf{d}\}^{\mathrm{T}} \mathbf{f} \tag{12.17}$$

with the subscript *e* standing for *equivalent*.

The virtual work done by the internal stress resultants $\boldsymbol{\sigma}$ (which exist in the element prior to the application of the virtual displacement, and are a result of the actual loads acting over the element), and the actual loads acting within the volume of the element and over its surface, is given by

$$\delta \mathbf{W}_a = \{\delta \boldsymbol{\varepsilon}\}^{\mathrm{T}} \boldsymbol{\sigma} - \{\delta \mathbf{u}\}^{\mathrm{T}} \mathbf{G} - \{\delta \mathbf{u}\}^{\mathrm{T}} \boldsymbol{\Phi} \tag{12.18}$$

with the subscript *a* standing for *actual*.

Making use of relations (12.16) to eliminate $\delta \boldsymbol{\varepsilon}$ and $\delta \mathbf{u}$, equation (12.18) becomes

$$\begin{aligned} \delta \mathbf{W}_a &= \{\delta \mathbf{d}\}^{\mathrm{T}} \mathbf{B}^{\mathrm{T}} \boldsymbol{\sigma} - \{\delta \mathbf{d}\}^{\mathrm{T}} \mathbf{N}^{\mathrm{T}} \mathbf{G} - \{\delta \mathbf{d}\}^{\mathrm{T}} \mathbf{N}^{\mathrm{T}} \boldsymbol{\Phi} \\ &= \{\delta \mathbf{d}\}^{\mathrm{T}} (\mathbf{B}^{\mathrm{T}} \boldsymbol{\sigma} - \mathbf{N}^{\mathrm{T}} \mathbf{G} - \mathbf{N}^{\mathrm{T}} \boldsymbol{\Phi}) \end{aligned} \tag{12.19}$$

Equating $\delta \mathbf{W}_e$ (which is already a node-by-node summation of virtual work) to the total of $\delta \mathbf{W}_a$ (which is obtained by integrating over the volume V or surface S of the element as appropriate), we obtain

$$\mathbf{f} = \int_V \mathbf{B}^{\mathrm{T}} \boldsymbol{\sigma} \, dV - \int_V \mathbf{N}^{\mathrm{T}} \mathbf{G} \, dV - \int_S \mathbf{N}^{\mathrm{T}} \boldsymbol{\Phi} \, dS \tag{12.20}$$

Using equation (12.2) to eliminate $\boldsymbol{\sigma}$, then equation (12.14) to eliminate $\boldsymbol{\varepsilon}$, we obtain

$$\begin{aligned} \mathbf{f} &= \left(\int_V \mathbf{B}^{\mathrm{T}} \mathbf{D} \mathbf{B} \, dV\right) \mathbf{d} - \int_V \mathbf{N}^{\mathrm{T}} \mathbf{G} \, dV - \int_S \mathbf{N}^{\mathrm{T}} \boldsymbol{\Phi} \, dS \\ &= \mathbf{k} \, \mathbf{d} - \int_V \mathbf{N}^{\mathrm{T}} \mathbf{G} \, dV - \int_S \mathbf{N}^{\mathrm{T}} \boldsymbol{\Phi} \, dS \end{aligned} \tag{12.21}$$

where

$$\mathbf{k} = \int_V \mathbf{B}^{\mathrm{T}} \mathbf{D} \mathbf{B} \, dV \tag{12.22}$$

is the *element stiffness matrix*.

For the static equilibrium of the element, the nodal forces must themselves, of course, be in equilibrium with each other. Setting **f** in equation (12.21) equal to zero, we obtain

$$\mathbf{k} \, \mathbf{d} = \int_V \mathbf{N}^{\mathrm{T}} \mathbf{G} \, dV + \int_S \mathbf{N}^{\mathrm{T}} \boldsymbol{\Phi} \, dS + \mathbf{Q} \tag{12.23}$$

where the third term $\mathbf{Q}$ did not feature in equation (12.21), and is being introduced to account for any concentrated loads that may exist at the nodes of the element (i.e. point loads that happen to be applied directly to the nodes of the element).

At this stage, it is convenient to redefine $\mathbf{f}$ so that it denotes *only* the terms on the right-hand side of equation (12.23). This allows equation (12.23) to be written in the final form

$$\mathbf{k}\,\mathbf{d} = \mathbf{f} \tag{12.24}$$

where

$$\mathbf{f} = \int_V \mathbf{N}^{\mathrm{T}}\,\mathbf{G}\,dV + \int_S \mathbf{N}^{\mathrm{T}}\,\boldsymbol{\Phi}\,dS + \mathbf{Q} \tag{12.25}$$

Equation (12.24) expresses the static equilibrium of the element. The nodal forces $\mathbf{f}$ are related to the nodal displacements $\mathbf{d}$ via the element stiffness matrix $\mathbf{k}$, in the same way the force applied to a spring is related to the extension of the spring via the spring stiffness. The set of nodal forces $\mathbf{f}$ is said to be consistent if the first two terms of the right-hand side of equation (12.25) are calculated from the same shape functions as $\mathbf{k}$.

For our example of the 2-noded conical shell element, the stiffness matrix $\mathbf{k}$ may readily be computed using the $\mathbf{B}$ matrix given by equation (12.15), and the $\mathbf{D}$ matrix given by equation (12.3). The result is a 6×6 matrix. The vector $\mathbf{f}$ is computed using the shape functions defined by equations (12.10) and (12.11), once the loading on the shell has been specified.

12.3.2 Assembly and solution of system equations

For the detailed procedure of assembling the stiffness matrix for the whole structure (in this case the whole shell), and of solving the ensuing set of system equations, reference should be made to texts devoted exclusively to the finite-element method.[2–7] Here it will suffice to summarise the key steps.

Let us assume the structure has already been discretised into a suitable mesh of finite elements, and the matrices $\mathbf{k}$ for the individual elements calculated in local element coordinates as per the considerations of the last section. We now need to define a coordinate system for the whole structure, so that all elements have a common reference system. By using appropriate transformation matrices, we may then express element stiffness matrices in terms of the global coordinates of the system.

The various finite elements making up the structure are joined to each other at certain points, or along certain lines or surfaces. The nodes of the structure (as opposed to the nodes of an individual element) are typically the locations at which the elements are connected together. The total number of degrees of freedom of the structure, which we will denote by the integer n, will generally be given by the number of degrees of freedom (i.e. possible displacements) at each structural node multiplied by the total number of structural nodes in the entire structure, less any freedoms that are constrained by supports or some other restraints.

The stiffness of a structural node may be envisaged as the sum of the node stiffnesses of all elements meeting at that node. Assembly of the system stiffness matrix $\mathbf{K}$ (also referred to as the global stiffness matrix) is then achieved by simply superimposing the element stiffness matrices (in the global coordinate system) in accordance with the way the elements are connected. The result is a relatively large matrix representing the stiffness of the structure as a whole, and of a size $n\times n$ that corresponds to the n degrees of freedom of the structure.

For the whole system, we may write down the equilibrium equations as follows:

$$\mathbf{K}\,\mathbf{X} = \mathbf{F} \tag{12.26}$$

where **X** is the vector of nodal freedoms (generalised displacements) of the system, and **F** is the vector of nodal loads (generalised forces) of the system. The unknowns are the generalised displacements **X**. Once the **X** for the whole system have been solved for, we transform the results back to element level, to get nodal displacements **d** in local coordinates. For the example of the axisymmetric shell element, with **d** now known, the internal actions (i.e. stress resultants and bending moments) in the element are evaluated from equations (12.2) and (12.14):

$$\sigma = \begin{Bmatrix} N_s \\ N_\theta \\ M_s \\ M_\theta \end{Bmatrix} = \mathbf{D}\varepsilon = \mathbf{D}\,\mathbf{B}\,\mathbf{d} \tag{12.27}$$

12.4. General formulation for dynamic analysis

For dynamic analysis, we may compute a consistent element mass matrix (based on the same shape functions that we used to obtain **k**) as follows:

$$\mathbf{m} = \int_V \rho \mathbf{N}^{\mathrm{T}}\,\mathbf{N}\,dV \tag{12.28}$$

where ρ is the mass per unit volume of the material. With **N** as given in equation (12.10), it is therefore a simple matter to calculate **m** for our conical shell element.

To derive equation (12.28), we note that, in dynamic problems, the distributed body forces **G** per unit volume (see equation (12.25)) are the inertial loads. The acceleration of points within the finite element may be expressed in terms of the accelerations of the nodes of the element as follows (compare with equation (12.11)):

$$\ddot{\mathbf{u}} = \mathbf{N}\ddot{\mathbf{d}} \tag{12.29}$$

The inertial forces per unit volume are given by the product of the mass per unit volume (ρ) and the accelerations:

$$\mathbf{G} = -\rho\ddot{\mathbf{u}} = -\rho\mathbf{N}\ddot{\mathbf{d}} \tag{12.30}$$

The minus sign denotes that inertial forces are opposite to the motion. The component of nodal forces due to distributed body forces (see the first term on the right-hand side of equation (12.25)) therefore becomes

$$\begin{aligned} \int_V \mathbf{N}^{\mathrm{T}}\,\mathbf{G}\,dV &= -\int_V \mathbf{N}^{\mathrm{T}}\rho\mathbf{N}\ddot{\mathbf{d}}\,dV \\ &= -\left(\int_V \rho\mathbf{N}^{\mathrm{T}}\,\mathbf{N}\,dV\right)\ddot{\mathbf{d}} \\ &= -\,\mathbf{m}\,\ddot{\mathbf{d}} \end{aligned} \tag{12.31}$$

where **m** represents the integral in brackets, and is the consistent mass matrix of the element. We have thus derived equation (12.28).

The equation of motion for the element may therefore be written, using equations (12.24), (12.25) and (12.31), as follows:

$$\mathbf{k}\,\mathbf{d} = \int_V \mathbf{N}^{\mathrm{T}}\,\mathbf{G}\,dV + \int_S \mathbf{N}^{\mathrm{T}}\,\boldsymbol{\Phi}\,dS + \mathbf{Q}$$

$$= -\,\mathbf{m}\,\ddot{\mathbf{d}} + \mathbf{f}_{\mathbf{ex}}(\mathrm{t}) \tag{12.32}$$

where $\mathbf{f}_{\mathbf{ex}}(\mathrm{t})$ is the sum of the second and third terms of the first expression, and represents excitation nodal forces due to time-varying distributed surface loads and point loads. Rearranging equation (12.32), we obtain the following form of the equation of motion of the element:

$$\mathbf{k}\,\mathbf{d} + \mathbf{m}\,\ddot{\mathbf{d}} = \mathbf{f}_{\mathbf{ex}}(\mathrm{t}) \tag{12.33}$$

For the whole structure, we may assemble the global mass matrix **M** using a procedure similar to the assembly of the global stiffness matrix **K**. The system dynamic equations are finally written down as follows:

$$\mathbf{K}\,\mathbf{X} + \mathbf{M}\,\ddot{\mathbf{X}} = \mathbf{F}_{\mathbf{ex}}(\mathrm{t}) \tag{12.34}$$

where **X** and $\ddot{\mathbf{X}}$ are vectors of system nodal displacements and system nodal accelerations, respectively, and $\mathbf{F}_{\mathbf{ex}}(\mathrm{t})$ is the vector of the time-varying system nodal excitation forces. In deriving the formulation represented by equation (12.34), we have ignored the effects of damping on the system, but these can easily be taken into account. If $\mathbf{F}_{\mathbf{ex}}(\mathrm{t})$ is set equal to zero, we get the eigenvalue problem of free vibration, the solution of which yields the natural frequencies of vibration of the system, and corresponding mode shapes.

12.5. General formulation for stability problems

We will limit our considerations to linear elastic buckling. We are interested in the relatively large out-of-plane (transverse) displacements w that are associated with buckling. Buckling was considered from a theoretical perspective in Chapter 11. Here, we want to develop a finite-element formulation of the problem for numerical implementation. Assuming a suitable displacement field for w has been chosen, we can express w in terms of element nodal freedoms via shape functions N_i as already explained in earlier sections. As an example, Figure 12.5 shows a 4-noded quadrilateral shell element with three translations $\{u, v, w\}$ and two rotations $\{\theta_x, \theta_y\}$ per node, as illustrated for node 1.

The transverse displacement w at any point $\{x, y\}$ may be written in the form

$$w = \mathbf{N}\mathbf{d} = \{N_1(x,y) \quad N_2(x,y) \quad \cdots \quad N_n(x,y)\} \begin{Bmatrix} u_1 \\ v_1 \\ w_1 \\ \theta_{x1} \\ \theta_{y1} \\ u_2 \\ v_2 \\ \vdots \end{Bmatrix} \tag{12.35}$$

Figure 12.5 A 4-noded curved quadrilateral shell element in the *xy* coordinate system. This has three translations and two rotations per node.

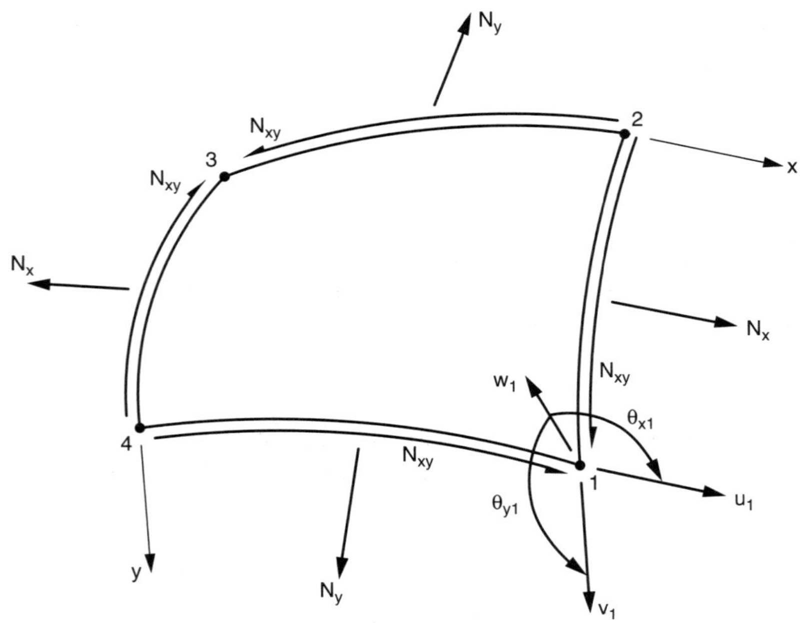

where n is the total number of degrees of freedom of the element (in this example $n = 20$). We may also express the first derivatives of w with respect to the x and y directions, in terms of the nodal displacements, as follows:

$$\begin{Bmatrix} \dfrac{\partial w}{\partial x} \\ \dfrac{\partial w}{\partial y} \end{Bmatrix} = \begin{bmatrix} \dfrac{\partial N_1(x,y)}{\partial x} & \dfrac{\partial N_2(x,y)}{\partial x} & \cdots & \dfrac{\partial N_n(x,y)}{\partial x} \\ \dfrac{\partial N_1(x,y)}{\partial y} & \dfrac{\partial N_2(x,y)}{\partial y} & \cdots & \dfrac{\partial N_n(x,y)}{\partial y} \end{bmatrix} \begin{Bmatrix} u_1 \\ v_1 \\ w_1 \\ \theta_{x1} \\ \theta_{y1} \\ u_2 \\ v_2 \\ \vdots \end{Bmatrix} = \mathbf{H}\,\mathbf{d} \qquad (12.36)$$

where $\mathbf{H}$ is the $2 \times n$ matrix of shape-function derivatives. Thus matrix $\mathbf{H}$ links the slopes of the shell element at any given interior point with the nodal displacements of the element.

The membrane stress resultants in the shell that exist before the onset of buckling are $\{N_x, N_y, N_{xy}\}$, as shown in Figure 12.5. Consistent with the considerations of Chapter 11, we assume these membrane stress resultants do not change during buckling, but the rotations of the shell (as a result of buckling displacements) will generate components of $\{N_x, N_y, N_{xy}\}$ in the transverse direction. These components of force are proportional to the slopes of the shell as defined by equation (12.36). Thus the displaced membrane stress resultants contribute to resistance of loads in the transverse direction, and hence to the stiffness of the shell element. The associated *initial-stress* or *geometric stiffness matrix* $\mathbf{k}_\sigma$ may readily be shown to be

$$\mathbf{k}_\sigma = \int_V \mathbf{H}^\mathrm{T} \begin{bmatrix} N_x & N_{xy} \\ N_{xy} & N_y \end{bmatrix} \mathbf{H}\, dV$$

$$= \int_{y1}^{y2} \int_{x1}^{x2} \mathbf{H}^\mathrm{T} \begin{bmatrix} N_x & N_{xy} \\ N_{xy} & N_y \end{bmatrix} \mathbf{H}\, dx\, dy \qquad (12.37)$$

where $\{x_1, x_2\}$ and $\{y_1, y_2\}$ define the bounding edges of the element. Since our $\mathbf{k}_\sigma$ here is based on the same shape functions as were used to derive the basic stiffness matrix $\mathbf{k}$, it is a consistent geometric stiffness matrix.

In linear buckling analysis, we may start off by solving for the applied loads via equation (12.26), and then evaluate the internal actions in the elements, which in the most general case comprise stress resultants $\{N_x, N_y, N_{xy}\}$ and moments $\{M_x, M_y, M_{xy}\}$. We then use the values of the obtained stress resultants $\{N_x, N_y, N_{xy}\}$ to compute element geometric stiffness matrices $\mathbf{k}_\sigma$ via equation (12.37). The various element matrices $\mathbf{k}_\sigma$ are then assembled to form the global geometric stiffness matrix $\mathbf{K}_\sigma$. The system equations are formulated as

$$[\mathbf{K} + \lambda\, \mathbf{K}_\sigma]\, \mathbf{X} = \mathbf{F} \qquad (12.38a)$$

where $\mathbf{K}_\sigma$ is based on some magnitude of the applied loading (which does not necessarily cause buckling), and λ is a scaling factor corresponding to the buckling load.

If an additional displacement $\Delta\mathbf{X}$ occurs without change of applied loads $\mathbf{F}$ (which is the definition of buckling), we may rewrite equation (12.38a) as

$$[\mathbf{K} + \lambda\, \mathbf{K}_\sigma]\{\mathbf{X} + \Delta\mathbf{X}\} = \mathbf{F} \qquad (12.38b)$$

Upon subtracting equation (12.38a) from equation (12.38b), we obtain the buckling equation in terms of the additional displacements only:

$$[\mathbf{K} + \lambda\, \mathbf{K}_\sigma]\, \Delta\mathbf{X} = \mathbf{0} \qquad (12.39)$$

We recognise equation (12.39) as an eigenvalue problem. The solution $\Delta\mathbf{X} = \mathbf{0}$ is trivial, in fact not legitimate, as buckling has already occurred. A non-trivial solution is obtained by setting the determinant of the main term to zero – that is,

$$|\mathbf{K} + \lambda\, \mathbf{K}_\sigma| = 0 \qquad (12.40)$$

The lowest solution for λ gives the critical buckling load. As noted in Chapter 11, buckling of shells of revolution is usually characterised by waves in the circumferential and/or meridional directions, and it may not be obvious what combination of wave numbers (the *m* and *n* parameters of Chapter 11) will be associated with the lowest buckling load. Several finite-element analyses may have to be carried out to establish the critical combination of wave numbers (i.e. critical mode), and hence the lowest buckling load.

With the general procedures for stress analysis, dynamic analysis and stability analysis now explained, we will devote the rest of the chapter to a discussion of some common finite elements for shell analysis. Effective use of FE programmes requires a good understanding of the strengths and limitations of various types of finite elements that are available. We will consider flat elements first, introduce the concept of isoparametric elements, and then apply this to shell analysis.

12.6. Flat shell elements

Curved shell elements require choice of an appropriate shell theory as a basis for the formulation of strain-displacement relationships and equilibrium equations, among other considerations. As we saw in Chapter 1, many versions of shell theory exist, and it is not always easy to make the best possible choice for a particular set of circumstances. To avoid this problem, we may resort to flat finite elements. In a flat element, the in-plane actions are decoupled from the bending actions (provided the displacements are small), whereas in a curved shell element, extensional and flexural actions are coupled together in the governing equations, which of course complicates the analysis. Clearly, flat facets will never model a curved surface exactly, but if the elements are small enough, the geometrical errors will diminish, and the behaviour of the assembly of tiny flat elements will closely match that of the curved surface.

The most versatile shape of facet for modelling arbitrary curved surfaces is the triangular element. Figure 12.6 shows a 3-noded flat triangular element with five degrees of freedom per node: $\{u, v, w\}$ are translations in the coordinate $\{x, y, z\}$ directions, while $\{\theta_x, \theta_y\}$ are rotations in the xz and yz planes respectively.

Owing to the uncoupled nature of in-plane and bending actions, the stiffness matrix of this element may be written down in block-diagonal form, as a combination of the stiffness matrices of a plane-stress element and a plate-bending element:

$$\mathbf{k} = \left[\begin{array}{c|c} \mathbf{k}^{\mathrm{m}} & \mathbf{0} \\ \hline \mathbf{0} & \mathbf{k}^{\mathrm{b}} \end{array}\right] \tag{12.41}$$

Figure 12.6 A 3-noded flat triangular element with five degrees of freedom per node.

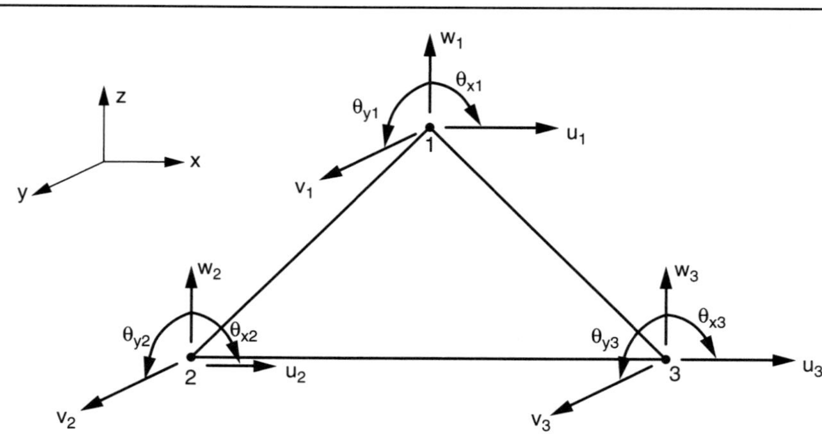

The matrix $\mathbf{k^m}$ is a 6×6 matrix corresponding to the membrane (i.e. in-plane) freedoms $\{u_i, v_i\}$ (for $i = 1, 2, 3$), while $\mathbf{k^b}$ is a 9×9 matrix corresponding to the bending (i.e. out-of-plane) freedoms $\{w_i, \theta_{xi}, \theta_{yi}\}$ (for $i = 1, 2, 3$). A variation of this element is possible, by adding a sixth degree of freedom θ_z (in-plane rotation about the z axis) at each node, to give an element with 18 degrees of freedom in total.

Flat rectangular elements may also be used. They are less versatile in modelling arbitrary curved surfaces, and are best suited for modelling general cylindrical surfaces (both open and closed profiles) of Chapter 9. Figure 12.7 shows a 4-noded flat rectangular element combining the properties of a 4-noded plane-stress element and a 4-noded plate-bending element, giving 20 degrees of freedom in total.

12.7. Isoparametric finite elements

12.7.1 General considerations

On account of their ability to incorporate curvature, isoparametric elements are particularly useful for modelling structures with curved features, such as shells. In the general (i.e. 3-dimensional) isoparametric formulation, the origin of the coordinate system is taken at the geometrical centroid of the element, and a special coordinate system $\{\xi, \eta, \mu\}$ is adopted. The usual $\{x, y, z\}$ coordinates are mapped onto a normalised coordinate system $\{\xi, \eta, \mu\}$ where

$$-1\le\xi\le+1;\quad -1\le\eta\le+1;\quad -1\le\mu\le+1 \tag{12.42}$$

Let $\mathbf{L}$ be a matrix that relates the normal coordinates $\{x, y, z\}$ of a point within an element to the coordinates of the nodes of the element. Let $\mathbf{N}$, as usual, be a matrix that relates the displacements $\{u, v, w\}$ of a point within an element to the nodal freedoms (i.e. the displacements of the nodes). Thus

$$\begin{Bmatrix} x \\ y \\ z \end{Bmatrix} = \mathbf{Lc};\quad \begin{Bmatrix} u \\ v \\ w \end{Bmatrix} = \mathbf{Nd} \tag{12.43}$$

Figure 12.7 A 4-noded flat rectangular element with five degrees of freedom per node.

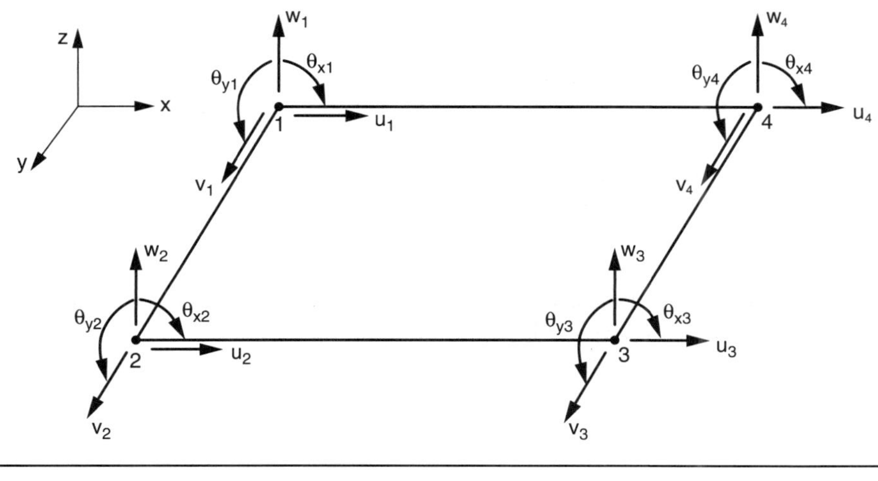

Figure 12.8 Isoparametric quadrilateral elements in 2-dimensional $\xi\eta$ space: (a) 4-noded linear element; (b) 8-noded quadratic element.

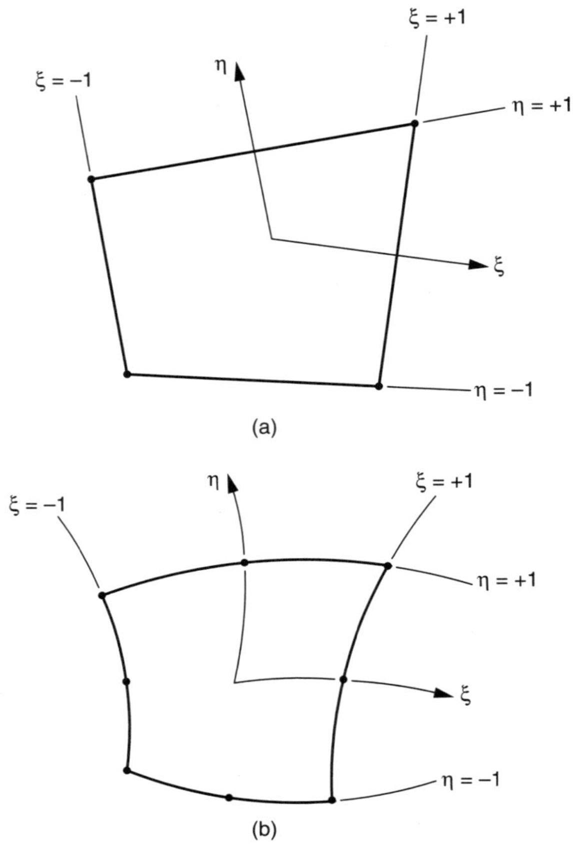

where **c** is the vector of nodal coordinates and, as usual, **d** is the vector of nodal displacements.

An element is said to be *isoparametric* if $\{u, v, w\}$ and $\{x, y, z\}$ are defined by the same set of nodes, and matrices **L** and **N** are identical.

For 2-dimensional elements, the $\{x, y\}$ normal coordinate system is mapped onto the $\{\xi, \eta\}$ special coordinate system. Examples of 2-dimensional isoparametric finite elements in the $\{\xi, \eta\}$ special coordinate system are shown in Figure 12.8.

To see more clearly how the isoparametric formulation works, let us consider a rectangular plane-stress finite element with four nodes, as shown in Figure 12.9(a). This element has two degrees of freedom $\{u_i, v_i\}$ at each node i $(i = 1, 2, 3, 4)$. The mapping of the element to the $\{\xi, \eta\}$ coordinate system is shown in Figure 12.9(b).

Figure 12.9 A 4-noded plane rectangular element: (a) actual element in *xy* space; (b) mapping of element to $\xi\eta$ coordinate system.

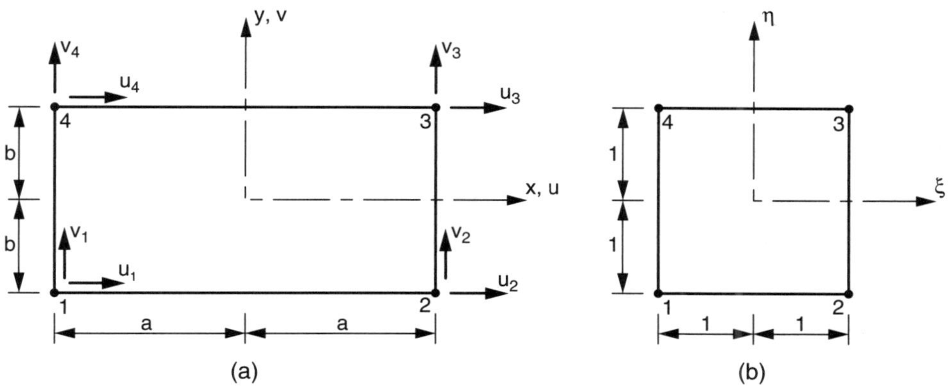

Considering the actual element (Figure 12.9(a)), we may write

$$\begin{Bmatrix} x \\ y \end{Bmatrix} = \mathbf{N}\,\mathbf{c}; \quad \begin{Bmatrix} u \\ v \end{Bmatrix} = \mathbf{N}\,\mathbf{d} \tag{12.44}$$

where

$$\mathbf{N} = \begin{bmatrix} N_1 & 0 & N_2 & 0 & N_3 & 0 & N_4 & 0 \\ 0 & N_1 & 0 & N_2 & 0 & N_3 & 0 & N_4 \end{bmatrix} \tag{12.45}$$

$$\mathbf{c} = \{x_1 \quad y_1 \quad x_2 \quad y_2 \quad x_3 \quad y_3 \quad x_4 \quad y_4\}^{\mathrm{T}} \tag{12.46a}$$

$$\mathbf{d} = \{u_1 \quad v_1 \quad u_2 \quad v_2 \quad u_3 \quad v_3 \quad u_4 \quad v_4\}^{\mathrm{T}} \tag{12.46b}$$

To obtain the shape functions $\{N_1, N_2, N_3, N_4\}$ in terms of the special coordinates ξ and η, let us assume displacement functions of the form

$$u = a_1 + a_2\xi + a_3\eta + a_4\xi\eta \tag{12.47a}$$

$$v = b_1 + b_2\xi + b_3\eta + b_4\xi\eta \tag{12.47b}$$

As the form of these functions is the same, we need only consider one of them, say u. The boundary conditions for u (at the corners) are as follows:

$$\text{At node 1 } (\xi = -1\ ;\ \eta = -1): \quad u = u_1 \quad \Rightarrow \quad a_1 - a_2 - a_3 + a_4 = u_1 \tag{12.48a}$$

$$\text{At node 2 } (\xi = +1\ ;\ \eta = -1): \quad u = u_2 \quad \Rightarrow \quad a_1 + a_2 - a_3 - a_4 = u_2 \tag{12.48b}$$

At node 3 $(\xi = +1 \ ; \eta = +1)$: $u = u_3 \quad \Rightarrow \quad a_1 + a_2 + a_3 + a_4 = u_3$ (12.48c)

At node 4 $(\xi = -1 \ ; \eta = +1)$: $u = u_4 \quad \Rightarrow \quad a_1 - a_2 + a_3 - a_4 = u_4$ (12.48d)

Solving for the a_i $(i = 1, 2, 3, 4)$, we obtain

$$a_1 = +\frac{1}{4}(u_1 + u_2 + u_3 + u_4) \tag{12.49a}$$

$$a_2 = -\frac{1}{4}(u_1 - u_2 - u_3 + u_4) \tag{12.49b}$$

$$a_3 = -\frac{1}{4}(u_1 + u_2 - u_3 - u_4) \tag{12.49c}$$

$$a_4 = +\frac{1}{4}(u_1 - u_2 + u_3 - u_4) \tag{12.49d}$$

Substituting these solutions back into equation (12.47a), we obtain

$$\begin{aligned} u &= \frac{1}{4}(u_1 + u_2 + u_3 + u_4) - \frac{1}{4}(u_1 - u_2 - u_3 + u_4)\xi \\ &- \frac{1}{4}(u_1 + u_2 - u_3 - u_4)\eta + \frac{1}{4}(u_1 - u_2 + u_3 - u_4)\xi\eta \\ &= \frac{1}{4}(1-\xi)(1-\eta)u_1 + \frac{1}{4}(1+\xi)(1-\eta)u_2 + \\ &+ \frac{1}{4}(1+\xi)(1+\eta)u_3 + \frac{1}{4}(1-\xi)(1+\eta)u_4 \\ &= [N_1 \quad N_2 \quad N_3 \quad N_4]\{u_1 \quad u_2 \quad u_3 \quad u_4\}^T \end{aligned} \tag{12.50}$$

where

$$N_1 = \frac{1}{4}(1-\xi)(1-\eta) \tag{12.51a}$$

$$N_2 = \frac{1}{4}(1+\xi)(1-\eta) \tag{12.51b}$$

$$N_3 = \frac{1}{4}(1+\xi)(1+\eta) \tag{12.51c}$$

$$N_4 = \frac{1}{4}(1-\xi)(1+\eta) \tag{12.51d}$$

For the displacement field $\{u, v\}$ taken together, we may write

$$\begin{Bmatrix} u \\ v \end{Bmatrix} = \begin{bmatrix} N_1 & 0 & N_2 & 0 & N_3 & 0 & N_4 & 0 \\ 0 & N_1 & 0 & N_2 & 0 & N_3 & 0 & N_4 \end{bmatrix} \begin{Bmatrix} u_1 \\ v_1 \\ u_2 \\ v_2 \\ u_3 \\ v_3 \\ u_4 \\ v_4 \end{Bmatrix} = \mathbf{N}(\xi, \eta)\, \mathbf{d} \tag{12.45}$$

which concludes our derivation of shape functions for the isoparametric element of Figure 12.9.

In calculations for **k**, the strain-displacement matrix **B** is needed. The evaluation of strains involves derivatives of displacements. If we denote any arbitrary function (whose derivatives we want) by Γ, then, for problems in two dimensions, we may write

$$\frac{\partial \Gamma}{\partial \xi} = \frac{\partial \Gamma}{\partial x}\frac{\partial x}{\partial \xi} + \frac{\partial \Gamma}{\partial y}\frac{\partial y}{\partial \xi} \tag{12.53a}$$

$$\frac{\partial \Gamma}{\partial \eta} = \frac{\partial \Gamma}{\partial x}\frac{\partial x}{\partial \eta} + \frac{\partial \Gamma}{\partial y}\frac{\partial y}{\partial \eta} \tag{12.53b}$$

or

$$\begin{Bmatrix} \dfrac{\partial \Gamma}{\partial \xi} \\ \dfrac{\partial \Gamma}{\partial \eta} \end{Bmatrix} = \mathbf{J} \begin{Bmatrix} \dfrac{\partial \Gamma}{\partial x} \\ \dfrac{\partial \Gamma}{\partial y} \end{Bmatrix} \tag{12.54}$$

where **J**, the Jacobian matrix, is given by

$$\mathbf{J} = \begin{bmatrix} \dfrac{\partial x}{\partial \xi} & \dfrac{\partial y}{\partial \xi} \\ \dfrac{\partial x}{\partial \eta} & \dfrac{\partial y}{\partial \eta} \end{bmatrix} \tag{12.55}$$

12.7.2 Isoparametric formulation for thick shells

For the thick-shell element shown in Figure 12.10, let the $\xi\eta$ plane coincide with the shell midsurface, and the μ direction be the direction normal to the shell midsurface. Furthermore, let the nodes on the surface $\mu = 1$ be denoted by subscript t (for top), and those on the surface $\mu = -1$ be denoted by subscript b (for bottom).

The coordinates $\{x, y, z\}$ at any point within the element may be expressed in terms of the nodal coordinates $\{x_{it}, y_{it}, z_{it}\}$ of the top nodes and $\{x_{ib}, y_{ib}, z_{ib}\}$ of the bottom nodes as follows:

Figure 12.10 A thick shell element with top and bottom nodes.

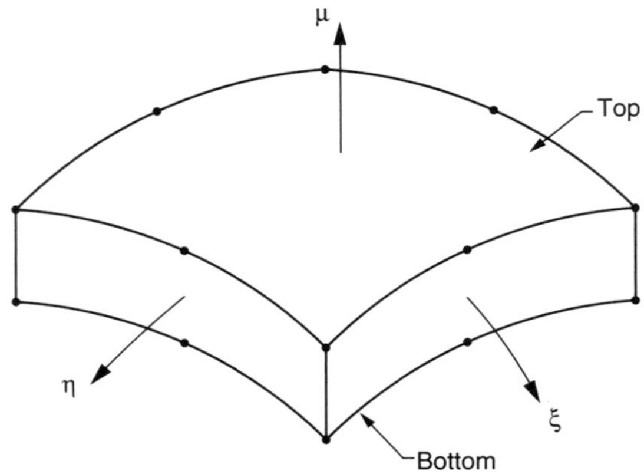

$$\begin{Bmatrix} x \\ y \\ z \end{Bmatrix} = \sum N_i \left(\frac{1+\mu}{2}\right) \begin{Bmatrix} x_{it} \\ y_{it} \\ z_{it} \end{Bmatrix} + \sum N_i \left(\frac{1-\mu}{2}\right) \begin{Bmatrix} x_{ib} \\ y_{ib} \\ z_{ib} \end{Bmatrix} \tag{12.56}$$

where the shape functions N_i are functions of ξ and η only.

Equation (12.56) may alternatively be expressed in terms of the coordinates of the middle surface of the element, $\{x_{im}, y_{im}, z_{im}\}$, as follows:

$$\begin{Bmatrix} x \\ y \\ z \end{Bmatrix} = \sum N_i \begin{Bmatrix} x_{im} \\ y_{im} \\ z_{im} \end{Bmatrix} + \sum N_i \frac{\mu}{2} \bar{\mathbf{n}}_\mathbf{i} \tag{12.57}$$

where $\bar{\mathbf{n}}_\mathbf{i}$ is a vector normal to the shell midsurface and of length equal to the shell thickness t_i:

$$\bar{\mathbf{n}}_\mathbf{i} = \begin{Bmatrix} x_{it} - x_{ib} \\ y_{it} - y_{ib} \\ z_{it} - z_{ib} \end{Bmatrix} \tag{12.58}$$

Considering an element with nodes lying on the middle surface, and without loss of generality, we may similarly write the displacement field throughout the element as a function of the nodal freedoms as follows:

$$\begin{Bmatrix} u \\ v \\ w \end{Bmatrix} = \sum N_i \begin{Bmatrix} u_i \\ v_i \\ w_i \end{Bmatrix} + \sum N_i \, \mu \, \frac{t_i}{2} \begin{Bmatrix} \theta_{xi} \\ \theta_{yi} \end{Bmatrix} \tag{12.59}$$

The first summation represents the contribution of the translational nodal freedoms $\{u_i, v_i, w_i\}$ to the displacement field, while the second summation represents the contribution of the rotations $\{\theta_{xi}, \theta_{yi}\}$ of the normal $\bar{\mathbf{n}}_i$ to the in-plane displacements u and v. To visualise the latter, consider Figure 12.11, which shows a portion of the element between the midsurface $(\mu = 0)$ and another surface a distance $\mu(t_i/2)$ above the midsurface. The vector $\bar{\mathbf{n}}_i$ intersects this second surface at point P. If the vector $\bar{\mathbf{n}}_i$ rotates by an amount θ_{xi} in the xz plane, the point P will displace by u' in the ξ (i.e. x) direction. Similarly, a rotation of θ_{yi} in the yz plane will give point P a displacement of v' in the η (i.e. y) direction.

From the sketched triangles in the figure, it is clear that u' and v' are given by

$$u' = \mu \frac{t_i}{2} \theta_{xi} \quad ; \quad v' = \mu \frac{t_i}{2} \theta_{yi} \tag{12.60}$$

This explains the second summation term of equation (12.59). Consistent with the first-order assumptions of shell theory, this treatment assumes zero strain in the thickness direction.

With the displacement field now defined, the strain vector is given by

$$\boldsymbol{\varepsilon} = \begin{Bmatrix} \varepsilon_x \\ \varepsilon_y \\ \gamma_{xy} \\ \gamma_{xz} \\ \gamma_{yz} \end{Bmatrix} = \begin{Bmatrix} \dfrac{\partial u}{\partial x} \\ \dfrac{\partial v}{\partial y} \\ \dfrac{\partial u}{\partial y} + \dfrac{\partial v}{\partial x} \\ \dfrac{\partial u}{\partial z} + \dfrac{\partial w}{\partial x} \\ \dfrac{\partial v}{\partial z} + \dfrac{\partial w}{\partial y} \end{Bmatrix} \tag{12.61}$$

Figure 12.11 Contribution of the rotations of $\bar{\mathbf{n}}_i$ to the in-plane displacements u and v.

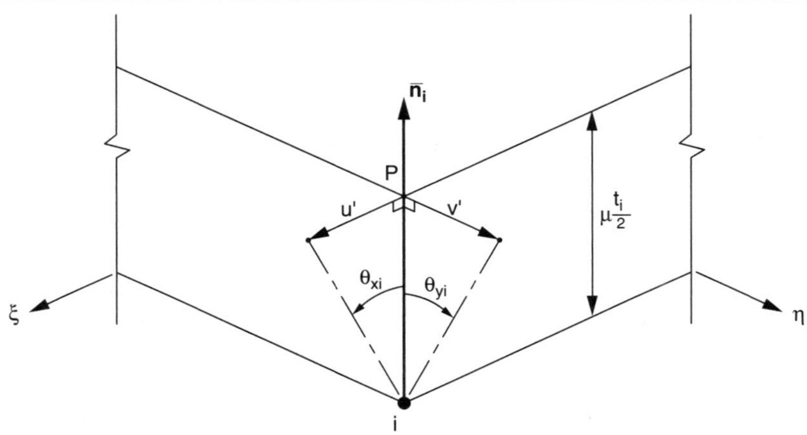

The stress vector $\boldsymbol{\sigma}$ and matrix of material stiffnesses $\mathbf{D}$ are given by

$$\boldsymbol{\sigma} = \begin{Bmatrix} \sigma_x \\ \sigma_y \\ \tau_{xy} \\ \tau_{xz} \\ \tau_{yz} \end{Bmatrix} = \mathbf{D}\,\boldsymbol{\varepsilon}; \quad \mathbf{D} = \frac{E}{1-\nu^2}\begin{bmatrix} 1 & \nu & 0 & 0 & 0 \\ \nu & 1 & 0 & 0 & 0 \\ 0 & 0 & \dfrac{1-\nu}{2} & 0 & 0 \\ 0 & 0 & 0 & \dfrac{1-\nu}{2\alpha} & 0 \\ 0 & 0 & 0 & 0 & \dfrac{1-\nu}{2\alpha} \end{bmatrix} \qquad (12.62, 12.63)$$

where α is a shear-distribution correction factor (normally taken as 1.2).

Equation (12.59) relates displacements $\{u, v, w\}$ to the $\{\xi, \eta, \mu\}$ coordinates. On the other hand, the strain-displacement matrix $\mathbf{B}$ is defined in terms of the derivatives of $\{u, v, w\}$ with respect to the $\{x, y, z\}$ coordinates. The conversion of derivatives with respect to $\{x, y, z\}$ to derivatives with respect to $\{\xi, \eta, \mu\}$ may readily be achieved via the relationship:

$$\begin{bmatrix} \dfrac{\partial u}{\partial x} & \dfrac{\partial v}{\partial x} & \dfrac{\partial w}{\partial x} \\ \dfrac{\partial u}{\partial y} & \dfrac{\partial v}{\partial y} & \dfrac{\partial w}{\partial y} \\ \dfrac{\partial u}{\partial z} & \dfrac{\partial v}{\partial z} & \dfrac{\partial w}{\partial z} \end{bmatrix} = \mathbf{J}^{-1} \begin{bmatrix} \dfrac{\partial u}{\partial \xi} & \dfrac{\partial v}{\partial \xi} & \dfrac{\partial w}{\partial \xi} \\ \dfrac{\partial u}{\partial \eta} & \dfrac{\partial v}{\partial \eta} & \dfrac{\partial w}{\partial \eta} \\ \dfrac{\partial u}{\partial \mu} & \dfrac{\partial v}{\partial \mu} & \dfrac{\partial w}{\partial \mu} \end{bmatrix} \qquad (12.64)$$

where the Jacobian matrix $\mathbf{J}$ is given by

$$\mathbf{J} = \begin{bmatrix} \dfrac{\partial x}{\partial \xi} & \dfrac{\partial y}{\partial \xi} & \dfrac{\partial z}{\partial \xi} \\ \dfrac{\partial x}{\partial \eta} & \dfrac{\partial y}{\partial \eta} & \dfrac{\partial z}{\partial \eta} \\ \dfrac{\partial x}{\partial \mu} & \dfrac{\partial y}{\partial \mu} & \dfrac{\partial z}{\partial \mu} \end{bmatrix} \qquad (12.65)$$

and calculated from equation (12.57). The calculation of the element stiffness matrix and other relevant matrices involves integration over the volume of the element. Integration with respect to $\{x, y, z\}$ is transformed to integration with respect to $\{\xi, \eta, \mu\}$ via the relationship

$$dx\, dy\, dz = |\mathbf{J}|\, d\xi\, d\eta\, d\mu \qquad (12.66)$$

In FE programmes, the integration is performed numerically. Assembly and solution of the system equations follow standard procedures as outlined in Sections 12.3 to 12.5. It may be remarked that, in real engineering problems involving thousands of solid elements each with a relatively large number of degrees of freedom (as when 20-node 60-degree-of-freedom brick elements are used), the numerical computations can be very expensive, but if solid elements with symmetry properties can be adopted (e.g. rectangular hexahedral elements with symmetric arrangements of nodes), group theory may be used to reduce the computational effort.[8]

12.7.3 Isoparametric formulation for shells of revolution

Figure 12.12(a) shows a finite element of a shell of revolution in the $\{R, z\}$ coordinate system, where the z direction is parallel to the axis of revolution of the shell, and the R plane is perpendicular to the axis of revolution. For ease of reference, we will regard the axis of revolution as vertical (and hence the R plane as horizontal). Figure 12.12(b) shows the isoparametric coordinate system adopted for the element. Owing to axisymmetry, only two coordinates $\{\xi, \eta\}$ are required to define locations within the element: ξ along the meridian and η normal to the midsurface.

In referring to the location of the nodes, we now replace subscripts t and b (for top and bottom) of the last section with o and i, which denote the outer and inner surfaces, respectively, of the shell of revolution. The first subscript of R and z denotes 'node i' as before, while the second subscript (either o, i or m) denotes the shell surface (outer, inner or middle). Reduction of equations (12.56) and (12.57) to the $\xi\eta$ space yields

$$\begin{Bmatrix} R \\ z \end{Bmatrix} = \sum N_i(\xi)\left(\frac{1+\eta}{2}\right)\begin{Bmatrix} R_{io} \\ z_{io} \end{Bmatrix} + \sum N_i(\xi)\left(\frac{1-\eta}{2}\right)\begin{Bmatrix} R_{ii} \\ z_{ii} \end{Bmatrix}$$

$$= \sum N_i(\xi)\begin{Bmatrix} R_{im} \\ z_{im} \end{Bmatrix} + \sum N_i(\xi)\,\frac{\eta}{2}\,\bar{\mathbf{n}}_{\mathbf{i}} \qquad (12.67)$$

where the vector $\bar{\mathbf{n}}_{\mathbf{i}}$ is now defined in terms of the meridional angle ϕ of the shell midsurface as follows (refer to the sketch in Figure 12.13):

$$\bar{\mathbf{n}}_{\mathbf{i}} = \begin{Bmatrix} t_i \sin\phi_i \\ t_i \cos\phi_i \end{Bmatrix} \qquad (12.68)$$

The parameter t_i (which is the magnitude of the vector $\bar{\mathbf{n}}_{\mathbf{i}}$) is the shell thickness at node i.

Figure 12.12 A finite element of a shell of revolution experiencing axisymmetric deformations: (a) element in the *Rz* coordinate system; (b) element in the $\xi\eta$ coordinate system.

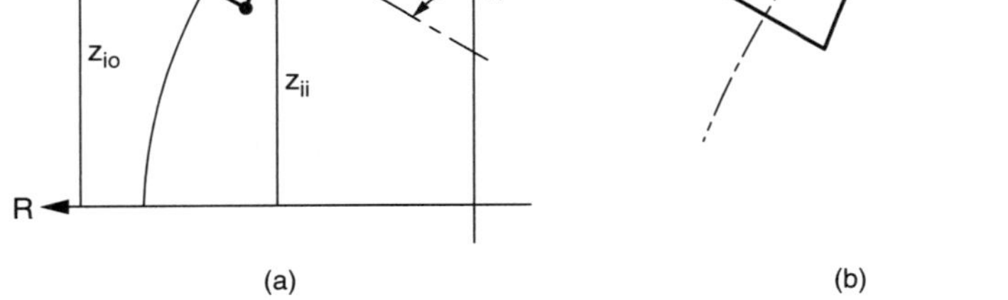

Figure 12.13 Components of vector $\bar{\mathbf{n}}_i$ of the element of the axisymmetric shell of revolution.

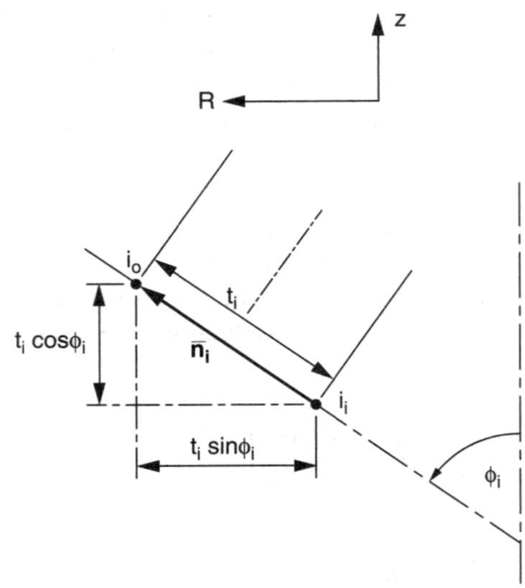

For axisymmetric deformations, only two displacements are relevant: v in the global R direction and w in the global z direction. Equation (12.59) reduces to

$$\begin{Bmatrix} v \\ w \end{Bmatrix} = \sum N_i \begin{Bmatrix} v_i \\ w_i \end{Bmatrix} + \sum N_i \; \eta \; \frac{t_i}{2} \; \varphi_i \begin{Bmatrix} \cos\phi_i \\ -\sin\phi_i \end{Bmatrix} \tag{12.69}$$

noting that a rotation of the normal by φ causes a point P, a distance $\eta(t/2)$ above the midsurface, to translate in the direction of s (i.e. the meridional direction) by an amount $\eta(t/2)\varphi$, which can in turn be resolved into components v' (in the direction of v) and w' (in the negative direction of w), where

$$v' = \eta \; \frac{t}{2} \; \varphi \cos\phi \tag{12.70a}$$

$$w' = \eta \; \frac{t}{2} \; \varphi \sin\phi \tag{12.70b}$$

(refer to Figure 12.14). In equation (12.69), the first summation is the contribution of the translational nodal freedoms $\{v_i, w_i\}$ of the element, while the second summation is the contribution of the rotational nodal freedoms φ_i of the element.

Thus, for the quadratic axisymmetric shell element with three midsurface nodes as shown in Figure 12.15, we may write

$$v = N_1 v_1 + N_2 v_2 + N_3 v_3 + N_1\left(\eta \; \frac{t_1}{2} \; \cos\phi_1\right)\varphi_1 + N_2\left(\eta \; \frac{t_2}{2} \; \cos\phi_2\right)\varphi_2 + N_3\left(\eta \; \frac{t_3}{2} \; \cos\phi_3\right)\varphi_3$$

$$w = N_1 w_1 + N_2 w_2 + N_3 w_3 - N_1\left(\eta \; \frac{t_1}{2} \; \sin\phi_1\right)\varphi_1 - N_2\left(\eta \; \frac{t_2}{2} \; \sin\phi_2\right)\varphi_2 - N_3\left(\eta \; \frac{t_3}{2} \; \sin\phi_3\right)\varphi_3 \tag{12.71a, b}$$

Figure 12.14 Contribution of the rotation of $\bar{\mathbf{n}}_i$ to the displacements v and w of the element of the axisymmetric shell of revolution.

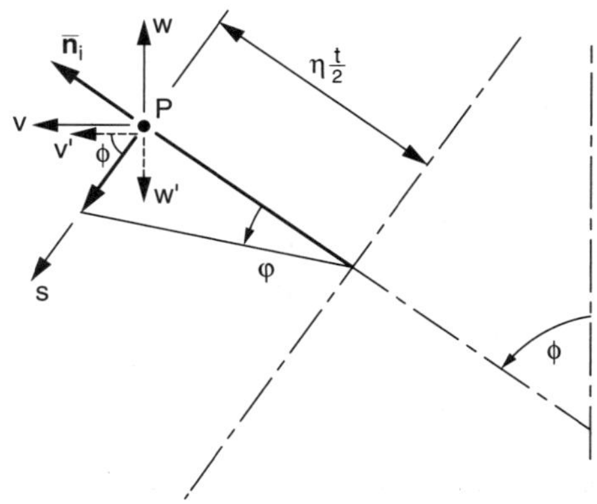

Figure 12.15 A quadratic element of the axisymmetric shell of revolution in the $\xi\eta$ coordinate system.

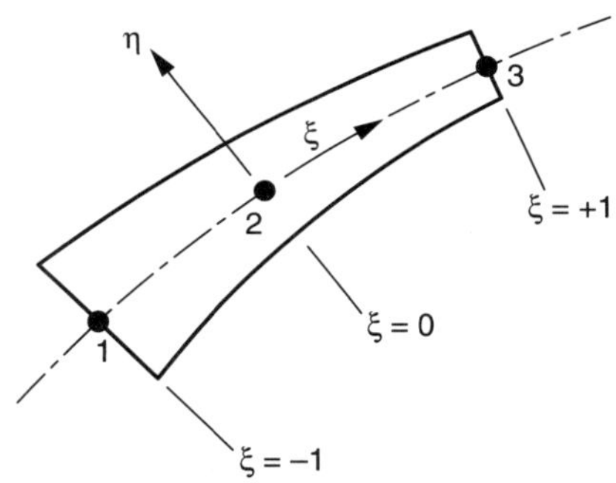

– that is,

$$\begin{Bmatrix} v \\ w \end{Bmatrix} = \mathbf{N}\,\mathbf{d} \tag{12.72}$$

where

$$\mathbf{N} = \left[\begin{array}{c:c:c:c:c:c:c:c:c} N_1 & 0 & \left(N_1\eta\frac{t_1}{2}\cos\phi_1\right) & N_2 & 0 & \left(N_2\eta\frac{t_2}{2}\cos\phi_2\right) & N_3 & 0 & \left(N_3\eta\frac{t_3}{2}\cos\phi_3\right) \\ \hdashline 0 & N_1 & \left(-N_1\eta\frac{t_1}{2}\sin\phi_1\right) & 0 & N_2 & \left(-N_2\eta\frac{t_2}{2}\sin\phi_2\right) & 0 & N_3 & \left(-N_3\eta\frac{t_3}{2}\sin\phi_3\right) \end{array}\right] \tag{12.73a}$$

$$\mathbf{d} = \{v_1 \quad w_1 \quad \varphi_1 \quad v_2 \quad w_2 \quad \varphi_2 \quad v_3 \quad w_3 \quad \varphi_3\}^{\mathrm{T}} \tag{12.73b}$$

If the loading is not axisymmetric, two additional freedoms come into play, namely u (translation in the circumferential direction) and β (rotation of the shell normal in a plane perpendicular to the meridional plane and containing the original position of the shell normal). The displacement field is expressed as a Fourier series in the circumferential (θ) direction, the jth term of the series being

$$\begin{Bmatrix} v_{(j)} \\ w_{(j)} \\ u_{(j)} \end{Bmatrix} = \begin{bmatrix} \cos j\theta & 0 & 0 \\ 0 & \cos j\theta & 0 \\ 0 & 0 & \sin j\theta \end{bmatrix} \times$$

$$\left(\sum N_i \begin{Bmatrix} v_{i(j)} \\ w_{i(j)} \\ u_{i(j)} \end{Bmatrix} + \sum N_i \, \eta \frac{t_i}{2} \begin{bmatrix} \cos \phi_i & 0 \\ -\sin \phi_i & 0 \\ 0 & 1 \end{bmatrix} \begin{Bmatrix} \varphi_{i(j)} \\ \beta_{i(j)} \end{Bmatrix} \right) \quad (12.74)$$

The loading is also similarly decomposed into Fourier components.

Let us pursue this more general case of non-axisymmetric loading. For the global coordinate system defined by $\{R, z, \theta\}$ as illustrated in Figure 12.16, the corresponding displacements are v (in the horizontal direction that passes through the axis of revolution of the shell), w (in the vertical direction) and u (in the horizontal direction tangential to the circle of latitude of the shell of revolution).

By definition, the direct strains are given by

$$\varepsilon_R = \frac{\partial v}{\partial R} \quad (12.75a)$$

$$\varepsilon_z = \frac{\partial w}{\partial z} \quad (12.75b)$$

Figure 12.16 Global coordinates $\{R, z, \theta\}$ and displacements $\{v, w, u\}$ of a shell of revolution under non-axisymmetric loading conditions.

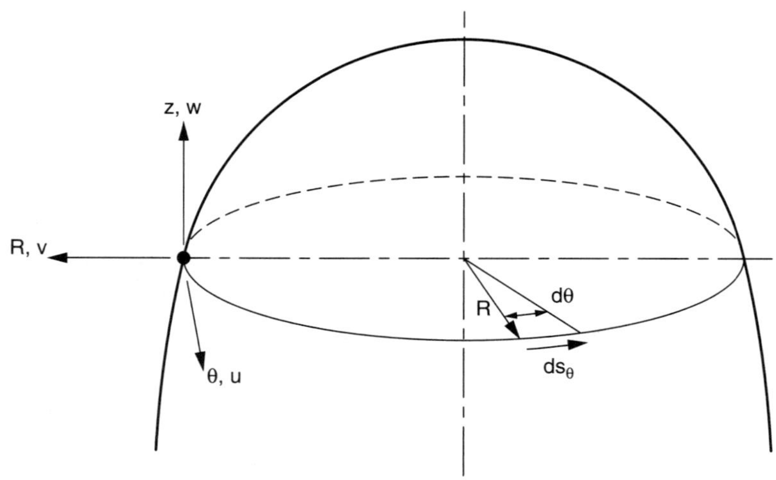

$$\varepsilon_\theta = \frac{\partial u}{\partial s_\theta} + \frac{v}{R} = \frac{\partial u}{R\partial\theta} + \frac{v}{R} \tag{12.75c}$$

noting that in the case of ε_θ, $ds_\theta = R\,d\theta$ (see Figure 12.16); ε_θ is made up of a contribution due to the purely circumferential displacement u and a contribution due to the lateral expansion of the radius R by an amount v.

Similarly, the shear strains with respect to these three global directions are

$$\gamma_{Rz} = \frac{\partial v}{\partial z} + \frac{\partial w}{\partial R} \tag{12.76a}$$

$$\gamma_{R\theta} = \frac{\partial v}{\partial s_\theta} + \frac{\partial u}{\partial R} + \frac{u}{R} = \frac{\partial v}{R\partial\theta} + \frac{\partial u}{\partial R} + \frac{u}{R} \tag{12.76b}$$

$$\gamma_{z\theta} = \frac{\partial w}{\partial s_\theta} + \frac{\partial u}{\partial z} = \frac{\partial w}{R\partial\theta} + \frac{\partial u}{\partial z} \tag{12.76c}$$

noting that, in the case of $\gamma_{R\theta}$, a movement through u of a point on the horizontal circle of latitude will cause the tangent at that point to rotate by an amount u / R, adding to the total relative change in angle (shear strain) of a lamina in the $R\theta$ plane.

Thus, the required strain matrix is

$$\boldsymbol{\varepsilon} = \begin{Bmatrix} \varepsilon_R \\ \varepsilon_z \\ \varepsilon_\theta \\ \gamma_{Rz} \\ \gamma_{R\theta} \\ \gamma_{z\theta} \end{Bmatrix} = \begin{Bmatrix} \dfrac{\partial v}{\partial R} \\ \dfrac{\partial w}{\partial z} \\ \dfrac{1}{R}\left(v + \dfrac{\partial u}{\partial\theta}\right) \\ \dfrac{\partial v}{\partial z} + \dfrac{\partial w}{\partial R} \\ \dfrac{1}{R}\left(u + \dfrac{\partial v}{\partial\theta}\right) + \dfrac{\partial u}{\partial R} \\ \dfrac{1}{R}\dfrac{\partial w}{\partial\theta} + \dfrac{\partial u}{\partial z} \end{Bmatrix} \tag{12.77}$$

In the case of axisymmetric deformations, the strain matrix reduces to

$$\boldsymbol{\varepsilon} = \begin{Bmatrix} \varepsilon_R \\ \varepsilon_z \\ \varepsilon_\theta \\ \gamma_{Rz} \end{Bmatrix} = \begin{Bmatrix} \dfrac{\partial v}{\partial R} \\ \dfrac{\partial w}{\partial z} \\ \dfrac{v}{R} \\ \dfrac{\partial v}{\partial z} + \dfrac{\partial w}{\partial R} \end{Bmatrix} \tag{12.78}$$

The derivation of the strain-displacement matrix **B** and the element stiffness matrix **k**, as well as the assembly of the system equations, all follow standard procedures. Our formulation may thus be regarded as complete.

12.8. Concluding remarks

As stated at the beginning, the Finite Element Method (FEM) is a vast area of research. This chapter has summarised the essence of the method as applied to the linear elastic analysis of shells. Historically, the earliest FEM developments were associated with the numerical prediction of the linear elastic behaviour of structures in instances where exact mathematical solutions were difficult or impossible to obtain, but over the years finite-element applications have extended to a wide range of non-linear problems in solid and structural mechanics,[9,10] computational fluid dynamics, electromagnetics and other problems in physics. Developments in these areas continue.

References

1. Oden J.T. and Reddy J.N. *An Introduction to the Mathematical Theory of Finite Elements*. John Wiley, New York, 1976.
2. Zienkiewicz O.C. and Taylor R.L. *The Finite Element Method*. McGraw-Hill Book Company, Maidenhead (UK), 1991.
3. Cook R.D. *Concepts and Applications of Finite Element Analysis*. John Wiley & Sons, New York, 1981.
4. Bathe K.J. *Finite Element Procedures in Engineering Analysis*. Prentice Hall, Englewood Cliffs (New Jersey), 1996.
5. Hughes T.J.R. *The Finite Element Method: Linear Static and Dynamic Finite Element Analysis*. Prentice Hall, Englewood Cliffs (New Jersey), 1987.
6. Gould P.L. *Finite Element Analysis of Shells of Revolution*. Pitman Publishing, London, 1985.
7. Yang T.Y. *Finite Element Structural Analysis*. Prentice Hall, Englewood Cliffs (New Jersey), 1986.
8. Zingoni A. A group-theoretic formulation for symmetric finite elements. *Finite Elements in Analysis and Design*, 2005, **41** (6), 615–635.
9. Crisfield M.A. *Non-Linear Finite Element Analysis of Solids and Structures*. John Wiley & Sons, New York, 1991.
10. Belytschko T., Liu W.K., Moran B. and Elkhodary K. *Nonlinear Finite Elements for Continua and Structures*. John Wiley & Sons, New York, 2013.

Shell Structures in Civil and Mechanical Engineering
ISBN 978-0-7277-6028-9

http://dx.doi.org/10.1680/ssicame.60289.421

Chapter 13
Design considerations for shell structures

13.1. Introduction

In design, not only do we use the results of analysis to inform us of the magnitude of the effects to be countered, but we also need to incorporate sufficient margins of safety against various uncertainties and errors incurred in the assessment of loading and other environmental conditions, the evaluation of the properties of the material, the modelling of the real behaviour of the structure, the manufacture of the structural elements and the assembly of the structure. Codes of practice, design standards and authoritative reports are available to assist the designer in complying with national requirements of safety and with what is generally considered to be good professional practice based on experience.

In this final chapter, we will consider practical aspects of the design of steel shells and the design of thin reinforced concrete shells, by reference to some European and American standards, including technical publications of international associations and national institutes. Where appropriate, the discussion will include analytical procedures that are recommended for the purposes of design, and some simple design formulae will be presented. It will be noted that the recommendations of the European and American standards are essentially similar (or even identical) on certain aspects, since the scientific basis of the recommendations is the same, but there are also significant differences with regard to other aspects. In general, the notation that will be used in each section is that of the standard or code of practice being discussed.

13.2. Steel shells

13.2.1 A brief review of Eurocode 3

13.2.1.1 Introduction

In the past five years, the countries of the European Union, as well as several other countries in Asia, the Middle East and Africa, have shifted towards the use of the suite of standards called the Eurocodes, for the design of buildings, bridges and other structures, including plate and shell structures. The Eurocode dealing with the design of steel structures is Eurocode 3, formally referred to as EN 1993. The design recommendations summarised below are those of the Eurocode standard that deals with the design of steel shells for strength and stability, which is formally referred to as EN 1993-1-6 (i.e. Eurocode 3 Part 1-6).[1]

Figure 13.1 illustrates typical configurations of steel tanks and silos. According to Eurocode 3, the shell structure should be designed to satisfy overall stability, the equilibrium between the applied loads and the internal actions, as well as all specified serviceability requirements. Testing is permitted as a tool for design. The considerations in this chapter are mainly concerned with ensuring that the shell can carry the applied loads.

Figure 13.1 Examples of steel containment shell structures: (a) cylindrical grain silo with conical roof and conical hopper; (b) cylindrical oil tank with spherical roof.

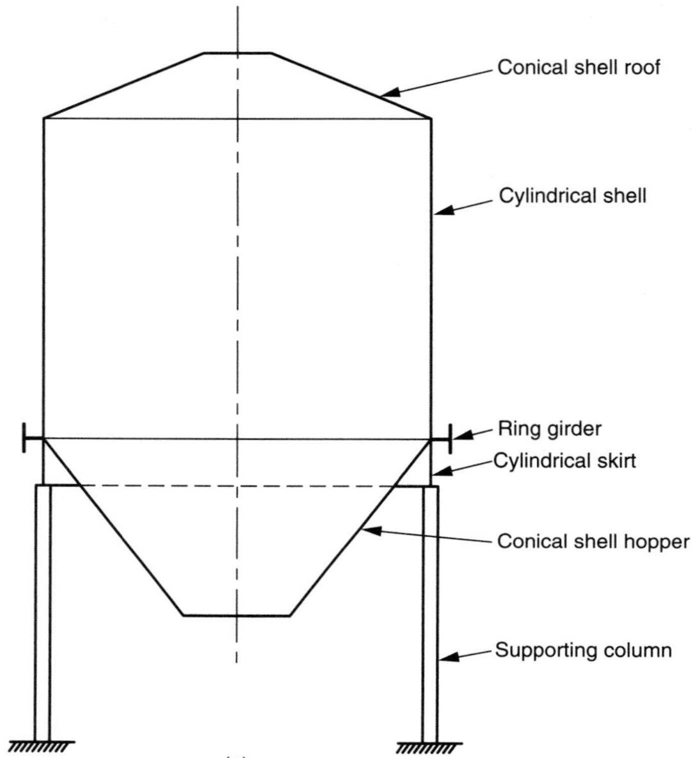

(a)

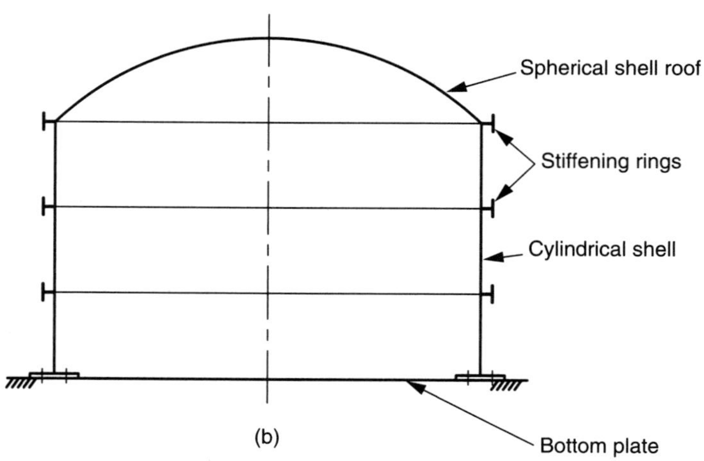

(b)

13.2.1.2 Analysis

Depending on the limit state being considered, any one or more of the following types of analyses, listed more or less in increasing level of complexity, may be used:

(*i*) membrane theory (for membrane forces)
(*ii*) linear elastic shell analysis (for membrane and bending stresses)
(*iii*) linear elastic bifurcation analysis (for the lowest buckling load, assuming perfect geometry)
(*iv*) geometrically non-linear elastic analysis (takes account of changes in geometry of the structure caused by loading)
(*v*) materially non-linear analysis (for the plastic limit load)
(*vi*) geometrically and materially non-linear analysis
(*vii*) geometrically non-linear elastic analysis taking into account imperfections
(*viii*) geometrically and materially non-linear analysis taking into account imperfections.

Methods (*i*)–(*iii*) may be executed via analytical closed-form solutions of the types that have been developed in Chapters 2 to 11 of this book. Methods (*iv*)–(*viii*) are usually executed numerically via the Finite Element Method (dealt with in Chapter 12).

The annexes to EN 1993-1-6 contain membrane-theory results for various load cases of circular-cylindrical shells and conical frusta, formulae for bending stresses in the vicinity of clamped and pinned edges of cylindrical shells under various loadings, as well as discontinuity stresses associated with abrupt changes in loading and wall thickness, or with the presence of ring stiffeners. Many of these results were derived in Chapters 3 and 5.

Annex D, which gives expressions for theoretical buckling stresses for variously loaded cylindrical and conical shells prescribed with clamped, pinned or free-edge boundary conditions, is particularly useful, given the time-consuming nature of analytical buckling calculations. For cylinders, the code defines a dimensionless length parameter ω:

$$\omega = \frac{l}{\sqrt{at}} \tag{13.1}$$

where l is the length of the cylinder, a is the radius and t is the shell thickness. Formulae for critical buckling stresses are then presented for various ranges of ω, for the three load cases of axial compression, uniform external pressure and shear (these were analytically treated in Chapter 11), as well as combinations of these. The key results may be stated as follows.

For cylinders subjected to uniform axial compression, when the boundary conditions are either fixed or pinned at both edges, the critical buckling stress is meridional (i.e. axial), and given by the formula

$$\sigma_{xcr} = 0.605\, EC_x \frac{t}{a} \tag{13.2}$$

where

$$C_x = 1.36 - \frac{1.83}{\omega} + \frac{2.07}{\omega^2} \text{ for } \omega \le 1.7 \text{ (short cylinders)} \tag{13.3a}$$

$$C_x = 1.0 \text{ for } 1.7 \le \omega \le 0.5\frac{a}{t} \text{ (medium-length cylinders)} \tag{13.3b}$$

$$C_x = \text{the greater of: } 0.60 \text{ or } 1 + \frac{0.2}{C_{xb}}\left(1 - 2\omega\frac{t}{a}\right) \text{ for } \omega > 0.5\frac{a}{t} \text{ (long cylinders)} \tag{13.3c}$$

and C_{xb} is a parameter that depends on the edge conditions and is given in a table. For medium-length cylinders (where $C_x = 1.0$), equation (13.2) coincides with the result that was derived in Chapter 11, when $\nu = 0.3$ (for steel) is substituted into equation (11.26).

When the loading becomes uniform external pressure, the critical buckling stress is now circumferential (i.e. hoop), and is given by the formulae:

$$\sigma_{\theta cr} = 0.92\, E\left(\frac{C_{\theta s}}{\omega}\right)\frac{t}{a} \text{ for } \frac{\omega}{C_\theta} < 20 \text{ (short cylinders)} \tag{13.4a}$$

$$\sigma_{\theta cr} = 0.92\, E\left(\frac{C_\theta}{\omega}\right)\frac{t}{a} \text{ for } 20 \le \frac{\omega}{C_\theta} \le 1.63\frac{a}{t} \text{ (medium-length cylinders)} \tag{13.4b}$$

$$\sigma_{\theta cr} = E\left(\frac{t}{a}\right)^2\left[0.275 + 2.03\left(\frac{C_\theta}{\omega}\frac{a}{t}\right)^4\right] \text{ for } \frac{\omega}{C_\theta} > 1.63\frac{a}{t} \text{ (long cylinders)} \tag{13.4c}$$

where C_6 and $C_{\theta s}$ depend on the edge conditions, and are given in tables. In the case of very long cylinders (e.g. pipelines), the second term in the square brackets of equation (13.4c) may be neglected in comparison with 0.275, and the expression reduces to

$$\sigma_{\theta cr} = 0.275\, E\frac{t^2}{a^2} \tag{13.4d}$$

which, of course, is the same result as equation (11.43) of the chapter on buckling, when we substitute the value $\nu = 0.3$ into that equation, and replace ap_{cr}/t with the stress $\sigma_{\theta cr}$.

Reduction factors are also given to allow for meridional and circumferential imperfections in shell geometry, as are buckling interaction parameters for cases where axial force, external pressure and torsional shear co-exist.

For conical frusta of constant shell thickness and with a vertical angle α (refer to Figure 11.8) not exceeding 65° (i.e. cones that are not too shallow), Annex D of EN 1993-1-6 defines an equivalent cylinder length l_e and an equivalent cylinder radius a_e, which depend on the type of loading applied on the cone. This allows the conical shell to be replaced by a cylinder, and the usual formulae for cylindrical shells then used. For cones under meridional compression, the equivalent parameters are

$$l_e = L; \quad a_e = \frac{r}{\cos\alpha} \tag{13.5a}$$

while for cones under circumferential (i.e. hoop) compression, the equivalent parameters are

$$l_e = L; \quad a_e = \frac{r_1 + r_2}{2\cos\alpha} \tag{13.5b}$$

where L is the length of the conical frustum measured along the sloping meridian, r_1 and r_2 are the radii of the smaller and the larger opening of the cone (refer to Figure 11.8), and r ($r_1 \le r \le r_2$) is the horizontal radius at any general location, assuming the axis of rotation of the shell is vertical.

If we substitute the equivalent parameters of equation (13.5b) into equation (13.4b) for the medium-length cylinder, with C_6 taken as 1.0 (pinned–pinned edges), we get

$$\sigma_{\theta cr} = 0.92\,E\,\frac{\sqrt{a_e t}}{L}\,\frac{t}{a_e} = 0.92\,E\,\frac{1}{L\sqrt{a_e}}\,t^{3/2} \qquad (13.6)$$

which is almost the same as the result that was given in Chapter 11 as equation (11.63), when the R of that equation is replaced with a_e (these symbols mean the same), and noting that $\sigma_{\theta cr} = a_e p_{cr} / t$. The only difference is the factor γ in equation (11.63), which depends on the ratio r_1 / r_2 of the cone, and is only 1.0 when r_1 / r_2 is close to 1.0. Equivalent parameters for conical shells under shear and torsion are also given.

13.2.1.3 Design

For design purposes, the thickness t of the shell may be taken as the nominal dimension less any anticipated reduction due to corrosion. Deviations of the geometry of the shell midsurface from the perfect shape are inevitable, and tolerance values for these must be specified beforehand and used as a basis for estimating geometrical imperfections, as required in the non-linear buckling analysis accounting for imperfections. Ultimate limit states that need to be considered are:

(*i*) *Plastic Limit*: This is the situation when the shell fails to carry the applied loading as a result of yielding of the material. It includes tensile rupture, when the shell wall separates into two as a result of high tensile stresses.

(*ii*) *Cyclic Plasticity Limit*: This is the situation when repeated cycles of loading and unloading take the material stresses into the yield range, the plastic work repeatedly done on the shell eventually leading to local cracking. Failure occurs after a relatively small number of cycles.

(*iii*) *Fatigue Limit*: This is the situation when repeated cycles of loading variations, not necessarily taking the material stresses into the yield range, eventually lead to the development of fatigue cracks. Failure typically occurs after a relatively large number of cycles.

(*iv*) *Buckling Limit*: This is the situation when, as a result of excessive compressive membrane or shear membrane stresses, all or part of the shell suddenly experiences large displacements transverse to the shell surface, resulting in immediate collapse of the shell, or failure of the shell to carry any further increase in the applied load.

The design stresses or stress resultants required to check the above limit states may be calculated using one or more of the methods of analysis listed in Section 13.2.1.2. In checking the plastic limit state, it is permitted to calculate the stresses via an elastic analysis, which provides a conservative estimate of the plastic resistance. When membrane theory is used to check the plastic limit state for a cylindrical shell, EN 1993-1-6 allows the resulting 2-dimensional field of stress resultants $\{N_x, N_\theta, N_{x\theta}\}$ to be represented by an equivalent design stress given by

$$\sigma_{eq} = \frac{1}{t}\sqrt{N_x^2 + N_\theta^2 - N_x N_\theta + 3N_{x\theta}^2} \qquad (13.7)$$

Where a linear elastic analysis or a geometrically non-linear elastic analysis is used, the resulting 2-dimensional field of stress may be represented by the von Mises equivalent design stress:

$$\sigma_{eq} = \sqrt{\sigma_x^2 + \sigma_\theta^2 - \sigma_x\sigma_\theta + 3\left(\tau_{x\theta}^2 + \tau_{xn}^2 + \tau_{\theta n}^2\right)} \tag{13.8}$$

where

$$\sigma_x = \frac{N_x}{t} \pm \frac{4M_x}{t^2}; \quad \sigma_\theta = \frac{N_\theta}{t} \pm \frac{4M_\theta}{t^2} \tag{13.9}$$

$$\tau_{x\theta} = \frac{N_{x\theta}}{t} \pm \frac{4M_{x\theta}}{t^2}; \quad \tau_{xn} = \frac{Q_{xn}}{t}; \quad \tau_{\theta n} = \frac{Q_{\theta n}}{t} \tag{13.10}$$

The check for the plastic limit state is satisfied if the following condition is satisfied:

$$\frac{\sigma_{eqEd}}{f_{yd}} \leq 1 \tag{13.11}$$

where σ_{eqEd} is the equivalent design stress calculated using the design values of the applied loads (i.e. applied loads multiplied by the appropriate partial load factors for the load combination in question), and f_{yd} is the design yield strength of the material (equal to the characteristic yield strength divided by the partial factor for resistance). In general, the Eurocode system uses the subscript *Ed* to denote 'design effects' (i.e. the internal actions or stresses that we are designing for, as obtained from an analysis), and the subscript *Rd* to denote 'design resistance' (i.e. the strength of the material or capacity of the structure).

For the buckling limit state, the design values of the stresses σ_{xEd}, $\sigma_{\theta Ed}$ and $\tau_{x\theta Ed}$ should be taken as the maximum values of compressive and shear membrane stresses as obtained from linear elastic shell analysis, using the design values of the applied loads. If the loading conditions are axisymmetric, membrane theory (instead of linear elastic shell analysis) may be used to obtain the maximum values of σ_{xEd} and $\sigma_{\theta Ed}$.

The buckling resistance σ_{Rd} (or buckling strength) is the critical buckling stress reduced by an appropriate partial factor as specified in the relevant national annex. The buckling limit state may be checked on the basis of the calculated design values of membrane stresses. If there is only one membrane-stress component with the potential to cause buckling at a particular location, the criterion to be satisfied will be one of the following:

$$\frac{\sigma_{xEd}}{\sigma_{xRd}} \leq 1; \quad \frac{\sigma_{\theta Ed}}{\sigma_{\theta Rd}} \leq 1; \quad \frac{\tau_{x\theta Ed}}{\tau_{x\theta Rd}} \leq 1 \tag{13.12}$$

If at a particular location there is more than one membrane-stress component with the potential to cause buckling, the following interaction formula should be used:

$$\left(\frac{\sigma_{xEd}}{\sigma_{xRd}}\right)^{k_x} - k_i\left(\frac{\sigma_{xEd}}{\sigma_{xRd}}\right)\left(\frac{\sigma_{\theta Ed}}{\sigma_{\theta Rd}}\right) + \left(\frac{\sigma_{\theta Ed}}{\sigma_{\theta Rd}}\right)^{k_\theta} + \left(\frac{\tau_{x\theta Ed}}{\tau_{x\theta Rd}}\right)^{k_\tau} \leq 1 \tag{13.13}$$

where k_x, k_θ, k_τ and k_i are buckling interaction parameters given in Annex D.

13.2.2 Design guidance of the American Petroleum Institute

Guidance for the design of oil-storage cylindrical steel vessels has been provided by the American Petroleum Institute through its publication, API Standard 650.[2] This standard states that the required shell thickness shall be the greater of the design shell thickness, inclusive of an allowance for corrosion, or the thickness as dictated by a hydrostatic test. In either case, the nominal shell thickness has to be at least:

- 5 mm if the nominal tank diameter is less than 15m
- 6 mm if the nominal tank diameter is between 15m and 36m
- 8 mm if the nominal tank diameter is between 36m and 60m
- 10 mm if the nominal tank diameter is greater than 60m

The permissible design stress S_d (in MPa) is tabulated for various shell-plate specifications; this is generally taken as the smaller of 2 / 3 of the yield strength or 2 / 5 of the tensile strength. The permissible hydrostatic-test stress S_t (in MPa) is also included in the tables; this is generally taken as the smaller of 3 / 4 of the yield strength or 3 / 7 of the tensile strength. The calculated stress in each shell course should not exceed the permissible stress.

Two approaches are indicated for obtaining the required shell thickness. The basic approach is the 1-Foot Method, where the thickness is calculated based on design points a distance of 0.3m (i.e. 1 foot) above the bottom of each shell course. The required shell thickness is taken as the greater of:

$$t_d = \frac{4.9D(H-0.3)G}{S_d} + CA \tag{13.14a}$$

$$t_t = \frac{4.9D(H-0.3)}{S_t} \tag{13.14b}$$

where t_d is the design shell thickness (mm), t_t is the hydrostatic-test shell thickness (mm), D is the nominal tank diameter (m), H is the design liquid head (m), G is the specific gravity of the liquid to be stored, and CA is the corrosion allowance (mm), with S_d and S_t (allowable stresses) being read from the tables. The method is not permitted for tanks larger than 61m in diameter.

The other approach, called the Variable-Design-Point Method, gives a more accurate estimate of shell thickness, but the benefits of this approach only become significant in the case of large-diameter tanks. Tanks whose diameter is greater than 61m may be designed using this second approach, or on the basis of a full elastic analysis. The calculated circumferential shell stresses must not exceed the permissible stresses.

In addition to the above, the shell needs to be checked for stability against buckling under the design wind speed. Measures against wind buckling include the provision of intermediate stiffening girders (see Figure 13.2), or the enhancement of shell thickness.

Some guidance is also provided on the choice of shell thickness for tank roofs. These include conical roofs and spherical domes supported only at the edges. For conical roofs, it is recommended that the slope α (angle between the inclined meridian and the horizontal) is chosen within the range $9.5° \leq \alpha \leq 37°$. The nominal shell thickness of the cone should neither be less than 5mm nor less than the greater of:

$$\frac{D}{4.8\sin\alpha}\sqrt{\frac{G_L}{2.2}} + CA; \quad \frac{83D}{\sin\alpha}\sqrt{\frac{P_r}{1.72E}} \tag{13.15}$$

Figure 13.2 Steel tank with stepwise-varied shell thickness and intermediate stiffening girders.

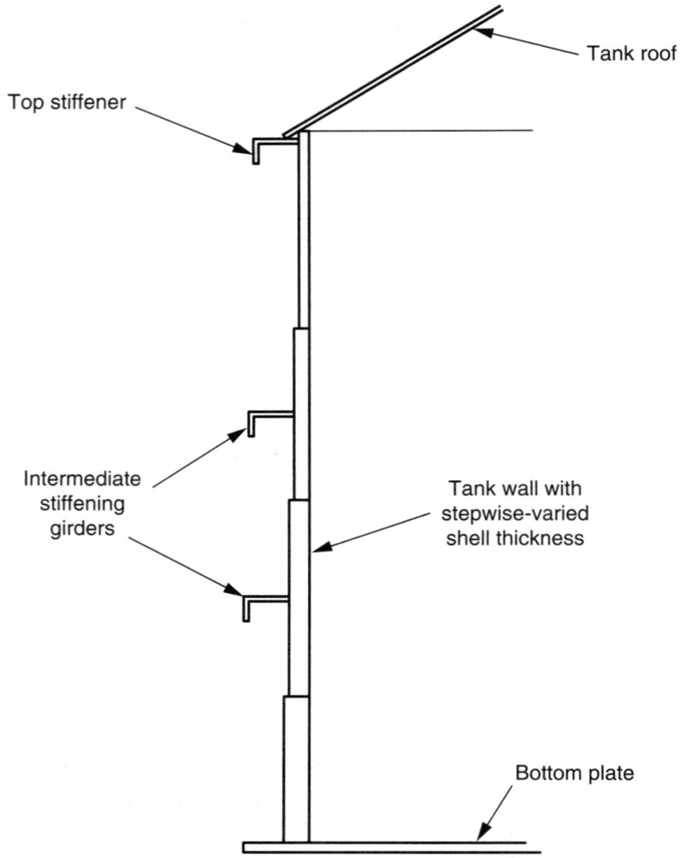

where D (in metres) is the nominal tank diameter, G_L (in kPa) is a load combination for gravity loads and external pressure, P_r (in kPa) is the total design external pressure on the roof, and E (in MPa) is Young's modulus.

For spherical roof domes, the recommended range for the radius of the dome is $0.8D \leq r_r \leq 1.2D$. The nominal shell thickness of the dome should not be less than 5mm or the greater of:

$$\frac{r_r}{2.4}\sqrt{\frac{G_L}{2.2}} + CA; \quad 141\, r_r \sqrt{\frac{P_r}{E}} \tag{13.16}$$

where r_r (in metres) is the dome radius, CA is the corrosion allowance, and E, G_L and P_r are as already defined.

13.3. Concrete shells

13.3.1 Introduction

Although the use of thin concrete shells for roofing has declined over the past 50 years, concrete shells still find widespread application as cooling towers at power stations, industrial chimneys, silos, water reservoirs and water treatment structures. The often-cited reason for the decline in architectural applications of thin concrete shells is the high cost of construction (formwork in particular). However, the simpler shell geometries (such as cylindrical shells and shells of revolution) have more moderate construction costs, as they permit modular units of formwork to be shifted progressively during construction, or the same formwork to be reused on multiple projects, which considerably reduces the unit cost of formwork. Here, we will not concern ourselves with construction aspects, but rather, with issues of the practical design of concrete shells. We will first consider guidelines of the American Concrete Institute, followed by those of the International Association for Shell and Spatial Structures.

13.3.2 Guidelines of ACI Committee 334

The report of Committee 334 of the American Concrete Institute (ACI)[3] provides some useful guidelines on practical aspects of the design of thin concrete shell structures. The determination of stresses in thin concrete shells, as well as the assessment of stability, is usually carried out on the basis of an elastic analysis, where the concrete is assumed to be uncracked, homogeneous and isotropic.

With regard to stresses, shells generally possess a considerable reserve of strength, with the calculated compressive stresses in the shell usually being only a fraction of the permissible compressive stress for concrete. The thickness of the concrete shell is usually dictated by cover requirements and construction considerations, with the result that the load-carrying capacity of the shell is substantially larger than that required to carry the calculated stresses. However, as the size of the shell is increased, or as the curvature is reduced (i.e. as the shell is made flatter), the compressive stresses in the shell become larger, and the reserve capacity of the shell reduces.

In stability assessments of concrete shells, consideration should be given to the gradual reduction of the buckling resistance of the shell caused by the creeping of concrete. Concrete shells that satisfy buckling criteria at the time of construction may experience buckling failure after a long period of time, not because of changes in the loading conditions, but due to creep. The effect of creep is equivalent to a lowering of the Young's modulus of concrete, so creep can be accounted for in buckling calculations by simply using a suitably reduced value of E.

Given that only a few geometric shell shapes have been studied in depth with regard to buckling behaviour, the ACI report suggests that, in the absence of better information, an estimate of the critical buckling stress in synclastic shells of arbitrary shape may be obtained by comparing the arbitrary shell with a cylindrical shell and a spherical shell, since the buckling resistance of synclastic shells of arbitrary shape usually lies somewhere in between that of a spherical shell and a cylindrical shell of similar proportions. This guidance might apply, for instance, to parabolic domes and elliptic paraboloids.

On the other hand, the buckling resistance of anticlastic shells (such as hyperbolic paraboloidal roofs and hyperbolic cooling towers) is significantly greater than that of synclastic shells of the same proportions, because the opposed pair of principal curvatures of an anticlastic shell at any given point tends to generate stabilising tensile action in one of the principal directions. Resistance to buckling is increased if reinforcement is provided on both faces of the shell wall instead of just one face.

For the concrete, the report specifies a minimum cylinder strength f_{cy} of 3000 psi, which converts to 21 MPa, or an equivalent cube strength of $f_{cu} = 26$ MPa. In practice, reinforced concrete shells are constructed with concrete of strength considerably higher than this value. As for other concrete structures, the principal tensile stresses in the shell are assumed to be fully resisted by the steel reinforcement. These stresses are also assumed to act in the middle surface of the shell, so the reinforcement to resist these should be placed in the middle of the shell section, preferably in the directions of the principal stresses.

Reinforcement should be spaced at not more than five times the thickness of the shell, but where the calculated principal tensile stress (expressed in *psi* units) exceeds $2\sqrt{f_{cy}}$ (where f_{cy} is also expressed in *psi* units), the spacing of reinforcement should not exceed three times the thickness of the shell. Reinforcement to resist bending moments should be provided in accordance with normal practice (as in slabs). Any direct tensile forces co-existing with the bending moments must also be allowed for. In the edge zones of thin concrete shells, where bending effects exist, top and bottom (i.e. inner- and outer-face) reinforcement should be provided.

When the concrete shell is connected to an edge member (e.g. a spherical dome with a ring beam at the base, or a cylindrical barrel shell with longitudinal edge beams), the normal practice is to gradually increase the thickness of the shell as the edge is approached. For barrel shells and domes, the recommended distance at which the gradual thickening should commence is ten times the normal shell thickness, while the recommended final shell thickness at the edge is twice the normal shell thickness, unless otherwise dictated by the results of the analysis. Typical proportioning of the shell and arrangement of reinforcement are illustrated in Figure 13.3 for a dome with a ring beam and for a cylindrical tank.

13.3.3 Guidelines of the International Association for Shell and Spatial Structures

The International Association for Shell and Spatial Structures (IASS), through its Working Group No. 5,[4] has compiled some useful guidelines on the design of reinforced concrete shells for roofing applications. Strength design, according to the IASS, consists of analysis of the shell and proportioning of the concrete and the reinforcement so that permissible stresses are not exceeded. It is recommended that all concrete shells be designed for strength at working loads, with elastic behaviour being the basis of the analysis.

Depending on the type of concrete shell and service environment, a number of loadings and load combinations need to be considered. These include dead load, live load, snow load, wind load, earthquake load, temperature and shrinkage effects, erection loads and prestressing loads. According to the IASS guidelines, the 28-day cylinder strength of the concrete, f_{cy}, must not be less than 20 MPa (practically the same as the ACI value). The value of Poisson's ratio ν for concrete may be taken as zero for the purposes of the analysis of the thin shell. The specified yield strength of the reinforcement, f_y, should not exceed 420 MPa. The permissible stresses in concrete and steel should comply with the standard or code of practice being used.

With regard to proportioning, and as also noted by the ACI,[3] the IASS guidelines make the point that shell thickness is usually dictated not by strength requirements, but by other factors such as construction and in-service considerations. For thin concrete shell roofs, the shell thickness typically lies in the range 50 – 100 mm, which, while relatively small, still permits the reinforcement and the concrete to be placed without too much difficulty, and which also ensures the cover to reinforcement is adequate. However, where bending reinforcement is required (typically near edges and supporting members), the thickness of the shell should be increased in a gradual manner, the recommended width of the thickened zone being *up to ten times the additional shell thickness*, and the maximum thickness at the edge being *two to three times the normal thickness of the shell.* These recommendations are quite similar to those of the ACI.[3]

Three types of reinforcement are distinguished: membrane, bending and special. Membrane reinforcement is usually placed in two orthogonal layers near the midsurface of the shell, while bending reinforcement is typically placed near the outer and inner surfaces of the shell; special reinforcement is usually required at corners or around shell openings. The use of high-strength bars (f_y greater than 280 MPa) is discouraged because of the increased risk of cracking.

Figure 13.3 Typical proportioning and reinforcement details for concrete shells: (a) dome with a ring beam at the base; (b) cylindrical tank wall and base.

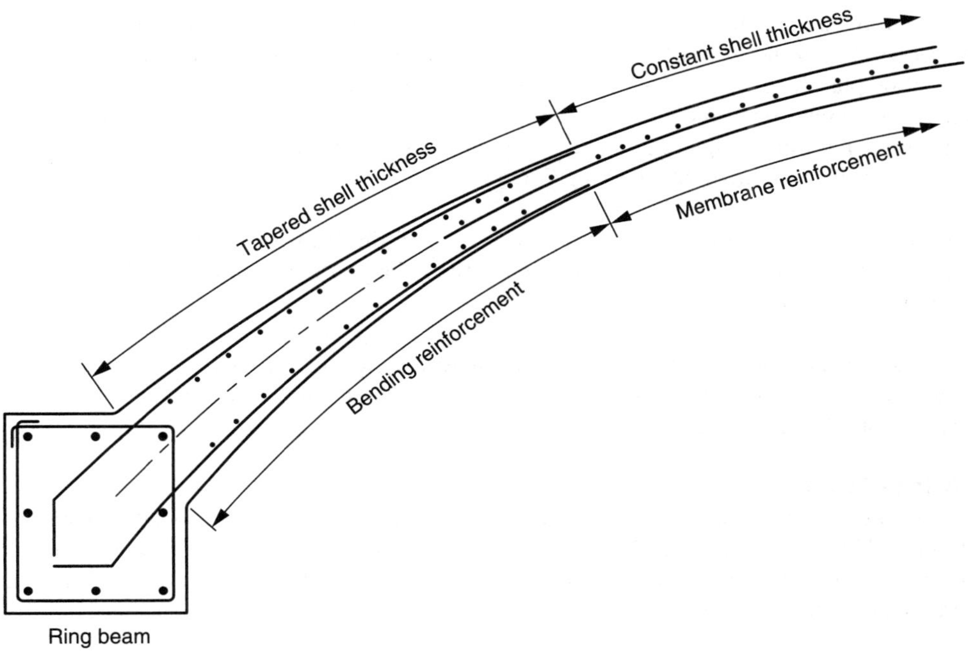

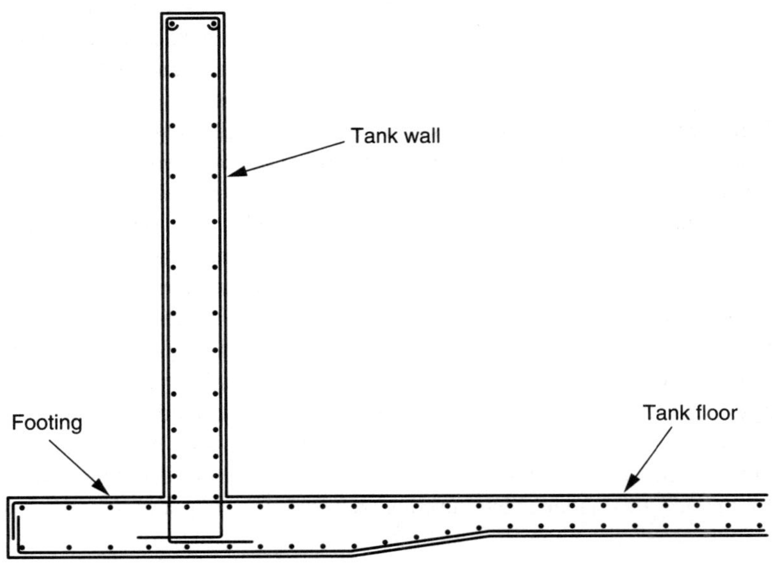

Membrane reinforcement should be provided to carry 100% of the calculated membrane tensile forces in the shell (i.e. concrete is assumed to contribute no strength). Instead of placing the reinforcement along the calculated directions of principal tensile stresses (which may be impractical), the two orthogonal layers of membrane reinforcement, placed close to the midsurface of the shell, may be oriented in more convenient directions. However, in zones subject to relatively high tensile stresses, it is recommended that the membrane reinforcement be placed as close as possible to the directions of principal tensile stresses. Alternatively, a third layer of membrane reinforcement may be added in the principal directions locally over such areas. A detailed approach for the design of reinforcement in thin concrete shells may be seen in a technical paper written by two members of the IASS Working Group.[5]

Spacing of bars should not be more than five times the shell thickness or 300 mm, whichever is smaller. Where the calculated principal tensile stresses are relatively large, the membrane reinforcement should be spaced at not more than three times the shell thickness.

Bending reinforcement should be provided in zones experiencing edge effects, on both the top and bottom faces. In calculating the required amount of bending reinforcement, any axial forces present in the area in question should also be catered for.

13.3.4 Miscellaneous guidelines

The following simple design guidelines for concrete shells are drawn from sources other than the report of ACI Committee 334 and the recommendations of IASS Working Group No. 5. Some of these are essentially similar to those of the ACI and the IASS, but others cover additional aspects that may be relevant to the designer of concrete shell structures.

For a concrete shell, the first step in the design process is to choose a shell geometry that satisfies the functional and architectural requirements. Determination of form for shells of unusual geometry can be aided by physical models. A trial set of dimensions for the shell may be chosen on the basis of functional considerations, available guidelines, previous experience, or observation of similar existing shell structures.

The stresses in the shell are then calculated on the basis of the linear elastic theory for isotropic shells, using either the membrane theory on its own, or the membrane theory in conjunction with an appropriate bending theory as presented in the chapters of this book. Once the stresses are available, one may proceed to choose the thickness of the shell and the amount of steel reinforcement required in the various regions of the shell.

From a material-strength point of view, any compressive membrane stresses that may occur in the shell are usually several times smaller than the crushing strength of concrete, so are unlikely to govern shell thickness requirements. On the other hand, if tensile membrane stresses occur in certain areas, we may use their magnitude as a criterion for choosing shell thickness – that is, the shell thickness is chosen such that the permissible *tensile* strength of the concrete is not exceeded anywhere on the shell. In the absence of specific values of the tensile strength of concrete, 10% of the compressive strength may be adopted as the tensile strength. One must also ensure that the chosen shell thickness meets minimum cover requirements for reinforcement as specified by the design standard being used. Construction considerations may also dictate the minimum shell thickness.

Where significant compressive actions occur in the shell, choice of shell thickness may be governed by buckling considerations rather than material strength. For concrete domes of double curvature, and in the absence of more exact information, the critical buckling load per unit area of shell surface may be taken as approximately

$$q_{cr} = \frac{2E}{\sqrt{3(1-\nu^2)}} \frac{t^2}{r_1 r_2} \tag{13.17}$$

where r_1 and r_2 are the principal radii of curvature at a given point on the dome, t is the shell thickness, E the Young modulus, and ν the Poisson ratio. Taking Poisson's ratio for concrete as 0.15, the above expression simplifies to

$$q_{cr} = 1.17E\frac{t^2}{r_1 r_2} \tag{13.18}$$

For a spherical dome with $r_1 = r_2 = a$, the critical buckling load intensity becomes

$$q_{cr} = 1.17E\left(\frac{t}{a}\right)^2 \tag{13.19}$$

while for a hyperbolic paraboloid, for which

$$z = \frac{h}{ab}xy \Rightarrow \frac{1}{r_1 r_2} = \left(\frac{h}{ab}\right)^2 \tag{13.20}$$

we get the result

$$q_{cr} = 1.17E\left(\frac{th}{ab}\right)^2 \tag{13.21}$$

Parametric studies show that the stresses in domes and other shell roofs are generally of low magnitude, and rarely dictate the choice of shell thickness, except in the edge zones. Over the interior regions of shell roofs, the shell thickness is usually made to lie in the range 60–120 mm in order to satisfy cover requirements and practical considerations during construction. Towards the supports, the shell thickness is usually gradually increased in order to cater for the localised bending stresses in the edge zone.

With shell thickness decided, reinforcement is provided primarily to resist the calculated tensile stresses. In calculating the amount of reinforcement to be provided in resisting the tensile stresses in the shell, the usual practice of neglecting the tensile strength of concrete must be followed. Note that, while we may possibly have used the peak value of tensile stresses to help us in choosing the shell thickness, we completely disregard the tensile strength of the concrete in determining the required amount of reinforcement, and assume all the tensile resistance is provided by the reinforcement.

The normal approach is to first provide primary reinforcement throughout the shell for resisting membrane stresses, and then add localised secondary reinforcement afterwards to cater for any edge effects. Because shell roofs are relatively thin, membrane reinforcement is usually placed as a single layer at the midsurface of the shell, in the two principal directions. On the other hand, the shell thickness for liquid-retaining vessels is generally greater, allowing membrane reinforcement to be placed in two layers, one near the outer surface of the shell and the other near the inner surface, each layer comprising two orthogonal systems of steel. Membrane reinforcement in the interior regions of the shell not only resists calculated tensile stresses, but also caters for temperature and shrinkage cracking.

Near shell edges and other sources of bending effects, two layers of reinforcement are always provided, irrespective of the type of shell structure. One layer is placed close to the outer surface of the shell, while the other layer is placed close to the inner surface, each layer comprising two orthogonal systems of steel. In these edge zones, it is recommended that steel be shared equally between the inner and outer layers, to cater for the generally oscillating nature of the bending-disturbance stresses. The area of reinforcement required to resist the calculated bending moment at a given location and in a specified direction is simply given by dividing the bending moment by the permissible stress for steel and the distance between the two layers of steel (i.e. the 'lever arm').

In providing reinforcement in any region of a given concrete shell, the minimum requirements for crack control must be satisfied. This minimum reinforcement may be taken as 0.3% in each of the two orthogonal directions, unless otherwise indicated by the design code being used.

Where a ring beam is provided along an edge of a shell of revolution, axial reinforcement must be provided in the ring beam. This should be placed at two levels, one close to the upper surface of the beam and the other close to the lower surface, so as to resist the net effect of *(i)* the extensional stresses due to the hoop force in the ring beam, and *(ii)* the flexural stresses due to the torque on the ring beam, as calculated in accordance with the shell-ring considerations of Chapter 7.

References

1. EN 1993-1-6. *Eurocode 3: Design of Steel Structures: Part 1-6: Strength and Stability of Shell Structures*. European Committee for Standardisation (CEN), Brussels, 2007.
2. API Standard 650. *Welded Tanks for Oil Storage* (12th edn). The American Petroleum Institute, Denver, 2013.
3. ACI Committee 334. *Concrete Shell Structures: Practice and Commentary*. American Concrete Institute, Michigan, 2002.
4. IASS Working Group No. 5. *Recommendations for Reinforced Concrete Shells and Folded Plates*. International Association for Shell and Spatial Structures, Madrid, 1979.
5. Medwadowski S.J. and Samartin A. Design of reinforcement in concrete shells: A unified approach. *Journal of the International Association for Shell and Spatial Structures*, 2004, **45** (144), 41–50.

Shell Structures in Civil and Mechanical Engineering
ISBN 978-0-7277-6028-9

http://dx.doi.org/10.1680/ssicame.60289.435

Author index

Shell Structures in Civil and Mechanical Engineering
ISBN 978-0-7277-6028-9

http://dx.doi.org/10.1680/ssicame.60289.437

Subject index